Petroleum Reservoir Modeling and Simulation

About the Authors
Sanjay Srinivasan, Ph.D., is a professor of petroleum and natural gas engineering at the Pennsylvania State University and holds the John and Willie Leone Family Chair in Energy and Mineral Engineering. He is a member of the International Association for Mathematical Geosciences, the Society of Exploration Geophysicists, and the Society of Petroleum Engineers.

Juliana Y. Leung, Ph.D., P.Eng., is a professor in the Civil and Environmental Engineering Department at the University of Alberta. She is a registered professional engineer with the Association of Professional Engineers and Geoscientists of Alberta.

Petroleum Reservoir Modeling and Simulation

Geology, Geostatistics, and Performance Prediction

Sanjay Srinivasan, Ph.D.

Juliana Y. Leung, Ph.D., P.Eng.

New York Chicago San Francisco
Athens London Madrid
Mexico City Milan New Delhi
Singapore Sydney Toronto

Library of Congress Cataloging-in-Publication Data

Names: Srinivasan, Sanjay, author. | Leung, Juliana, author.
Title: Petroleum reservoir modeling and simulation : geology, geostatistics,
  and performance prediction / Sanjay Srinivasan, Juliana Leung.
Description: New York : McGraw Hill, [2022] | Includes bibliographical
  references and index.
Identifiers: LCCN 2021049779 | ISBN 9781259834295 (hardback) | ISBN
  9781259834301 (ebook)
Subjects: LCSH: Oil reservoir engineering—Mathematical models. |
  Petroleum—Geology—Mathematical models.
Classification: LCC TN870.57 .S75 2022 | DDC 622/.3382—dc23/eng/20211103
LC record available at https://lccn.loc.gov/2021049779

McGraw Hill books are available at special quantity discounts to use as premiums and sales promotions or for use in corporate training programs. To contact a representative, please visit the Contact Us page at www.mhprofessional.com.

**Petroleum Reservoir Modeling and Simulation:
Geology, Geostatistics, and Performance Prediction**

1 2 3 4 5 6 7 8 9   CCD   26 25 24 23 22 21

ISBN 978-1-259-83429-5
MHID 1-259-83429-8

The pages within this book were printed on acid-free paper.

Sponsoring Editor Robin Najar	**Copy Editor** Neha Priya, MPS Limited
Editorial Supervisor Stephen M. Smith	**Proofreader** Subashree, MPS Limited
Production Supervisor Lynn M. Messina	**Indexer** Ariel Tuplano
Acquisitions Coordinator Elizabeth M. Houde	**Art Director, Cover** Jeff Weeks
Project Managers Jyoti Shaw and Rishabh Gupta, MPS Limited	**Composition** MPS Limited

Contents

CHAPTER 1

Introduction

The authors have envisioned this book to be a reference and a repository of algorithms and tools for practicing reservoir modelers and engineers. The book is written from the perspective of a practitioner of geostatistics and flow simulation. There are certainly many books currently available in the market that focus on specific topics in geostatistical reservoir modeling or numerical flow simulation, but it is difficult to find a book that covers the basic concepts relevant to both realms of study. In fact, it can be argued that geostatistical reservoir modeling and flow modeling are so inseparably intertwined that talking at length about one without extensively talking about the other would render advanced topics such as model calibration or history matching and upscaling/scale-up of models inaccessible to all but a select few. It is to remedy this relative deficiency that the authors conceived of this reference with the fond hope that both students in introductory courses on geostatistical modeling and numerical flow simulation, as well as advanced researchers working at the interface between reservoir modeling and numerical flow modeling, would find something of value in the book. The various chapters are structured to cover a broad range of relevant topics. In addition, some concepts that have emerged from research on some of the more advanced topics are also presented that can provide some fuel for future research. Besides having worked numerical examples and case studies embedded throughout the text, software modules (e.g., EXCEL® and MATLAB® codes) are also provided, on this book's website (www.mhprofessional.com/PRMS), for accomplishing the modeling and analysis examples presented in the text. These, we hope, will be of some value to the practicing engineer.

The chapters in this book can be grouped into two parts. The first four chapters focus on various aspects of geostatistics and stochastic reservoir modeling. An inspiration for this first part is the concise and yet fairly comprehensive review that Prof. Andre Journel presented in his *Fundamentals of Geostatistics in Five Lessons* (Journel, 1989). While that short-course reference starts with an initial discussion of basic concepts from probability and statistics, this work skips that initial discussion under the premise that a student intending to work in this area would have sufficient background in probability

1

and statistics. The discussion on spatial correlations and the inference of spatial continuity measures included in the first chapter of the short-course notes are discussed at length in Chap. 2 of this book. As in Andre's reference, kriging and spatial estimation are the focus of Chap. 3. Rather than splitting the discussion on kriging into a chapter on simple kriging (and its dual interpretation) and another chapter on kriging under constraints, both these variants of kriging are included in one chapter in this book. Data integration through co-kriging, as well as a discussion on the non-parametric form of kriging, i.e., indicator kriging, is also thrown in for good measure. This results in a rather long chapter but hopefully one that will provide a good foundation to the reader for exploring additional concepts in subsequent chapters. Stochastic simulation is the subject of Chap. 4, where the discussion starts with a simple decomposition of the covariance matrix and the use of those decompositions for stochastic simulation. Then sequential simulation techniques are discussed and the chapter culminates with techniques for data integration within non-parametric schemes for sequential simulation. Some of the details of the Markov–Bayes algorithm for accomplishing the integration can be found in App. B. Much of the development in geostatistics over the past decade has been in the area of pattern simulation or multiple point geostatistics. Chapter 5 presents a discussion on multiple point geostatistics starting from the concept of indicator basis function. The single extended normal equation is presented as a special case of the indicator basis function and some of the key techniques based on the concept of the single normal equation are discussed. Some tricks to make these simulations efficient are also discussed. The chapter ends with a musing on the use of the extended indicator basis function for stochastic simulation. Admittedly, this is still a nascent idea, but one that may become much more mainstream in the coming years with the advent of faster computers, etc. There have also been many developments in the area of spectral simulation and simulation using cumulants. These will hopefully make their way into future editions of the book. Similarly, the area of spatiotemporal geostatistics has also attracted much research. Some perspectives regarding that development would also hopefully make their way into future versions of this book.

The last four chapters focus on various aspects of flow simulation. Chapter 6 presents the basic mass conservation and momentum balance equations for multi-phase flow, relevant auxiliary equations, and the corresponding initial and boundary conditions. Formulation of an isothermal black-oil model based on the control volume finite difference scheme is emphasized. More advanced topics are presented in Chap. 7, including the issues of accuracy and stability of numerical schemes and unstructured meshes. Finally, numerical schemes for modeling media with dual- and triple-porosity systems, as in the case of fractured media, are discussed. Chapter 8 will focus specifically on upscaling, which is an essential link between geological modeling and reservoir flow modeling. Various approaches, including statistical (static) upscaling, as well as flow-based (dynamic) upscaling, are introduced in order to derive the effective medium properties for both flow (e.g., permeability) and transport (e.g., dispersivity) properties. Finally, Chap. 9 discusses the topic of dynamic flow data integration or history matching, which refers to the process of calibrating reservoir models to reflect the observed production characteristics.

There are many existing techniques for perturbing uncertain reservoir model parameters such that the simulation prediction becomes consistent with the actual production observations. Data inversion techniques, including both local (gradient-based) and global schemes, are widely adopted. Various data-assimilation techniques are also discussed to handle continuous data integration. Probabilistic history matching methods, which aim to sample an ensemble of realistic models from the posterior probability distribution, are emphasized.

As they say, "smartness is in standing on the shoulders of giants" in order to make limited forays into the frontiers of knowledge. While presenting this book, one would be remiss if the contributions of these giants are not respectfully acknowledged. These giants have not only added significantly to the body of knowledge in the areas of geostatistical reservoir modeling and numerical flow simulation, but also have nurtured a future generation of researchers who have further extended the knowledge base in those areas. The authors of this book owe an eternal debt to these teachers who have nurtured them and brought them to this point. It is also very important to point out that any defects in this rendition are entirely attributable to the authors and their limitations rather than a reflection on the teachers who imparted this knowledge to them. To the students picking up this book, it is the sincere hope that they are as spellbound by the inner details of the algorithms and methods discussed in this book as the authors themselves are.

1.1 Introductory Concepts

1.1.1 Stochastic Reservoir Modeling

A situation commonly encountered in natural resource development is the need to estimate the amount or quality of reserve at a particular location or the properties of the surrounding media that would facilitate the recovery of the resource. In petroleum reservoir development, it might be necessary to estimate the value of porosity and permeability at a candidate well location based on data at a few surrounding wells. An important aspect of this problem is that since the objective is to assess the resource characteristic at a location where direct measurement is unavailable, there is likely to be uncertainty associated with our estimation of that characteristic. It is for the description of that uncertainty that the notion of *probability* is most useful for. A second aspect of the problem is the process by which knowledge at nearby wells is utilized to make the estimation at the candidate location. This implies that the observations at the nearby locations have some bearing on the process of estimation that is to be carried out. Since several observations might be available at a number of locations where data were collected, this also implies that rather than quantifying the relationship between any specific data location and the candidate well location based on that single well observation, it might be useful to pool the information observed at several wells and compute an average or general relationship between the data and the unknown. In other words, we would like to *statistically* quantify the relationship between the data and the unknown. There is an intimate relationship between probability and statistics that can be gleaned from the above scenario. We would like to model the *probability* that quantifies the uncertainty in resource characteristics at a location, and in order to do so we would need to compute some *statistics*, based on the observations at wells that quantify the relationship between the data and the unknown.

There are some other important concepts that emerge out of the above discussion. It is clear that *estimation* or *prediction* is an important objective that motivates resource modeling. It also motivates data acquisition and analysis. Estimation utilizes the available data and also relevant statistics. On the other hand, the statistics and the estimation that they result in have to be used ultimately to present a clear picture of the uncertainty associated with the estimation of the unknown characteristic at the target location. So, one could argue that it is not the estimation or prediction that is the ultimate objective of the analysis/modeling process. Rather, it is the quantification of uncertainty that is the ultimate goal. Instead of engaging in an extended debate over this issue, suffice it to say that data analysis, inference of relevant statistics, utilization of the statistics to estimate/predict an outcome, and, subsequently, extension of the estimation to quantify the uncertainty in resource characteristics are the core steps of geostatistical modeling and will be presented in that sequence through the first few chapters of the book.

1.1.2 Flow Modeling in Highly Heterogeneous Reservoirs

The conservation of mass, momentum, and energy equations are the primary governing equations for flow and transport modeling. These equations are mathematical representations of the fundamental principles concerning the underlying physics. These balance equations are often expressed in the form of partial differential equations, where the dependent variable (e.g., fluid pressure) is a function of both spatial coordinates and time. Several auxiliary equations, such as phase behavior descriptions (e.g., equation of state), reaction kinetics, etc., are needed to fully define the state of the system. The purpose of computational flow modeling is to solve this set of equations numerically. Therefore, the first of numerical modeling should be the formulation of the appropriate governing and auxiliary equations, which can be written in many different forms. The equation forms, choice of discretization methods, and formulation of solution schemes should reflect careful consideration of the relevant physical mechanisms; for example, what types of heterogeneity are involved? What are the modeling scale and dimensions? Is the flow direction linear, radial, or spherical? A balance between computational efficiency and accuracy is often needed.

1.1.3   Linking Geostatistical Modeling to Flow Modeling

Well logs are calibrated using core data and they inform the spatial variation of reservoir petrophysical properties in the near well bore region (scale: feet). Well test informs a much larger reservoir volume (scale: hundreds of feet). The well test response is induced by the flow of hydrocarbons through the reservoir rock. The movement of fluid through the reservoir is a complex phenomenon that depends non-linearly on the spatial distribution of the reservoir petrophysical properties. Thus, the well test data inform indirectly the spatial variations of the reservoir attributes. Reservoir heterogeneity occurring at the even larger inter-well scale (thousands of feet) is informed by the production data. The relation between production data measured at wells and reservoir heterogeneity is again non-linear and is influenced by several other factors, such as reservoir fluid properties, boundary conditions, and rock-fluid interactions. Accurate and efficient reservoir modeling thus requires an understanding of the complex interplay between reservoir geology and the flow of fluids through the reservoir.

Since flow and future productivity are controlled by the spatial connectivity of permeability in the reservoir, the flow-related data $rp(\mathbf{u}_{w_1}, t)$ is particularly valuable for updating prior models for permeability variation in the reservoir. In other words, one hopes that conditioning to the well-specific data would result in permeability models $k(\mathbf{u})$, $\mathbf{u} \in$ Reservoir which would predict accurately the future reservoir performances $rpf(\mathbf{u}_{w_2}, t)$. In other words, we would like the response predicted by applying a transfer function model (flow simulator) TF to a well located at $\mathbf{u}_{w_2}$ to be as close to the actual response $rpf(\mathbf{u}_{w_2}, t)$ at a future instant in time.

This brings to fore, some interesting challenges associated with reservoir performance prediction:

- In the mapping between the permeability variations in the reservoir $k(\mathbf{u})$, $\mathbf{u} \in$ Reservoir and the response data $rp(\mathbf{u}_{w_1}, t)$, there are many more unknowns $k(\mathbf{u})$ than there are data $rp(\mathbf{u}_{w_1}, t)$. This implies that the application of the inverse transfer function TF^{-1} in order to model the permeability variations will not result in a unique solution. There are at least two ways to address the non-uniqueness of this inverse problem:

1. Constrain the permeability field $k(\mathbf{u})$ to as much different data as possible, for example, well logs, seismic, production data, etc., to reduce the number of degrees of freedom. This requires an understanding of geostatistical methods for reservoir modeling and data assimilation.

2. Accept that there are several solutions, say L realizations of the permeability field $k^\ell(\mathbf{u})$, $\mathbf{u} \in$ Reservoir, $\ell = 1, \ldots, L$ such that they yield flow response that match the data $rp(\mathbf{u}_{w_1}, t)$ to an acceptable level and reflect some prior geological information. This again entails the application of stochastic methods for reservoir modeling.

The preceding discussion motivates the ensuing discussion on stochastic modeling of reservoirs, modeling flow in such reservoirs, and finally the interplay between static and dynamic reservoir modeling.

1.2   Reference

Journel, A. G. (1989). *Fundamentals of Geostatistics in Five Lessons*. Washington DC: Americal Geophysical Union.

CHAPTER 2

Spatial Correlation

The following broad steps can be envisioned for modeling a petroleum reservoir:

- Define a population/area A that involves invoking the decision of stationarity.
- Infer statistics pertaining to spatial variability within A.
- Synthesize a stationary random function model representative of the phenomenon in A.
- Sample realizations of the reservoir from the multivariate joint distribution characterizing the random function model.

In this chapter, we will first begin with a discussion related to the inference and modeling of spatial covariances and semi-variograms. It is important to realize that these pair-wise correlation measures are adequate (optimal) to represent the variability exhibited by multi-Gaussian phenomena. Representing non-Gaussian phenomena using these pair-wise measures could result in the higher-order moments remaining unconstrained and that may impart some unintended patterns of spatial variability in the random function (RF) model. That will motivate our discussion about multiple point statistics in a later chapter to better constrain the RF models. Finally, we will make the argument that spatial statistics in the form of covariances, semi-variograms, or even multiple point statistics are but measures summarizing the bivariate or, more generally, the n-variate joint distributions characterizing the spatial variability of the phenomenon. Knowing the moments, the multivariate distribution cannot be constructed uniquely, and this is one of the motivations for a stochastic framework for modeling reservoirs. Such a framework will yield multiple, equi-probable representations of the RF model. The topic of stochastic simulation of reservoirs is also the focus of a separate chapter.

Let us begin our discussion with a brief look at the basic measures for spatial variability/similarity.

2.1 Spatial Covariance

Analogous to the covariance between a pair of variables, the spatial covariance between pairs of locations in a reservoir is defined as:

$$Cov\{Z(\mathbf{u}), Z(\mathbf{u} + \mathbf{h})\} = E\{Z(\mathbf{u})Z(\mathbf{u} + \mathbf{h})\} - E\{Z(\mathbf{u})\}$$
$$\cdot E\{Z(\mathbf{u} + \mathbf{h})\} \qquad (2\text{-}1)$$

The notations $\mathbf{u}$ and $\mathbf{u} + \mathbf{h}$ represent a pair of locations separated by a lag vector $\mathbf{h}$. The bold $\mathbf{u}$ implies a vector of coordinates $\mathbf{u} = [x, y, z]$ for a spatial process or equivalently $\mathbf{u} = [x, y, z; t]$ for a spatiotemporal process.

A number of samples are required at the locations $\mathbf{u}$ and $\mathbf{u} + \mathbf{h}$ in order to infer the covariance defined by Eq. (2-1). However, in most cases of reservoir modeling, only one pair of observations is available at the two locations. As in other statistical inference cases, we need to invoke *stationarity* in order to be able to infer the spatial covariance. Specifically, we make the decision that the relationship between pairs of locations a lag $\mathbf{h}$ apart is invariant regardless of where the pairs are located within the stationary reservoir domain. This means that the spatial covariance can be computed as:

$$\hat{C}(\mathbf{h}) = \frac{1}{N_h}\sum_{N_h}\sum_{N_h}z(\mathbf{u}_i) \cdot z(\mathbf{u}_j) - \frac{1}{N_h}\sum_{N_h}z(\mathbf{u}_i) \cdot \frac{1}{N_h}\sum_{N_h}z(\mathbf{u}_j) \qquad (2\text{-}2)$$

In Eq. (2-2), as the averaging is done over all locations making up the N_h pairs separated by a lag $\mathbf{h}$, the resultant spatial covariance is only a function of the lag. Thus, stationarity implies that the spatial covariance is a function only of the two-point spatial template $\mathbf{h}$. The covariance is computed by specifying a spatial template (Fig. 2-1) and subsequently, translating the spatial template over the entire stationary domain. This yields several outcomes for $Z(\mathbf{u})$ (tail value) and $Z(\mathbf{u} + \mathbf{h})$ (head value), and using these outcomes the spatial covariance $\hat{C}(\mathbf{h})$ can be computed.

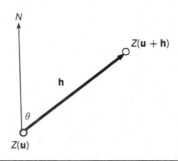

FIGURE 2-1 A two-spatial template characterized by a lag vector $\mathbf{h}$ separating pairs of points. The lag vector subtends an angle θ to the azimuthal north direction.

Remarks

- The spatial covariance, as defined in Eq. (2-3), measures the similarity or continuity in the RF value at two locations in space. If we write the spatial covariance as $C(\mathbf{h}) = E\{(Z(\mathbf{u}) - m) \cdot (Z(\mathbf{u} + \mathbf{h}) - m)\}$, where the mean at the two locations is the same under stationarity, it is evident that the covariance value is maximized when the outcomes at the two locations behave similarly with respect to the mean.

- The spatial covariance is in units of the square of the data when $Z(\mathbf{u})$ and $Z(\mathbf{u} + \mathbf{h})$ are the same variables at two locations or in the units of the product of data when the two variables are different. In order to render the covariance unitless, it is customary to work with the spatial correlogram, which is defined as:

$$\hat{\rho}(\mathbf{h}) = \frac{\hat{C}(\mathbf{h})}{\sqrt{Var(Z(\mathbf{u})) \cdot Var(Z(\mathbf{u} + \mathbf{h}))}} \qquad (2\text{-}3)$$

- The spatial covariance can also be interpreted as the moment of the bivariate scatterplot between $Z(\mathbf{u})$ and $Z(\mathbf{u} + \mathbf{h})$. The presence of outliers will influence the moment of the $\mathbf{h}$-scatterplot.

2.2 Semi-Variogram

The inference of a spatial covariance or correlogram is possible when the reservoir characteristic being modeled is stationary. An alternate measure of spatial variability is the semi-variogram defined as:

$$\gamma_Z = \frac{1}{2}E\{(Z(\mathbf{u}) - Z(\mathbf{u} + \mathbf{h}))^2\} \qquad (2\text{-}4)$$

If $E\{Z(\mathbf{u})\} = E\{Z(\mathbf{u} + \mathbf{h})\}$ then $Var\{Z(\mathbf{u}) - Z(\mathbf{u} + \mathbf{h})\} = E\{(Z(\mathbf{u}) - Z(\mathbf{u} + \mathbf{h}))^2\}$ and then the variogram γ_Z can be interpreted as the semi-variance of increments $\{Z(\mathbf{u}) - Z(\mathbf{u} + \mathbf{h})\}$. The inference of the variogram requires stationarity of the increments—referred to as intrinsic stationary (Matheron, 1973). A phenomenon may not be second-order stationarity (i.e., variance and covariance of the RF may not exist), but it may be intrinsically stationary; i.e., the increments of the RF may be stationary and its variance may exist. A small example to illustrate this point follows.

Consider the following layered system. The values tabulated may be indicative of variations in some rock property (e.g., permeability). In order that the calculated statistics are not unduly affected by the lack of statistical mass, a hundred values were used for each layer, and the table only lists ten of those values per layer.

5.55	4.28	4.70	4.72	4.86	....	4.35	4.51	5.24	5.03	4.77		4.81
10.22	8.25	8.12	7.45	8.47	....	7.76	8.29	6.83	7.85	8.06		7.85
12.38	11.40	9.82	10.46	9.22	....	12.71	11.83	12.35	12.25	10.32		11.03
13.46	14.46	15.04	15.76	9.75	....	13.29	14.02	13.28	13.33	12.94		14.05

The last column indicates the mean value for each layer, and it can be observed that there is a trend exhibited by the mean. If we use the tabulated data to compute the covariance over a unit lag in the vertical direction using Eq. (2-2), the variation in the covariance values computed using the values in adjacent layers is tabulated below:

C(h)
0.22
−0.10
0.49

The first value is the spatial covariance calculated using the values in the first two rows of the table, the second value is the covariance calculated using the values in the second and third rows of the table, and so on. It can be seen that the spatial covariance calculated over the same lag exhibits a systematic variation. This, in turn, calls into question the validity of pooling data over the entire domain (all combination of points separated by a unit lag in the vertical direction) in order to compute a spatial covariance.

The semi-variogram values computed using Eq. (2-4) for the same unit lag in the vertical direction are summarized below:

γ(h)
2.61
2.59
2.54

The first value is the semi-variance of increments computed using the values in the first two rows, the second value between the second and third rows, and so on. The semi-variance behaves in a much more stable manner, and thus the semi-variogram computed by pooling all the data together is much more interpretable. This would thus be an example of a phenomenon that is not second-order stationary but is intrinsically stationary.

In the spatial context, the semi-variogram between a pair of locations $\mathbf{u}$ and $\mathbf{u}'$ is defined as:

$$2\gamma(\mathbf{u}, \mathbf{u}') = E\{(Z(\mathbf{u}) - Z(\mathbf{u}'))^2\} \qquad (2\text{-}5)$$

Under the intrinsic hypothesis:

$$\gamma(\mathbf{u}, \mathbf{u}') \equiv \gamma(\mathbf{u} - \mathbf{u}') = \gamma(\mathbf{h})$$

The lag $\mathbf{h} = (\mathbf{u} - \mathbf{u}')$ is the vector distance between pairs of locations within the stationary domain; i.e., the semi-variogram inferred by invoking intrinsic stationarity loses any specificity to the location $\mathbf{u}$ and instead represents the average variability exhibited by pairs of points separated by a lag $\mathbf{h}$.

2.2.1 Variogram Inference

Consider the bivariate scatterplot (Fig. 2-2) between pairs of values $Z(\mathbf{u})$ and $Z(\mathbf{u} + \mathbf{h})$ separated by a lag $\mathbf{h}$. Suppose there are $N(\mathbf{h})$ pairs of observations pooled together under the decision of intrinsic stationarity.

The projection of any point on the scatterplot to the 45° bisector is denoted as d_i. From the figure:

$$d_i = |z(\mathbf{u}) - z(\mathbf{u} + \mathbf{h})| \cdot \text{Sin}(45°)$$

$$d_i^2 = \frac{1}{2} \cdot [z(\mathbf{u}) - z(\mathbf{u} + \mathbf{h})]^2$$

The average d_i^2 is:

$$E\{d_i^2\} = \frac{1}{2 \cdot N(\mathbf{h})} \cdot \sum_{i=1}^{N(\mathbf{h})} [z(\mathbf{u}) - z(\mathbf{u} + \mathbf{h})]^2 \tag{2-6}$$

By comparing Eq. (2-6) with Eq. (2-4), it can be concluded that $E\{d_i^2\} = \hat{\gamma}(\mathbf{h})$. The quantity d_i can also be interpreted as the moment of the data value $Z(\mathbf{u})$ about the 45° bisector. By definition, the moment of inertia is the sum of the products of the mass of each particle in the body with the square of its distance from the axis of rotation. In the context of the semi-variogram, the particles (points) all have the same mass $\frac{1}{N(\mathbf{h})}$, and the square distance of each particle

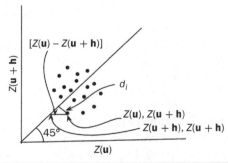

FIGURE 2-2   The h-scatterplot between pairs of values separated by a lag $\mathbf{h}$. Along the 45° bisector $Z(\mathbf{u}) = Z(\mathbf{u} + \mathbf{h})$.

from the axis of rotation (the 45° bisector) is $d_i^2 = \dfrac{1}{2} \cdot [z(\mathbf{u}) - z(\mathbf{u} + \mathbf{h})]^2$.

The semi-variogram is thus inferred as the moment of inertia of the h-scatterplot.

Inference of the experimental variogram requires:

- h-scatterplot obtained by systematically varying $|\mathbf{h}|$ in several spatial directions.

- For each lag, computing the corresponding moment of inertia of the scatterplot.

- Plotting the computed moment of inertia versus the lag $\mathbf{h}$ in specific directions.

Remarks

- There must be a sufficient number of pairs in each scatterplot (statistical mass).

- Outliers (abnormal high or low values) can cause the moment of inertia to fluctuate wildly.

- If the phenomenon is deemed second-order stationary, then we can work out a relationship between the spatial covariance and the semi-variogram:

$$
\begin{aligned}
2\gamma(\mathbf{h}) &= E\left\{[Z(\mathbf{u}) - Z(\mathbf{u} + \mathbf{h})]^2\right\} \\
&= E\left\{Z^2(\mathbf{u})\right\} - 2 \cdot E\left\{Z(\mathbf{u}) \cdot Z(\mathbf{u} + \mathbf{h})\right\} + E\left\{Z^2(\mathbf{u} + \mathbf{h})\right\} \\
&= \left[E\left\{Z^2(\mathbf{u})\right\} - \mu^2\right] - 2 \cdot \left[E\left\{Z(\mathbf{u}) \cdot Z(\mathbf{u} + \mathbf{h})\right\} - \mu^2\right] \quad\quad (2\text{-}7) \\
&\quad + \left[E\left\{Z^2(\mathbf{u} + \mathbf{h})\right\} - \mu^2\right] \\
&= 2 \cdot Cov(0) - 2Cov(\mathbf{h})
\end{aligned}
$$

The aforementioned equation is based on the fact that under stationarity, $E\left\{Z(\mathbf{u})\right\} = E\left\{Z(\mathbf{u} + \mathbf{h})\right\} = \mu$ and $Cov(0) = Var\left\{Z(\mathbf{u})\right\}$. As $|\mathbf{h}| \to \infty$, $Cov(\mathbf{h}) \to 0$ which in turn implies that $\gamma(\mathbf{h}) \to C(0)$. For a second-order stationary phenomenon, the semi-variogram thus tends to stabilize at the prior variance value. Actually, the semi-variogram will tend to stabilize at a *sill value* equivalent to the variance of the data that contribute to the calculation of variogram values at all lags. The subset of data contributing to the variogram values at all lags is also referred to as the common eroded set.

2.2.2   Variogram Characteristics

It has already been shown that for a second-order stationary phenomenon, $\displaystyle\lim_{|\mathbf{h}| \to \infty} \gamma(\mathbf{h}) = Cov(0)$. But there do exist exceptional cases when the semi-variogram instead of stabilizing at a sill (variance) value oscillates around the variance value. This happens in the case

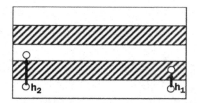

Figure 2-3   A layered rock sequence made of alternating sand and shale layers.

of a spatial (or temporal) periodic phenomenon, such as a repeat sequence of sand-shale, as shown in Fig. 2-3.

As the lag $\mathbf{h}_1$ increases within sand, the semi-variogram picks up the variability within the sand. When the lag vector is such that the tail value is in sand while the head value is in shale, the semi-variogram attains its maximum value, indicating maximum variability. But then as the lag increases to $\mathbf{h}_2$, the tail value in sand may correlate with another location $\mathbf{u} + \mathbf{h}_2$ also in sand, but in a different sequence. Thus, the semi-variogram value decreases from its previous maximum. But then as the lag keeps increasing, the semi-variogram rises again because of the variability between the sand in the two different sequences. Thus, the semi-variogram at large lags shows a periodic behavior also referred to as a *hole effect*.

As the lag spacing $\mathbf{h}$ decreases, the similarity between the attribute value at a pair of locations increases, and correspondingly, the semi-variogram values that reflect spatial variability also decrease. For a phenomenon exhibiting second-order stationarity, as $\mathbf{h} \to 0$, then $Cov(\mathbf{h}) = Cov(0)$, the prior variance exhibited by the phenomenon. Consequently:

$$\gamma(0) = Cov(0) - Cov(0) = 0$$

The semi-variogram does start at the origin for any natural phenomenon. For a non-second-order stationary RF, the semi-variogram can be represented as (Agterberg, 1994):

$$\gamma(\mathbf{h}) = |\mathbf{h}|^{\omega} \qquad (2\text{-}8)$$

Even in this case at $|\mathbf{h}| = 0$, the semi-variogram value is 0. However, very close to origin when $|\mathbf{h}| = \varepsilon$, the semi-variogram may actually exhibit a discontinuity or a jump termed as the nugget effect. There are two possible reasons for this discontinuity:

1. Measurement errors: Due to measurement error or, in many cases, position error of the measuring device, repeat measurements at the same spatial location may yield some uncertainty or variability in the measured value. The semi-variogram value at a lag infinitesimally close to 0 is thus non-zero and equal to the *nugget effect* (Clark, 2010).

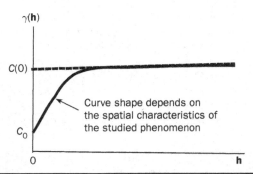

FIGURE 2-4   A typical semi-variogram plotted as a function of lag **h** in a particular direction.

2. Sub-scale variability: Rock attributes such as porosity exhibit variability at very small length scales. For example, if porosity is measured using an X-ray probe, a particular spatial location may be assigned 100% porosity if the X-ray penetrates a pore space, or 0% if it penetrates a rock grain. This dramatic change in porosity occurs over the size of a typical pore-body, say at a length scale of a few microns. In the case of reservoir modeling, the semi-variogram may be computed corresponding to a lag spacing of tens or hundreds of feet. The sub-scale variability at a lag spacing of microns, thus, shows up as a nugget effect when looking at the semi-variogram at the reservoir scale (Pitard, 1994).

Putting all these things together, a typical semi-variogram may be denoted as:

$$\gamma(\mathbf{h}) = C_o + A(\mathbf{h}) \tag{2-9}$$

The continuous part $A(\mathbf{h}) \to 0$ as $\mathbf{h} \to 0$ and $A(\mathbf{h}) \to C(0) - C_o$ as $\mathbf{h} \to \infty$. This characteristic of the semi-variogram is depicted in Fig. 2-4. The semi-variogram shown is for a second-order stationary phenomenon for which the variance is defined. For a non-stationary (but intrinsically stationary) phenomenon, the semi-variogram may not stabilize at a stationary value [as implied by Eq. (2-9)]. Another variation that is possible is of a semi-variogram that may exhibit an oscillatory characteristic referred to as a hole effect above.

2.2.3   Pure Nugget Effect

When there is no spatial correlation at any lag $|\mathbf{h}| > 0$, then $\gamma(\mathbf{h}) = C(0)$; i.e., the variogram is a straight line at the value approximately equal to the prior variance of the data, as shown in Fig. 2-5. This is referred to as the pure nugget effect model (Journel & Huijbregts, 1978).

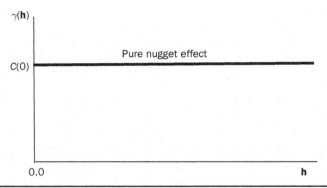

FIGURE 2-5   A semi-variogram representing pure nugget effect or lack of correlation at all lags.

The pure nugget effect model implies a lack of correlation at all lags. Of course, the semi-variogram value at lag $|\mathbf{h}| = 0$ is identically 0, but at a lag distance infinitesimally close to 0, the semi-variogram abruptly jumps to the variance value. In the case of Gaussian RFs, a pure nugget effect implies that $F(Z(\mathbf{u}) \mid Z(\mathbf{u} + \mathbf{h})) = F_Z(z)$ and, consequently, $E(Z(\mathbf{u}) \mid Z(\mathbf{u} + \mathbf{h})) = E(Z(\mathbf{u}))$; i.e., the data is of no value for estimating values at unsampled locations. The best estimate corresponding to a pure nugget scenario is simply the arithmetic average of data:

$$E(Z(\mathbf{u}) \mid Z(\mathbf{u} + \mathbf{h})) = E(Z(\mathbf{u})) = \frac{1}{N}\sum_{i=1}^{N} z(\mathbf{u}_i) \qquad (2\text{-}10)$$

i.e., no resolution of spatial highs and lows is possible. A pure nugget effect model is a drastic one that is predicated on a natural geologic system exhibiting absolute disorder, in which case data does not inform the predictions. As such drastic situations rarely arise as manifestation of geologic processes, a pure nugget effect behavior of the semi-variogram is more likely due to sparse data and the difficulty it poses for inferring a reliable semi-variogram.

2.2.4   Behavior Next to the Origin

A semi-variogram may exhibit different characteristics at short lags depending on the characteristics of the geologic phenomenon being studied. Some of these characteristics could be:

Parabolic: The continuous portion of the semi-variogram function; i.e., $A(\mathbf{h})$ exhibits the following characteristic:

$$\lim_{|\mathbf{h}|\to 0} A(\mathbf{h}) = b \cdot \mathbf{h}^2$$

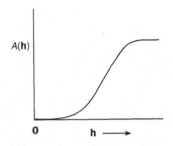

Figure 2-6 A semi-variogram exhibiting a parabolic structure near the origin.

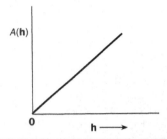

Figure 2-7 A semi-variogram exhibiting a linear structure near the origin.

The function $A(\mathbf{h})$ is twice differentiable near the origin; i.e., it exhibits curvature, as seen in Fig. 2-6. It can be observed in the figure that the rate of climb of the semi-variogram near the origin is very slow, implying that the phenomenon exhibits a great deal of continuity (i.e., variability is low) at shorter lag distances.

Linear: In this case, the continuous part of the semi-variogram function can be written as:

$$\lim_{|\mathbf{h}| \to 0} A(\mathbf{h}) = b \cdot |\mathbf{h}|$$

The function $A(\mathbf{h})$ is differentiable only once but is continuous at the origin, as depicted in Fig. 2-7. The rise of the semi-variogram is steeper in this case and would be indicative of a phenomenon exhibiting more variability at short lags.

Remark
- The linear or parabolic description of $A(\mathbf{h})$ near the origin has to be combined with the observed nugget effect in order to arrive at the complete model for the semi-variogram $\gamma(\mathbf{h})$, as depicted in Fig. 2-8.

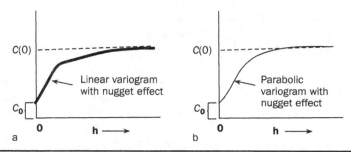

Figure 2-8   **a)** A semi-variogram exhibiting linear characteristics at short lags combined with a nugget effect. **b)** Semi-variogram exhibiting parabolic structure at short lags combined with a nugget effect.

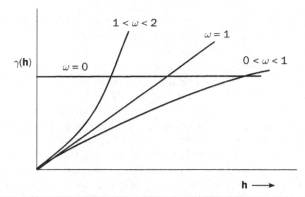

Figure 2-9   Power law semi-variogram corresponding to different values of the exponent.

Behavior at Large Lags

Matheron (1973) demonstrates that a second-order continuous intrinsic RF $Z(\mathbf{u})$ is characterized by the following property:

$$\lim_{|\mathbf{h}| \to \infty} \frac{\gamma(\mathbf{h})}{|\mathbf{h}|^2} \to 0$$

This leads to the following general formulation for the variogram of a continuous, intrinsic phenomenon:

$$\gamma(\mathbf{h}) \propto |\mathbf{h}|^\omega \text{ for } \omega \in [0, 2] \tag{2-11}$$

The power law semi-variogram corresponding to various exponents is depicted in Fig. 2-9. It is clear that the semi-variogram corresponding to $\omega = 0$ is the pure nugget effect model.

Remarks

- For a second-order stationary phenomenon, $\lim_{|\mathbf{h}| \to \infty} \gamma(\mathbf{h}) \to C(0)$; i.e., the semi-variogram has a stable sill.

- The power law model $\gamma(\mathbf{h}) \propto |\mathbf{h}|^{\omega}$ with $0 < \omega < 2$ in general does not converge to a sill and is hence indicative of a non-second-order stationary phenomenon.

- If $\gamma(\mathbf{h}) \propto |\mathbf{h}|^{\omega}$, then the semi-variogram corresponding to a linearly rescaled lag $\mathbf{h}' = r \cdot \mathbf{h}$ is $\gamma(\mathbf{h}') \propto r^{\omega}|\mathbf{h}|^{\omega} = r^{\omega}\gamma(\mathbf{h})$ that exhibits the same (rescaled) characteristic of the original semi-variogram $\gamma(\mathbf{h})$. That is, the modeled phenomenon is self-similar at all scales, which is the definition of a fractal phenomenon (Cheng & Agterberg, 1996).

2.2.5   Outliers and Semi-Variogram Inference

Outliers can cause the semi-variogram to fluctuate and appear to be noisy. As shown in Fig. 2-10, the presence of an outlier in a particular **h**-scatterplot can result in the moment of inertia of the scatterplot to be high as it is affected by the square difference between the outlier value and the value on the bisector.

One way to mitigate influence of outliers would be to exclude outlier pairs from the scattergram for a particular **h**. However, it is important to note that a datum that acts as an outlier for a particular lag **h** could be representative of the spatial variability at another lag **h′**. For example, if the semi-variogram within a sand bed is being computed, the presence of an occasional low value (corresponding to a shale) would be an outlier in the **h**-scatterplot, but that same value would not be an outlier when calculating the semi-variogram corresponding to lags within the shale. This implies that the deliberate deletion of outlier data from datasets is probably a bad idea; instead, it might be more useful to devise other tools that may be used to infer the spatial structure of the phenomenon despite the presence of outliers.

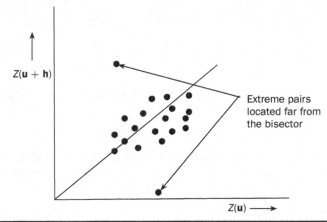

FIGURE 2-10   Depicting the influence of outliers on the semi-variogram value at a particular lag.

The sensitivity to outliers is exaggerated by squaring the differences $[Z(\mathbf{u}) - Z(\mathbf{u} + \mathbf{h})]$ when computing the semi-variogram value. By computing the semi-variogram as (Journel, 1988):

$$2\gamma(\mathbf{h}) = E\left\{[Z(\mathbf{u}) - Z(\mathbf{u} + \mathbf{h})]^{\omega}\right\} \quad \omega \in [0, 2]$$

the effect of outliers can be dampened. This alternate semi-variogram measure can be estimated as:

$$\hat{\gamma}_{\omega}(\mathbf{h}) = \frac{1}{2 \cdot N(\mathbf{h})} \sum_{i=1}^{N(\mathbf{h})} [z(\mathbf{u}_i) - z(\mathbf{u}_i + \mathbf{h})]^{\omega}$$

when $\omega = 2$, we are back to the *traditional semi-variogram*. When $\omega = 1$, we refer to the *madogram* (Bacro, Bel & Lantuejoul, 2008) defined as:

$$2\gamma_1(\mathbf{h}) = E\{|E(\mathbf{u}) - Z(\mathbf{u} + \mathbf{h})|\} \tag{2-12}$$

The *madogram* mitigates the influence of the outliers by accounting for only the first-order difference between the head and tail values of pairs making up the **h**-scatterplot. It can thus be estimated as:

$$\hat{\gamma}_1(\mathbf{h}) = \frac{1}{2 \cdot N(\mathbf{h})} \sum_{i=1}^{N(\mathbf{h})} |z(\mathbf{u}_i) - z(\mathbf{u}_i + \mathbf{h})|$$

By making $\omega = \frac{1}{2}$, the influence of outliers can be further dampened, in which case, we obtain the *rodogram* (Cressie & Hawkins, 1980):

$$\hat{\gamma}_{1/2}(\mathbf{h}) = \frac{1}{2 \cdot N(\mathbf{h})} \sum_{i=1}^{N(\mathbf{h})} \sqrt{|z(\mathbf{u}_i) - z(\mathbf{u}_i + \mathbf{h})|} \tag{2-13}$$

The best possible way for mitigating the influence of outliers is perhaps by taking the logarithm of the values making up the **h**-scatterplot. This results in the *semi-variogram of logarithms* (Goovaerts, 1997):

$$\hat{\gamma}_{\log}(\mathbf{h}) = \frac{1}{2 \cdot N(\mathbf{h})} \sum_{i=1}^{N(\mathbf{h})} [\ln(z(\mathbf{u}_i)) - \ln(z(\mathbf{u}_i + \mathbf{h}))]^2 \tag{2-14}$$

It needs to be noted that these alternate measures can be used to detect features such as the range and anisotropy of the spatial phenomena in the presence of noisy data. However, these alternate semi-variogram measures do not have a variance interpretation associated with them and as such cannot be used directly within a variance-minimization scheme for making predictions such as in kriging.

So, the standard practice is to use these measures to infer the relevant parameters for the semi-variogram and subsequently use the traditional semi-variogram within spatial estimation/prediction schemes.

2.2.6 Proportion Effect

Consider a spatial phenomenon that exhibits different degrees of variability (variance) in different regions of the domain of interest. Furthermore, let us assume that the variance computed in different regions is proportional to the corresponding mean value of RF in that region. Such heteroscedasticity is observed frequently in the case of variables that are characterized by a skewed distribution, such as a lognormal distribution. The change in variance as the mean of a lognormally distributed RV increases is shown in Fig. 2-11.

If, in addition to this heteroscedastic characteristic, the reservoir is preferentially sampled in the high-value regions, the mean of these samples would be high and consequently the variance would be high. Moreover, wells in these regions may be typically tightly spaced in order to maximize the recovery of hydrocarbons. This implies that the data pairs informing the short lag **h**-scattergrams would consist predominantly of high values and consequently will be characterized by high variance (because of heteroscedasticity). The moment of inertia of the scatterplots will consequently be high, and correspondingly, the semi-variogram value $\gamma(\mathbf{h})$ at short lags will be high. Figure 2-12 shows the behavior of the semi-variogram in the presence of proportion effect. The spatial phenomenon (e.g., the distribution of permeability) can be misinterpreted as pure nugget effect.

A strategy to mitigate the effects of proportion effect would be to standardize the computed semi-variogram by the locally varying variance. Furthermore, because of heteroscedasticity, the locally

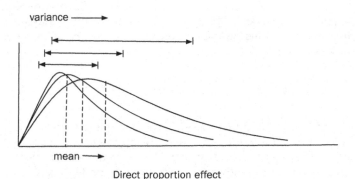

Direct proportion effect

Figure 2-11   An illustration of the heteroscedasticity exhibited by a lognormal random variable.

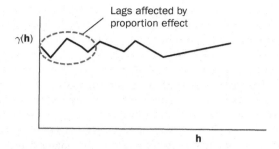

FIGURE 2-12 Semi-variogram exhibiting the effect of proportion effect. (The color version of this figure is available at www.mhprofessional.com/PRMS.)

varying variance can be made a function of the lag mean m(h). Consequently, the generalized relative semi-variogram is defined as:

$$\hat{\gamma}_{GR}(\mathbf{h}) = \frac{\hat{\gamma}(\mathbf{h})}{f(m(\mathbf{h}))} \tag{2-15}$$

The function $f(m(\mathbf{h}))$ may be derived by plotting the variance of values in different regions versus the corresponding mean. A typical form of the function is $f(m(\mathbf{h})) = m^2(\mathbf{h})$, in which case, the corresponding statistic is called *pair-wise relative semi-variogram* (Srivasatava & Parker, 1989):

$$\hat{\gamma}_{PR}(\mathbf{h}) = \frac{1}{2 \cdot N(\mathbf{h})} \sum_{i=1}^{N(\mathbf{h})} \frac{\left[z(\mathbf{u}_i) - z(\mathbf{u}_i + \mathbf{h})\right]^2}{\left[\dfrac{z(\mathbf{u}_i) + z(\mathbf{u}_i + \mathbf{h})}{2}\right]^2} \tag{2-16}$$

The general relative variograms permit inference of semi-variogram characteristics when the data is masked by proportion effect. It will cause the semi-variogram values corresponding to short lags to be decreased, thereby conforming to the typical semi-variogram characteristics. However, the pair-wise relative semi-variogram cannot be directly used in estimation/simulation. After identifying parameters such as the range and the structures, the desired model for the traditional variogram is developed and used in estimation.

2.3 Variogram Modeling

In most applications, the data available to infer a semi-variogram is sparse, and consequently, inference is possible only for few lags $\mathbf{h}_k$. However, during estimation, especially when we are trying to estimate values on a grid (see "kriging" in Chap. 3), semi-variogram values may be needed for arbitrary lags $\mathbf{h}$. These vectors may be at orientations different from those for which the experimental semi-variogram may be computed and may be of magnitude different from the lag separation exhibited by the data.

But there are some other, more powerful reasons for modeling semi-variograms. As will be made clear during the subsequent discussion on kriging, spatial estimation consists of assigning weights to the available data values in order to combine them and make predictions at un-informed locations. We would like to assign unique weights to the available data values so that the resultant map of estimates will be unique. At the same time, we would like the estimated values to satisfy some conditions, such as exhibiting minimum error variance that is positive. It turns out that in order to meet these requirements posed by the estimation procedures, the semi-variogram or under stationarity, the covariance function model has to satisfy some mathematical properties. Our objective is therefore to understand and apply these properties in order to develop a model that yields semi-variogram values for all lags.

Before embarking on a discussion of the essential mathematical properties, it is useful to revisit some semi-variogram properties:

- The type of spatial/temporal phenomena modeled is assumed to be mean square continuous, i.e.,

$$\underset{|\mathbf{h}| \to 0}{\text{Lim}} \left[Z(\mathbf{u}) - Z(\mathbf{u} + \mathbf{h}) \right]^2 = 0$$

Because the semi-variogram $\gamma(\mathbf{h})$ is the variance of stationary increments, it follows that the RF $Z(\mathbf{u})$ is mean square continuous only when the semi-variogram is continuous at $\mathbf{h} = 0$.

- The variogram model is developed as $\gamma(\mathbf{h}) = C_o + A(\mathbf{h})$, where $A(\mathbf{h})$ has to be a continuous function of the lag $\mathbf{h}$.

- The increase in $\gamma(\mathbf{h})$ with $\mathbf{h}$ is regulated by: $\underset{|\mathbf{h}| \to \infty}{\text{Lim}} \dfrac{\gamma(\mathbf{h})}{|\mathbf{h}|^2} = 0$ for a second-order RF. (Strang, 2016)

2.3.1 Positive Definiteness

Consider the variance of a linear combination of random variables $Y(\mathbf{u}) = \sum_{i=1}^{N} \lambda_i \cdot Z(\mathbf{u}_i)$. This represents the linear decomposition of a random function $Y(\mathbf{h})$ using a set of basis RFs $Z(\mathbf{u}_i)$ (Chiles & Delfiner, 1999). This could be, e.g., the decomposition of permeability heterogeneity in a reservoir into a long-range component, such as a channel form and a short-range component, e.g., due to trough cross-bed features. Assume that the basis RFs $Z(\mathbf{u}_i)$ are described by a common covariance function $C(\mathbf{h})$. The variance of such a linear combination under second-order stationarity can be written as:

$$Var\{Y(\mathbf{u})\} = \sum_{j=1}^{N}\sum_{i=1}^{N} \lambda_i \lambda_j \cdot Cov\{\mathbf{h}_{ij}\} \qquad (2\text{-}17)$$

The variance has to be positive, and consequently, the covariance $Cov(\mathbf{h}_{ij})$ in Eq. (2-17) has to be such that the right-hand side double summation is positive no matter what the values of λ_i, N, and $\mathbf{h}_{ij} = \mathbf{u}_i - \mathbf{u}_j$.

Definition (Strang, 2016)   A function $\varphi(x, y)$ is said to be positive definite if and only if $\sum_{j=1}^{n}\sum_{i=1}^{n} a_i a_j \cdot \varphi\{x_i, y_j\} \geq 0 \; \forall \; a_i$ and n.

Furthermore, it is said to be strictly positive definite if and only if:

$$\sum_{j=1}^{n}\sum_{i=1}^{n} a_i a_j \cdot \varphi\{x_i, y_j\} = 0 \text{ only if } a_i = 0 \; \forall \; i$$

The implications of the necessary and sufficient condition for positive definiteness are:

- If $\varphi(x, y)$ is a positive definite function, then the double summation $\sum_{j=1}^{n}\sum_{i=1}^{n} a_i a_j \cdot \varphi\{x_i, y_j\}$ is guaranteed to be positive, no matter the value of a_i.

- If the double summation $\sum_{j=1}^{n}\sum_{i=1}^{n} a_i a_j \cdot \varphi\{x_i, y_j\}$ is positive, then the function $\varphi(x, y)$ has to be positive definite.

In the case of the linear combination of RVs, the double summation corresponds to the variance of a random function $Y(\mathbf{u})$. According to the previous definition, for this variance to be positive, the covariance $Cov(\mathbf{h}_{ij})$ therefore has to be positive definite. Conversely, if there exists a stationary RF $Y(\mathbf{u}) = \sum_{i=1}^{N}\lambda_i \cdot Z(\mathbf{u}_i)$ characterized by a positive variance, then the covariance function $Cov(\mathbf{h}_{ij})$ exists and is positive definite.

In the case of a second-order stationary RF,

$$\gamma(\mathbf{h}) = C(0) - C(\mathbf{h})$$

Substituting this in the expression for variance of $Y(\mathbf{u})$:

$$Var\{Y(\mathbf{u})\} = \sum_{j=1}^{N}\lambda_j\sum_{i=1}^{N}\lambda_i\, C(0) - \sum_{j=1}^{N}\sum_{i=1}^{N}\lambda_i\lambda_j\gamma\left(\mathbf{u}_i - \mathbf{u}_j\right)$$

We would like to be able to infer variance of Y regardless of whether $C(0)$ exists [i.e., when $Z(\mathbf{u})$ is intrinsically stationary]. This is possible by ensuring that $\sum_{i=1}^{N}\lambda_i = 0$. Putting all this together, for the variance of the linear combination $Y(\mathbf{u}) = \sum_{i=1}^{N}\lambda_i \cdot Z(\mathbf{u}_i)$ to be positive with weights, such that $\sum_{i=1}^{N}\lambda_i = 0$, we need to ensure that $\sum_{j=1}^{N}\sum_{i=1}^{N}\lambda_i\lambda_j\gamma(\mathbf{u}_i - \mathbf{u}_j)$ be negative definite, i.e.,

$$-\sum_{j=1}^{N}\sum_{i=1}^{N}\lambda_i\lambda_j\gamma(\mathbf{u}_i - \mathbf{u}_j) \geq 0$$

This implies that the variogram function $\gamma(\mathbf{h}_{ij})$ should be conditionally negative definite, with the accompanying condition $\sum_{i=1}^{N}\lambda_i = 0$. There is another approach to arrive at the conditionally negative condition for the semi-variogram even in the case of an intrinsically stationary RF.

2.3.2 Allowable Linear Combinations

For intrinsic RF, an allowable linear combination is defined as (Chiles & Delfiner, 1999):

$$Y(\mathbf{u}) = \sum_{i=1}^{N}\lambda_i \cdot Z(\mathbf{u}_i) \text{ such that } \sum_{i=1}^{N}\lambda_i = 0$$

Consider the variance of such an RF $Y(\mathbf{u})$:

$$Var\{Y(\mathbf{u})\} = Cov\{Y(\mathbf{u}), Y(\mathbf{u})\} = Cov\left\{\sum_{i=1}^{N}\lambda_i \cdot Z(\mathbf{u}_i), \sum_{i=1}^{N}\lambda_i \cdot Z(\mathbf{u}_i)\right\}$$

Introducing an arbitrary reference RF $Z(\mathbf{u}_o)$ (or Z_o), the right-hand side of the above equation becomes: $\sum_{i=1}^{N}\sum_{j=1}^{N}\lambda_i\lambda_j Cov\{Z(\mathbf{u}_i) - Z(\mathbf{u}_o), Z(\mathbf{u}_j) - Z(\mathbf{u}_o)\}$. Furthermore:

$$(Z_i - Z_j)^2 = \left[(Z_i - Z_o) - (Z_j - Z_o)\right]^2$$
$$= (Z_i - Z_o)^2 + (Z_j - Z_o)^2 - 2(Z_i - Z_o)(Z_j - Z_o)$$

Taking the expectation of both sides and recognizing that $E\{(Z_i - Z_j)^2\} = 2 \cdot \gamma(\mathbf{h}_{ij})$ under intrinsic stationarity:

$$\gamma(Z_i - Z_j) = \gamma(Z_i - Z_o) + \gamma(Z_j - Z_o) - Cov\{Z_i - Z_o, Z_j - Z_o\}$$

This yields:

$$\sum_{i=1}^{N}\sum_{j=1}^{N}\lambda_i\lambda_j Cov\{Z(\mathbf{u}_i) - Z(\mathbf{u}_o), Z(\mathbf{u}_j) - Z(\mathbf{u}_o)\}$$
$$= \sum_{i=1}^{N}\sum_{j=1}^{N}\lambda_i\lambda_j\gamma(Z_i - Z_o) + \sum_{i=1}^{N}\sum_{j=1}^{N}\lambda_i\lambda_j\gamma(Z_j - Z_o) - \sum_{i=1}^{N}\sum_{j=1}^{N}\lambda_i\lambda_j\gamma(Z_i - Z_j)$$
$$= \sum_{j=1}^{N}\lambda_j\sum_{i=1}^{N}\lambda_i\gamma(Z_i - Z_o) + \sum_{i=1}^{N}\lambda_i\sum_{j=1}^{N}\lambda_j\gamma(Z_j - Z_o) - \sum_{i=1}^{N}\sum_{j=1}^{N}\lambda_i\lambda_j\gamma(Z_i - Z_j)$$

Recognizing that $\sum_{i=1}^{N} \lambda_i = 0$ in the above equation (the first two terms become 0), we obtain:

$$Var\{Y(\mathbf{u})\} = Cov\left\{\sum_{i=1}^{N}\lambda_i \cdot Z(\mathbf{u}_i), \sum_{i=1}^{N}\lambda_i \cdot Z(\mathbf{u}_i)\right\}$$

$$= -\sum_{i=1}^{N}\sum_{j=1}^{N}\lambda_i\lambda_j\gamma(Z_i - Z_j) \qquad (2\text{-}18)$$

Imposing the condition that the variance be necessarily positive results in the same condition that the semi-variogram $\gamma(Z_i - Z_j)$ be conditionally negative definite.

Remarks
- In light of the above discussion, it is useful to examine what some of the attributes of a positive definite function are. The following is excerpted from Buescu & Paixão (2012):
  - A positive semi-definite function of a real argument x is, in general, a complex valued function f, such that for any set of real numbers $x_1, x_2, ..., x_n$, the $n \times n$ matrix: $A = [a_{ij}]_{i,j=1}^{n}$, where $a_{ij} = f(x_i - x_j)$ is positive semi-definite. The function f in this case is the covariance that is a function of lag $\mathbf{h}_{ij} = |\mathbf{u}_i - \mathbf{u}_j|$. The implication is that the *matrix A* has to be Hermitian [i.e., the complex-square matrix is equal to its complex conjugate or $f(-x) = \overline{f(x)}$], or in the case of a real valued function, the matrix A has to be symmetric and $f(-x) = f(x)$.
  - It is also necessary (but not sufficient) that $f(0) \geq 0$ and $|f(x)| \leq f(0)$. These conditions stem from the previous requirement for A to be positive definite (which requires that all the sub-determinants also be positive definite—the above two conditions arise out of positive definiteness of the $n = 1, 2$ sub-matrices).
- A function is deemed negative semi-definite if the inequalities are reversed, i.e., $f(0) \leq 0$ and $|f(x)| \geq f(0)$.
- For a strictly positive or negative definite function, the weak inequalities ($\leq$, $\geq$) will be replaced by the corresponding strong inequality.

2.3.3 Legitimate Structural Models

One way to get a symmetric positive semi-definite function is by taking the Fourier transform of a positive function (Bochner, 1941) defined in $\Re^n$:

$$C(|\mathbf{h}|) = \int_{\Re^n} e^{ih\cdot\omega}G(\omega)d^n\omega \qquad (2\text{-}19)$$

By Fourier inversion:

$$G(\omega) = (2\pi)^{-n} \int_{\Re^n} e^{-ih\cdot\omega} C(|\mathbf{h}|) d^n\omega \qquad (2\text{-}20)$$

Note that in the above-mentioned equation, ω is a vector of frequencies belonging to $\Re^n$. If $G(\omega)d\omega = G(d\omega)$ is absolutely continuous, then $G(\omega)$ is referred to as the spectral density and $G(d\omega)$ as the spectral measure. Because the RF $Z(\mathbf{u})$ or more generally $Y(\mathbf{u})$ in our cases are real-valued, the spectral density is an even, symmetric function, i.e., $G(\omega) = G(-\omega)$ and:

$$G(\omega) = (2\pi)^{-n} \int_{\Re^n} \cos(\omega \cdot |\mathbf{h}|) C(|\mathbf{h}|) d^n\omega \qquad (2\text{-}21)$$

If the RF $Y(\mathbf{u})$ is isotropic, then the spectral density is only a function of the magnitude of ω, i.e., $g(|\omega|)$. We denote the magnitude of ω as r, i.e., $r = |\omega|$.

In order to evaluate the integrals associated with the Fourier transform, it is useful to perform a polar transformation and work with spherical coordinates (Bracewell, 1986). If we denote by σ, the arc of a unit circle in $\Re^2$ or equivalently, the sector of a sphere in $\Re^3$ and designate a uniform probability density for σ: $d\sigma$, we can transform Cartesian coordinates (x_1, x_2) to spherical coordinates (r, σ). The derivative dx^2 becomes $2\pi r\, dr\, d\sigma$ in $\Re^2$. In $\Re^3$, the transformation results in $dx^3 \rightarrow 4\pi r^2$. In general, the transformation in $\Re^n$ results in $dx^n \rightarrow \dfrac{(2\pi)^{n/2} r^{n-1}}{\Gamma(n/2)} dr\, d\sigma.$

Now getting back to the vector of values ω, as noted above, the vector is only positive valued for an isotropic $Y(\mathbf{u})$. In the transformed space, each outcome of σ; that is, σ_j has the corresponding probability density $d\sigma_j$. The multivariate joint distribution of σ_j^2 (positive valued) is thus described by the Dirichlet distribution. The domain of a Dirichlet distribution is itself a set of probability distributions, just like the distribution of each σ_j^2. The Dirichlet distribution is parameterized by $[\alpha_i]$. As pointed out by Frigyik et al. (Frigyik, Kapila & Gupta, 2010), if the parameters $[\alpha_i]$ are identically 1, then the distribution is uniform over the entire range of its support. On the other hand, if the parameters $[\alpha_i]$ are all less than 1, then the joint distribution has a spike corresponding to each outcome in its support. Clearly, the latter is what we want and so the parameters of the Dirichlet distribution are set to $\left[\dfrac{1}{2}\right]$. The marginal of the Dirichlet distribution is the Beta distribution $\text{Beta}(\alpha_i, \alpha_0 - \alpha_i)$, where $\alpha_0 = \sum_n \alpha_i$. In our case, the marginal distribution is described by $\text{Beta}\left(\dfrac{1}{2}, \dfrac{(n-1)}{2}\right)$. This is the marginal for σ_j^2. From this distribution, an outcome for σ_j can be obtained as the square root of a Beta$\left[\dfrac{1}{2}, \dfrac{(n-1)}{2}\right]$ random variable.

Now, we are ready to develop a strategy for obtaining positive semi-definite covariance functions $C(\mathbf{h})$. In the following, we recognize that $r = |\omega| \geq 0$ and $\sigma = \dfrac{\omega}{|\omega|}: \sigma \in Sphere^{n-1}$. We define the component of σ in the direction of $\mathbf{h}$ as σ_h: $\sigma_h = \sigma \cdot \dfrac{\mathbf{h}}{|\mathbf{h}|}$. Furthermore, we designate $|\mathbf{h}| = \rho$.

$$C(\rho) = \int\limits_{\Re^n} \cos(\omega \cdot |\mathbf{h}|) g(|\omega|) d^n\omega$$

$$= \int\limits_{\Re^+} \int\limits_{S^{n-1}} \cos(r \cdot \rho \cdot \sigma_h) \cdot g(r) \frac{(2\pi)^{n/2} r^{n-1}}{\Gamma(n/2)} dr\, d\sigma$$

Now, getting back to our argument regarding σ being the square root of a Beta variable, u, the Beta distribution is written as (Nadarajah & Gupta, 2004):

$$\beta\left(u : \frac{1}{2}, \frac{(n-1)}{2}\right) = \frac{\Gamma\left(\dfrac{1}{2} + \dfrac{(n-1)}{2}\right)}{\Gamma\left(\dfrac{1}{2}\right) \cdot \Gamma\left(\dfrac{(n-1)}{2}\right)} u^{1-1/2}(1-u)^{1-(n-1)/2}$$

Substituting in the above integral:

$$C(\rho) = \int\limits_{\Re^+} \int\limits_0^1 \cos(r \cdot \rho \cdot \sqrt{u}) \cdot g(r) \frac{(2\pi)^{n/2} r^{n-1}}{\Gamma\left(\dfrac{1}{2}\right) \cdot \Gamma\left(\dfrac{(n-1)}{2}\right)} u^{1/2}(1-u)^{(3-n)/2} dr\, du$$

Substituting $\cos\theta = \sqrt{u}$ and recognizing that the Bessel's function of first order is:

$$J_v(x) = \frac{(x/2)^v}{\sqrt{\pi}\Gamma(v+1/2)} \int\limits_0^\pi \cos(z\cos\theta)\sin(\theta)^{2v} d\theta$$

In the above expression: $v = \dfrac{n}{2} - 1$. The following expression for a positive definite covariance function is obtained:

$$C(\rho) = \int\limits_0^\infty \rho(2\pi r/\rho)^{v+1} J_v(r\rho) g(r) dr \tag{2-22}$$

The forms of Eq. (2-22) for different values of n are given below:

$$n = 1: C(\rho) = \int_0^\infty 2\cos(r\rho)g(r)dr$$

$$n = 2: C(\rho) = \int_0^\infty 2\pi r J_0(r\rho)g(r)dr \qquad (2\text{-}23)$$

$$n = 3: C(\rho) = \int_0^\infty \rho(2\pi r/\rho)^{3/2} J_{1/2}(r\rho)g(r)dr$$

Equation (2-22) provides a recipe for developing various covariance models that will be positive semi-definite and will thus be conducive to modeling second-order RFs. This is achieved by specifying different models for the spectral density $g(r)$. More details regarding permissible models for covariances and semi-variograms can be found in Christakos (1984).

The Exponential Family
In addition to the robust procedure for developing positive definite covariance models described above, there are some other mathematical functions that are guaranteed to be positive definite. One such function is the exponential function (Deutsch, 1998). Models in this family can be broadly written as:

$$C(\rho) = \exp\left[-\left(\frac{\rho}{a}\right)^p\right], \quad 0 < p \le 2 \qquad (2\text{-}24)$$

The parameter a is the correlation range of the phenomenon being studied. When $p = 1$, we get the *exponential model*:

$$C(\rho) = \exp\left[-\left(\frac{\rho}{a}\right)\right] \qquad (2\text{-}25)$$

The spectral density corresponding to this exponential model is:

$$g(r) = \frac{(a^2/2\pi)}{(1 + r^2 a^2)^{3/2}}$$

This is of a form very similar to the bivariate Cauchy distribution:

$$f(x,y;x_o,y_o,\gamma) = \frac{1}{2\pi} \cdot \frac{\gamma^2}{\left(\gamma^2 + \rho^2\right)^{1.5}}$$

where
$$\rho^2 = (x - x_o)^2 + (y - y_o)^2$$

When $p = 2$ in Eq. (2-24), we get the *Gaussian model* (Dubrule, 2017). The reference specifically talks about conditions when the

Gaussian models may not be permissible to model a spatial phenomenon. The model is written as:

$$C(\rho) = \exp\left(-\left[\frac{\rho}{a}\right]^2\right)$$ (2-26)

The corresponding spectral density is similar to a normal density:

$$g(r) = \frac{a^2}{4\pi} e^{-\frac{r^2 a^2}{4}}$$

If the range specified for an exponential model is a, then the actual range of the covariance or the semi-variogram is $3 \cdot a$, i.e., the distance at which semi-variogram reaches 95% of sill. In order to represent the covariance converging asymptotically to a range a, it is customary to write the model as:

$$C(\rho) = \exp\left(-3 \cdot \left[\frac{\rho}{a}\right]\right)$$ (2-27)

The Gaussian model is similarly written as:

$$C(\rho) = \exp\left(-3 \cdot \left[\frac{\rho}{a}\right]^2\right)$$ (2-28)

If the exponential covariance function is expanded in the form of a Taylor's series:

$$C(\rho) = 1 - 3 \cdot \frac{\rho}{a} + \frac{1}{2!}\left(3 \cdot \frac{\rho}{a}\right)^2 - \ldots$$

That is, in the limit that the lag $\rho \to 0$:

$$\lim_{\rho \to 0} C(\rho) = 1 - 3 \cdot \frac{\rho}{a}$$

i.e., the higher-order terms converge rapidly to 0, and so the exponential covariance model behaves in a linear fashion close to the origin (Fig. 2-13).

The Gaussian covariance function [Eq. (2-26)] is infinitely differentiable and, hence, representative of a highly continuous phenomenon. The Taylor's series expansion for the Gaussian covariance function is:

$$C(\rho) = 1 - 3 \cdot \left(\frac{\rho}{a}\right)^2 + \frac{3^2}{2!}\left(\frac{\rho}{a}\right)^4 - \ldots$$

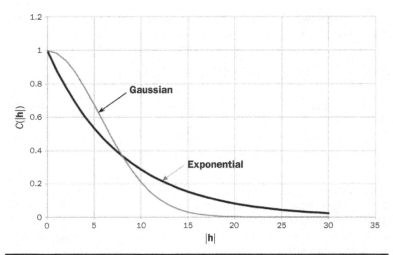

FIGURE 2-13 Positive semi-definite covariance models of the exponential family.

Again, at the limit when $\rho \to 0$:

$$\operatorname*{Lim}_{\rho \to 0} C(\rho) = 1 - 3 \cdot \left(\frac{\rho}{a}\right)^2$$

This suggests that the covariance function exhibits a parabolic behavior near the origin (Fig. 2-13). At short lags, the covariance experiences a very slight decrease indicating a quasi-deterministic phenomenon.

Matern Model
A general formulation for positive definite covariance models results corresponding to the spectral density:

$$g(r) = \frac{a^n}{\Gamma(s)\pi^{n/2}}\left(1 + a^2 r^2\right)^{-s-n/2} \tag{2-29}$$

The spectral density has an additional parameter s that lends shape to the resultant covariance function and is referred to as the shape parameter. The corresponding covariance function expression is (Diggle & Ribeiro, 2007):

$$C(\rho) = \frac{1}{2^{s-1}\Gamma(s)}\left(\frac{\rho}{2a}\right)^s K_s\left(\frac{\rho}{a}\right) \tag{2-30}$$

K_s is the modified Bessel function of the third kind of order s. The covariance function corresponding to shape factor $s = 2$ is shown in Fig. 2-14.

As can be seen, the Matern model bears resemblance to the exponential models described above. In fact, as $s \to \infty$, the Matern model converges to the Gaussian model and when $s = \dfrac{1}{2}$, the Matern

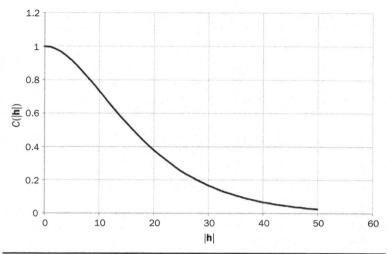

FIGURE 2-14   The Matern model for covariance corresponding to a shape factor of 2 and a covariance range of 8.

model behaves like the exponential covariance model. These limiting characteristics of the Matern model are arrived at by recognizing the similarity of the spectral density function to a Student t-distribution with $2s$ degrees of freedom and variance that scales as $\sigma^2 = \dfrac{1}{2a^2s}$. Indeed, when $s \to \infty$, the variance scales to 0; i.e., the phenomenon acts deterministic just as in a Gaussian model.

Spherical Model
A positive definite covariance model can be defined by looking at the intersection between two discs of radii a separated by a distance ρ. Consider a cluster of points distributed in $\Re^3$ such that each coordinate of the point is uniformly sampled (Wackernagel, 2003). The average point density per unit volume is designated as θ, and N_V is the number of such points within a volume V. If we consider a sphere S of radius a centered at the location $\mathbf{u}$, then the random variable $Z(\mathbf{u})$ is defined as:

$$Z(\mathbf{u}) = N(S_{\mathbf{u}})$$

The random variable measures the volume of influence of the point $\mathbf{u}$. Next, we define a binary indicator variable:

$$I_S(\mathbf{u}') = \begin{cases} 1, & \mathbf{u}' \in S_{\mathbf{u}} \\ 0, & \text{otherwise} \end{cases}$$

The indicator or geometric covariance measures the joint probability that two points $\mathbf{u}$ and $\mathbf{u}' = \mathbf{u} + \rho$ fall within $S_{\mathbf{u}}$. This can be computed as:

$$K(\rho) = \int\limits_{-\infty}^{\infty} \int\limits_{-\infty}^{\infty} \int\limits_{-\infty}^{\infty} I_S(\mathbf{u}) \cdot I_S(\mathbf{u} + \rho) d\mathbf{u}$$

Alternatively, defining two spheres, S_u and $S_{u+\rho}$, the indicator covariance can be calculated as the probability that the point u falls within the intersection of S_u and $S_{u+\rho}$:

$$K(\rho) = \int\limits_{-\infty}^{\infty} \int\limits_{-\infty}^{\infty} \int\limits_{-\infty}^{\infty} I_{S_u}(u) \cdot I_{S_{u+\rho}}(u) du$$

i.e.,

$$K(\rho) = \left| S_u \cap S_{u+\rho} \right|$$

The covariance function for the RV $Z(u)$ is simply the centered form of $K(\rho)$ and can be written as (Wackernagel, 2003):

$$C(\rho) = \theta \cdot K(\rho)$$

Because $K(\rho)$ is the volume of intersection of two spheres separated by a lag ρ, an expression can be analytically developed. The resultant covariance function is:

$$C(\rho) = \theta \cdot |B| \cdot \left[1 - \frac{3}{2} \cdot \frac{\rho}{a} + 0.5 \left(\frac{\rho}{a} \right)^3 \right] \qquad (2\text{-}31)$$

$|B|$ is the volume of the sphere: $|B| = \frac{4}{3}\pi a^3$, θ is the density of points in the sphere and so the product $\theta \cdot |B|$ represents the prior variance of $Z(u) : C(0)$. At lags greater than the range a, the volume of the intersection goes to 0 and so does the covariance.

The spectral density corresponding to the spherical covariance (and semi-variogram) can be developed but does not reveal any interesting insights. Instead, it is useful to think of the spherical covariance as a convolution or moving average of Gaussian white noise. Positive definite covariance models in lower dimensions can also be derived as the circular covariance ($\Re^2$) or the triangular covariance ($\Re^1$) as well in higher dimensions in terms of intersection of n-dimensional hyper-spheres (Chiles & Delfiner, 1999).

The spherical model is bounded in that:

$$\gamma(\rho > a) = 1 - C(\rho > a) = 1$$

At the limit when $\rho \rightarrow 0$:

$$\lim_{\rho \rightarrow 0} C(\rho) = 1 - \frac{3}{2} \left(\frac{\rho}{a} \right)$$

Similar to the exponential model, the spherical model also exhibits a linear characteristic at short lags. However, the slope of the exponential covariance model is 3 which is double the slope of the spherical model $\frac{3}{2}$. This implies that the exponential model is representative of phenomena exhibiting much more variability.

Power Law Model

As discussed before, the semi-variogram corresponding to a second-order intrinsic RF can be developed as:

$$\gamma(\mathbf{h}) = |\mathbf{h}|^\omega \quad 0 < \omega < 2$$

The reason for the constraint on the value of ω is detailed in Chiles & Delfiner (1999). Corresponding to $\omega = 2 : \dfrac{d\gamma(\mathbf{h})}{d\mathbf{h}} = 2 \cdot |\mathbf{h}|$, i.e., linear with a slope depending on $|\mathbf{h}|$.

Power law variogram represents a self-similar (fractal) phenomena, i.e., phenomena for which a characteristic scale cannot be defined.

Remarks

- If the covariance function is strictly positive definite, then an equation system such as the kriging system results in a unique solution for the weights λ_β:

$$\sum_{\beta=1}^{N} \lambda\beta \cdot Cov\left(\mathbf{h}_{\alpha\beta}\right) = Cov(\mathbf{h}_{\alpha o})$$

This implies that a unique estimate for the RV $Y(\mathbf{u})$ can be determined.

- Positive definite function (conditional or not) defined in $\Re^n$ will exhibit positive definiteness in $\Re^m$, $\forall m \leq n$ but not for $m > n$ (Posa & De Iaco, 2018). For example,

$$C(r) = \begin{cases} 1 - \dfrac{r}{a}, & \text{if } r \leq a, \quad a > 0 \\ 0 & \text{otherwise} \end{cases}$$

$C(r)$ is positive definite in $\Re^1$ but not in $\Re^2$ and for all higher dimensions.

The implications of this are that when modeling a 3D phenomenon, modeling the semi-variograms in individual directions will not ensure that a positive definite 3D model would result. On the other hand, if a 3D model is developed, then semi-variograms in individual directions computed using that model would be positive definite.

2.3.4 Positive Combinations

A single correlation range a characterizes the positive definite models for stationary random processes discussed previously. Several geological phenomena exhibit variability at multiple length scales. For example, within a deltaic sequence, the sandstone beds may exhibit long range continuity defined by sea-level advances. However, within such massive sandstone beds, there may be finer scale laminar structures produced by local variations in sea levels caused by tidal influence.

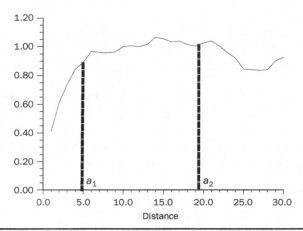

FIGURE 2-15   Semi-variogram corresponding to a spatial phenomenon exhibiting two length scales. (The color version of this figure is available at www.mhprofessional.com/PRMS.)

Figure 2-15 depicts the semi-variogram exhibited by such a multi-scale phenomenon. It can be observed that the semi-variogram exhibits a nugget effect of 0.10 and rises normally to a range of 5 and a sill value of 0.85. The semi-variogram then continues to rise to a sill value of 1.0 over a range of around 19. It is clear that attempting to model such a multi-scale phenomenon using a traditional positive definite model for a semi-variogram would be inadequate (Gringarten & Deutsch, 2001).

Figure 2-16 Depicts the fitted model using a single spherical structure of range 5. It can be concluded that modeling such multi-scale variograms may not be feasible using a single positive-definite structure, such as those described earlier.

In order to model spatial phenomena exhibiting multiple length scale structures, we make use of the following theorem related to positive combinations of legitimate semi-variogram structures. Note that legitimacy in this context implies positive semi-definite covariance (or equivalently, negative semi-definite semi-variogram) structures (Grzebyk & Wackernagel, 1994).

Theorem   Any positive linear combination or product of covariances is also a licit covariance. Mathematically, consider a basic, legitimate covariance model $C_k(\mathbf{h})$, such as one of the models discussed earlier. Suppose the K in such models is positive semi-definite in $\Re^n$. Then, according to the theorem stated above, $C(\mathbf{h}) = \sum_{k=1}^{K} \lambda_k \cdot C_k(\mathbf{h})$, $\Re^n$, provided $\lambda_k > 0$, $\forall k$.

Proof   Consider the following linear decomposition of an RF:

$$Z(\mathbf{u}) = \sum_{k=1}^{K} \lambda_k \cdot Z_k(\mathbf{u}) \,\forall\, \lambda_k \tag{2-32}$$

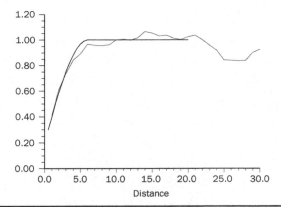

Model fit using a single legitimate variogram structure for a spatial phenomenon exhibiting two length scales. (The color version of this figure is available at www.mhprofessional.com/PRMS.)

As we saw earlier, the covariance of a linear sum can be written as:

$$\text{Cov}\{Z(\mathbf{u}), Z(\mathbf{u} + \mathbf{h})\} = \sum_{k=1}^{K} \sum_{k'=1}^{K'} \lambda_k \lambda_{k'} \cdot \text{Cov}\{Z_k(\mathbf{u}), Z_{k'}(\mathbf{u} + \mathbf{h})\}$$

Next, we make an assumption that the K random functions $Z_k(\mathbf{u})$ are independent of each other, i.e.,

$$Cov\{Z_k(\mathbf{u}), Z_{k'}(\mathbf{u} + \mathbf{h})\} = 0 \ \forall \ k \neq k'$$

Each RF exhibits spatial correlation described by $C_k(\mathbf{h})$. Using this assumption in the previous double summation results in:

$$C(\mathbf{h}) = \sum_{k=1}^{K} \lambda_k^2 \cdot C_k(\mathbf{h}) \tag{2-33}$$

Thus, the linear combination of covariance functions $C_k(\mathbf{h})$ yields a positive definite function $C(\mathbf{h})$, provided the weights $\mu_k = \lambda_k^2 > 0$.

Continuing with the previous example in Fig. 2-15, the semi-variogram of the phenomenon seems to exhibit a nugget effect of 0.10. Subsequently, two structures are evident, one with a contribution of 0.75 (0.85 − 0.10) to the total variance and with a range of approximately 5 units. The second structure exhibits a contribution of 0.15 (1 − 0.15) to the total variance and has a range of approximately 20 units. The sum of the three contributions (0.10—Nugget; 0.75—Structure 1; 0.15—Structure 2) adds up to 1. Note also that because these are contributions to the variance, we are assured that these will always be positive. The remaining task is to pick the most appropriate

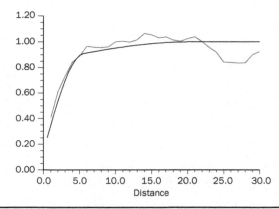

Figure 2-17   Semi-variogram model fit obtained using nested structures. Two spherical structures have been used to model the spatial structure exhibited by the phenomenon. (The color version of this figure is available at www.mhprofessional.com/PRMS.)

semi-variogram functions to represent the two structures. We persist with the previous choice of spherical models for both the structures. Figure 2-17 shows the model obtained using this scheme of nested structures and how it fits the experimental semi-variogram.

As is evident, a much better fit to the experimental semi-variogram is achieved using the nested model. More interestingly, the nest model suggests that the spatial phenomenon being modeled exhibits two length scale structures of correlation lengths 5 units and 20 units, respectively. Figure 2-18 shows the image corresponding to which the original experimental semi-variogram was computed. As is evident in this image, there are darker-shaded regions of smaller correlation length embedded within the gray-shaded regions that exhibit longer continuity. This confirms that what we see in the semi-variogram in terms of nested structures indeed conforms to the physical characteristics of the phenomenon being modeled.

Myers & Journel (1990) point out the problem posed by insisting on a positive combination of nested structures for the semi-variograms. They point out the singular matrix system that results when a positive combination of semi-variograms is used to model a zonal anisotropy. Posa & De Iaco (2018) also point out the difficulties that may arise when a positive liner combination is used to model oscillatory phenomena, such as a hole effect. They provide some details about a scheme for using arbitrary combinations of semi-variogram that can still preserve the positive definite condition.

2.3.5   Geometric Anisotropy

The previous discussions on basic positive-definite semi-variogram structures and nested semi-variogram modeling presumed that the spatial phenomenon being modeled exhibits the same range

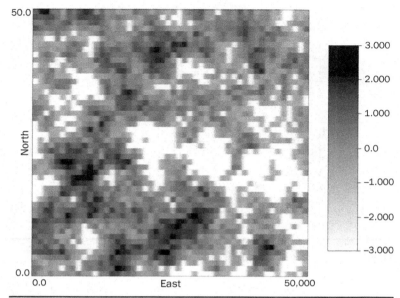

FIGURE 2-18 The spatial model corresponding to the experimental semi-variogram depicted in Fig. 2-15.

of correlation in all directions. That isotropic range shows up as the parameter a in the semi-variogram functions. If the range corresponding to the semi-variogram in different directions is plotted, the corresponding rose diagram will appear as shown in Fig. 2-19(a). In contrast, if the spatial phenomenon exhibits different ranges of correlation in different directions, then the corresponding rose diagram will appear as shown in Fig. 2-19(b). In this case, instead of the semi-variogram being characterized by a single parameter a, the set of ranges a_1, a_2, $\rightleftharpoons$ in different directions characterizes the semi-variogram. In that case, how can we represent such an anisotropic phenomenon using semi-variogram functions that are parameterized using a single range a? This question is addressed in the ensuing discussion.

Remarks

- The spatial phenomenon will be assumed to exhibit anisotropy described by a pair of orthogonal directions. This is consistent with how we perceive patterns—we see continuity in a particular direction, and orthogonal to it we see contrast, i.e., least continuity.

- The variability exhibited by the spatial phenomenon (indicated by the semi-variogram sill) is assumed to be the same in all directions. It is only the range over which the sill is attained that is assumed to be different in different directions.

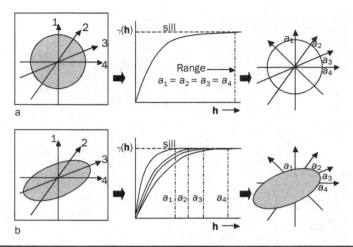

Figure 2-19   Rose diagram corresponding to an **a)** isotropic spatial phenomenon and **b)** anisotropic spatial phenomenon.

- The phenomenon may exhibit multiple scales of variability that each may exhibit different directions of anisotropy [see examples in Gringarten & Deutsch (2001)]. Our strategy will be to identify the nested structures and subsequently model the anisotropy exhibited by each structure.

- Convention: Anisotropy (azimuth) angles are with reference to North axis or y direction. For 3D phenomena, dip angles are with reference to the horizontal plane.

The inference of geometric anisotropy is possible by computing the semi-variogram along pairs of orthogonal directions. If the ratio of the range in the minimum direction of continuity to that in the direction of maximum continuity (i.e., the anisotropy ratio) is computed, the pair of directions along which that ratio is the largest would be indicative of the anisotropy direction. Once the anisotropy direction is identified, in order to compute semi-variogram values that account for the anisotropy direction, some transformations are necessary, and these are discussed next.

2.3.6   Coordinate Transform in 2D

Consider the coordinate system shown in Fig. 2-20. In this figure, a vector $\mathbf{h}$ (in $\Re^2$) specified in the original Cartesian space (with projections h_x, h_y on the original x and y axes) corresponds to an equivalent vector $\mathbf{h}'$ (with projections $h_{x'}$, $h_{y'}$) relative to a rotated direction of anisotropy. The spatial phenomenon exhibits anisotropy at an azimuth angle θ. Suppose the semi-variogram value is desired corresponding to a lag vector $\mathbf{h}$ specified either in terms of the projections h_x, h_y or alternatively by specifying the magnitude of the vector and the azimuthal angle α it subtends.

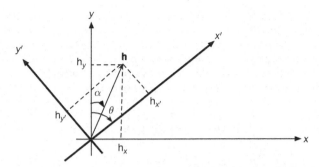

FIGURE 2-20 Rotational transformation brought about due to anisotropy exhibited by the spatial phenomenon.

The vector **h** makes an angle α with the Cartesian y direction and angle $(\theta - \alpha)$ with the anisotropy axis x'. The projection of **h** on the rotated system of coordinates is $\mathbf{h}_{x'}$ and $\mathbf{h}_{y'}$:

$$\mathbf{h}_{x'} = |\mathbf{h}| \cdot \cos(\theta - \alpha)$$

and

$$\mathbf{h}_{y'} = |\mathbf{h}| \cdot \sin(\theta - \alpha)$$

The projections of **h** on the original system of axes are:

$$\mathbf{h}_{x} = |\mathbf{h}| \cdot \sin\alpha$$

and

$$\mathbf{h}_{y} = |\mathbf{h}| \cdot \cos\alpha$$

Combining the two representations:

$$\mathbf{h}_{x'} = |\mathbf{h}| \cdot \cos(\theta - \alpha) = |\mathbf{h}| \cos\theta \cos\alpha + |\mathbf{h}| \sin\theta \sin\alpha$$
$$= h_y \cos\theta + h_x \sin\theta$$

Similarly,

$$\mathbf{h}_{y'} = h_y \sin\theta - h_x \cos\theta$$

That is, the projection of **h** coordinates can be written as:

$$\begin{bmatrix} \mathbf{h}_{x'} \\ \mathbf{h}_{y'} \end{bmatrix} = \begin{bmatrix} \sin\theta & \cos\theta \\ -\cos\theta & \sin\theta \end{bmatrix} \cdot \begin{bmatrix} \mathbf{h}_{x} \\ \mathbf{h}_{y} \end{bmatrix} \tag{2-34}$$

Denoting the rotation matrix: $R_\theta = \begin{bmatrix} \sin\theta & \cos\theta \\ -\cos\theta & \sin\theta \end{bmatrix}$

$$\mathbf{h'} = \begin{bmatrix} h'_x \\ h'_y \end{bmatrix} = R_\theta \cdot \mathbf{h} \tag{2-35}$$

Equation (2-35) depicts the transformation of a vector **h** defined with respect to the original coordinate frame of reference to an equivalent vector **h'** defined with respect to the anisotropic coordinate frame of reference. In this rotated frame of reference, the projections $\mathbf{h}_{x'} \neq \mathbf{h}_{y'}$. Extending this to the range of the semi-variogram, $a_{x'} \neq a_{y'}$. But in order to utilize the previous formulas for positive definite semi-variograms, the range should be the same in all directions. This implies that the ellipse that depicts the anisotropy (with major axis $a_{x'} > a_{y'}$) has to be transformed into a circle such that $a_{x''} = a_{y''}$. In order to accomplish this, one strategy would be to shrink the major axis $a_{x'}$ into $a_{x''}$ such that $a_{x''} = a_{y'}$. This is accomplished by further multiplying $\mathbf{h}'$ as given in Eq. (2-35) by an anisotropy matrix:

$$\begin{bmatrix} \mathbf{h}_{x''} \\ \mathbf{h}_{y''} \end{bmatrix} = \begin{bmatrix} \lambda & 0 \\ 0 & 1 \end{bmatrix} \cdot \begin{bmatrix} \mathbf{h}_{x'} \\ \mathbf{h}_{y'} \end{bmatrix} \tag{2-36}$$

The anisotropy ratio λ is defined as $\lambda = \frac{a_{minor}}{a_{major}}$. The resultant vector $\mathbf{h}'' = \begin{bmatrix} \mathbf{h}_{x''} \\ \mathbf{h}_{y''} \end{bmatrix}$ is in terms of an isotropic set of coordinates with range equal to a_{minor} or $a_{y'}$. The two-step transformation process can therefore be written as:

$$\mathbf{h}'' = \begin{bmatrix} \mathbf{h}_{x''} \\ \mathbf{h}_{y''} \end{bmatrix} = D_\lambda \cdot R_\theta \cdot \mathbf{h} \tag{2-37}$$

Example 2-1

Consider the pattern of spatial heterogeneity exhibited in Fig. 2-21(a). It is clear that the random field exhibits continuity in a direction oblique to the EW or NS directions. In actuality, however, the exhaustive spatial field will never ever be accessible, and instead, the spatial random field will be probed at a few discrete sampling locations, as shown in Fig. 2-21(b).

Using the sparse conditioning data, experimental semi-variograms were computed in several different directions. The ranges of the semi-variograms in several directions are depicted in the rose

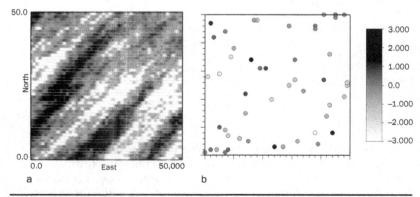

Figure 2-21   **a)** Random field exhibiting anisotropy in the 45° azimuth direction. **b)** Conditioning data sampled from the reference model on the left.

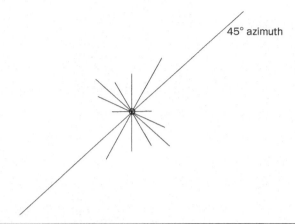

45° azimuth

FIGURE 2-22   Rose diagram reflecting the geometric anisotropy observed in Fig. 2-21.

Semi-Variogram Type	Gaussian
Nugget effect	0.04
Sill component of structure	0.96
Range in 45° direction	17.0
Range in 135° direction	4.0

TABLE 2-1   Semi-Variogram Parameters Used to Model the Random Field Shown in Fig. 2-21

diagram shown in Fig. 2-22. Based on this result, it is evident that the data exhibits a strong anisotropy in the azimuth 45° direction.

Based on the rose diagram observed in Fig. 2-22, the anisotropic characteristics of the random field are modeled using the semi-variogram model shown in Table 2-1.

The semi-variogram fitted using the parameters in Table 2-1 is shown in Fig. 2-23.

In order to demonstrate the computation of semi-variogram values using the parameters specified above, consider the computation of a semi-variogram value corresponding to a lag of 5 units in the 30° azimuth direction.

The rotation matrix in this case, given that the anisotropy direction $\theta = 45°$, is:

$$\begin{bmatrix} 0.707 & 0.707 \\ -0.707 & 0.707 \end{bmatrix}$$

The anisotropy matrix is: $\begin{bmatrix} 0.235 & 0.0 \\ 0.0 & 1.0 \end{bmatrix}$

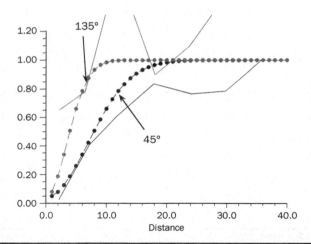

FIGURE 2-23 Anisotropic semi-variogram model with a single structure fitted to the experimental semi-variogram inferred using the conditioning data in Fig. 2-21(b). (The color version of this figure is available at www.mhprofessional.com/PRMS.)

The application of the rotation matrix and the anisotropy matrix will yield a vector in an isotropic frame of reference with range 4.0 (equal to the minor range of the anisotropic semi-variogram). The target vector can be resolved on to the original Cartesian space, given that it is of length 5 units in the azimuth 30° direction:

$$\begin{bmatrix} h_x \\ h_y \end{bmatrix} = \begin{bmatrix} |\mathbf{h}| \sin\alpha \\ |\mathbf{h}| \cos\alpha \end{bmatrix} = \begin{bmatrix} 2.5 \\ 4.33 \end{bmatrix}$$

Applying the anisotropy and the rotation transformation to the above vector yields:

$$\begin{bmatrix} h_x'' \\ h_y'' \end{bmatrix} = \begin{bmatrix} 1.14 \\ 1.29 \end{bmatrix}$$

The magnitude of this transformed vector is $\sqrt{h_x''^2 + h_y''^2} = 1.72$. The semi-variogram value corresponding to this lag within an isotropic structure of range 4.0 is:

$$\gamma(\mathbf{h}) = e^{-3\left(\frac{|h|}{a}\right)^2} = e^{-3\left(\frac{1.72}{4}\right)^2} = 0.43$$

Recalling the discussion regarding positive combination of legitimate semi-variogram structures, the calculated semi-variogram value has to be combined with the nugget effect in order to yield the final semi-variogram value. Because the standardized sill of the semi-variogram is 1.0 and the nugget effect is specified to be 0.04, the calculated semi-variogram value has to be multiplied by the sill contribution of the Gaussian structure, which is $(1 - 0.04 = 0.96)$, and then combined with the nugget effect. This yields the final semi-variogram value:

$$\gamma(\mathbf{h}) = 0.04 + 0.96 \cdot 0.43 = 0.45$$

Nested Anisotropic Model

It can be observed in Fig. 2-23 that though the single structure aniso-tropic model captures the variability at short lags well, the fit is rather poor at larger lags especially in the direction of maximum continuity (45° azimuth). In order to improve the fit, it is necessary to consider a nested semi-variogram model.

Recall the theorem on positive combinations of legitimate covariance functions (Eq. 2-33), which states that positive linear combinations of covariance functions will yield a legitimate cova-riance function. The strategy to fit a nested semi-variogram model to the experimental semi-variogram would therefore be to identify multiple anisotropic structures and the sill contribution of those structures.

It is noted from the experimental semi-variogram depicted in Fig. 2-23 that the semi-variogram value at the shortest lag exhibits an offset from the origin, indicating a Gaussian structure. The Gaussian model exhibits a good fit up to a sill contribution of 0.40, as seen in Fig. 2-23. The semi-variogram in the minor direction (45° azimuth) rises steeply and so will have a shorter range for this first structure. The Gaussian structure is combined with a spherical structure that depicts continuity in the major direction almost up to a range of 40 units. The range of this structure in the minor direction is smaller in order to fit the experimental semi-variogram in that direction. The parameters of the fitted nested semi-variogram model are presented in Table 2-2.

The semi-variogram model fit obtained using the parameters in Table 2-2 is shown in Fig. 2-24. It can be observed that a much better fit is now obtained at all lags.

Again, let us go back to the problem of computing the semi-var-iogram value corresponding to a lag of 5 units in the 30° azimuth direction. The rotation matrix corresponding to both the structures is the same as before, because both structures are anisotropic in the 45° azimuth direction. The anisotropy matrices for the two structures are different, however. We perform the calculations for the semi-variogram contribution from each structure.

Structure 1

Rotation matrix: $\qquad R_\theta = \begin{bmatrix} 0.707 & 0.707 \\ -0.707 & 0.707 \end{bmatrix}$

Anisotropy matrix: $\qquad D_\lambda = \begin{bmatrix} 0.5 & 0.0 \\ 0.0 & 1.0 \end{bmatrix}$

Nugget effect	0.01
Structure 1	
Type	Gaussian
Sill component of structure	0.40
Anisotropy direction	45°
Range in 45° direction	15.0
Range in 135° direction	7.5
Structure 2	
Type	Spherical
Sill component of structure	0.59
Anisotropy direction	45°
Range in 45° direction	44.0
Range in 135° direction	13.5

TABLE 2-2 Parameters of the Nested Anisotropic Semi-Variogram Model Obtained by Fitting to the Experimental Semi-Variogram

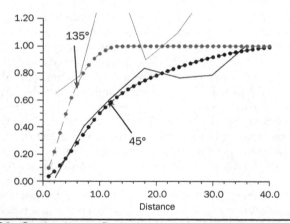

FIGURE 2-24 Semi-variogram fit obtained by using nested structures. In this case, two nested structures were used—a Gaussian structure combined with a spherical structure. (The color version of this figure is available at www.mhprofessional.com/PRMS.)

Applying these transformations to the target vector yields:

$$\begin{bmatrix} h_x'' \\ h_y'' \end{bmatrix} = \begin{bmatrix} 2.41 \\ 1.29 \end{bmatrix}$$

The magnitude of the vector is 2.74. The isotropic range of the structure after the transformation process is 7.5. The semi-variogram value corresponding to a vector of magnitude 4.51 is therefore:

$$\gamma(\mathbf{h}) = e^{-3\left(\frac{|h|}{a}\right)^2} = e^{-3\left(\frac{2.74}{7.5}\right)^2} = 0.33$$

Because the structure contributes a fraction 0.4 of the total variability, the contribution of the above semi-variogram value to the total at the specified lag is $0.4 \times 0.33 = 0.132$.

Structure 2

Rotation matrix: $R_\theta = \begin{bmatrix} 0.707 & 0.707 \\ -0.707 & 0.707 \end{bmatrix}$

Anisotropy matrix $(\frac{a_{\min}}{a_{\text{maj}}} = \frac{13.5}{44} = 0.31):$ $D_\lambda = \begin{bmatrix} 0.31 & 0.0 \\ 0.0 & 1.0 \end{bmatrix}$

Applying these transformations to the target vector yields:

$$\begin{bmatrix} h_x'' \\ h_y'' \end{bmatrix} = \begin{bmatrix} 1.48 \\ 1.29 \end{bmatrix}$$

The magnitude of the vector is 1.97. The isotropic range of the structure after the transformation process is 13.5. This structure is given to be spherical, and thus the semi-variogram value corresponding to a vector of magnitude 4.4 is:

$$\gamma(\mathbf{h}) = 1.5 \cdot \left(\frac{|\mathbf{h}|}{13.5}\right) - 0.5 \cdot \left(\frac{|\mathbf{h}|}{13.5}\right)^3 = 0.217$$

Because the structure contributes a fraction 0.59 to the total variability, the contribution of the above semi-variogram value to the total at the specified lag is $0.59 \times 0.217 = 0.13$.

Combining the nugget effect and the contributions from Structures 1 and 2, the semi-variogram value at a lag of 5 units in the $30°$ azimuth direction is calculated to be:

$$\gamma(\mathbf{h}) = 0.01 + 0.4 \cdot 0.33 + 0.59 \cdot 0.217 = 0.27$$

Remarks
- The nested structures may have different directions and ranges of continuity.
- The nested structures may be interpreted as a sequence of nested geologic structures exhibiting different correlation lengths in different directions.

2.3.7   Modeling Anisotropy in 3D

Most geological phenomena exhibit heterogeneity in all three dimensions. In addition to the azimuthal anisotropy, a 3D phenomenon may exhibit continuity in the dip direction also. As an extension to 2D modeling, three-dimensional anisotropy is assumed to be defined by an ellipsoid.

The inference of 3D anisotropy proceeds by first inferring the anisotropy in the azimuth direction. Subsequently, the dip is varied in the azimuthal direction of anisotropy in order to determine the major direction of continuity. The minor axis of the ellipsoid is

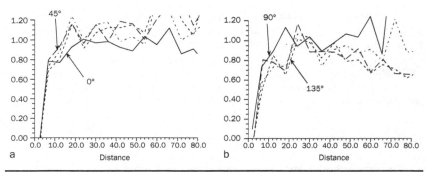

a   b

FIGURE 2-25 Semi-variograms computed in several directions in order to determine **a)** main direction of continuity and **b)** minor direction of continuity.

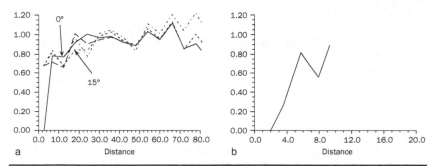

a   b

FIGURE 2-26 **a)** Semi-variograms computed in different dip directions keeping the azimuth direction fixed at 0°, the direction of maximum continuity. **b)** Semi-variogram in the direction perpendicular to the dip plane.

in the direction perpendicular to the major direction along the dip plane. Finally, the third axis of the ellipsoid is aligned in the direction perpendicular to the dip plane. The angle convention used in the subsequent discussion is that the dip 0° direction corresponds to the horizontal plane. Dip is subsequently measured clockwise from the dip 0° direction.

Figure 2-25(a) and (b) depicts the semi-variogram inferred in several azimuthal directions corresponding to a sparse 3D dataset of permeabilities. These semi-variograms have been computed assuming the dip to be 0°. It can be inferred that the 3D phenomenon exhibits maximum anisotropy in the azimuth 0° (or north) direction.

Next, the dip is varied in the direction of maximum continuity. The resultant variation in the range of continuity in the azimuth 0° direction, as the dip angle is systematically varied, is shown in Fig. 2-26(a). It is evident by looking at the figure that the maximum continuity is observed corresponding to a dip angle of 15°. The continuity in the direction perpendicular to the dip plane is shown in Fig. 2-26(b).

The parameters for the 3D semi-variogram model can be retrieved from the experimental semi-variograms shown in Figs. 2-25 and 2-26. The model parameters are summarized in Table 2-3.

Nugget effect	0.01
Structure 1	
Type	Gaussian
Sill component of structure	0.59
Anisotropy direction	90°
Dip direction	15°
Range in 90° direction	20.0
Range in 0° direction	10.0
Range in direction ⊥ to dip plane	8.0
Structure 2	
Type	Spherical
Sill component of structure	0.40
Anisotropy direction	0°
Dip direction	15°
Range in 0° direction	34.0
Range in 90° direction	20.0
Range in direction ⊥ to dip plane	10.0

TABLE 2-3   Parameters of the 3D Semi-Variogram Model Obtained from the Experimental Semi-Variograms Shown in Figs. 2-25 and 2-26

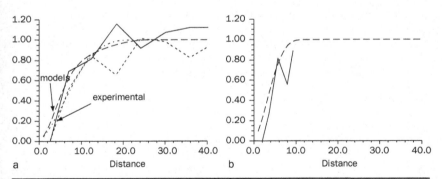

FIGURE 2-27   Semi-variogram model fits in **a)** the major and minor directions of continuity and **b)** the direction perpendicular to the dip plane.

The semi-variogram fits obtained using this model are shown in Fig. 2-27.

2.3.8   Semi-Variogram Computation in 3D

In order to compute the semi-variogram value corresponding to a vector when the phenomenon being modeled is anisotropic in 3D, the calculation procedure is basically the same as in 2D with an additional rotation about the dip axis. As mentioned previously, the dip angle is measured as rotation in clockwise sense around the O_y axis.

Similar to the 2D transformation, a three-step transformation procedure is implemented that takes into consideration the azimuthal and dip angles subtended by the main axis of continuity, the range in that direction, and the ranges in the direction perpendicular to that direction along the dip plane as well as in the plane perpendicular to the dip plane. These three ranges specify the axes of the ellipsoid. The three steps of the transformation are:

1. Projection of vector **h** on to the rotated system of coordinates x', y', z keeping the vertical axis as the original z axis (i.e., dip angle equal to $0°$). This will yield the transformed vector $\mathbf{h}' = \begin{bmatrix} \mathbf{h}_{x'} \\ \mathbf{h}_{y'} \\ \mathbf{h}_z \end{bmatrix}$ corresponding to the original vector $\mathbf{h} = \begin{bmatrix} \mathbf{h}_x \\ \mathbf{h}_y \\ \mathbf{h}_z \end{bmatrix}$.

2. Projection of vector **h** on to the rotated system of coordinates x'', y', z' keeping the minor axis same as the rotated y' axis. This will yield the transformed vector $\mathbf{h}'' = \begin{bmatrix} \mathbf{h}_{x''} \\ \mathbf{h}_{y'} \\ \mathbf{h}_{z'} \end{bmatrix}$.

3. Now we have the original vector **h** projected on to the space defined by the anisotropy ellipsoid. The next step is to squeeze the ellipsoid into a sphere (isotropic) and find the vector $\mathbf{h}'''$ consistent with the isotropic sphere.

These rotational transformations are illustrated in Fig. 2-28(a) and (b).

Consider the phenomenon exhibiting maximum continuity along azimuth angle θ, dip angle φ, and a vector $\mathbf{h} = \begin{bmatrix} \mathbf{h}_x \\ \mathbf{h}_y \\ \mathbf{h}_z \end{bmatrix}$. The first rotation

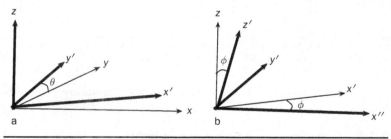

FIGURE 2-28 **a)** Rotation of original Cartesian axes about the vertical axis consistent with the azimuth angle of anisotropy. **b)** Subsequent rotation about the minor axis consistent with the dip angle of anisotropy.

around the vertical axis is the same as the earlier rotation in the 2D case:

$$\mathbf{h}' = \begin{bmatrix} \mathbf{h}_{x'} \\ \mathbf{h}_{y'} \\ \mathbf{h}_z \end{bmatrix} = \begin{bmatrix} \sin(\theta) & \cos(\theta) & 0 \\ -\cos(\theta) & \sin(\theta) & 0 \\ 0 & 0 & 1 \end{bmatrix} \cdot \begin{bmatrix} \mathbf{h}_x \\ \mathbf{h}_y \\ \mathbf{h}_z \end{bmatrix} \tag{2-38}$$

The transformation above results in the vertical projection h_z remaining unaltered. Suppose that the vector $\mathbf{h}'$ subtends an angle ω with the z axis as shown in Fig. 2-29, then:

$$\mathbf{h}_{z'} = |\mathbf{h}'| \cdot \cos(\omega - \varphi)$$

But

$$\mathbf{h}_z = |\mathbf{h}'| \cdot \cos\omega$$

$$\mathbf{h}_{x'} = |\mathbf{h}'| \cdot \sin\omega$$

Therefore:

$$\mathbf{h}_{z'} = h_z \cos\varphi + h_{x'} \sin\varphi$$

The projection along the dip axis x'' is similarly:

$$\mathbf{h}_{x''} = |\mathbf{h}'| \cdot \sin(\omega - \varphi)$$

$$= h_{x'} \cos\varphi - h_z \sin\varphi$$

The projection along the y' direction remains unaltered. The projections on the rotated z' and x'' can therefore be written as:

$$\mathbf{h}'' = \begin{bmatrix} \mathbf{h}_{x''} \\ \mathbf{h}_{y'} \\ \mathbf{h}_{z'} \end{bmatrix} = \begin{bmatrix} \cos\varphi & 0 & -\sin\varphi \\ 0 & 1 & 0 \\ \sin\varphi & 0 & \cos\varphi \end{bmatrix} \cdot \begin{bmatrix} \mathbf{h}_{x'} \\ \mathbf{h}_{y'} \\ \mathbf{h}_z \end{bmatrix}$$

The complete transformation can be written as:

$$\mathbf{h}'' = \begin{bmatrix} \cos\varphi & 0 & -\sin\varphi \\ 0 & 1 & 0 \\ \sin\varphi & 0 & \cos\varphi \end{bmatrix} \cdot \begin{bmatrix} \sin\theta & \cos\theta & 0 \\ -\cos\theta & \sin\theta & 0 \\ 0 & 0 & 1 \end{bmatrix} \cdot \begin{bmatrix} \mathbf{h}_x \\ \mathbf{h}_y \\ \mathbf{h}_z \end{bmatrix} \tag{2-39}$$

The transformation $\quad \mathbf{h}'' = R_{\theta,\varphi} \cdot \mathbf{h}$

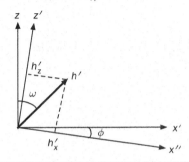

Figure 2-29 Projections corresponding to the dip rotation.

The next step is to transform **h″** to an isotropic set of axes. If we define the anisotropy ratios:

$$\lambda_1 = \frac{a_{\text{minor}}}{a_{\text{major}}} \text{ and } \lambda_2 = \frac{a_{\text{minor}}}{a_{\text{vertical}}}$$

The transformation:

$$\mathbf{h'''} = \begin{bmatrix} \lambda_1 & 0 & 0 \\ 0 & 1 & 0 \\ 0 & 0 & \lambda_2 \end{bmatrix} R_{\theta,\varphi} \cdot \mathbf{h} \tag{2-40}$$

The resultant vector $\mathbf{h'''} = R_{\theta,\varphi} \cdot \mathbf{h}$ is isotropic with range a_{minor}.

Notes

- In many cases, $a_{\text{vert}} < a_{\text{minor}}$ in which case $\lambda_2 > 1$. This implies that the transformation effectively stretches the vertical range in order to form the isotropic spheroid.

- As mentioned earlier, the angle convention for dip is positive angle measured counter-clockwise from the horizontal. Consequently, when a geologic structure dips downwards, the dip angle is considered to be negative.

Example 2-2

Let us use the transformation matrix to calculate the semi-variogram value corresponding to a lag vector of length 5 units in the azimuth 45° and 10° dip direction using the semi-variogram parameters listed in Table 2-3.

The lag vector specified can be projected to the original Cartesian axes to yield:

$$\mathbf{h} = \begin{bmatrix} \mathbf{h}_x \\ \mathbf{h}_y \\ \mathbf{h}_z \end{bmatrix} = \begin{bmatrix} 5 \cdot \sin 45 \cdot \cos 10 \\ 5 \cdot \cos 45 \cdot \cos 10 \\ 5 \cdot \sin 10 \end{bmatrix}$$

The transformation matrices corresponding to the first structure are:

$$D_\lambda = \begin{bmatrix} 0.50 & 0.00 & 0.00 \\ 0.00 & 1.00 & 0.00 \\ 0.00 & 0.00 & 1.25 \end{bmatrix} \quad D_\theta = \begin{bmatrix} 1.00 & 0.00 & 0.00 \\ 0.00 & 1.00 & 0.00 \\ 0.00 & 0.00 & 1.00 \end{bmatrix} \quad D_\varphi = \begin{bmatrix} 0.966 & 0.00 & 0.259 \\ 0.00 & 1.00 & 0.00 \\ -0.259 & 0.00 & 0.966 \end{bmatrix}$$

$$D_{\lambda,\theta,\varphi} = \begin{bmatrix} 0.50 & 0.00 & 0.00 \\ 0.00 & 1.00 & 0.00 \\ 0.00 & 0.00 & 1.25 \end{bmatrix} \cdot \begin{bmatrix} 1.00 & 0.00 & 0.00 \\ 0.00 & 1.00 & 0.00 \\ 0.00 & 0.00 & 1.00 \end{bmatrix} \cdot \begin{bmatrix} 0.966 & 0.00 & 0.259 \\ 0.00 & 1.00 & 0.00 \\ -0.259 & 0.00 & 0.966 \end{bmatrix}$$

The transformed vector is thus:

$$\mathbf{h}'' = \begin{bmatrix} 1.79 \\ 3.48 \\ -0.078 \end{bmatrix}$$

The contribution of a Gaussian structure of range 10 units corresponding to a vector of length $\|\mathbf{h}''\| = \sqrt{1.79^2 + 3.48^2 + (-0.078)^2} = 3.92$ is 0.369.

Repeating this exercise for the second structure, the transformation matrix in this case is:

$$D_{\lambda,\theta,\phi} = \begin{bmatrix} 0.59 & 0.00 & 0.00 \\ 0.00 & 1.00 & 0.00 \\ 0.00 & 0.00 & 2.00 \end{bmatrix} \cdot \begin{bmatrix} 0.00 & 1.00 & 0.00 \\ -1.00 & 0.00 & 0.00 \\ 0.00 & 0.00 & 1.00 \end{bmatrix} \cdot \begin{bmatrix} 0.966 & 0.00 & 0.259 \\ 0.00 & 1.00 & 0.00 \\ -0.259 & 0.00 & 0.966 \end{bmatrix}$$

The change in the transformation matrix is because the azimuthal anisotropy direction for this structure is in the $0°$ azimuth direction and the anisotropic ranges listed in Table 2-3. Applying this transformation matrix to the previous lag vector results in:

$$\mathbf{h}'' = \begin{bmatrix} 2.11 \\ -3.48 \\ -0.125 \end{bmatrix}$$

The resultant semi-variogram value computed assuming a spherical structure of range 20 units is 0.301. The total semi-variogram value is therefore $0.01 + 0.40 \cdot 0.301 + 0.59 \cdot 0.369 = 0.348$, using the sill contributions tabulated in Table 2-3.

2.3.9   Zonal Anisotropy

Consider a spatial phenomenon exhibiting strong periodic characteristics (layered media) such that within each period (layer), the properties are very homogenous, as seen in Fig. 2-30.

For the layered medium shown in Fig. 2-30, the vertical semi-variogram exhibits more variance (higher sill) than horizontal

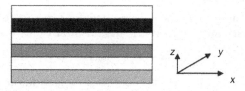

FIGURE 2-30 A layered medium exhibiting strong contrast in layer properties in the vertical direction and exhibiting a great deal of homogeneity within the layers.

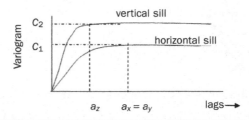

Figure 2-31   A semi-variogram exhibiting zonal anisotropy with sills different in the vertical and horizontal directions.

semi-variograms—since the property values within a layer are very homogenous. This is shown in Fig. 2-31.

Such phenomena that exhibit different variability (sill) in different directions are said to exhibit *zonal anisotropy* (Journel & Huijbregts, 1978; Zimmerman, 1993).

2.3.10   Modeling Strategy

An observation is that if a particular semi-variogram structure has infinite range in a particular direction, then the semi-variogram does not register any variability; i.e., the variogram function does not exhibit any rise in that direction. The semi-variogram structure does not consequently contribute to the sill in that direction.

In Fig. 2-31, the 3D semi-variogram model may be assumed to be made of two structures where the first structure is assumed to have an infinite range in the vertical direction (in other words, it does not contribute in the vertical direction), while the second structure is assumed to have an infinite range in the horizontal directions and only contributes in the vertical direction. That is the composite semi-variogram can be modeled as:

$$\gamma(h) = C_1 \cdot \gamma(a_{x'}, a_{y'}, \infty) + C_2 \cdot \gamma(\infty, \infty, a_z)$$

Remarks

- Infinite range simply implies that the range in that direction is set to be several times the size of the domain so that the semi-variogram value computed with that range in that direction will essentially work out to be 0.0.

- Zonal anisotropy is thus modeled as a particular form of geometric anisotropy where the anisotropy ratio is essentially 0. ($\frac{a_{min}}{a_{max}} = 0$ when $a_{max} \rightarrow \infty$).

- Myers & Journel (1990) point out the condition when zonal anisotropy can result in non-invertible kriging matrices. This is because the semi-variogram model in that condition becomes conditionally semi-definite instead of being positive definite.

Example 2-3

An example of a semi-variogram exhibiting zonal anisotropy is presented in Fig. 2-32. It is apparent that the semi-variogram in the x direction stabilizes at a lower value of sill than the semi-variogram in the y direction, indicating that the variability in the x direction is lower.

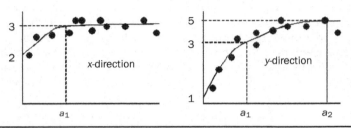

FIGURE 2-32   An example of a semi-variogram exhibiting zonal anisotropy.

Some additional specific observations are:

- Nugget effect is 0.2 in x direction and 1.0 in y direction indicating an anisotropic nugget effect.
- Sill is 3.0 in x direction and 5.0 in y direction—zonal anisotropy.
- Variogram in x direction has a single structure of range a_1, while in y direction, it has two structures with ranges a_1 and a_2.

Based on these observations, we can develop uni-directional semi-variograms.

Variogram in x direction: $\gamma(\mathbf{h}_x) = 2 + 1 \cdot \gamma\left(\dfrac{\mathbf{h}_x}{a_1}\right)$. The sill contributions of the nugget effect and the one structure (at distances greater than a_1) add up to 3, as it should.

Variogram in y direction: $\gamma(\mathbf{h}_y) = 1 + 2 \cdot \gamma\left(\dfrac{\mathbf{h}_y}{a_1}\right) + 2 \cdot \gamma\left(\dfrac{\mathbf{h}_y}{a_2}\right)$. The sill contributions of the nugget effect and the two structures [at distances greater than $\max(a_1, a_2)$] add up to 5.0.

However, our goal is to develop a single 3D model such that when it is queried in the x and y directions, it gives back the uni-directional models identified above. In other words, if we substitute $\mathbf{h}_y = 0$ in the 3D model, we get back $\gamma(\mathbf{h}_x)$.

2.3.11   3D Model

Our basic principle is to utilize all structures in all directions and then eliminate the contribution of structure in some directions by setting the variogram range to infinity. We will apply this principle to model each of the structures identified above—the nugget effect and the two remaining structures.

Nugget Effect   In order to accommodate the anisotropic nugget effect, we stipulate two structures. We make the first structure isotropic with a sill contribution of 1 and an infinitesimally small range of ε. This takes care of the nugget effect in the y direction. But then there is a remaining nugget effect of 1.0 in the x direction. We therefore make the second structure anisotropic with a range of infinity in the y direction (so this structure does not contribute in that direction) and a range of ε (infinitesimally small) in the x direction. This structure will therefore contribute in the x direction and also toward the semi-variograms computed for orientation close to an azimuth of 90° (x direction). The nugget effect model can thus be written as:

$$C_o = 1 \cdot Sph\left(\frac{|\mathbf{h}|}{\varepsilon}\right) + 1 \cdot Sph\left(\sqrt{\left(\frac{h_x}{\varepsilon}\right)^2 + \left(\frac{h_y}{\infty}\right)^2}\right)$$

We then start to work on the remaining structures.

Structure 1   We noted that there is a single structure of range a_1 and sill contribution 1.0 in the x direction. We see a structure with the same range in the y direction but that structure has a sill contribution of 2.0 in the y direction. In order to accommodate this, let us first propose an isotropic structure of range a_1:

$$\gamma_1(\mathbf{h}) = 1 \cdot Sph\left(\frac{|\mathbf{h}|}{a_1}\right)$$

Structure 2   However, we need to accommodate a remaining sill of 1.0 in the y direction from this structure of range a_1. In order to do this, we propose a structure that has a sill contribution of 1.0 with a range of a_1 in the y direction. This structure will have a range of infinity in the x direction, implying that it will not contribute in that direction.

$$\gamma_2(\mathbf{h}) = 1 \cdot Sph\left(\sqrt{\left(\frac{\mathbf{h}_x}{\infty}\right)^2 + \left(\frac{\mathbf{h}_y}{a_1}\right)^2}\right)$$

Structure 3   There is now the remaining structure that has a range a_2 in the y direction. This structure is not visible in the x direction. Hence, we propose the following model:

$$\gamma_3(\mathbf{h}) = 2 \cdot Sph\left(\sqrt{\left(\frac{\mathbf{h}_x}{\infty}\right)^2 + \left(\frac{\mathbf{h}_y}{a_2}\right)^2}\right)$$

The combined 3D semi-variogram model therefore can be written as:

$$\gamma(\mathbf{h}) = 1 \cdot Sph\left(\frac{|\mathbf{h}|}{\varepsilon}\right) + 1 \cdot Sph\left(\sqrt{\left(\frac{h_x}{\varepsilon}\right)^2 + \left(\frac{h_y}{\infty}\right)^2}\right) + 1 \cdot Sph\left(\frac{|\mathbf{h}|}{a_1}\right)$$

$$+ 1 \cdot Sph\left(\sqrt{\left(\frac{\mathbf{h}_x}{\infty}\right)^2 + \left(\frac{\mathbf{h}_y}{a_1}\right)^2}\right) + 2 \cdot Sph\left(\sqrt{\left(\frac{\mathbf{h}_x}{\infty}\right)^2 + \left(\frac{\mathbf{h}_y}{a_2}\right)^2}\right)$$

Note that there are other possible interpretations of the anisotropic semi-variogram structures that may be possible. However, a good modeling practice would be to select the 3D model that is the simplest and represents the anisotropy with the fewest number of structures.

2.4 References

Agterberg, F. P. (1994). Fractals, Multifractals, and Change of Support. In R. Dimitrakopoulos (Ed.), *Geostatistics for the Next Century. Quantitative Geology and Geostatistics* (Vol. 6). Dordrecht: Springer.

Bacro, J.-N., Bel, L., & Lantuejoul, C. (2008). Testing for Spatial Asymptotic Independence Of Extremes Using The Madogram. *GEOSTATS 2008: 8th International Geostatistics Congress.* Santiago: FCFM Gecamin 2008.

Bochner, S. (1941, July). Hilbert Distances And Positive Definite Functions. *Annals of Mathematics, 42*(3).

Bracewell, R. (1986). *The Fourier Transform and Its Applications.* New York: McGraw-Hill Book Company.

Buescu, J., & Paixão, A. C. (2012). Real And Complex Variable Positive Definite Functions. *São Paulo Journal of Mathematical Sciences, 6*(2), 155-169.

Cheng, Q., & Agterberg, F. (1996). Muitifractai Modeling And Spatial Statistics. *Mathematical Geology, 28*(1).

Chiles, J. P., & Delfiner, P. (1999). *Geostatistics Modeling Spatial Uncertainty.* New York: Wiley Interscience.

Christakos, G. (1984, February). On The Problem Of Permissible Covariance And Variogram Models. *Water Resources Research, 20*(2), 251-265.

Clark, I. (2010, June). Statistics or geostatistics? Sampling Error Or Nugget Effect? *Journal of the Southern African Institute of Mining and Metallurgy, 110*(6).

Cressie, N., & Hawkins, D. M. (1980). Robust Estimation of the Variogram: 1. *Mathematical Geology, 12*(2), 115-125.

Deutsch, C. V. (1998). *GSLIB: Geostatistical Software Library and User's Guide.* New York, New York, U.S.A.: Oxford University Press.

Diggle, P. J., & Ribeiro, P. J. (2007). *Model-based Geostatistics.* New York: Springer.

Dubrule, O. (2017). Indicator Variogram Models: Do We Have Much Choice? *Mathematical Geosciences, 49*, 441-465.

Frigyik, B. A., Kapila, A., & Gupta, M. R. (2010). *Introduction to the Dirichlet Distribution and Related Processes.* UWEE Technical Report, University of Washington, Department of Electrical Engineering, Seattle.

Goovaerts, P. (1997). *Geostatistics for Natural Resources Evaluation.* Oxford, U.K.: Oxford University Press.

Gringarten, E., & Deutsch, C. V. (2001). Teacher's Aide Variogram Interpretation and Modeling. *Mathematical Geosciences, 33*, 507-534.

Grzebyk, M., & Wackernagel, H. (1994). Multivariate Analysis and Spatial/Temporal Scales: Real and Complex Models. *Proceedings of XVIIth International Biometric Conference, 1*, pp. 19-33. Hamilton.

Journel, A. G. (1988). New Distance Measures: The Route Toward Truly Non-Gaussian Geostatistics. *Mathematical Geology, 20*(4), 459-475.

Journel, A. G., & Huijbregts, C. J. (1978). *Mining Geostatistics.* London: Academic Press.

Matheron, G. (1973). The Intrinsic Random Functions And Their Applications. *Advances in Applied Probability, 5*(3), 439-468.

Myers, D., & Journel, A. G. (1990). Variograms with Zonal Anisotropies and Noninvertible Kriging Systems. *Mathmatical Geosciences, 22*(7).

Nadarajah, S., & Gupta, A. K. (2004). Beta Function and the Incomplete Beta Function. In A. K. Gupta, & S. Nadarajah (Eds.), *Handbook of Beta Distribution and its Applications.* New York: Marcell Dekker.

Pitard, F. F. (1994). Exploration of the "Nugget Effect". In R. Dimitrakopoulos (Ed.), *Geostatistics for the Next Century. Quantitative Geology and Geostatistics* (Vol. 6, pp. 124-136). Springer.

Posa, D., & De Iaco, S. (2018). Strict positive definiteness in geostatistics. *Stochastic Environmental Research and Risk Assessment volume, 32*, 577-590.

Srivasatava, R. M., & Parker, H. M. (1989). Robust Measures of Spatial Continuity: Geostatistics. In M. Armstrong (Ed.), *Quantitative Geology and Geostatistics* (Vol. 4, pp. 295-308). Dordrecht.

Strang, G. (2016). *Introduction to Linear Algebra.* Wellesley Cambridge Press.

Wackernagel, H. (2003). *Multivariate Geostatistics.* Berlin: Springer-Verlag.

Zimmerman, D. L. (1993). Another Look at Anisotropy in Geostatistics. *Mathematical Geology, 25*(4), 453-470.

CHAPTER 3

Spatial Estimation

3.1 Linear Least Squares Estimation or Interpolation

We have thus far focused on devising a technique for interpolating contours or values of reservoir properties based on the availability of information or data at a few locations. The discussion thus far has been restricted to techniques that account for the physical distance of an estimation node to the data locations. These include spline-based contouring, distance-based interpolation, etc. In all these interpolation techniques, the goal is to consider the available data and subsequently decide what the reservoir property value will be at a location where data is unavailable. In the contour-based approaches, the magnitude of the mapped variable at a location is controlled by the fact whether the location is on one side or the other of a contour.

The interpolation or estimation of spatial attributes away from data locations can be viewed from a more statistical perspective. The goal then is to fit an interpolant or an estimator taking into consideration the statistics of the data and the error. This is the objective of regression. In the subsequent section, a brief exposition of linear least squares (regression) as applied in traditional statistics is presented. This is followed by a discussion on how this regression can be performed using spatial data for the purpose of interpolation.

Note The estimation examples in this chapter are included as an Excel Workbook on this book's website, www.mhprofessional.com/PRMS.

3.2 Linear Regression

Given data (y_i, x_i), $i = 1, \ldots, n$, the objective in linear regression is to fit a straight line through the data and subsequently use the line for prediction. Write the equation of such a predictor in linear form as (Weisberg, 2005; Montgomery, 2012):

$$y^* = \beta_o + \beta_1 \mathbf{x} \tag{3-1}$$

The objective is to determine the parameters (β_o, β_1) using the available data. In order to accomplish this objective, an error is defined:

$$e = (y_i - y_i^*) \tag{3-2}$$

This error can be squared and the sum over all the available n data can be computed:

$$\begin{aligned} S(\beta_o, \beta_1) &= \sum_{i=1}^{n} e_i^2 \\ &= \sum_{i=1}^{n} [y_i - (\beta_o + \beta_1 \cdot x_i)]^2 \end{aligned} \tag{3-3}$$

Our objective is to fit a best-fit straight line through the available data such that this squared error is minimized – hence the terminology *linear least squares* (Montgomery, 2012). Our goal is therefore to find the combination of parameters (β_o, β_1) such that $S(\hat{\beta}_o, \hat{\beta}_1)$ is minimized. This is accomplished by the usual operation of taking a derivative of S with respect to the parameters setting the derivatives to 0:

$$\frac{\partial S}{\partial \hat{\beta}_o} = -2\sum_{i=1}^{n}\left[y_i - \left(\hat{\beta}_o + \hat{\beta}_1 \cdot x_i\right)\right] = 0$$

$$\frac{\partial S}{\partial \hat{\beta}_1} = -2\sum_{i=1}^{n}x_i\left[y_i - \left(\hat{\beta}_o + \hat{\beta}_1 \cdot x_i\right)\right] = 0$$

$$\Rightarrow \hat{\beta}_o = \frac{\sum\limits_{i=1}^{n}x_i^2 \cdot \sum\limits_{i=1}^{n}y_i - \sum\limits_{i=1}^{n}x_i \cdot \sum\limits_{i=1}^{n}x_iy_i}{n\sum\limits_{i=1}^{n}x_i^2 - \left[\sum\limits_{i=1}^{n}x_i\right]^2} \quad \text{and} \tag{3-4}$$

$$\hat{\beta}_1 = \frac{n\sum\limits_{i=1}^{n}x_iy_i - \sum\limits_{i=1}^{n}x_i \cdot \sum\limits_{i=1}^{n}y_i}{n\sum\limits_{i=1}^{n}x_i^2 - \left[\sum\limits_{i=1}^{n}x_i\right]^2}$$

3.2.1   Linear Least Squares—An Interpretation

Adopting the following variable definitions:

- Variance of X as: $S_{xx} = \dfrac{1}{n}\sum\limits_{i=1}^{n}x_i^2 - \left[\dfrac{1}{n}\sum\limits_{i=1}^{n}x_i\right]^2$

- Variance of Y as: $S_{yy} = \dfrac{1}{n}\sum\limits_{i=1}^{n}y_i^2 - \left[\dfrac{1}{n}\sum\limits_{i=1}^{n}y_i\right]^2$

- Covariance between X and Y as: $S_{xy} = \dfrac{1}{n}\sum_{i=1}^{n} x_i y_i - \dfrac{1}{n}\sum_{i=1}^{n} x_i \cdot \dfrac{1}{n}\sum_{i=1}^{n} y_i$

- Correlation coefficient as: $r_{XY} = \dfrac{S_{xy}}{\sqrt{S_{xx} \cdot S_{yy}}}$

We can write the coefficient (Weisberg, 2005) (Beale, 2010) as:

$$\hat{\beta}_1 = \frac{n\sum_{i=1}^{n} x_i y_i - \sum_{i=1}^{n} x_i \cdot \sum_{i=1}^{n} y_i}{n\sum_{i=1}^{n} x_i^2 - \left[\sum_{i=1}^{n} x_i\right]^2} = \frac{S_{xy}}{S_{xx}} \qquad (3\text{-}5)$$

Therefore:

$$r_{XY} = \frac{S_{xy}}{\sqrt{S_{xx} \cdot S_{yy}}} = \hat{\beta}_1 \sqrt{\frac{S_{xx}}{S_{yy}}} \qquad (3\text{-}6)$$

The best-fit line $y_i = \hat{\beta}_o + \hat{\beta}_1 x_i$ will also pass through the mean $(\bar{x}, \bar{y})$; i.e., $\bar{y} = \hat{\beta}_o + \hat{\beta}_1 \bar{x}$

Working with the standardized form of the variables:

$$u_i = \frac{x_i - \bar{x}}{\sqrt{S_{xx}}}; \qquad v_i = \frac{y_i - \bar{y}}{\sqrt{S_{yy}}}$$

This implies that $S_{uu} = S_{vv} = 1$ and $S_{uv} = r = \hat{\beta}_1$. Because we already said that the best-fit line passes through the mean $(\bar{x}, \bar{y})$, it will also pass through the mean of the transformed variables. This implies:

$$\hat{\beta}_o = \bar{v} - \hat{\beta}_1 \bar{u}$$

But because $\bar{u} = \bar{v} = 0$ (given the above definitions for u and v), this entails that $\hat{\beta}_o = 0$ and consequently, the vector of estimates $\hat{v}$ can be expressed as:

$$\hat{v} = \hat{\beta}_1 \hat{u} = r \cdot \hat{u} \qquad (3\text{-}7)$$

i.e., the standardized estimate $\hat{v}$ can be viewed as a linear combination of the standardized data u_i weighted by the correlation coefficient r. In the original variable space:

$$\frac{\hat{y} - \bar{y}}{\sqrt{S_{yy}}} = r \cdot \frac{x_i - \bar{x}}{\sqrt{S_{xx}}}$$

Linear regression can thus be viewed as an estimate $\hat{y}$ obtained as a linear combination of x_i weighted by the correlation coefficient r.

3.2.2 Application to Spatial Estimation

If the spatial correlation describing the variability of reservoir properties is known, then the estimate at any location where a sampled
value is not available can be computed as a weighted combination of
the available data (Beale, 2010). According to the preceding discussion regarding linear regression [Eq. (3-1)], the weights are equal to
the spatial covariance between the data and the unsampled value.
Inferring such a covariance would require several outcomes or sample sets at the unsampled location and the data location. Pooling the
data together, one might be able to compute the required measure.
However, in actuality, no such mechanism for drawing multiple outcomes is available, as there is only one "true" value available at each
of the data locations. Consequently, a decision of statistical stationarity is necessary in order to pool data at all locations that are separated
by a similar lag vector (distance and orientation). This decision of stationarity implies that the computed covariance value is not specific
to the particular estimation location or the location of the data. Rather
it is a value that describes on the average, the relationship between
all locations separated by the same lag distance and orientation. Such a
covariance function can be conceived of as a function of the lag distance
in a particular direction, and anisotropy of the modeled phenomenon
would be represented by covariance functions that exhibit variation in
different directions. The details of how the covariance or its counter-
part, the semi-variogram, is computed and the process of inferring and
modeling anisotropy is presented in the previous chapter.

Consider the thickness variations in a reservoir represented by a
covariance function shown in Fig. 3-1(a). For simplicity, the thickness
variations are assumed to be isotropic implying that the variability
in all directions is approximately the same. The formation thickness
at a few locations in the reservoir is assumed to be available. These
locations are shown in Fig. 3-1(b). Using the thickness data at these
locations and the covariance function, estimates for the thickness at
all locations in the reservoir can be computed.

In order to develop a map showing variability of thickness in
the reservoir, one approach could be to overlay a grid on the available data. The size of the grid imposed should be consistent with the
support of the data. For example, if the horizontal resolution of the
well logs is 8 ft, the selected horizontal dimension of each grid cell
should at the minimum be 8 ft. Of course, the regression estimation
outlined earlier can be performed directly at any location within the
domain, regardless of the superimposed grid, by directly computing
the distance between the estimation point and all locations with data.
Consider, for example, the estimation at a location with coordinates
(100,100) when the data is available at the four locations marked in
Fig. 3-1(b). The origin is located at the bottom left. The correlation
function has a range of 800 ft. The location of the four data values and

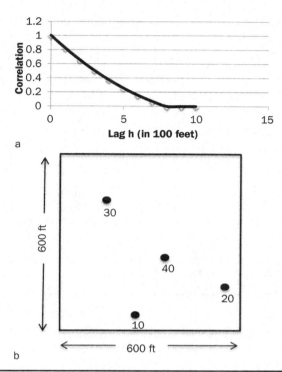

Figure 3-1 a) Covariance function C(**h**) assumed for the application example. **b)** Data configuration for the same example.

Data Location (in 100 ft) Estimation Node	(2,5)	(4,3)	(6,2)	(3,1)
(1,1)	4.12	3.61	5.10	2.00

Table 3-1 Spatial Lag between the Data and the Estimation Node Corresponding to the Conditioning Data in Fig. 3-1(b)

the corresponding lag (distance) separating the data locations from the estimation location are tabulated in Table 3-1.

The correlation function shown in Fig. 3-1(a) is fitted using a function:

$$\rho(\mathbf{h}) = 1.5\frac{|\mathbf{h}|}{80} - 0.5\left(\frac{|\mathbf{h}|}{80}\right)^2$$

The function is in terms of the magnitude of lag vector **h** and irrespective of the direction of the vector. The correlation coefficients (*r*) corresponding to the lag distances tabulated in Table 3-1 are presented in Table 3-2.

As stated in Table 3-2, these correlation coefficients act as weights applied on the standardized data obtained by subtracting the mean and dividing it by the standard deviation of the data. The estimate at (1,1) is thus obtained as:

$$thickness = 0.649 \cdot \frac{(10 - 25)}{12.91} - 0.116 \cdot \frac{(20 - 25)}{12.91} + 0.0169 \cdot \frac{(40 - 25)}{12.91}$$
$$+ 0.147 \cdot \frac{(30 - 25)}{12.91} = 16.83$$

The estimate for formation thickness is thus 16.83 ft. This process of estimation can be performed at all other locations on the superimposed grid, and the resultant interpolated map for thickness is shown in Fig. 3-2.

Data Location (in 100 ft) Estimation Node	(2,5)	(4,3)	(6,2)	(3,1)
(1,1)	0.649	−0.116	0.0169	0.147

TABLE 3-2 Weights Assigned to the Conditioning Data in Fig. 3-1(b) Computed by Linear Regression

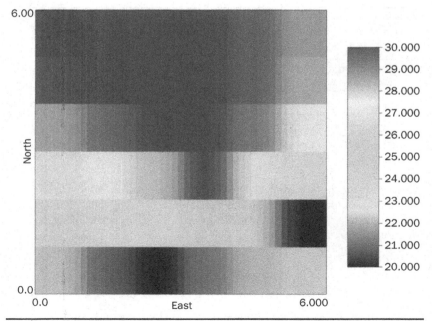

FIGURE 3-2 Estimated map of thickness obtained by applying linear regression. The available conditioning data and the covariance model are shown in Fig. 3-1. (The color version of this figure is available at www.mhprofessional.com/PRMS.)

Looking at the map in Fig. 3-2, its resemblance to a contour map drawn to reflect the variability exhibited by the data is obvious. Just like a contour map, the map shows an orderly transition from high to low thickness values. This map is exact in the sense that at the locations where conditioning data is available, the estimated value is equal to the conditioning data value.

Several cautionary points regarding the application of regression in a spatial setting have to be considered (Beale, 2010). Linear regression is predicated on the notion that the estimation errors are uncorrelated or independent. This is what allows us to simply sum the square errors corresponding to each datum, as seen in Eq. (3-3). However, that notion may not hold true (Beale, 2010) if the extrinsic cause for the pattern of variability of y is not fully explained by the spatial structure of the data x. The notion of independence of errors may also not hold true if there is an intrinsic cause for the auto-correlation of y, such as any non-stationary trends.

3.3   Estimation in General

In general, estimation implies the combination of available information into the estimated outcome of a random variable (RV) such that a criteria or objective is satisfied. In linear regression, for example, given points on an x–y plot, the objective is to fit a straight line such that the mean-squared error between data points and the corresponding projection on to the fitted straight line are minimized (see Fig. 3-3). We saw an application of such a regression-based approach to reservoir modeling (contouring) earlier in this chapter. The equation of the regression line is:

$$Y^* = b_o + \sum_{i=1}^{n} b_i \cdot X_i$$

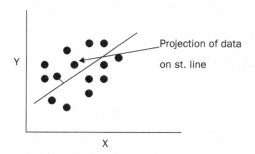

FIGURE 3-3   A regression example.

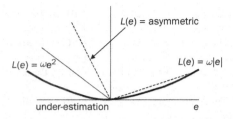

FIGURE 3-4   Examples of loss functions.

And the coefficients b_0 and $b_1, i = 1, \ldots, n$ are obtained such that the mean squared error

$$S = \left| \frac{1}{N} \sum_{i=1}^{N} \{Y_i - Y_i^*\}^2 \right|$$

is minimized. This notion can be generalized using the concepts of utility theory (Longford, 2013; Weiss, 1961).

Consider an attribute x whose outcome is uncertain and hence denoted as an RV X. The RV is characterized by the probability function $F_X(x)$. An estimate X^* is constructed using the available data/information. We can compute an error of estimation as:

$$E = X - X^*$$

A specific outcome of error is e. If $e > 0$, we have an over-estimation, and when $e < 0$, we have an under-estimation. Corresponding to the error, let us define a loss function $L(e)$ that assesses the impact of the error. In the case of the regression estimator above, the loss function is quadratic:

$$L(e) = \left| \frac{1}{N} \sum_{i=1}^{N} e^2 \right| = S$$

where the error $e = x_i - x_i^*$. In that case, the loss function is symmetric as it is equal to the error squared. But in general, the error function could be asymmetric. Some examples are shown in Fig. 3-4. Loss functions could also be continuous or discrete with jumps at set thresholds.

In Fig. 3-4, asymmetric loss function such as:

$$L(e) = \begin{cases} \omega_1 \cdot e & \text{when } e < 0 \\ \omega_2 \cdot e & \text{when } e > 0 \end{cases}$$

implies that the impact of an under-estimation is different from that corresponding to an over-estimation. The error RV is described by the random variable X and consequently by the distribution function

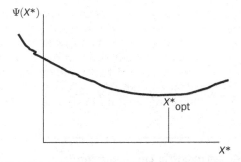

FIGURE 3-5 Expected loss as a function of the estimate value X*.

$F_X(x)$ or alternatively by the density function $f_x(x)$. The expected loss is given by:

$$E\{L(e)\} = \int\limits_{-\infty}^{\infty} L(X - X^*) f_X(x) dx$$
$$= \Psi(X^*)$$

As is evident from the above development, the expected loss is a function of the estimator X^*. This is depicted in Fig. 3-5. Conceivably, there could be an optimal value of the estimator corresponding to which the expected loss function will be minimized. That optimal estimator will be specific to the loss function selected.

As long as a loss function and the distribution of X: $F_X(x)$ is specified, an optimal estimate can always be retrieved.

The above discussion provides an avenue for constructing different optimal estimators corresponding to specific loss functions and forms the basis for estimation methods. Some analytical estimators corresponding to different loss functions are presented next.

3.3.1 Loss Function—Some Analytical Results

Consider a univariate *cdf* $F_X(x)$ describing a random variable X. Now consider the following specific examples of loss functions.

Example 3-1: Loss Function $L(e) = we^2$ Where $e = x^* - x$

If $w = 1$, the loss function is the same as the function S we saw earlier for linear regression. The expected loss can be calculated as:

$$E\{L(e)\} = \int\limits_{-\infty}^{\infty} (X^* - X)^2 dF_X(x)$$

The integral in the above expression is the same as that for the variance provided X^* is identified with the mean. The optimal estimate, per our earlier discussion, is one that minimizes the expected loss, which in this case would tantamount to minimizing the variance. By the characteristic property of the conditional mean, among all possible estimates that can be created using the available data, the mean is the one that is guaranteed to minimize the error variance. Consequently, the optimal estimate $X^*_{optimal}$ in the case of a quadratic loss function we^2 is the conditional mean denoted as $X^* = E\{X\,|\,X_\alpha\}$ where X_α are conditioning data.

Example 3-2: An Asymmetric Loss Function

$$L(e) = \begin{cases} w_1 e, & \text{if } e \geq 0 \\ w_2 |e|, & \text{if } e \leq 0 \end{cases}$$

When $e \geq 0$, then the estimate X^* is greater than the true value X: $X \leq X^*$. When $e \leq 0$, then $X \geq X^*$. The expected loss is computed by performing integration over the ranges $X \leq X^*$ and $X \geq X^*$:

$$E\{L(e)\} = \int_{-\infty}^{x^*} w_1(X^* - X)dF_X(x) + \int_{x^*}^{\infty} w_2(X - X^*)dF_X(x)$$

$$= w_1 x^* \cdot F_X(x^*) - w_1 \int_{-\infty}^{x^*} X\,dF_X(x) + \int_{x^*}^{\infty} w_2 X\,dF_X(x)$$

$$- w_2 x^*(1 - F_X(x^*))$$

In order to find the optimal estimate $X^*_{optimal}$, we take the derivative with respect to x^* and set it equal to 0. Recognizing that:

$$\frac{d\left(\int_{-\infty}^{x^*} X\,dF_X(x)\right)}{dx^*} = x^* f_X(x^*) \qquad \text{and} \qquad \frac{dF_X(x^*)}{dx^*} = f_X(x^*)$$

$$\int_{x^*}^{\infty} X\,dF_X(x) = 1 - \int_{-\infty}^{x^*} X\,dF_X(x) \Rightarrow \frac{d\left(\int_{x^*}^{\infty} X\,dF_X(x)\right)}{dx^*} = -x^* f_X(x^*)$$

The derivative $\frac{dE\{L(e)\}}{dx^*}$ becomes:

$$\frac{dE\{L(e)\}}{dx^*} = w_1 x^* f_X(x^*) + w_1 F_X(x^*) - w_1 x^* f_X(x^*) - w_2 x^* f_X(x^*)$$

$$- w_2 + w_2 x^* f_X(x^*) + w_2 F_X(x^*)$$

$$\Rightarrow \frac{dE\{L(e)\}}{dx^*} = w_1 F_X(x^*) - w_2 + w_2 F_X(x^*)$$

Setting to 0, we get: $x^*_{opt} = F_X^{-1}\left(\dfrac{w_2}{w_1 + w_2}\right)$

If $w_1 = w_2$ then $x^*_{optimal} = F_X(0.5) =$ Median

The optimal estimate corresponding to symmetric linear loss function; i.e., $E = |X^* - X|$ is thus the median. In that case, E is also referred to as the *mean absolute deviation* (MAD), and so the median is the optimal MAD estimator.

Example 3-3: A Loss Function Independent of Error

In this case, the loss function is given by:

$$L(e) = \begin{cases} 0, & if\ e = 0 \\ C, & if\ \text{not} \end{cases}$$

The expected loss in this case is given as

$$E\{L(e)\} = \int\limits_{-\infty}^{x^*-} c\,dF_X(x) + \int\limits_{x^*+}^{\infty} c\,dF_X(x)$$

because the loss function is the same for over- or under-estimation. Minimizing the expected loss results in:

$$\frac{d(E\{L(e)\})}{dx^*} = 0 \Rightarrow c \cdot f_X(x^*+) = c \cdot f_X(x^*-)$$

This implies that the optimal estimate x^* is such that the pdf approaching from smaller values is the same as the pdf value approaching from values greater than x^*, that is, the optimal estimate corresponds to a point of inflection in the pdf. The mode corresponds to a location where there is an inflection in the pdf; i.e., it corresponds to a peak in the probability distribution. Consequently, the optimal estimate:

$$x^*_{opt} = Mode$$

Longford (2013) presents the application of an asymmetric loss function for the estimation of the mean of a Gaussian distribution. He postulates that the error $\tilde{\mu} - \mu$ associated with the estimator $\tilde{\mu}$ of the true mean μ may be characterized by a loss $(\tilde{\mu} - \mu)^2$ when $\tilde{\mu}$ underestimates the mean, but the loss could be $R(\tilde{\mu} - \mu)^2$ when $\tilde{\mu}$ overestimates the mean. R is a constant greater than 1. In this setting, he investigates estimators $\tilde{\mu} = \bar{\mu} - c$, with c being a constant and $\bar{\mu}$ being the sample mean and derives a closed form expression for the expected loss corresponding to this estimator $\tilde{\mu} \sim N(\mu - c, 1/n)$ in terms of the parameter R.

In a spatial context, Kilmartin (2009) applied a loss function to estimate the permeability vaiations in a petroleum reservoir. The loss function in that case depicts the economic impact of an over- or under-estimation of permeability at a location and retrieves the optimal simulated value that minimizes the expected loss. Both an asymmetric linear loss function model and a parabolic loss function model are investigated. The key results of this modeling procedure are reservoir realizations that exhibit the correct spatial characteristics (i.e., variogram reproduction), while, at the same time, yielding the minimum expected loss. A potential extension of this method would be to alter the loss function (so as to emphasize either under- or over-estimation) and generate other realizations at the extremes of the global uncertainty distribution, thereby eliminating the necessity for the generation of a large suite of realizations to locate the global extremes of the uncertainty distribution.

The loss function-based approach presents an avenue to retrieve an optimal estimate corresponding to any specified loss criteria while taking into consideration the uncertainty associated with the RV. However, in typical reservoir modeling, several response variables $r_1, r_2, \ldots, r_N$ may contribute to loss. These response variables may be influenced by a set of parameters $x_1, x_2, \ldots, x_N$; i.e., the loss function is on the basis of several reservoir parameters taken jointly: $L(e) = h(x_1, x_2, \ldots, x_N)$. Consequently, the joint distribution $F_{X_1 X_2, \ldots, X_N}(x_1, x_2, \ldots, x_N)$ is needed for computing expected loss and this may be hard to specify at least in an analytical form.

3.4 Kriging

Based on the characteristic of conditional expectations, the best estimate in a least error variance sense is:

$$z_0^* = \varphi(z_\alpha, \alpha = 1, \ldots, n)$$

taking all the data jointly. The function $\phi(\bullet)$ in the general case for an arbitrary multivariate distribution could be non-linear. Restricting our discussion to a linear estimator of the following form:

$$z_0^* = \lambda_o \cdot 1 + \sum_{\alpha=1}^{n} \lambda_\alpha z_\alpha \tag{3-8}$$

The estimator expression bears a resemblance to the linear regression estimator discussed earlier.

Remark Consider the bivariate Gaussian distribution:

$$f_{XY} = \frac{1}{2\pi\sigma_X\sigma_Y\sqrt{1-\rho^2}} e^{-\frac{1}{2(1-\rho^2)}\left[\left(\frac{x-\mu_X}{\sigma_X}\right)^2 - 2\rho\left(\frac{x-\mu_X}{\sigma_X}\right)\left(\frac{y-\mu_Y}{\sigma_Y}\right) + \left(\frac{y-\mu_Y}{\sigma_Y}\right)^2\right]}$$

defined by five parameters:

$$-\infty \le \mu_X \le \infty, \ -\infty \le \mu_Y \le \infty, \ \sigma_X > 0, \ \sigma_Y > 0, \ -1 < \rho_{XY} < 1$$

The conditional distribution $f_{X|Y}(x|y)$ can be derived as (Mardia, 2000):

$$f_{X|Y(x|y)} = \frac{1}{\sigma_X\sqrt{2\pi(1-\rho_{XY}^2)}} e^{-\frac{\left[x-\left[\mu_X+\rho_{XY}\frac{\sigma_X}{\sigma_Y}(y-\mu_Y)\right]\right]^2}{2\sigma_X^2(1-\rho_{XY}^2)}}$$

Comparing the above expression for the conditional *pdf* to the general expression for a Gaussian density, it can be concluded that (Rao, 1973):

$$\text{Mean} = E\{E\,|\,Y\} = \mu_X + \rho_{XY}\frac{\sigma_X}{\sigma_Y}(y-\mu_Y) \tag{3-9}$$

$$\text{Variance} = Var\{X\,|\,Y\} = \sigma_X^2\left(1-\rho_{XY}^2\right) \tag{3-10}$$

The expression for the conditional mean is linear similar to the expression in Eq. (3-8). By the characteristic of the mean estimator that it minimizes the error variance, it can be concluded that a linear estimator of the form in Eq. (3-8) can be constructed that minimizes the error variance.

Consider the estimation error:

$$z_0 - z_0^* = z_0 - \lambda_0 - \sum_{\alpha=1}^{n}\lambda_\alpha z_\alpha$$

The error is unknown because the true value z_0 is unknown. We therefore treat the error as an RV:

$$e = z_0 - z_0^* \rightarrow Z_0 - Z_0^* = E$$

The goal is then to model the probability distribution characterizing the random variable E: $\text{Prob}\{E = Z_0 - Z_0^* \leq e\}$. Since the linear estimator z_0^* is optimal in a Gaussian setting and the "true" value is deterministic, the error distribution in the optimal case is Gaussian. The implication is that the error distribution is completely characterized by the first- and second-order moments.

$$\text{Mean error} - E\{Z_0 - Z_0^*\}$$
$$\text{Error variance} - Var\{Z_0 - Z_0^*\}$$

Our strategy is therefore to play with weights λ_α so as to impart some suitable properties to the moments of the error distribution.

3.4.1   Expected Value of the Error Distribution

The error is defined as $E = Z_o - Z_o^*$. Substituting the expression for the linear estimate yields:

$$E\{Z_o - Z_o^*\} = E\left\{Z_o - \lambda_o - \sum_{\alpha=1}^{n} \lambda_\alpha Z_\alpha\right\} = E\{Z_o\} - \lambda_o - \sum_{\alpha=1}^{n} \lambda_\alpha E\{Z_\alpha\}$$

In this expression for the first moment of the error distribution, $E\{Z_o\}$ is the population mean m_o; $E\{Z_\alpha\}$ computed using the data in the vicinity of the data location $\mathbf{u}_\alpha$ is m_α that may be different from m_o to account for local fluctuations in the mean. Therefore:

$$E\{Z_o - Z_o^*\} = m_o - \lambda_o - \sum_{\alpha=1}^{n} \lambda_\alpha \cdot m_\alpha$$

The first desired property of the estimate Z_o^* is that it be *unbiased* and that implies:

$$E\{Z_o^*\} = E\{Z_o\} = m_o$$

That in turn implies: $E\{Z_o - Z_o^*\} = 0$, and therefore:

$$m_o - \lambda_o - \sum_{\alpha=1}^{n} \lambda_\alpha \cdot m_\alpha = 0$$

that yields: $\lambda_o = m_o - \sum_{\alpha=1}^{n} \lambda_\alpha \cdot m_\alpha$ and consequently the updated form of the estimator:

$$Z_o^* - m_o = \sum_{\alpha=1}^{n} \lambda_\alpha (Z_\alpha - m_\alpha) \tag{3-11}$$

We are thus attempting to estimate the residue $(Z_o - m_o)$ at the unsampled location using a weighted combination of residue at the conditioning data locations $(Z_\alpha - m_\alpha)$.

Before looking at the variance of the error distribution, let us perform some algebraic manipulations. The error $Z_o - Z_o^*$ can be written as:

$$Z_o - Z_o^* = Z_o - m_o - (Z_o^* - m_o)$$

$$= Z_o - m_o - \sum_{\alpha=1}^{n} \lambda_\alpha \cdot (Z_\alpha - m_\alpha)$$

$$= \sum_{\alpha=0}^{n} a_\alpha \cdot (Z_\alpha - m_\alpha)$$

where $a_\alpha = 1$ when $\alpha = 0$ and $a_\alpha = -\lambda_\alpha \, \forall \, \alpha \neq 0$. Now we are ready to work with the error variance.

3.4.2 Error Variance

The variance is a second-order moment:

$$Var\left\{Z_o - Z_o^*\right\} = E\left\{\left[Z_o - Z_o^*\right]^2\right\} - \left\{E\left\{Z_o - Z_o^*\right\}\right\}^2 \qquad (3\text{-}12)$$

We have already established that the estimator is unbiased, implying that the second term on the RHS of Eq. (3-12) is 0. Furthermore, we have derived the corresponding expression for λ_o in terms of λ_α at the data locations. This results in:

$$\begin{aligned}
Var\left\{Z_o - Z_o^*\right\} &= E\left\{\left[Z_o - Z_o^*\right]^2\right\} \\
&= E\left\{\left[\sum_{\alpha=0}^{n} a_\alpha Z_\alpha\right]^2\right\} \qquad (3\text{-}13) \\
&= \sum_{\alpha=0}^{n}\sum_{\beta=0}^{n} a_\alpha \cdot a_\beta \cdot Cov\left\{Z_\alpha, Z_\beta\right\}
\end{aligned}$$

In order to compute the covariance $Cov\{Z_\alpha, Z_\beta\}$, we need several outcomes of the random variable Z at locations $\mathbf{u}_\alpha$ and $\mathbf{u}_\beta$. However, at those locations, only one outcome, i.e., the data value at those locations, is available. Therefore, in order to obtain the covariance value, we need to invoke *stationarity* in order to pool data at all pairs of locations that are separated by the lag $\mathbf{h}_{\alpha\beta}$ separating the locations $\mathbf{u}_\alpha$ and $\mathbf{u}_\beta$. Of course, when all the data separated by the same lag $\mathbf{h}_{\alpha\beta}$ is pooled together, the resultant covariance is not specific to the locations $\mathbf{u}_\alpha$ and $\mathbf{u}_\beta$. Instead, it is only dependent on the lag $\mathbf{h}_{\alpha\beta}$ separating the pair of locations. Thus, under second-order stationarity:

$$Cov\left\{Z_\alpha, Z_\beta\right\} = C\left(\mathbf{h}_{\alpha\beta}\right) \qquad (3\text{-}14)$$

and consequently,

$$\begin{aligned}
Var\left\{Z_o - Z_o^*\right\} &= \sum_{\alpha=0}^{n}\sum_{\beta=0}^{n} a_\alpha a_\beta \cdot C\left(\mathbf{h}_{\alpha\beta}\right) \\
&= \sum_{\alpha=0}^{n} a_\alpha^2 \cdot Var\left\{Z_\alpha\right\} + \sum_{\alpha=0}^{n}\sum_{\beta\alpha} a_\alpha a_\beta \cdot C(\mathbf{h}_{\alpha\beta})
\end{aligned}$$

We have already said that the linear estimator is optimal in the Gaussian case or in other words, it is the conditional expectation that has the characteristic property of minimizing the error or residual variance. Consequently, our next task is to minimize the previous expression for error variance by varying the parameters a_α. Taking the derivative of the error variance with respect to the parameters yields:

$$\frac{\partial Var\left\{Z_o - Z_o^*\right\}}{\partial a_\alpha} = 2 \cdot a_\alpha \cdot Var\left\{Z_\alpha\right\} + 2 \cdot \sum_{\beta \neq \alpha} a_\beta \cdot C(\mathbf{h}_{\alpha\beta})$$

$$= 2 \cdot \sum_{\beta=0}^{n} a_\beta \cdot C(\mathbf{h}_{\alpha\beta})$$

$$= a_0 \cdot C(\mathbf{h}_{\alpha 0}) \sum_{\beta=1}^{n} a_\beta \cdot C(\mathbf{h}_{\alpha\beta})$$

Setting the above to 0 and using $a_0 = 1$: $a_\alpha = -\lambda_\alpha \ \forall \ \alpha \neq 0$, we get:

$$C(\mathbf{h}_{\alpha 0}) + \sum_{\beta=1}^{n} -\lambda_\beta \cdot C(\mathbf{h}_{\alpha\beta}) = 0$$

that results in the following normal or kriging system (Olea, 1999; Chiles, 1999; Journel & Huijbregts, 1978):

$$\sum_{\beta=1}^{n} \lambda_\beta \cdot C(\mathbf{h}_{\alpha\beta}) = C(\mathbf{h}_{\alpha 0}) \text{ for } \alpha = 1, 2, \ldots, n \qquad (3\text{-}15)$$

Equation (3-15) is a system of n equations that can be solved for the n unknowns λ_β. It can be established that the solution of the system of equations does yield weights that minimize the error variance. The linear estimator Z_o^* is now fully determined, and it retains the following characteristics—it is linear, unbiased-, and best in a minimum error variance sense. In other words, the kriging estimator Z_o^* is **BLUE** (Lewis-Beck, 2004). Summarizing the results thus far:

Simple kriging (SK) estimator: $Z_o^* - m_o = \sum_{\alpha=1}^{n} \lambda_\alpha (Z_\alpha - m_\alpha)$

Kriging system: $\sum_{\beta=1}^{n} \lambda_\beta \cdot C(\mathbf{h}_{\alpha\beta}) = C(\mathbf{h}_{\alpha 0}), \ \forall \alpha = 1, \ldots, n$

The corresponding minimum error variance is:

$$Min. \ Var\left\{Z_o - Z_o^*\right\} = \sigma_{SK}^2 = \sum_{\alpha=0}^{n} \sum_{\beta=0}^{m} a_\alpha \cdot a_\beta \cdot C(\mathbf{h}_{\alpha\beta})$$

$$= a_o^2 C(0) + 2 \cdot \sum_{\alpha=1}^{n} a_\alpha a_o C(\mathbf{h}_{\alpha o}) + \sum_{\alpha=1}^{n} \sum_{\beta=1}^{n} a_\alpha \cdot a_\beta \cdot C(\mathbf{h}_{\alpha\beta})$$

$$= C(0) - 2 \cdot \sum_{\alpha=1}^{n} \lambda_\alpha C(\mathbf{h}_{\alpha o}) + \sum_{\alpha=1}^{n} -\lambda_\alpha \sum_{\beta=1}^{n} -\lambda_\beta \cdot C(\mathbf{h}_{\alpha\beta})$$

Substituting the kriging system:

$$\sigma_{SK}^2 = C(0) - 2 \cdot \sum_{\alpha=1}^{n} \lambda_\alpha C(\mathbf{h}_{\alpha 0}) + \sum_{\alpha=1}^{n} \lambda_\alpha C(\mathbf{h}_{\alpha 0})$$

$$= C(0) - \sum_{\alpha=1}^{n} \lambda_\alpha C(\mathbf{h}_{\alpha 0})$$

i.e., the minimum error variance is homoscedastic, dependent only on the data configuration and not on the absolute value of the data themselves. Again, this is consistent with the optimality of the linear kriging estimator in a Gaussian framework, because then the conditional variance is indeed homoscedastic, as seen in Eq. (3-10).

Remarks

- The requisites for SK are all means m_o, m_α, $\alpha = 1, \ldots, n$ and a covariance model $C(\mathbf{h})$ using which all the data-to-data covariances $C(\mathbf{h}_{\alpha\beta})$ and data-to-unknown covariances $C(\mathbf{h}_{\alpha 0})$, $\alpha = 1, \ldots, n$; $\beta = 1, \ldots, n$ can be computed.

- The kriging expression allows for the mean to vary locally within neighborhoods. While the random variable Z is assumed globally stationary (i.e., its global mean and variance exist), its mean is allowed to vary locally within the neighborhoods containing Z_o and Z_α.

- In case the means remain invariant throughout the entire domain, i.e., $m_\alpha = m_o = m \,\forall\, \alpha$, then the SK estimator reduces to:

$$Z_o^* = \left(1 - \sum_{\alpha=1}^{n} \lambda_\alpha\right) \cdot m + \sum_{\alpha=1}^{n} \lambda_\alpha Z_\alpha \tag{3-16}$$

The externally specified mean m thus gets a weight equivalent to $\lambda_m = \left(1 - \sum_{\alpha=1}^{n} \lambda_\alpha\right)$ in the SK procedure.

- The kriging system accounts for the information in the data toward the unknown through the $C(\mathbf{h}_{\alpha 0})$ term and the data-to-data redundancy through the $C(\mathbf{h}_{\alpha\beta})$ term.

- The minimum estimation variance can be derived as:

$$\sigma_{SK}^2 = \sum_{\alpha=0}^{n} \sum_{\beta=0}^{n} a_\alpha \cdot a_\beta \cdot C(\mathbf{h}_{\alpha\beta}) \tag{3-17}$$

For this variance to be guaranteed positive, the combination of parameters a_α and the covariance $C(\mathbf{h})$ has to be such that the previous double summation is always greater than or equal to 0; in other words, the covariance model $C(\mathbf{h})$ has to be positive definite. This notion of a positive definite covariance function is discussed further a little later.

- The kriging system can also be interpreted as a system that ensures the reproduction of the target covariance between the data Z_α and the kriged value Z_o^*. The LHS of the kriging system [Eq. (3-15)] states that the weights λ_β are computed such that they in combination with the data-to-data covariances equate the RHS of the equation, i.e., the data-to-unknown covariance.

- The solution of the kriging system may yield weights λ_β that are negative. Negative weights can result in estimates Z_o^* that fall outside the range of data values, i.e., $Z_o^* \notin [z_{min}, z_{max}]$. Kriging is a non-convex estimator (Olea, 1999).

- The minimum error variance [Eq. (3-17)] reflects the reduction in prior uncertainty [$C(0)$—the prior variance] due to the conditioning influence of the data $C(\mathbf{h}_{\alpha 0})$. If a number of data receive negative weights $\lambda_\beta < 0$, then it may turn out that the estimation uncertainty $\sigma_{SK}^2 > C(0)$. The implication is that if data from multiple sources is integrated into kriging and if it turns out that the data is redundant with each other (resulting in some data receiving negative weights), the uncertainty may actually be compounded (contrary to the notion that uncertainty decreases with integration of more data).

- In the kriging system, [Eq. (3-15)], if the data-to-data covariance is 0 for all $\alpha \neq \beta$, the weights are such that $\lambda_\alpha = \dfrac{C(\mathbf{h}_{\alpha 0})}{C(\mathbf{h}_{\alpha\alpha})}$. The estimate in that case becomes:

$Z_o^* - m_0 = \displaystyle\sum_{\alpha=1}^{n} \dfrac{C(\mathbf{h}_{\alpha 0})}{C(\mathbf{h}_{\alpha\alpha})}(Z_\alpha - m_\alpha)$. Defining standardized vari-

ables $Y_o^* = \dfrac{Z_o^* - m_o}{\sqrt{C(\mathbf{h}_{oo})}}$ and $Y_\alpha^* = \dfrac{Z_\alpha^* - m_\alpha}{\sqrt{C(\mathbf{h}_{\alpha\alpha})}}$, then:

$$Y_o^* = \sum_{\alpha=1}^{n} \rho(\mathbf{h}_{\alpha 0})Y_\alpha$$

a form identical to the linear regression estimator: $\hat{v} = r \cdot u_i$ where r is the correlation between the variables v and u. This implies that kriging performed by ignoring the data-to-data correlation or redundancy amounts to linear regression. In fact, a special aspect of kriging or generalized linear regression is accounting for the data-to-data redundancy when arriving at an estimate.

3.4.3 An Example

Consider the following estimation scenario with data available at two locations $\mathbf{u}_1$ and $\mathbf{u}_2$ for estimation at location $\mathbf{u}_o$, as shown in Fig. 3-6(b). The spatial dependence between the values at the data locations, as well as between the data and the estimation node, is assumed to be given by the stationary covariance function $C(\mathbf{h})$, shown in Fig. 3-6(a). As the prior variance indicated by the covariance at a lag distance of 0 is 1 in Fig. 3-6(a), the covariance $C(\mathbf{h})$ is the same as the correlogram $\rho(\mathbf{h})$ and so we will develop the kriging expressions in terms of the correlogram $\rho(\mathbf{h})$.

The stationary phenomenon $Z(\mathbf{u})$ is assumed to be characterized by a population mean m. The SK estimate at location $\mathbf{u}_o$ given the data $Z(\mathbf{u}_1)$ and $Z(\mathbf{u}_2)$ can be written as:

$$Z^*(\mathbf{u}_o) = \left(1 - \sum_{\alpha=1}^{n}\lambda_\alpha\right)\cdot m + \sum_{\alpha=1}^{n}\lambda_\alpha Z(\mathbf{u}_\alpha)$$
$$= (1 - \lambda_1 - \lambda_2)\cdot m + \lambda_1\cdot Z(\mathbf{u}_1) + \lambda_2\cdot Z(\mathbf{u}_2)$$

The weights are obtained by solving the SK system:

$$\sum_{\beta=1}^{n}\lambda_\beta\rho(\mathbf{h}_{\alpha\beta}) = \rho(\mathbf{h}_{\alpha0})$$

leading to the following simultaneous equation system:

$$\lambda_1\rho(\mathbf{h}_{11}) + \lambda_2\rho(\mathbf{d}) = \rho(\mathbf{h}_1)$$
$$\lambda_1\rho(\mathbf{d}) + \lambda_2\rho(\mathbf{h}_{22}) = \rho(\mathbf{h}_2)$$

The weights come out to be:

$$\lambda_1 = \frac{\rho(\mathbf{h}_1) - \rho(\mathbf{h}_2)\cdot\rho(\mathbf{d})}{1 - \rho^2(\mathbf{d})} \qquad \lambda_2 = \frac{\rho(\mathbf{h}_2) - \rho(\mathbf{h}_1)\cdot\rho(\mathbf{d})}{1 - \rho^2(\mathbf{d})}$$

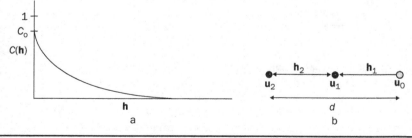

FIGURE 3-6 **a)** Covariance function $C(\mathbf{h})$ assumed for the application example. **b)** Data configuration for the same example.

The weight assigned to the mean comes out to be:

$$\lambda_m = 1 - \frac{\rho(h_1) - \rho(h_2) \cdot \rho(d)}{1 - \rho^2(d)} - \frac{\rho(h_2) - \rho(h_1) \cdot \rho(d)}{1 - \rho^2(d)}$$

$$= 1 - \frac{\rho(h_1) + \rho(h_2)}{1 + \rho(d)}$$

The estimation variance is given as:

$$\sigma_K^2 = \rho(0) - \sum_{\alpha=1}^{n} \lambda_\alpha \rho(h_{\alpha 0})$$

$$= 1 - \frac{\rho(h_1) - \rho(h_2)\rho(d)}{1 - \rho^2(d)} \cdot \rho(h_1) - \frac{\rho(h_2) - \rho(h_1)\rho(d)}{1 - \rho^2(d)} \cdot \rho(h_2)$$

$$= \frac{1 - (\rho^2(d) + \rho^2(h_1) + \rho^2(h_2) - 2\rho(h_1)\rho(h_2)\rho(d))}{1 - \rho^2(d)}$$

When the data is located at a distance larger than the variogram range from the estimation node, then $\rho(h_1) = \rho(h_2) = 0$ and $\sigma_K^2 = 1$ regardless of $\rho(d)$.

Case 1: Data location u_2 is pushed to infinity   In this case, $\rho(h_2) = \rho(d) = 0$ and so the weights $\lambda_1 = \rho(h_1)$, $\lambda_2 = 0$, and $\lambda_m = 1 - \rho(h_1)$. The estimate is therefore $Z_0^* = (1 - \rho(h_1)) \cdot m + \rho(h_1) \cdot Z(u_1)$, which is of a form similar to linear regression, where the available data receives a weight equal to the correlation coefficient between the data and the unknown.

Case 2: Both data locations u_1 and u_2 are moved to infinity   In this case, $\rho(h_1) = \rho(h_2) = \rho(d) = 0$ and so the weights $\lambda_1 = 0$, $\lambda_2 = 0$, and $\lambda_m = 1$. This implies that the estimate $Z_0^* = m$; that is, the prior mean is the optimal estimate when the data is too far to exert any influence on the estimate. As seen previously, the estimation variance is simply the prior variance $\rho(0) = 1$.

Case 3: Data location u_1 is relocated to be concurrent with u_0   In this case, $\rho(h_1) = \rho(0) = 1$ and $\rho(h_2) = \rho(d)$ so the weights $\lambda_1 = 1$, $\lambda_2 = 0$, and $\lambda_m = 0$. This implies that the estimate $Z^*(u_1) = 1 \cdot Z(u_1) = Z(u_1)$; that is, the estimate is exactly the data value $Z(u_1)$. This is a demonstration of the *data exactitude* of kriging.

Case 4: Data at their original locations, estimation node u_0 within a ε-neighborhood around u_1   The covariance model specified has a nugget effect of magnitude c_o; that is, a location within a ε neighborhood of a data location will not have an estimated value exactly equal to the data value. The extent to which the estimated value will differ from the data value is controlled by the extent of the nugget effect.

In the case of the above example, the weights assigned to the data become:

$$\lambda_1 = \frac{\rho(\mathbf{h}_1) - \rho(\mathbf{h}_2) \cdot \rho(\mathbf{d})}{1 - \rho^2(\mathbf{d})} = \frac{(1 - c_o) - \rho^2(\mathbf{d})}{1 - \rho^2(\mathbf{d})}$$

$$= 1 - \frac{c_o}{1 - \rho^2(\mathbf{d})}$$

$$\lambda_2 = \frac{\rho(\mathbf{h}_2) - \rho(\mathbf{h}_1) \cdot \rho(\mathbf{d})}{1 - \rho^2(\mathbf{d})} = \frac{c_o \cdot \rho(\mathbf{d})}{1 - \rho^2(\mathbf{d})}$$

In the preceding example, we have assumed that the distance between the two data locations is sufficiently large so that $\rho(\mathbf{h}_2) = \rho(\mathbf{d} \pm \varepsilon) \cong \rho(\mathbf{d})$.

The weight assigned to the mean is:

$$\lambda_m = 1 - \left(1 - \frac{c_o}{1 - \rho^2(\mathbf{d})} + \frac{c_o \cdot \rho(\mathbf{d})}{1 - \rho^2(\mathbf{d})}\right) = \frac{c_o}{1 - \rho^2(\mathbf{d})}$$

For simplicity, let us assume that the two data values are located at a distance greater than the range of the covariance function. In that case, $\rho(\mathbf{d}) = 0$ and the resultant estimate at the location $\mathbf{u}_o$ becomes:

$$Z^*(\mathbf{u}_o) = c_o \cdot m + (1 - c_o) \cdot Z(\mathbf{u}_1) \neq Z(\mathbf{u}_1)$$

In fact, the deviation of the estimate at $\mathbf{u}_o$ that is located within a ε neighborhood of $Z(\mathbf{u}_1)$ from the data value is:

$$\Delta = Z(\mathbf{u}_1) - Z^*(\mathbf{u}_o) = Z(\mathbf{u}_1) - c_o \cdot m - (1 - c_o) \cdot Z(\mathbf{u}_1)$$
$$= c_o \cdot (Z(\mathbf{u}_1) - m)$$

The kriged map will be data exact; i.e., the kriged value at the data location will be equal to the data value, but at an arbitrarily small distance from the data location, there will be a discontinuity whose magnitude will be controlled by the magnitude of the nugget effect and the difference between the data value and the mean specified for kriging. These results can be depicted visually, as shown in Fig. 3-7. For convenience, the figure only shows the case of one data (the other data location is assumed outside the range of the correlation function).

Case 5: Characteristics of the weights depending on the choice of the covariance function   Going back to the original configuration of two data values informing the estimation node, the weight assigned to the first data location is given as:

$$\lambda_2 = \frac{\rho(\mathbf{h}_1) - \rho(\mathbf{h}_2) \cdot \rho(\mathbf{d})}{1 - \rho^2(\mathbf{d})}$$

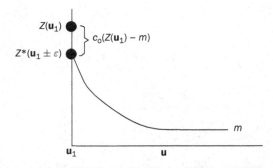

FIGURE 3-7   Illustration of the variation in the kriging estimate as the estimation node is moved away from a conditioning data location.

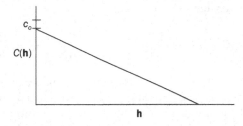

FIGURE 3-8   A linear covariance function.

Given that the denominator is guaranteed to be positive, the weight will come out to be negative if: $\rho(\mathbf{h}_2) \cdot \rho(\mathbf{d}) > \rho(\mathbf{h}_1)$

Similarly: $\lambda_2 < 0$ if $\rho(\mathbf{h}_1) \cdot \rho(\mathbf{d}) > \rho(\mathbf{h}_2)$

In order to explore conditions when the weight assigned to one of the data may become negative, let us alter the covariance structure, as shown in Fig. 3-8:

This new covariance model is linear and can be written as:

$$C(\mathbf{h}) = 1 - |\mathbf{h}|$$

In that case, $\rho(\mathbf{h}_1) \cdot \rho(\mathbf{d})$ is:

$$\rho(\mathbf{h}_1) \cdot \rho(\mathbf{d}) = (1 - |\mathbf{h}_1|) \cdot (1 - |\mathbf{d}|)$$
$$= 1 - (|\mathbf{h}_1| + |\mathbf{d}|) + |\mathbf{h}_1| \cdot |\mathbf{d}|$$

Given the data configuration in Fig. 3-6 $(|\mathbf{h}_1| + |\mathbf{d}|) = |\mathbf{h}_2|$. Therefore:

$$\rho(\mathbf{h}_1) \cdot \rho(\mathbf{d}) = 1 - |\mathbf{h}_2| + |\mathbf{h}_1| \cdot |\mathbf{d}|$$

But $1 - |\mathbf{h}_2| = \rho(\mathbf{h}_2)$ and therefore:

$$\rho(\mathbf{h}_1) \cdot \rho(\mathbf{d}) = \rho(\mathbf{h}_2) + |\mathbf{h}_1| \cdot |\mathbf{d}|$$
$$> \rho(\mathbf{h}_2)$$

Because $\rho(\mathbf{h}_1) \cdot \rho(\mathbf{d}) > \rho(\mathbf{h}_2)$, this implies that $\lambda_2 < 0$; that is, corresponding to the data configuration assumed in Fig. 3-6(b) when one datum is directly behind another and the covariance model describing the relationship between the data and the unknown is linear, the influence of the second datum toward the estimation is screened by the nearer data value. This screening is observed in the form of the negative weight assigned to the datum farther away.

Let us re-compute the weight λ_1 when the linear covariance model is replaced with an exponential model, as shown in Fig. 3-9.

In this case: $C(\mathbf{h}) = e^{-|\mathbf{h}|}$

Therefore: $\rho(\mathbf{h}_1) \cdot \rho(\mathbf{d}) = e^{-|\mathbf{h}_1|} \cdot e^{-|\mathbf{d}|} = e^{-(|\mathbf{h}_1| + |\mathbf{d}|)} = e^{-|\mathbf{h}_2|}$

The last equality is because of the particular configuration of the data shown in Fig. 3-6(b). Therefore, the weight assigned to the second datum in this case is:

$$\lambda_2 = \frac{\rho(\mathbf{h}_2) - \rho(\mathbf{h}_1) \cdot \rho(\mathbf{d})}{1 - \rho^2(\mathbf{d})}$$
$$= \frac{e^{-|\mathbf{h}_2|} - e^{-|\mathbf{h}_2|}}{1 - e^{-2|\mathbf{d}|}} = 0$$

i.e., for the same data configuration in Fig. 3-6(b), given an exponential model for the correlation between pairs of locations, the weight ascribed to the farther-away datum is exactly 0. This is an example of the Markov (screening) property of the exponential covariance structure. Indeed,

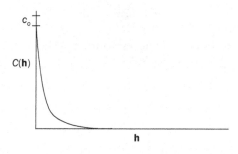

FIGURE 3-9   An exponential covariance function.

phenomena described by an exponential covariance structure are often described to be "memory-less"; i.e., estimation or prediction is based only on the data in the immediate neighborhood of the estimated quantity and data events in the distant "memory" are ignored.

Case 6: Case when the data act independent of each other   This case is tantamount to a data configuration where the lags h_1, h_2 are smaller than the range of the correlation function but the lag separating the data d is greater than the range.

In that case, the weights assigned to the two data are:

$$\lambda_1 = \frac{\rho(h_1) - \rho(h_2) \cdot \rho(d)}{1 - \rho^2(d)} \qquad \text{and} \qquad \lambda_2 = \frac{\rho(h_2) - \rho(h_1) \cdot \rho(d)}{1 - \rho^2(d)}$$
$$= \rho(h_1) \qquad\qquad\qquad\qquad\qquad = \rho(h_2)$$

because $\rho(d) = 0$ in this case. The estimate is then:

$$Z^*(u_o) = \rho(h_1) \cdot Z(u_1) + \rho(h_2) \cdot Z(u_2) + (1 - \rho(h_1) - \rho(h_2)) \cdot m$$

This is exactly the form of the estimator in linear regression in terms of the standardized data Eq. (3-7). It can therefore be concluded that kriging performed by ignoring the correlation between the data is equivalent to linear regression.

Example 3-4: Numerical Example
The conditioning data and covariance model in Fig. 3-6 are used to demonstrate the implementation of SK. Using the configuration of data and the covariance model, the following matrix of covariance between the data is computed:

$$K = \begin{bmatrix} 1.00 & 0.49 & 0.62 & 0.36 \\ 0.49 & 1.00 & 0.62 & 0.26 \\ 0.62 & 0.62 & 1.00 & 0.53 \\ 0.36 & 0.26 & 0.53 & 1.00 \end{bmatrix}$$

The covariance between the data and the kriged node remains the same as for the previous regression example.

$$k = \begin{bmatrix} 0.66 \\ 0.25 \\ 0.43 \\ 0.36 \end{bmatrix}$$

The weights are calculated as:

$$\lambda = K^{-1} \cdot k = \begin{bmatrix} 1.68 & -0.29 & -0.81 & -0.10 \\ -0.29 & 1.69 & -0.97 & 0.18 \\ -0.81 & -0.97 & 2.53 & -0.81 \\ -0.10 & 0.18 & -0.81 & 1.42 \end{bmatrix} \begin{bmatrix} 0.66 \\ 0.25 \\ 0.43 \\ 0.36 \end{bmatrix} = \begin{bmatrix} 0.65 \\ -0.12 \\ 0.02 \\ 0.15 \end{bmatrix}$$

The kriged estimate is thus obtained as:

$$z^*_{SK} = 0.65 \cdot \frac{(10 - 25)}{12.91} - 0.12 \cdot \frac{(20 - 25)}{12.91} + 0.02 \cdot \frac{(40 - 25)}{12.91}$$
$$+ 0.15 \cdot \frac{(30 - 25)}{12.91} = 16.83$$

Performing kriging at all un-informed nodes in the spatial domain for the example, we obtain the following kriged map (Fig. 3-10).

Comparing the results of regression shown in Fig. 3-1 and kriging shown in Fig. 3-10, it can be observed that the regression results are smoother, and the transition from higher to lower values is generally more gradual. The kriging estimation variance map can also be calculated using Eq. (3-17). The result is shown in Fig. 3-11.

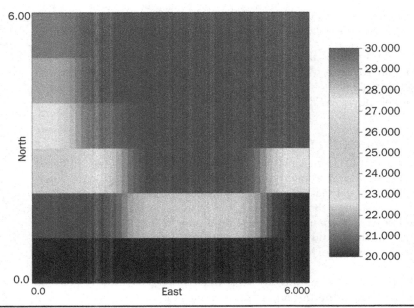

FIGURE 3-10 Estimated map of thickness obtained by kriging. The available conditioning data and the covariance model are shown in Fig. 3-6. (The color version of this figure is available at www.mhprofessional.com/PRMS.)

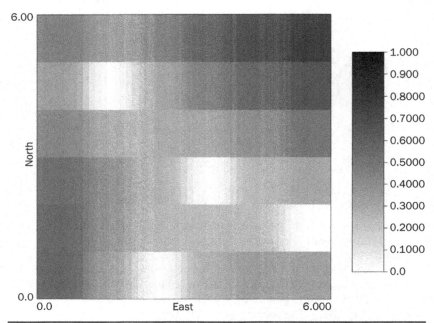

Figure 3-11 Map of simple kriging estimation variances. The variance is 0 at the locations of available conditioning data and exhibits spatial variation consistent with the covariance model shown in Fig. 3-6. (The color version of this figure is available at www.mhprofessional.com/PRMS.)

The location of conditioning data can be discerned in the estimation variance map. Away from the data location, the estimation variance exhibits isotropic variability consistent with the covariance model shown in Fig. 3-6(a).

3.5 Universal Kriging

The previous discussion of SK was centered on the concept of linear interpolation or a linear estimator where the weights for the interpolation or estimation were obtained as a solution to a linear system that considered the data-to-data covariances, as well as the data-to-unknown covariances. The key requirement for SK is a model for the covariance and the prior means at each estimation location. This prior mean can exhibit some spatial variation (denoted by $m(\mathbf{u}_\alpha)$ but is considered stationary and known. What happens in the case where the prior mean is unknown? More interestingly, how do we perform kriging when the mean is unknown and non-stationary, e.g., when the natural system being modeled exhibits a trend?

In order to address these practical questions, let us consider the following additive model for the random function $Z(\mathbf{u})$ being modeled:

$$Z(\mathbf{u}) = m(\mathbf{u}) + R(\mathbf{u}) \tag{3-18}$$

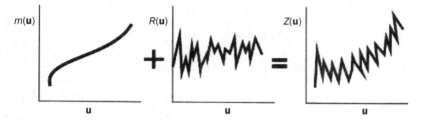

FIGURE 3-12 Conceptual rendering of the linear decomposition of a random function $Z(\mathbf{u})$ into a trend function $m(\mathbf{u})$ and a residue $R(\mathbf{u})$.

$m(\mathbf{u}) = E\{Z(\mathbf{u})\}$	is a smooth, continuous function depicting the spatial variation in the prior mean
$R(\mathbf{u})$	is a stationary residual or variability around the mean
$E\{R(\mathbf{u}), R(\mathbf{u} + \mathbf{h})\} = C_R(\mathbf{h})$	is the stationary spatial covariance characterizing the stationary residual

This decomposition can be visualized as shown in Fig. 3-12 and is quite commonly used for modeling phenomena whose variability is in the form of a time series.

There are several options for modeling the smooth trend function. One option would be to use an analytical function that describes how the mean varies in different directions. Another option could be to use a smooth map that reflects the variability of the mean. Yet another option could be to relate the smooth variability exhibited by a measurement such as seismic and relate it to the spatial variations in the mean. We will discuss the implementation of kriging using some of these options for describing the spatial variations in the mean next.

3.5.1   Kriging with a Trend Function

Consider the following analytical model for the trend (Journel & Huijbregts, 1978; Chiles, 1999):

$$m(\mathbf{u}) = E\{Z(\mathbf{u})\} = \sum_{\ell=0}^{L} a_\ell f_\ell(\mathbf{u}) \tag{3-19}$$

This functional form is similar to a series representation, for example, a Fourier series that has basis functions (sine or cosine) and coefficients that are estimated using the data. Similarly, in the above formulation, the basis $f_\ell(\mathbf{u})$ is assumed to be a known function of the coordinates and the $(L + 1)$ coefficients a_ℓ are deemed unknown and to be determined using the available data. By convention, the leading basis $f_0(\mathbf{u})$ is set to one. In order to understand how the function in Eq. (3-19) models the mean, consider the following examples:

Example   If trend is assumed to be linear in the x direction:

$$m(\mathbf{u}) = a_o + a_1 \cdot x$$

Example   If trend is quadratic in space coordinates $\mathbf{u}$, then:

$$m(\mathbf{u}) = a_0 + a_1 \cdot \mathbf{u} + a_2 \cdot \mathbf{u}^2$$

Another example of a smooth second-order variability of the mean could be:

$$m(\mathbf{u}) = a_o + a_1 \cdot x + a_2 \cdot y + a_3 \cdot x \cdot y$$

A key idea that we will explore in the subsequent sections is to use the analytical formulation to filter out the mean $m(\mathbf{u})$ from the kriging expression and construct an unbiased, minimum error variance estimate.

Remarks   In the random function model $Z(\mathbf{u}) = m(\mathbf{u}) + R(\mathbf{u})$, the trend model $m(\mathbf{u}) = E\{Z(\mathbf{u})\} = \sum_{\ell=0}^{L} a_\ell f_\ell(\mathbf{u})$ is deterministic, as the value of the mean at every location in the domain is known once we have estimates for the parameters a_ℓ. The residue $R(\mathbf{u}) = Z(\mathbf{u}) - m(\mathbf{u})$ is stochastically characterized by the covariance $C_R(\mathbf{h})$. The implication is that in order to develop an estimate for $Z(\mathbf{u})$, it is the stationary covariance $C_R(\mathbf{h})$ that is required.

If the mean is non-stationary and unknown *a priori*, it is hopeless to try and compute the residual covariance based on measured values $Z(\mathbf{u}_\alpha)$ at data locations. This is because $m(\mathbf{u})$ changes with the location $\mathbf{u}$ and is unknown *a priori*. It is likely that if the mean is expected to vary systematically in a particular direction, the random function in the direction perpendicular to this trend exhibits stationarity (Armstrong, 1984). This implies that the spatial characteristics of $R(\mathbf{u}) = Z(\mathbf{u}) - m$ have to be inferred in the direction perpendicular to the mean and so, a possibility could be $C_R(\mathbf{h}) = C_Z(\mathbf{h})$. An approximation therefore is to infer the covariance using the available data in the direction perpendicular to the direction of the trend and identify that to the residual covariance $C_R(\mathbf{h})$. Note that the residual covariance inferred in this fashion will necessarily be isotropic. The anisotropy exhibited by the random function $Z(\mathbf{u})$ is accounted for by combining the isotropic covariance with a trend model extending in the direction of the anisotropy.

The decomposition $Z(\mathbf{u}) = m(\mathbf{u}) + R(\mathbf{u})$ implies that the large-scale correlation of Z is captured through $m(\mathbf{u})$ and the smaller-scale (high frequency) variability is captured by $R(\mathbf{u})$. The correlation range of $C_R(\mathbf{h})$ should therefore be expected to be smaller. This is true if $C_R(\mathbf{h})$ is inferred in the direction perpendicular to the trend.

Unbiasedness

Consider the following linear estimator:

$$Z_o^* = \sum_{\alpha=1}^{n} \lambda_\alpha Z(\mathbf{u}_\alpha) \tag{3-20}$$

The estimator is of a form similar to the SK estimator [Eq. (3-8)] without the additional weight λ_0. It is important to note that the mean does not explicitly appear in the estimator expression. Unbiasedness of the estimator implies:

$$E\{Z_o - Z_o^*\} = 0 \text{ which implies that } \sum_{\alpha=1}^{n} \lambda_\alpha m(\mathbf{u}_\alpha) = m(\mathbf{u}_o)$$

Substituting the functional form for the trend function, we can write:

$$\sum_{\alpha=1}^{n} \lambda_\alpha \sum_{\ell=0}^{L} a_\ell f_\ell(\mathbf{u}_\alpha) = \sum_{\ell=0}^{L} a_\ell f_\ell(\mathbf{u}_o)$$

or

$$\sum_{\ell=0}^{L} a_\ell \sum_{\alpha=1}^{n} \lambda_\alpha f_\ell(\mathbf{u}_\alpha) = \sum_{\ell=0}^{L} a_\ell f_\ell(\mathbf{u}_o)$$

Equating the functional basis multiplying each coefficient a_ℓ on the left- and right-hand side, we get:

$$\sum_{\alpha=1}^{n} \lambda_\alpha f_\ell(\mathbf{u}_\alpha) = f_\ell(\mathbf{u}_o) \,\forall\, \ell = 0, \dots, L \tag{3-21}$$

The unbiasedness condition thus gives rise to a system of equations. These equations state that a weighted linear combination of the coordinate function at the data locations equals the coordinate function at the estimation location. Some examples of these unbiased conditions for various forms of the coordinate function are as follows:

Example   If the mean is stationary but unknown then:

$$m(\mathbf{u}) = a_o \text{ i.e., mean is not a function of } \mathbf{u}$$

This implies that in this case, $L = 0$, and

$$\sum_{\alpha=1}^{n} \lambda_\alpha \cdot f_0(\mathbf{u}_\alpha) = f_0(\mathbf{u}_o)$$

Because $f_0(\mathbf{u})$ is by convention equal to one, the unbiasedness condition yields:

$$\sum_{\alpha=1}^{n} \lambda_\alpha = 1$$

Example If the mean or trend function is linear in the x and y coordinates:

$$m(\mathbf{u}) = a_o + a_1 \cdot x + a_2 \cdot y$$

In this case $f_0(\mathbf{u}) = 1, f_1(\mathbf{u}) = x, f_2(\mathbf{u}) = y$ and the corresponding unbiasedness conditions are:

$$\sum_{\alpha=1}^{n} \lambda_\alpha = 1$$

$$\sum_{\alpha=1}^{n} \lambda_\alpha \cdot x_\alpha = x_o$$

$$\sum_{\alpha=1}^{n} \lambda_\alpha \cdot y_\alpha = y_o$$

Estimation Variance

The estimation variance corresponding to the estimator in Eq. (3-20) can be written as:

$$Var\left\{Z_o - Z_o^*\right\} = E\left\{\left[Z_o - Z_o^*\right]^2\right\} = E\left\{\left[Z_o - \sum_{\alpha=1}^{n} \lambda_\alpha Z_\alpha\right]^2\right\}$$

$$= E\left\{\left[Z_o^2 - 2\sum_{\alpha=1}^{n} \lambda_\alpha \cdot Z_o \cdot Z_\alpha + \sum_{\alpha=1}^{n}\sum_{\beta=1}^{n} \lambda_\alpha \cdot \lambda_\beta \cdot Z_\alpha \cdot Z_\beta\right]\right\}$$

$$= C_R(\mathbf{h}_{oo}) - 2\sum_{\alpha=1}^{n} C_R(\mathbf{h}_{\alpha o}) + \sum_{\alpha=1}^{n}\sum_{\beta=1}^{n} \lambda_\alpha \cdot \lambda_\beta \cdot C_R(\mathbf{h}_{\alpha\beta})$$

Because the means are unknown, we have used the decomposition $Z(\mathbf{u}) - m(\mathbf{u}) = R(\mathbf{u})$ to develop the estimation variance expression in terms of the residual variance and the residual covariance instead of the variance and covariance of $Z(\mathbf{u})$. This estimation variance has to be minimized under the constraint $\sum_{\alpha=1}^{n} \lambda_\alpha f_\ell(\mathbf{u}_\alpha) = f_\ell(\mathbf{u}_o) \ \forall \ \ell = 0,\ldots,L$.

In order to solve minimization problem, we can use the method of Lagrange multipliers (Rockafellar, 1993). Assigning $L+1$ Lagrange multipliers μ_ℓ, one for each constraint, we can write the minimization problem as:

$$\text{min.} \left[\begin{array}{l} C_R(\mathbf{h}_{oo}) - 2\displaystyle\sum_{\alpha=1}^{n} \lambda_\alpha \cdot C_R(\mathbf{h}_{\alpha o}) + \displaystyle\sum_{\alpha=1}^{n}\sum_{\beta=1}^{n} \lambda_\alpha \cdot \lambda_\beta \cdot C_R(\mathbf{h}_{\alpha\beta}) \\[2ex] + \displaystyle\sum_{\ell=0}^{L} 2 \cdot \mu_\ell \left(\displaystyle\sum_{\alpha=1}^{n} \lambda_\alpha f_\ell(\mathbf{u}_\alpha) - f_\ell(\mathbf{u}_o) \right) \end{array} \right]$$

The parameters for the optimization problem are the n weights λ_α and the $L + 1$ Lagrange parameters μ_ℓ. Performing the usual minimization by taking the derivative with respect to the weights λ_α and setting the derivative to 0, we get a system of n equations:

$$\sum_{\beta=1}^{n} \lambda_\beta \cdot C_R(\mathbf{h}_{\alpha\beta}) + \sum_{\ell=0}^{L} \mu_\ell \cdot f_\ell(\mathbf{u}_\alpha) = C_R(\mathbf{h}_{\alpha o}) \ \forall \ \alpha \qquad (3\text{-}22)$$

Taking the derivative with respect to the $L + 1$ Lagrange parameters and setting equal to 0, we get an auxiliary set of $L + 1$ equations:

$$\sum_{\alpha=1}^{n} \lambda_\alpha f_\ell(\mathbf{u}_\alpha) = f_\ell(\mathbf{u}_o) \ \forall \ \ell = 0,\dots,L \qquad (3\text{-}23)$$

These set of auxiliary equations are the same as the unbiasedness conditions we derived earlier. The first one of these unbiasedness conditions is [because $f_0(\mathbf{u}) = 1$]:

$$\sum_{\alpha=1}^{n} \lambda_\alpha = 1 \qquad (3\text{-}24)$$

Putting all this together, we write the UK/KT estimator as:

$$Z_o^* = \sum_{\alpha=1}^{n} \lambda_\alpha Z(\mathbf{u}_\alpha)$$

where the weights are obtained by solving a system:

$$\sum_{\beta=1}^{n} \lambda_\beta \cdot C_R(\mathbf{h}_{\alpha\beta}) + \sum_{\ell=0}^{L} \mu_\ell \cdot f_\ell(\mathbf{u}_\alpha) = C_R(\mathbf{h}_{\alpha o}) \ \forall \ \alpha$$

$$\sum_{\alpha=1}^{n} \lambda_\alpha f_\ell(\mathbf{u}_\alpha) = f_\ell(\mathbf{u}_o) \ \forall \ \ell = 0,\dots,L$$

The solution of $(L + n + 1) \times (L + n + 1)$ system of equations is guaranteed to yield a unique solution for the weights λ_α and the $L + 1$ Lagrange parameters μ_ℓ, provided $C_R(\mathbf{h})$ is positive definite and $\sum_{\alpha=1}^{n} \lambda_\alpha f_\ell(\mathbf{u}_\alpha) = 0$ if and only if $\mu_\ell = 0 \ \forall \ \ell$.

The expression for the error variance is:

$$\sigma_K^2 = \sum_{\alpha=0}^{n} \sum_{\beta=0}^{n} a_\alpha a_\beta C(\mathbf{h}_{\alpha\beta})$$

where $a_\alpha = \begin{cases} 1, & \text{for } \alpha = 0 \\ -\lambda_\alpha & \text{for } \alpha > 0 \end{cases}$

This is equivalent to:

$$\sigma_K^2 = \sum_{\alpha=0}^{n} \sum_{\beta=0}^{n} a_\alpha a_\beta C_R(\mathbf{h}_{\alpha\beta})$$

$$= C(0) - 2 \cdot \sum_{\alpha=1}^{n} \lambda_\alpha C_R(\mathbf{h}_{\alpha o}) + \sum_{\alpha=1}^{n} \sum_{\beta=1}^{n} \lambda_\alpha \lambda_\beta C_R(\mathbf{h}_{\alpha\beta})$$

substituting $\sum_{\beta=1}^{n} \lambda_\beta C_R(\mathbf{h}_{\alpha\beta}) = C_R(h_{\alpha o}) - \sum_{\ell=1}^{L} \mu_\ell f_\ell(\mathbf{u}_\alpha)$, we get:

$$\sigma_K^2 = \sum_{\alpha=0}^{n} \sum_{\beta=0}^{n} a_\alpha a_\beta C_R(\mathbf{h}_{\alpha\beta}) = C(0) - \sum_{\alpha=1}^{n} \lambda_\alpha C_R(\mathbf{h}_{\alpha o}) - \sum_{\ell=0}^{L} \mu_\ell f_\ell(\mathbf{u}_\alpha) \sum_{\alpha=1}^{n} \lambda_\alpha$$

$$= C(0) - \sum_{\alpha=1}^{n} \lambda_\alpha C_R(\mathbf{h}_{\alpha o}) - \sum_{\ell=0}^{L} \mu_\ell f_\ell(\mathbf{u}_o)$$

$$(3\text{-}25)$$

The last equality comes about because $\sum_{\alpha=1}^{n} \lambda_\alpha = 1$ and $\sum_{\alpha=1}^{n} \lambda_\alpha f_\ell(\mathbf{u}_\alpha) = f_\ell(\mathbf{u}_o) \; \forall \; \ell = 0, \ldots, L$. The expression for the estimation variance suggests that lack of knowledge about the mean could add to the uncertainty associated with the predictions if the Lagrange parameter comes out to be negative. On the other hand, if the data is abundant and consistent with the trend and residual covariance model, then the uncertainty may decrease (as the Lagrange parameters in this case will be positive).

3.5.2 Ordinary Kriging

The first example for kriging with a trend is the case of a stationary, constant mean. That case is equivalent to truncating the trend function [Eq. (3-19)] at the constant term:

$$m(\mathbf{u}) = a_o$$

As noted previously, under this condition, the unbiasedness condition is:

$$\sum_{\alpha=1}^{n} \lambda_\alpha = 1$$

The kriging system is:

$$\sum_{\beta=1}^{n} \lambda_\beta \cdot C_R(\mathbf{h}_{\alpha\beta}) + \mu_o = C_R(\mathbf{h}_{\alpha o}) \; \forall \; \alpha = 1, \ldots, n \qquad (3\text{-}26)$$

The system of equations and the accompanying estimator is referred to as the *ordinary kriging* (OK) system and estimator. The corresponding OK variance is:

$$\sigma_{OK}^2 = C(0) - \sum_{\alpha=1}^{n} \lambda_\alpha C_R(\mathbf{h}_{\alpha o}) - \mu_o \qquad (3\text{-}27)$$

Remarks

- If the unbiasedness condition $\sum_{\alpha=1}^{n} \lambda_\alpha = 1$ is substituted in the SK estimator:

$$Z_o^* = \left(1 - \sum_{\alpha=1}^{n} \lambda_\alpha\right) \cdot m + \sum_{\alpha=1}^{n} \lambda_\alpha \cdot Z(\mathbf{u}_\alpha)$$

  the mean m is filtered out of the estimator. So the unbiasedness condition provides an avenue to perform kriging without specifying a value for the mean.

- In the case of a stationary but unknown mean, the covariance inferred using the data and modeled using a positive definite function suffices for the kriging system. Inference of the covariance in the direction orthogonal to the trend is unnecessary. This is because there is no systematic drift that will render the estimation of covariance biased.

Example 3-5: Ordinary Kriging Example

Considering the same example depicted in Fig. 3-6, the corresponding OK system given the data configuration and the covariance model can be written as:

$$\begin{bmatrix} 1.00 & 0.49 & 0.62 & 0.36 & 1 \\ 0.49 & 1.00 & 0.62 & 0.26 & 1 \\ 0.62 & 0.62 & 1.00 & 0.53 & 1 \\ 0.36 & 0.26 & 0.53 & 1.00 & 1 \\ 1 & 1 & 1 & 1 & 0 \end{bmatrix} \begin{bmatrix} \lambda_1 \\ \lambda_2 \\ \lambda_3 \\ \lambda_4 \\ \mu \end{bmatrix} = \begin{bmatrix} 0.66 \\ 0.25 \\ 0.43 \\ 0.36 \\ 1.00 \end{bmatrix}$$

The above system is for the estimation at node (1,1) at the lower-left corner of the grid. Solution of the system yields the following solution:

$$\begin{bmatrix} \lambda_1 \\ \lambda_2 \\ \lambda_3 \\ \lambda_4 \\ \mu \end{bmatrix} = \begin{bmatrix} 0.73 \\ -0.01 \\ 0.01 \\ 0.27 \\ -0.17 \end{bmatrix}$$

It can be verified that the sum of the weights (λ's) is one. By looking at the estimated weights and Lagrange parameter and the expression for the estimation variance [Eq. (3-27)], it can be concluded that lack of knowledge about the mean adds to the uncertainty in estimation through the negative value for the Lagrange parameter. The kriged estimate for thickness at the estimation location is:

$$z_{OK}^* = 0.73 \cdot \frac{(10 - 25)}{12.91} - 0.01 \cdot \frac{(20 - 25)}{12.91} + 0.01 \cdot \frac{(40 - 25)}{12.91}$$
$$+ 0.27 \cdot \frac{(30 - 25)}{12.91} = 15.50$$

The spatial map with OK estimates at all locations in the 6 × 6 grid is presented in Fig. 3-13.

The kriging estimation variance can be computed using Eq. (3-27). This is plotted in Fig. 3-14.

Comparing the estimation variance map in Fig. 3-14 to that corresponding to SK in Fig. 3-11, it is observed that due to the lack of data in the top-right corner, the estimation variance obtained by OK tends to be higher in that region. The estimation variances at locations close to data are fairly similar by both the methods.

Example 3-6: Trend Function

Now let us consider an example of a linear trend function. Specifying the trend as:

$$m(\mathbf{u}) = a_o + a_1 \cdot x + a_2 \cdot y$$

The corresponding unbiasedness conditions for this case have been previously established as:

$$\sum_{\alpha=1}^{n} \lambda_\alpha = 1$$
$$\sum_{\alpha=1}^{n} \lambda_\alpha \cdot x_\alpha = x_o$$
$$\sum_{\alpha=1}^{n} \lambda_\alpha \cdot y_\alpha = y_o$$

The kriging system in this case is:

$$\sum_{\alpha=1}^{n} \lambda_\beta \cdot C_R(\mathbf{h}_{\alpha\beta}) + \mu_o + \mu_1 \cdot x_\alpha + \mu_1 \cdot y_\alpha = C_R(\mathbf{h}_{\alpha o})$$

In order to demonstrate the application of this linear trend model for estimation, the covariance model for the previous estimation

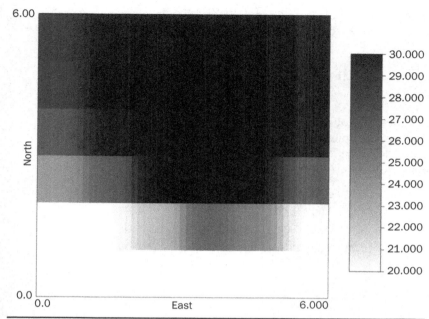

FIGURE 3-13 Estimates for reservoir thickness obtained by ordinary kriging. The calculations use the configuration of data and the covariance model in Fig. 3-6. (The color version of this figure is available at www.mhprofessional.com/PRMS.)

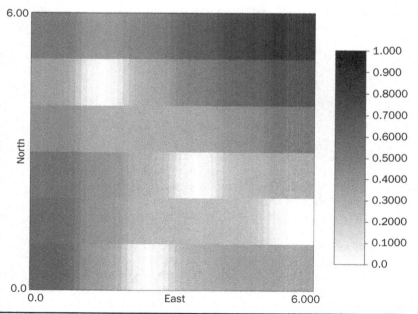

FIGURE 3-14 The map of ordinary kriging estimation variance corresponding to the configuration of data and the covariance model in Fig. 3-6. (The color version of this figure is available at www.mhprofessional.com/PRMS.)

examples (Fig. 3-6) is modified and the following isotropic covariance model is specified:

$$C(\mathbf{h}) = \begin{cases} 1.5 \cdot \left(\dfrac{|\mathbf{h}|}{4}\right) - 0.5 \cdot \left(\dfrac{|\mathbf{h}|}{4}\right)^2 , \forall\, |\mathbf{h}| \le 4 \\ 0 \quad \text{otherwise} \end{cases}$$

In other words, the range of the correlation function is half that in the previous examples, with the idea being that the trend function acts as an additional factor that adds correlation in the resultant spatial model. The following is the kriging system at the node (1,1):

$$\begin{bmatrix} 1.00 & 0.13 & 0.32 & 0.00 & 1 & 3 & 1 \\ 0.13 & 1.00 & 0.32 & 0.00 & 1 & 6 & 2 \\ 0.32 & 0.32 & 1.00 & 0.19 & 1 & 4 & 3 \\ 0.00 & 0.00 & 0.19 & 1.00 & 1 & 2 & 5 \\ 1 & 1 & 1 & 1 & 0 & 0 & 0 \\ 3 & 6 & 4 & 2 & 0 & 0 & 0 \\ 1 & 2 & 3 & 5 & 0 & 0 & 0 \end{bmatrix} \begin{bmatrix} \lambda_1 \\ \lambda_2 \\ \lambda_3 \\ \lambda_4 \\ \mu_0 \\ \mu_1 \\ \mu_2 \end{bmatrix} = \begin{bmatrix} 0.38 \\ 0.00 \\ 0.05 \\ 0.00 \\ 1.00 \\ 1.00 \\ 1.00 \end{bmatrix}$$

The solution to this linear system yields the following weights and Lagrange parameters at the node (1,1):

$$\begin{bmatrix} \lambda_1 \\ \lambda_2 \\ \lambda_3 \\ \lambda_4 \\ \mu_0 \\ \mu_1 \\ \mu_2 \end{bmatrix} = \begin{bmatrix} 1.50 \\ -0.49 \\ -0.26 \\ 0.25 \\ -2.35 \\ 0.36 \\ 0.28 \end{bmatrix}$$

As expected, the weights sum up to one. The large negative number for the Lagrange parameter μ_o indicates that the data is incompatible with the assumption of a linear trend in x and y directions. This contributes to the higher estimation variance observed for this case. The map of kriged estimates and the corresponding estimation variance map are shown in Fig. 3-15.

Comparing the estimation variances obtained by OK in Fig. 3-14 to the one in Fig. 3-15(b), it can be observed that at location away from data, the estimation variances are much larger and in fact at some nodes, the estimation variance is greater than the prior variance. This may be because of the additional uncertainty introduced by not knowing the mean and additionally, the inconsistency between the assumed model for the trend and the conditioning data available.

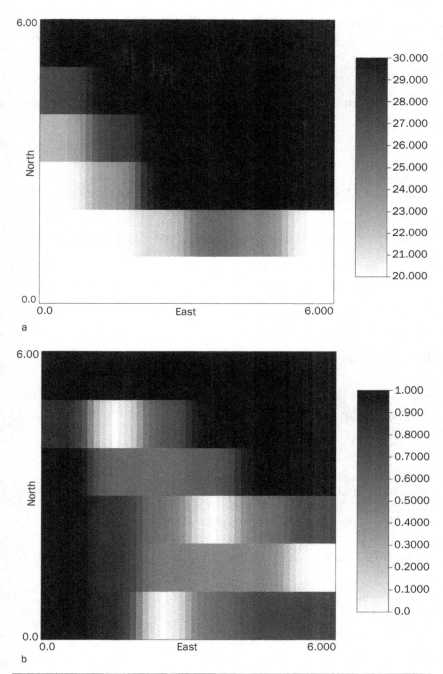

Figure 3-15   a) Map of kriged estimates obtained by universal kriging assuming a linear trend model in both x and y directions. The data configuration is the same as in Fig. 3-6(b). **b)** Map of estimation variance for the same example. (The color version of this figure is available at www.mhprofessional.com/PRMS.)

3.5.3   Universal Kriging Estimate for Trend

In Ordinary Kriging

The estimator in universal kriging (UK) is written as: $Z^*(\mathbf{u}_o) = \sum_{\alpha=1}^{n} \lambda_\alpha Z(\mathbf{u}_\alpha)$. Compared with the SK estimator in Eq. (3-16), the unknown mean has been filtered out by specifying the unbiasedness condition $\sum_{\alpha=1}^{n} \lambda_\alpha = 1$. While the mean does not formally appear in the estimation expression, nevertheless there is a mean influencing the estimate and its functional form in terms of the coordinates of the estimation node is known. So, a relevant question could be *"How to model spatial variations in the underlying trend (mean)?"* Before we address this question in the general framework for UK, let us first look at the mean in OK and then attempt to generalize our results to the case with non-stationary and unknown mean.

Like the expression for estimate of $Z(\mathbf{u}_o)$, the unknown mean is expressed as a linear function of the data:

$$m_{OK}^*(\mathbf{u}_0) = \sum_{\alpha=1}^{n} v_\alpha \cdot Z(\mathbf{u}_\alpha)$$

In order for this estimate to be unbiased, the requirement is that:

$$\sum_{\alpha=1}^{n} v_\alpha = 1$$

The estimation variance can be written as:

$$\sigma^2_{OK_m} = Var\{m - m^*\} = Var\{m\} + Var\{m^*\} - 2 \cdot Cov\{m, m^*\}$$

In OK, the assumption is that there is a single stationary mean that exists but is unknown for the estimation. The true mean m is therefore deterministic. Consequently: $Var\{m\} = 0$ and $Cov\{m, m^*\}$ is indeterminate. The estimation variance consequently reduces to:

$$\sigma^2_{OK_m} = Var\{m^*\} = Var\left(\sum_{\alpha=1}^{n} v_\alpha Z(\mathbf{u}_\alpha)\right) = \sum_{\alpha=1}^{n}\sum_{\beta=1}^{n} v_\alpha v_\beta Cov(Z_\alpha, Z_\beta)$$

This estimation variance has to be a minimized subject to the condition $\sum_{\alpha=1}^{n} v_\alpha = 1$.

Again applying the Lagrange multiplier principle:

$$S = \sum_{\alpha=1}^{n}\sum_{\beta=1}^{n} v_\alpha \cdot v_\beta \cdot C(\mathbf{h}_{\alpha\beta}) + 2 \cdot \mu_{OK_M} \cdot \left(\sum_{\alpha=1}^{n} v_\alpha - 1\right)$$

$$\frac{\partial S}{\partial v_\alpha} = 2 \cdot v_\alpha \cdot Var\{Z_\alpha\} + 2 \cdot \sum_{\beta\neq\alpha} v_\beta \cdot C(\mathbf{h}_{\alpha\beta}) + 2 \cdot \mu_{OK_M}$$

$$= 2 \cdot \sum_{\beta=1}^{n} v_\beta \cdot C(\mathbf{h}_{\alpha\beta}) + 2\mu_{OK_M}$$

Setting the derivative to 0 yields the following system of equations:

$$\sum_{\alpha=1}^{n} v_\beta \cdot C(\mathbf{h}_{\alpha\beta}) + \mu_{OK_M} = 0, \quad \alpha = 1,\ldots,n$$

Taking the derivative with respect to the Lagrange parameter μ_{OK_M} yields the unbiasedness condition $\sum_{\alpha=1}^{n} v_\alpha = 1$. These results can be summarized as follows:

Estimator: $m^* = \sum_{\alpha=1}^{n} v_\alpha Z(\mathbf{u}_\alpha)$

System: $\sum_{\beta=1}^{n} v_\beta \cdot C(\mathbf{h}_{\alpha\beta}) + \mu_{OK_M} = 0 \qquad \forall \alpha$ (3-28)

$$\sum_{\alpha=1}^{n} v_\alpha = 1$$

The system is very similar to that for the estimation of Z except that the RHS of the system is 0. The corresponding expression for the minimum estimation variance can be derived as:

$$\sigma^2_{KT_m} = \sum_{\alpha=1}^{n}\sum_{\beta=1}^{n} v_\alpha v_\beta Cov(Z_\alpha, Z_\beta)$$

$$= \sum_{\alpha=1}^{n} v_\alpha \sum_{\beta=1}^{n} v_\beta Cov(\mathbf{h}_{\alpha\beta})$$

$$= \sum_{\alpha=1}^{n} v_\alpha \cdot (-\mu_{OK_M}) = -\mu_{OK_M} \cdot \sum_{\alpha=1}^{n} v_\alpha = -\mu_{OK_M}$$

In order for the variance to be guaranteed positive, the Lagrange parameter μ_{OK_M} has to be necessarily less than 0.

Remark   Based on the preceding discussion, it can be interpreted that the OK estimate can be simply reconstituted by performing SK with the known mean m in SK replaced by the local mean m^*_{OK} estimated using the data within a local neighborhood of the estimation point. This can be seen as:

$$Z^*_{0_{OK}} = \left[1 - \sum_{\alpha=1}^{n} \lambda_{\alpha_{SK}}\right] \cdot m^*_{OK} + \sum_{\alpha=1}^{n} \lambda_{\alpha_{SK}} Z(\mathbf{u}_\alpha)$$

In comparison: $$Z^*_{0_{SK}} = \left[1 - \sum_{\alpha=1}^{n} \lambda_{\alpha_{SK}}\right] \cdot m + \sum_{\alpha=1}^{n} \lambda_{\alpha_{SK}} Z(\mathbf{u}_\alpha)$$

Therefore: $Z^*_{0_{OK}} - Z^*_{0_{SK}} = \lambda_{m_{SK}} \cdot (m^*_{OK} - m)$,   where $\lambda_{m_{SK}} = \left[1 - \sum_{\alpha=1}^{n} \lambda_{\alpha_{SK}}\right]$. If the weight assigned to the mean in SK is greater than 0; that is, $\lambda_{m_{SK}} > 0$, then in areas where $m^*_{OK} > m$, the OK estimate will be greater than the SK estimate, $Z^*_{0_{OK}} > Z^*_{0_{SK}}$.

Example 3-7
Going back to the previous example shown in Fig. 3-6, the system for estimating the mean at coordinate location (1,1) can be written as:

$$\begin{bmatrix} 1.00 & 0.49 & 0.62 & 0.36 & 1 \\ 0.49 & 1.00 & 0.62 & 0.26 & 1 \\ 0.62 & 0.62 & 1.00 & 0.53 & 1 \\ 0.36 & 0.26 & 0.53 & 1.00 & 1 \\ 1 & 1 & 1 & 1 & 0 \end{bmatrix} \begin{bmatrix} v_1 \\ v_2 \\ v_3 \\ v_4 \\ \mu_{OK_M} \end{bmatrix} = \begin{bmatrix} 0 \\ 0 \\ 0 \\ 0 \\ 1 \end{bmatrix}$$

The solution is:

$$\begin{bmatrix} v_1 \\ v_2 \\ v_3 \\ v_4 \\ \mu_{OK_M} \end{bmatrix} = \begin{bmatrix} 0.28 \\ 0.36 \\ -0.03 \\ 0.40 \\ -0.57 \end{bmatrix}$$

As expected, the Lagrange parameter $\mu_{OK_M} < 0$ and the weights sum to one. The resultant value of the mean is 20.62. In comparison, the average of the four conditioning data values (Fig. 3-6) is 25 and that was the mean used in the SK calculations. Let us verify the result presented earlier regarding the link between SK and OK. At node (1,1), the calculated estimates are:

$$Z^*_{SK} = 16.83 \text{ and } Z^*_{OK} = 15.50$$

The SK weights for the estimation at (1,1) are:

$$\lambda = \begin{bmatrix} 0.65 \\ -0.12 \\ 0.02 \\ 0.15 \end{bmatrix}$$

The weight assigned to the mean is therefore:

$$\lambda_{m_{SK}} = 1 - (0.65 - 0.12 + 0.02 + 0.15) = 0.30$$

$$\lambda_{m_{SK}} \cdot (m_{OK}^* - m) = 0.3 \cdot (20.62 - 25.0) = -1.3 \approx (Z_{o_{OK}}^* - Z_{o_{SK}}^*)$$

The slight discrepancy in the results is because of round off errors.

Estimation of the Trend in Universal Kriging

Similar to the derivation for OK estimate of the trend, the UK estimator for the trend can be written as:

$$Z^*(\mathbf{u}_0) = \sum_{\alpha=1}^{n} v_{UK_{M_\beta}} Z(\mathbf{u}_\alpha)$$

The system of equations for the weights can be written as:

$$\sum_{\beta=1}^{n} v_{UK_{M_\beta}} C_R(\mathbf{h}_{\alpha\beta}) + \sum_{\ell=1}^{L} \mu_{UK_{M_\ell}} f_\ell(\mathbf{u}_\alpha) = 0, \quad \alpha = 1,\ldots,n$$

$$\sum_{\alpha=1}^{n} v_{UK_{M_\alpha}} f_\ell(\mathbf{u}_\alpha) = f_\ell(\mathbf{u}_0), \quad \ell = 0,\ldots,L$$

The corresponding expression for the estimation variance is:

$$\sigma_{UK_M}^2(\mathbf{u}_0) = -\sum_{\ell=0}^{L} \mu_{UK_{M_\ell}} f_\ell(\mathbf{u}_0)$$

That estimation variance has to be greater than 0.

Remarks

- Similar to the relationship between OK and SK, a link between SK and UK also exists:

$$Z_{UK}^*(\mathbf{u}_0) - m_{UK}^*(\mathbf{u}_0) = \sum_{\alpha=1}^{n} \lambda_{\alpha_{SK}} [Z(\mathbf{u}_\alpha) - m_{UK}^*(\mathbf{u}_\alpha)]$$

$\lambda_{\alpha_{SK}}$ are the SK weights corresponding to the same data configuration.

- In the previous discussion, the trend model was assumed to be deterministic but unknown. We can also render the trend model stochastic by making the coefficients a_ℓ stochastic. We can develop a UK estimator and system for each individual coefficient a_ℓ similarly. Each coefficient is written as a linear combination of the data and a UK system similar to the one for the mean can be derived.

Example 3-8

Calculation of the trend implied in the UK results shown in Fig. 3-15 is demonstrated. This calculation assumes the conditioning data depicted in Fig. 3-6. As discussed for the previous UK model, an isotropic covariance with a range of 4 is used. The range is smaller than that shown in Fig. 3-6(a) because the linear trend in the x and y directions introduce additional correlation.

The UK system for the estimation of the mean at node (1,1) is:

$$
\begin{bmatrix}
1.00 & 0.13 & 0.32 & 0.00 & 1 & 3 & 1 \\
0.13 & 1.00 & 0.32 & 0.00 & 1 & 6 & 2 \\
0.32 & 0.32 & 1.00 & 0.19 & 1 & 4 & 3 \\
0.00 & 0.00 & 0.19 & 1.00 & 1 & 2 & 5 \\
1 & 1 & 1 & 1 & 0 & 0 & 0 \\
3 & 6 & 4 & 2 & 0 & 0 & 0 \\
1 & 2 & 3 & 5 & 0 & 0 & 0
\end{bmatrix}
\begin{bmatrix}
v_{UK_{M1}} \\
v_{UK_{M2}} \\
v_{UK_{M3}} \\
v_{UK_{M4}} \\
\mu_{UK_{M0}} \\
\mu_{UK_{M1}} \\
\mu_{UK_{M2}}
\end{bmatrix}
=
\begin{bmatrix}
0 \\
0 \\
0 \\
0 \\
1 \\
1 \\
1
\end{bmatrix}
$$

This yields the solution:

$$
\begin{bmatrix}
v_{UK_{M1}} \\
v_{UK_{M2}} \\
v_{UK_{M3}} \\
v_{UK_{M4}} \\
\mu_{UK_{M0}} \\
\mu_{UK_{M1}} \\
\mu_{UK_{M2}}
\end{bmatrix}
=
\begin{bmatrix}
1.55 \\
-0.36 \\
-0.56 \\
0.37 \\
-2.30 \\
0.29 \\
0.28
\end{bmatrix}
$$

The estimate for the mean is therefore:

$$
m^*_{UK} = 1.55 \cdot \frac{(10 - 25)}{12.91} - 0.36 \cdot \frac{(20 - 25)}{12.91} - 0.56 \cdot \frac{(40 - 25)}{12.91}
$$
$$
+ 0.37 \cdot \frac{(30 - 25)}{12.91} = -2.93
$$

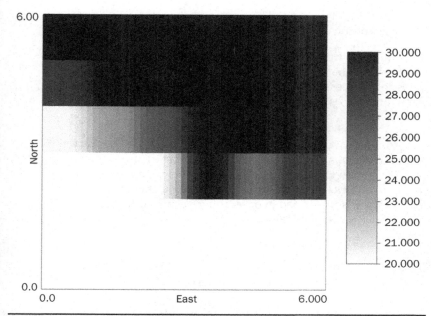

6.00

North

0.0

0.0   East   6.000

Figure 3-16   Map of mean estimated by universal kriging when a linear trend model in x and y directions is assumed. (The color version of this figure is available at www.mhprofessional.com/PRMS.)

As can be seen, the estimate for the mean is negative and is well outside the range of the data (10 to 40). This points to the non-convex nature of the kriging interpolator that can sometimes result in estimates outside the range of data. In some cases, the kriged value can be unphysical. The map of kriged estimates for the mean in Fig. 3-16 is consistent with that of the estimates for Z in Fig. 3-15(a).

The estimation variance associated with the estimation of the mean at (1,1) is:

$$\sigma^2_{UK_M}(\mathbf{u}_0) = -\sum_{\ell=0}^{L} \mu_{UK_{M\ell}} f_\ell(\mathbf{u}_0)$$
$$= -(-2.30 + 0.29 \cdot 1 + 0.28 \cdot 1)$$
$$= 1.73$$

The results indicate that the assumed trend model is inconsistent with the conditioning data, and this results in a compounding of uncertainty at locations farther away from the conditioning data.

3.6   Kriging with an External Drift

Let us consider the same decomposition:

$$Z(\mathbf{u}) = m(\mathbf{u}) + R(\mathbf{u})$$

But instead of considering an explicit trend model $m(\mathbf{u}) = \sum_{\ell=0}^{L} a_\ell f_\ell(\mathbf{u})$, where the basis functions that define the trend are assumed known, let us consider that an external (secondary) variable $t(\mathbf{u})$ that linearly relates to the mean of the variable to be estimated is available. In other words, let us write:

$$m(\mathbf{u}) = a_0 + a_1 \cdot t(\mathbf{u}) \tag{3-29}$$

Such a model is also referred to as an implicit model because the weights a_0 and a_1 assigned to the external variable $t(\mathbf{u})$ are calculated implicitly within kriging or regression. In order to implement such a model, the external variable $t(\mathbf{u})$ should be available at all locations where kriging is performed in order to arrive at an estimate.

Consider the KT estimator:

$$Z^*(\mathbf{u}_0) = \sum_{\alpha=1}^{N} \lambda_\alpha \cdot Z(\mathbf{u}_\alpha)$$

In order for this estimator to be unbiased:

$$E\{Z^*(\mathbf{u}_0)\} = \sum_{\alpha=1}^{n} \lambda_\alpha \cdot E\{Z(\mathbf{u}_\alpha)\} = \sum_{\alpha=1}^{n} \lambda_\alpha \cdot (a_0 + a_1 \cdot t(\mathbf{u}_\alpha))$$

This must be equal to $E\{Z(\mathbf{u}_0)\} = a_0 + a_1 \cdot t(\mathbf{u}_0)$. This results in:

$$\sum_{\alpha=1}^{n} \lambda_\alpha \cdot (a_0 + a_1 \cdot t(\mathbf{u}_\alpha)) = a_0 + a_1 \cdot t(\mathbf{u}_0)$$

This implies the following two unbiasedness conditions:

$$\sum_{\alpha=1}^{n} \lambda_\alpha = 1$$

$$\sum_{\alpha=1}^{n} \lambda_\alpha \cdot t(\mathbf{u}_\alpha) = t(\mathbf{u}_0)$$

Minimizing the error variance while satisfying the two unbiasedness equations yields the requisite system of equations for the unknown weights λ_α. The system is:

$$\sum_{\beta=1}^{n} \lambda_\beta C_R(\mathbf{h}_{\alpha\beta}) + \mu_{0_{ED}} + \mu_{1_{ED}} t(\mathbf{u}_\alpha) = C_R(\mathbf{h}_{\alpha 0}), \quad \forall \alpha$$

$$\sum_{\alpha=1}^{n} \lambda_\alpha = 1 \tag{3-30}$$

$$\sum_{\alpha=1}^{n} \lambda_\alpha \cdot t(\mathbf{u}_\alpha) = t(\mathbf{u}_0)$$

Remarks

- The external drift model can be considered as a simple but crude approach to integrate secondary/auxiliary data in our models. A linear relationship between $m(\mathbf{u})$ and $t(\mathbf{u})$ is assumed. The information in $t(\mathbf{u})$ toward the direct prediction of $Z(\mathbf{u})$ is ignored. Instead, a linear relationship between the mean of $Z(\mathbf{u})$ and the secondary attribute is calibrated. Calibration of the information in the secondary variable to the direct prediction of $Z(\mathbf{u})$ is the focus of co-kriging, which will be discussed later in this chapter.

- The second unbiasedness condition $\sum_{\alpha=1}^{n} \lambda_\alpha \cdot t(\mathbf{u}_\alpha) = t(\mathbf{u}_0)$ implies that the seismic value at any location $\mathbf{u}_0$ can be expressed as a linear combination of the seismic values in the neighborhood of that location. This in turn implies that the secondary map $t(\mathbf{u})$ has to be smooth.

Example 3-9

The data configuration in Fig. 3-6 is used for this example. The secondary variable is computed by applying a non-linear transformation to the kriged map, shown in Fig. 3-10. The sum of all kriged values within a 2×2 window are subject to a logarithmic transformation. This process is repeated by translating the 2×2 window over the whole domain. The resultant map of secondary values is shown in Fig. 3-17.

Unlike in the UK cases discussed earlier, the range of the covariance is restored back to 8 units [consistent with the covariance model in Fig. 3.6(a)]. The resultant kriging system for the estimation at node (1,1) is:

$$
\begin{bmatrix}
1.00 & 0.49 & 0.62 & 0.00 & 1 & 5.84 \\
0.49 & 1.00 & 0.62 & 0.00 & 1 & 5.78 \\
0.62 & 0.62 & 1.00 & 0.53 & 1 & 6.52 \\
0.00 & 0.00 & 0.53 & 1.00 & 1 & 6.22 \\
1 & 1 & 1 & 1 & 0 & 0 \\
5.84 & 5.78 & 6.52 & 6.22 & 0 & 0
\end{bmatrix}
\begin{bmatrix}
\lambda_1 \\ \lambda_2 \\ \lambda_3 \\ \lambda_4 \\ \mu_{0_{ED}} \\ \mu_{1_{ED}}
\end{bmatrix}
=
\begin{bmatrix}
0.66 \\ 0.25 \\ 0.43 \\ 0.36 \\ 1.00 \\ 5.23
\end{bmatrix}
$$

The solution is:

$$
\begin{bmatrix}
\lambda_1 \\ \lambda_2 \\ \lambda_3 \\ \lambda_4 \\ \mu_{0_{ED}} \\ \mu_{1_{ED}}
\end{bmatrix}
=
\begin{bmatrix}
1.178 \\ 0.530 \\ -1.044 \\ 0.337 \\ -4.428 \\ 0.715
\end{bmatrix}
$$

This yields the estimate at (1,1) as:

$$
m_{UK}^* = 1.178 \cdot \frac{(10-25)}{12.91} + 0.530 \cdot \frac{(20-25)}{12.91} - 1.044 \cdot \frac{(40-25)}{12.91}
$$
$$
+ 0.337 \cdot \frac{(30-25)}{12.91} = -9.30
$$

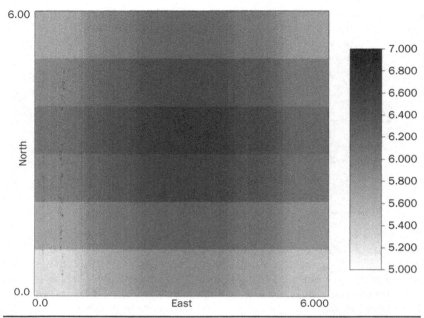

FIGURE 3-17 Map of external drift values obtained by applying a non-linear transformation procedure to the kriged estimates in Fig. 3-10. (The color version of this figure is available at www.mhprofessional.com/PRMS.)

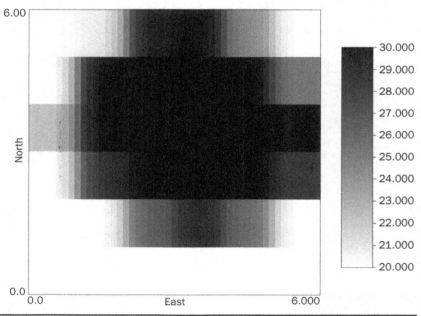

FIGURE 3-18 Map of kriging estimates obtained corresponding to the external drift map depicted in Fig. 3-17. (The color version of this figure is available at www.mhprofessional.com/PRMS.)

As noted before, the estimate is well outside the range of the data indicating that the kriging with external drift procedure is non-convex. The map of kriging estimates is shown in Fig. 3-18.

The specified external drift map serves to restrict the region of high kriged values in the north-east and north-west corners of the domain.

3.7   Indicator Kriging

3.7.1   Non-Parametric Approach to Modeling Distributions

Given some data $Z(\mathbf{u}_\alpha)$, $\alpha = 1, \ldots, n$, the experimental distribution $\hat{F}(\mathbf{u}; z_k)$ can be created by defining a series of thresholds $z_k, k = 1, \ldots, K$. We keep count of the number of data values less than a particular threshold z_k and then obtain the experimental *cdf* as:

$$\hat{F}(\mathbf{u}; z_k) = \frac{\text{count}(Z(\mathbf{u}_\alpha) \leq z_k; \alpha = 1, \ldots, n)}{n}$$

This same process of constructing the experimental *cdf* can be facilitated using an indicator random variable $I(\mathbf{u}; z_k)$:

$$I(\mathbf{u}; z_k) = \begin{cases} 1, \text{ if } Z(\mathbf{u}) \leq z_k \\ 0 \text{ otherwise} \end{cases} \tag{3-31}$$

The expected value of this RV is:

$$\begin{aligned} E\{I(\mathbf{u}; z_k)\} &= 1 \cdot \text{Prob}\{Z(\mathbf{u}) \leq z_k\} + 0 \cdot \text{Prob}\{Z(\mathbf{u}) > z_k\} \\ &= \text{Prob}\{Z(\mathbf{u}) \leq z_k\} = F(\mathbf{u}; z_k) \end{aligned} \tag{3-32}$$

If (n) conditioning data is available, then applying the same process, the conditional *cdf* can be obtained as:

$$F(\mathbf{u}; z_k \mid (n)) = E\{I(\mathbf{u}; z_k) \mid (n)\}, \forall k = 1, \ldots, k$$

If an estimator for the conditional expectation $E\{I(\mathbf{u}; z_k) \mid (n)\}$ denoted as $I^*(\mathbf{u}; z_k)$ can be constructed using the available data, then this can be equivalent to establishing the conditional probability at location $\mathbf{u}$ without resorting to any parametric assumptions regarding the probability distribution (Journel, 1983).

Remarks

- The indicator RV because of its binary characteristics is invariant under any non-linear transformation: $I^\omega(\mathbf{u}; z_k) = I(\mathbf{u}; z_k)$, $\forall \omega$.

- The indicator RV is immune to the presence of outliers of $Z(\mathbf{u})$. Regardless of the absolute value of the variable $Z(\mathbf{u})$, the resultant indicator transform always assumes binary outcomes.

- The variance of the indicator variable can be computed as:

$$
\begin{aligned}
Var\{I(\mathbf{u};z_k)\} &= E\{I^2(\mathbf{u};z_k)\} - [E\{I(\mathbf{u};z_k)\}]^2 \\
&= E\{I(\mathbf{u};z_k)\} - [F_Z(\mathbf{u};z_k)]^2 \\
&= F_Z(\mathbf{u};z_k) - [F_Z(\mathbf{u};z_k)]^2 \\
&= p \cdot (1-p)
\end{aligned}
\tag{3-33}
$$

The probability $F_Z(\mathbf{u}; z_k)$ is represented by p and the above equation suggests that the variance is the product of the probability and its compliment.

Prior to talking about a form for the estimator $I^*(\mathbf{u}; z_k)$ and the procedure to calculate it using the available data, let us take a step back and look at kriging (and its variations) in terms of projections.

3.7.2 Kriging in Terms of Projections

The interpretation of kriging in terms of projection to vector spaces is elucidated in Journel & Huijbregts (1978). A brief recap of the discussions in that text follows. Consider a set of RVs: $S = \{Z(\mathbf{u}), \mathbf{u} \in D\}$. Let us now define a vector space H comprising all finite linear combinations of the elements of S:

$$
H = \left\{ \sum_{\alpha} v_\alpha Z(\mathbf{u}_\alpha), Z(\mathbf{u}_\alpha) \in S \right\}
$$

The weights v_α are stipulated to be real. The random variables $Z(\mathbf{u})$ are assumed to be characterized by the covariance $Cov\{Z(\mathbf{u}_\alpha), Z(\mathbf{u}_\beta)\} = E\{Z(\mathbf{u}_\alpha)Z(\mathbf{u}_\beta)\}$. This is equivalent to the space H being equipped with the dot product $\langle Z(\mathbf{u}_\alpha), Z(\mathbf{u}_\beta) \rangle$.

Note   The RVs are assumed to be characterized by the non-centered covariance implying that the variables may be non-stationary.

The norm of a vector $Z(\mathbf{u})$ can also be computed in the vector space H:

$$
\|Z(\mathbf{u})\| = \sqrt{\|Z(\mathbf{u})\|^2} = \sqrt{\langle Z(\mathbf{u}), Z(\mathbf{u}) \rangle}
$$

The norm $\|Z(\mathbf{u}_\alpha) - Z(\mathbf{u}_\beta)\|$ measures the distance between the RVs $Z(\mathbf{u}_\alpha)$ and $Z(\mathbf{u}_\beta)$ that are elements of S.

Now consider a set of outcomes of the RVs $Z(\mathbf{u})$ in S. A space H' formed of linear combination of these outcomes will be a subspace of H. Consider an unknown random variable $Z(\mathbf{u}_0)$. There are various estimates of $Z(\mathbf{u}_0)$ that can be composed in the subspace H'. According to the projection theorem, there exists a unique element Z_0^* in H' that minimizes the distance $\|Z(\mathbf{u}_0) - Z_0^*\|$. This unique element

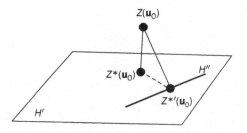

Figure 3-19   Projection of an unknown $Z(\mathbf{u}_o)$ to two vector sub-spaces.

Z_0^* is referred to as the projection of $Z(\mathbf{u}_o)$ onto the subspace H'. The different kriging procedures discussed earlier differ in the definition of the subspace H' on to which $Z(\mathbf{u}_o)$ is projected. In each case, the square of the distance $\left\| Z(\mathbf{u}_0) - Z_0^* \right\|^2 = E\{(Z(\mathbf{u}_0) - Z_0^*)^2\}$ is referred to as the kriging estimation variance.

In order to compare and contrast the various kriging estimators, consider another vector space H'' which is a subset of H'. The projection of $Z(\mathbf{u}_o)$ on H'' is denoted as $Z_0^{*'}$. As seen from Fig. 3-19, because $H'' \subset H' \subset H$, the distance $\left\| Z(\mathbf{u}_0) - Z_0^* \right\| \leq \left\| Z(\mathbf{u}_0) - Z_0^{*'} \right\|$. This implies that an estimate or projection of $Z(\mathbf{u}_o)$ to a space defined in higher dimensions is likely to result in smaller error variance. This notion is important when considering the relative merits and demerits of various kriging-based estimation schemes.

Simple Kriging
The estimator is given as:

$$Z_0^* = \lambda_0 \cdot \mathbf{1} + \sum_{\alpha=1}^{n} \lambda_\alpha \cdot Z(\mathbf{u}_\alpha)$$

i.e., the projection of $Z(\mathbf{u}_o)$ is to a space defined by the n data plus the unit vector $\mathbf{1}$. In order to preserve properties such as the unbiasedness of the estimator, we will restrict the subspace H_{n+1} to a linear manifold $L \subset H_{n+1}$.

As discussed earlier, the projection of $Z(\mathbf{u}_o)$ to the subspace H_{n+1} implies that the distance $\|Z(\mathbf{u}_o) - Z_0^*\|$ is minimized. As can be seen in Fig. 3-20, the distance is minimized when the vector $Z(\mathbf{u}_o) - Z_0^*$ is orthogonal to the subspace H_{n+1} or more precisely to the $(n+1)$ vectors that define the subspace H_{n+1}.

This yields the system:

$$\left\langle Z(\mathbf{u}_0) - Z_0^*, \mathbf{1} \right\rangle = 0 \quad \Rightarrow \lambda_0 = m(\mathbf{u}_0) - \sum_{\alpha=1}^{n} \lambda_\alpha \cdot m(\mathbf{u}_\alpha)$$

$$\left\langle Z(\mathbf{u}_0) - Z_0^*, Z(\mathbf{u}_\alpha) \right\rangle = 0 \Rightarrow \left\langle Z(\mathbf{u}_0) Z(\mathbf{u}_\alpha) \right\rangle = \sum_{\beta=1}^{n} \lambda_\beta \cdot \left\langle Z(\mathbf{u}_\alpha) Z(\mathbf{u}_\beta) \right\rangle$$

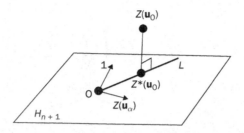

Figure 3-20   Orthogonality conditions for simple kriging.

We have substituted the linear estimator for Z_0^* in the last expression. Because the mean is assumed to be stationary and known in SK, this last expression can be written in terms of the centered covariance:

$$C(\mathbf{h}_{\alpha 0}) = \sum_{\beta=1}^{n} \lambda_\beta \cdot C(\mathbf{h}_{\alpha\beta})$$

Ordinary Kriging
In this case, the mean is assumed to be stationary but unknown. This necessitates that $\lambda_0 = 0$ in the previous conditions derived for SK; i.e., the mean receives no weight. The projection of $Z(\mathbf{u}_0)$ is thus to the sub-space $H_n \subset H_{n+1}$. In addition, the projection is restricted to a linear manifold $L_1 \subset H_n \subset H_{n+1}$. The condition $\lambda_0 = 0$ also results in the unbiasedness condition $\sum_{\alpha=1}^{n} \lambda_{\alpha=1}$. The projection $Z_0^* = \sum_{\alpha=1}^{n} \lambda_\alpha \cdot Z(\mathbf{u}_\alpha)$ is thus characterized by the conditions $Z_0^* \in L_1$ and $\sum_{\alpha=1}^{n} \lambda_\alpha = 1$.

Consider the projection shown in Fig. 3-21 compared to that shown in Fig. 3-20. The main difference is the absence of an additional vector 1 defining the linear manifold. The location of the mean signifying the centering value 0 in the projection plot in Fig. 3-21 is unknown. Consider an arbitrary reference location Y. It can be seen that the error vector $Z(\mathbf{u}_0) - Z_0^*$ must be orthogonal to any vector $Z(\mathbf{u}_\alpha) - Y$ in the linear manifold L_1. Thus:

$$\langle Z(\mathbf{u}_0) - Z_0^*, Z(\mathbf{u}_\alpha) - Y \rangle = 0$$

Denoting $\langle Z(\mathbf{u}_0) - Z_0^*, -Y \rangle = \mu_1$ and using $Z_0^* = \sum_{\alpha=1}^{n} \lambda_\alpha \cdot Z(\mathbf{u}_\alpha)$, we get the orthogonality condition:

$$C(\mathbf{h}_{\alpha 0}) = \sum_{\beta=1}^{n} \lambda_\beta \cdot C(\mathbf{h}_{\alpha\beta}) - \mu_1, \text{ for } = 1, \dots, n$$

together with the unbiasedness condition $\sum_{\alpha=1}^{n} \lambda_\alpha = 1$.

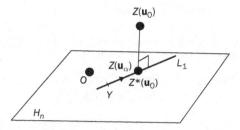

FIGURE 3-21   Orthogonality conditions for ordinary kriging.

Remark   An option for the reference location Y could be the arithmetic mean $Y = \frac{1}{n}\sum_{\alpha=1}^{n} Z(\mathbf{u}_\alpha)$ that is guaranteed to lie within the linear manifold L_1. The scalar μ_1 can be interpreted as a parameter linked to this arbitrary centering value Y that is uncertain.

In the case of UK, the estimator is $Z^*_{0_{UK}} = \sum_{\alpha=1}^{n}\lambda_{\alpha_{UK}} \cdot Z(\mathbf{u}_\alpha)$ which implies that the projection is to the sub-space $H_n \subset H_{n+1}$ under the condition $\sum_{\alpha=1}^{n}\lambda_{\alpha_{UK}} = 1$. In addition, there are L unbiasedness conditions of the form:

$$\sum_{\alpha=1}^{n}\lambda_{\alpha_{UK}} \cdot f_\ell(\mathbf{u}_\alpha) = f_\ell(\mathbf{u}_0), \quad \ell = 1,\ldots,L$$

This implies that the projection of $Z(\mathbf{u}_0)$ is to a linear manifold $L_{n-L} \subset H_n$. The system of equations required to solve for the weights is again obtained by stipulating that the error $Z(\mathbf{u}_o) - Z^*_0$ be orthogonal to vectors $Z(\mathbf{u}_\alpha) - Y_\ell$ for $\ell = 1,\ldots,L$. The arbitrary reference locations Y_ℓ correspond to the L unknown coefficients a_ℓ characterizing the non-stationary trend: $m(\mathbf{u}) = a_0 + \sum_{\ell=1}^{L}a_\ell \cdot f_\ell(\mathbf{u})$.

3.7.3   Indicator Basis Function

Now consider the projection manifold $\varphi(n)$ defined on the basis of n indicator random variables $I(\mathbf{u}_\alpha)$, $\alpha = 1,\ldots,n$:

$$\varphi(n) = a + \sum_{j_1=1}^{n} b^{(1)}_{j_1} \cdot I\left(\mathbf{u}_{j_1}\right) + \sum_{j_1=1}^{n}\sum_{j_2=1}^{n} b^{(2)}_{j_1,j_2} \cdot I\left(\mathbf{u}_{j_1}\right)I\left(\mathbf{u}_{j_2}\right)$$
$$+ \doteq + b^{(n)}_{j_1,j_2,\doteq,j_n} \prod_{i=1,n} I(\mathbf{u}_i) \tag{3-34}$$

The indicator estimator $I^*(\mathbf{u}_0)$, which as previously discussed is an estimator for the conditional expectation $E\{I(\mathbf{u}_0 \mid (n))\}$, is precisely the projection of the unknown indicator event $I(\mathbf{u}_0)$ on to $\varphi(n)$. By projecting $I(\mathbf{u}_0)$ to the space defined by the basis L_n: $\{1, I_{j_1}, I_{j_1}I_{j_2}, I_{j_1}I_{j_2}I_{j_3}, \ldots, I_{j_1}I_{j_2}I_{j_3}, \ldots, I_{j_n}\}$ and ensuring that the error vector $I - I^*(\mathbf{u}_0)$ is orthogonal to each basis vector defining the space L_n, we can solve for the 2^n coefficients $a, b_{j_1,j_2}, \ldots, b_{j_1,j_2,\ldots,j_n}$.

Remarks   Given the n indicator RVs, the expansion mentioned above is the most complete possible.

- There are $1 + \binom{n}{1} + \binom{n}{2} + \cdots + \binom{n}{n} = 2^n$ terms in the above expansion implying that complete characterization of the indicator random function is possible if all combinations of the RVs in the above expansion are known.

- Since we are dealing with binary indicator RVs, there are two outcomes for each $I(\mathbf{u}_\alpha)$ and since there are n RVs, there are a total of 2^n outcomes possible for the function $\varphi(n)$.

- Application of this complete indicator expansion in any practical algorithm for estimation would require inference of statistics corresponding to all combinations of indicator outcomes (2^n), which is infeasible as n increases. The indicator expansion is therefore truncated at order $k \ll n$.

The projection of $I(\mathbf{u}_0)$ on to the manifold of reduced order (L_k) can be expressed compactly as:

$$I^*_{0_k} = \sum_{\ell=1}^{n_k} \lambda_\ell \cdot V_\ell$$

$$V_1 = \{1\}; \ V_\ell = \{I_i\}, \ i = 1, \ldots, n, \ \ell = i + 1;$$

$$V_\ell = \{I_i \cdot I_j\}, \ i = 1, \ldots, n; \ j = 1, \ldots, n; \ell = (i - 1) \cdot n + j + 1 + \binom{n}{1}$$

and so on.

In the notation of the Hilbert projection theorem (Luenberger, 1969), L_k is a closed subspace of the Hilbert space L_n, I is a vector in L_n, and $I^*_{0_k}$ is a vector in L_k and it is unique, minimum norm projection if and only if $\left(I_0 - I^*_{0_k}\right)$ is orthogonal to L_k, or more precisely to all vectors defining the space L_k. The space L_k is represented by $L_k : \{V_\ell, \ \ell = 1, \ldots, n_k\}$. The implication of the projection theorem is that:

$$\left\langle \left(I_0 - I^*_{0_k}\right), V_\ell \right\rangle = 0 \Leftrightarrow \left\langle I_0 \cdot V_\ell \right\rangle = \left\langle I^*_{0_k} \cdot V_\ell \right\rangle$$

or in terms of expectations:

$$E\{I^*_{0_k} V_\ell\} = E\{I_0 V_\ell\}, \forall \ \ell = 1, \ldots, n_k \tag{3-35}$$

which is a system of n_k normal equations.

Examples $V_1 = 1 \Rightarrow E\{I_{0_k}^*\} = E\{I_0\}$, a statement of unbiasedness.

$$V_\ell = I_j, j = 1,\dots,n \Rightarrow E\{I_{0_k}^* I_j\} = \sum_{\ell=1}^{n_k} \lambda_\ell \cdot E\{V_\ell \cdot I_j\} = E\{I_0 \cdot I_j\}, \text{ a}$$

statement of reproduction of indicator covariances (Journel, 1983).

$$V_\ell = I_{j_1} \cdot I_{j_2}, j_1, j_2 = 1,\dots,n \Rightarrow E\{I_{0_k}^* I_{j_1} I_{j_2}\}$$
$$= \sum_{\ell=1}^{n_k} \lambda_\ell \cdot E\{V_\ell \cdot I_{j_1} I_{j_2}\} = E\{I_0 \cdot I_{j_1} I_{j_2}\}$$

a statement of reproduction of the third-order indicator covariances.

The orthogonality condition can be generally represented in the form:

$$\sum_{\ell=1}^{n_k} \lambda_\ell \cdot E\{V_\ell \cdot V_\ell'\} = E\{I_0 \cdot V_\ell'\}, \ell = 1,\dots,n_k$$

which requires knowledge of up to $k+1$ order indicator covariances $E\{I_0 \cdot V_\ell'\}$, $\ell = 1,\dots,n_k$

3.7.4 Indicator Kriging

Prior to taking an excursion into kriging in terms of projection, we had observed that if the conditional expectation $E\{I(\mathbf{u}; z_k) | (n)\}$ denoted as $I^*(\mathbf{u}; z_k)$ can be constructed using the available data, then this is equivalent to establishing the conditional probability at location $\mathbf{u}$ without resorting to any parametric assumptions regarding the probability distribution. We start with the indicator expansion truncated at the first order:

$$I_{0_k}^* = \sum_{\ell=1}^{n_k} \lambda_\ell \cdot V_\ell$$

Going back to $V_1 = 1 \Rightarrow \lambda_1 + \sum_{\ell=2}^{n_k} \lambda_\ell \cdot E\{V_\ell\} = E\{I\}$

$$\Rightarrow \lambda_\ell = E\{I\} - \sum_{\ell=2}^{n_k} \lambda_\ell \cdot m_\ell, \quad m_\ell = E\{V_\ell\}, \ell > 1$$

Substituting in the estimator $I_k^* = \sum_{\ell=1}^{n_k} \lambda_\ell \cdot V_\ell = \lambda_\ell + \sum_{\ell=2}^{n_k} \lambda_\ell \cdot V_\ell$, we get:

$$I_k^* - E\{I\} = \sum_{\ell=2}^{n_k} \lambda_\ell \cdot (V_\ell - m_\ell)$$

Putting all this together and recognizing that I_k^* is an approximation of $E\{I(\mathbf{u}) \mid (n)\}$ since $n_k \ll 2^n$, we can write the following indicator kriging estimator and the accompanying normal system:

Estimator: $E\{I \mid (n)\} - E\{I\} \cong \sum_{\ell=2}^{n_k} \lambda_\ell \cdot (V_\ell - m_\ell)$, for $n_k \ll 2^n$

As noted previously, the orthogonality condition results in:

$$\sum_{\ell=1}^{n_k} \lambda_\ell \cdot Cov\left\{V_\ell \cdot V_\ell'\right\} = Cov\left\{I \cdot V_\ell'\right\},\ \ell = 1,\ldots,n_k$$

In the specific case, $V_\ell = I_j, j = 1,\ldots,n \Rightarrow [E\{I \mid (n)\} - E\{I\}] =$

$\sum_{j=1}^{n} \lambda_j \cdot [I_j - E\{I_j\}]$ with the normal system: $\sum_{j=1}^{n} \lambda_\ell \cdot Cov\{I_j \cdot I_i\} =$

$Cov\{I \cdot I_i\},\ i = 1,\ldots,n$

These last two equations are commonly referred to as the *simple indicator kriging estimator* and the system for obtaining the corresponding kriging weights. The required statistics for performing the kriging are the indicator covariances $Cov\{I, I_j\}$, $Cov\{I_i, I_j\}$ and the prior proportions $E\{I\}$, $E\{I_j\}$.

Remarks

- The weights λ's are determined by solving the normal system of equations that utilize indicator covariances that in turn are computed employing the decision of stationarity (by averaging indicator data under some assumption of stationarity). Because the weights are computed using stationary statistics, they are not specific to any data values and are only influenced by the configuration of the data.

- One may wonder why it was necessary to talk about kriging from a projection perspective in order to arrive at the formulation of the indicator kriging estimator and the corresponding indicator kriging system. Why not simply derive the kriging system by defining an error variance and minimizing it? This is because the indicator kriging estimator yields a probability and it is awkward to talk about the variance of a probability. A second important reason is that indicator-based kriging is non-parametric and not restricted to Gaussian-type hypothesis where it is sufficient to investigate up to the second-order moment (variance) of the RVs.

Kriging/Projection with One Datum (n = 1)

In this case $I_o^* = E\{I_o \mid I_1\} = \varphi(I_1) = a \cdot 1 + b \cdot I_1$

Applying the first normal equation, orthogonality of the error vector $I_o^* - I_o$ with the unit vector **1** (this is the unbiasedness condition):

$$\left\langle \left(I_o^* - I_o\right), 1 \right\rangle = 0 \Leftrightarrow \left\langle I_o^* \right\rangle = \left\langle I_o \right\rangle$$
$$\Leftrightarrow a + b \cdot \left\langle I_1 \right\rangle = \left\langle I_o \right\rangle$$
$$\Leftrightarrow a + b \cdot E\{I_1\} = E\{I_o\}$$
$$\Leftrightarrow a + b \cdot p_1 = p_0$$

The second normal equation (orthogonality of the error with I_1):

$$\left\langle \left(I_o^* - I_o\right), I_1 \right\rangle = 0 \Leftrightarrow \left\langle I_o^*, I_1 \right\rangle - \left\langle I_o, I_1 \right\rangle = 0$$
$$\Leftrightarrow a \cdot E\{I_1\} + b \cdot E\{I_1^2\} = E\{I_o I_1\}$$

Since any function of the indicator RV (e.g., I^2) is simply the indicator RV I:

$$a \cdot E\{I_1\} + b \cdot E\{I_1^2\} = a \cdot p_1 + b \cdot p_1 = E\{I_o I_1\}$$

Substituting from the first normal equation: $b \cdot p_1 = p_o - a$

$$a = \frac{p_o - E\{I_o I_1\}}{1 - p_1}; \qquad b = \frac{E\{I_o I_1\} - p_o p_1}{p_1 \cdot (1 - p_1)}$$

This implies that $E\{I_o \mid I_1 = 1\} = a + b \cdot 1 = \dfrac{E\{I_o I_1\} \cdot (1 - p_1)}{p_1 \cdot (1 - p_1)}$
$$= \frac{E\{I_o I_1\}}{E\{I_1\}}$$

This is the same as the result provided by Bayes' rule. Similarly when the conditioning event $I_1 = 0$, which is equivalent to event $J = 1 - I = 1$:

$$E\{I_o \mid I_1 = 0\} = E\{I_o \mid J = 1\} = \frac{E\{I_o J\}}{E\{J\}} = \frac{E\{I_o(1 - I_1)\}}{1 - E\{I_1\}}$$

which is again the result provided by Bayes' rule.

Note   The expressions for the weights a and b are general and not specific to any particular outcome of the conditioning event. However, when particular outcomes, e.g., $I_1 = 0$ or $I_1 = 1$, are considered, the results for $E\{I_o \mid I_1\}$ in terms of the projection identify those obtained by Bayes' rule.

Simple Indicator Kriging

$$I^*(\mathbf{u};z_k) = F^*(\mathbf{u};z_k \mid (n))$$

$$= \left[1 - \sum_{\alpha=1}^{n}\lambda_\alpha(\mathbf{u};z_k)\right]F(z_k) + \sum_{\alpha=1}^{n}\lambda_\alpha(\mathbf{u};z_k) \cdot I(\mathbf{u}_\alpha;z_k) \quad \text{(3-36)}$$

the weights $\lambda_\alpha(\mathbf{u};z_k)$ are solutions to the indicator system:

$$\sum_{\beta=1}^{n}\lambda_\beta(\mathbf{u};z_k) \cdot C_I(\mathbf{h}_{\alpha\beta};z_k) = C_I(\mathbf{h}_{\alpha o};z_k) \text{ for } \alpha = 1,\ldots,n \quad \text{(3-37)}$$

Unique solution to the kriging system requires positive definite $C_I(\mathbf{h};z_k)$ corresponding to each threshold z_k. The indicator-coded data at thresholds z_k is used for inferring the experimental variograms and these are subsequently modeled using a linear combination of basic structures. The general practices for selection of thresholds $z_k, k = 1,\ldots,K$ are:

- 9 deciles of the experimental histogram computed using available data and
- thresholds chosen corresponding to critical values such as the value above which a zone may be classified as a thief zone

The simple indicator kriging formulation also permits the usage of different covariance or variogram models for each threshold; i.e., the spatial continuity of extremes may be different from that of the median thresholds. The drawback with such a flexible model for accounting for spatial continuity is that the inference and modeling of variogram corresponding to different thresholds can be tedious and infeasible when data is sparse.

Median Indicator Kriging

In order to render simple indicator kriging faster and feasible when faced with sparse data, we can assume that the K indicator random functions $I(\mathbf{u};z_k)$ are correlated with each other such that the K indicator variograms or covariances are proportional to a common variogram model. This common variogram model is generally taken to be the variogram model at the median threshold $\gamma_{mI}(\mathbf{h})$. In that case, the median kriging estimate is:

$$F^*(\mathbf{u};z_k \mid (n)) = \left[1 - \sum_{\alpha=1}^{n}\lambda_\alpha(\mathbf{u};z_k)\right]F(z_k) + \sum_{\alpha=1}^{n}\lambda_\alpha(\mathbf{u};z_k) \cdot I(\mathbf{u}_\alpha;z_k)$$

where weights are solutions to:

$$\sum_{\beta=1}^{n}\lambda_\beta(\mathbf{u};z_k) \cdot C_{mI}(\mathbf{h}_{\alpha\beta}) = C_{mI}(\mathbf{h}_{\alpha o}) \text{ for } \alpha = 1,\ldots,n$$

Note the following relationship between the median indicator variogram and the covariance $\gamma_{ml}(\mathbf{h}) = 0.25 - C_{ml}(\mathbf{h})$, where the theoretical prior variance corresponding to the median threshold ($p = 0.5$) is 0.25. This median indicator kriging approach is faster although some of the flexibility built into SIK is lost (Wackernagel, 1994; Goovaerts, 1994).

Example 3-10

Let us go back to the data configuration in Fig. 3-6. The covariance model presented in that figure is assumed to be representative at the median threshold. The thresholds are selected to be [15, 25, 35, 45]. The prior *cdf* values corresponding to these thresholds are assumed to be [0.20, 0.52, 0.70, 0.87]. The indicator-coded data corresponding to the first threshold is:

$$I(\mathbf{u}_1) = 1$$
$$I(\mathbf{u}_2) = 0$$
$$I(\mathbf{u}_3) = 0$$
$$I(\mathbf{u}_4) = 0$$

The indicator kriging system corresponding to the first threshold at the location (1,1) is:

$$\begin{bmatrix} 0.25 & 0.12 & 0.15 & 0.09 \\ 0.12 & 0.25 & 0.15 & 0.06 \\ 0.15 & 0.15 & 0.25 & 0.13 \\ 0.09 & 0.06 & 0.13 & 0.25 \end{bmatrix} \begin{bmatrix} \lambda_1 \\ \lambda_2 \\ \lambda_3 \\ \lambda_4 \end{bmatrix} = \begin{bmatrix} 0.16 \\ 0.06 \\ 0.11 \\ 0.09 \end{bmatrix}$$

This yields the weights:

$$\begin{bmatrix} \lambda_1 \\ \lambda_2 \\ \lambda_3 \\ \lambda_4 \end{bmatrix} = \begin{bmatrix} 0.649 \\ -0.112 \\ 0.017 \\ 0.147 \end{bmatrix}$$

The estimate corresponding to the first threshold can be computed as:

$$\begin{aligned} I_0^* = F^*(\mathbf{u}_0 \,|\, (n); 25) &= (1 - 0.649 + 0.116 - 0.017 + 0.147) \cdot 0.20 \\ &\quad + 0.649 \cdot 1 - 0.116 \cdot 0 + 0.017 + 0.147 \cdot 0 \\ &= 0.710 \end{aligned}$$

Repeating this for the remaining three thresholds yields the following updated or estimated *ccdf* (conditional cumulative distribution function):

$$F(\mathbf{u}_0 \,|\, (n); 15) = [0.710; 0.690; 0.892; 0.961]$$

The four *ccdf* values correspond to the four threshold values. The calculated values present a problem. A legitimate *cdf* function should necessarily be non-decreasing. The calculated values reveal that the *cdf* value for the second threshold is actually lower than that for the previous threshold. Such a violation is termed an order-relations violation (Journel & Posa, 1990) and stems from the fact that the indicator estimator at a particular threshold does not take into consideration the estimated value at the adjacent thresholds. A fix for this problem is to explicitly consider the *ccdf* value at an adjacent threshold as a constraint when computing the optimal estimate at the target threshold. In other words, rather than arriving at the estimate by applying the kriging system, solve the problem using a constrained optimization procedure where the error variance is minimized together with the constraint posed by the *ccdf* value at the adjacent threshold. Such a procedure though feasible would detract from the analytical simplicity of the kriging procedure presented earlier.

The estimated *ccdf* values at the location (4,4) are:

$$F(\mathbf{u}_0 \mid (n)) = [-0.054; 0.042; 0.317; 0.994]$$

This presents another set of order-relations violations. The calculated *ccdf* values fall outside the bounds for a legitimate *cdf* (0,1). These violations come out because of the non-convex characteristics of kriging; i.e., values may result outside the range of values input as conditioning data. Again these problems can be fixed by implementing a constrained minimization procedure.

However, a more *ad hoc* procedure for fixing order-relations violations has been proposed by Deutsch (1998). In this procedure, the correction is applied by making a forward and a backward pass through the kriged values. The forward pass starts with the *cdf* value corresponding to the lowest threshold. If this is less than 0, then it is set to 0. We then proceed to the next threshold and if the *cdf* value at that threshold is less than the previous threshold value, it is set equal to the value at the previous threshold. This process is continued until the *cdf* value corresponding to the last threshold is reached. If that happens to be more than 1.0, it is set equal to 1.0. Next, a backward pass is implemented. Starting with the *cdf* value corresponding to the highest threshold, if that value is more than 1.0, then it is set equal to 1.0. We proceed to the next lower threshold and if that happens to be greater than the adjacent higher threshold value, it is set equal to that value. This is continued until the *cdf* value corresponding to the lowest threshold is reached. If that value happens to be more than the next highest threshold value, it is set equal to the value. If the calculated *cdf* value is less than 0.0, then it is set to equal to 0. The final corrected *cdf* is then obtained as the average of the forward and the backward pass values. The magnitude of the order-relations corrections is reported and in most cases is very small (Deutsch, 1998). Data scarcity and incompatibility of the data with the covariance model used for kriging contribute to the order-relations violations.

The results of the forward and backward pass corrections at the node (1,1) are summarized below:

Original *ccdf* values:	[0.710;0.690;0.892;0.961]
Forward pass correction:	[0.710;0.710;0.892;0.961]
Backward pass correction:	[0.690;0.690;0.892;0.961]
Final corrected *cdf*:	[0.700;0.700;0.892;0.961]

At the end of the kriging procedure, the *ccdf* at all locations in the estimation grid are available. These *ccdf* values can be used to generate maps summarizing the spatial variability of the attribute under consideration in this case, as well as the uncertainty associated with the estimation procedure. For example, multiplying the threshold value by the probability density associated with that threshold and summing that across all the thresholds results in the E-type (or the expected value) map. For example, at node (1,1), the E-type estimate is:

$$Z_E = 0.700 \cdot 15 + (0.700 - 0.700) \cdot 25 + (0.892 - 0.700) \cdot 35$$
$$+ (0.961 - 0.892) \cdot 45 = 20.30$$

Figure 3-22 shows the E-type map obtained by implementing the median indicator kriging procedure. The map resembles the SK results shown in Fig. 3-10. An important difference is that the SK results (based on Gaussian assumptions) exhibit a gradual transition from the high values to the low values. The non-parametric indicator result on the other hand exhibits abrupt transition from the high values to the low values.

Example 3-11: Full Indicator Kriging

In the previous example, a covariance model representative of the spatial variability of the median threshold is specified, and the covariance model corresponding to all the other thresholds is proportional to that covariance model. Indicator kriging permits a different covariance model to be specified for each threshold. In order to demonstrate full indicator kriging, let us consider the following variogram models for the four thresholds:

$$z_k = 15: \quad \gamma(\mathbf{h}) = 1 - \left(1.5\frac{|\mathbf{h}|}{12} - 0.5\left(\frac{|\mathbf{h}|}{12}\right)^2\right)$$

$$z_k = 25: \quad \gamma(\mathbf{h}) = 1 - \left(1.5\frac{|\mathbf{h}|}{8} - 0.5\left(\frac{|\mathbf{h}|}{8}\right)^2\right)$$

$$z_k = 35: \quad \gamma(\mathbf{h}) = 1 - \left(1.5\frac{|\mathbf{h}|}{4} - 0.5\left(\frac{|\mathbf{h}|}{4}\right)^2\right)$$

$$z_k = 45: \quad \gamma(\mathbf{h}) = 1 - \left(1.5\frac{|\mathbf{h}|}{4} - 0.5\left(\frac{|\mathbf{h}|}{4}\right)^2\right)$$

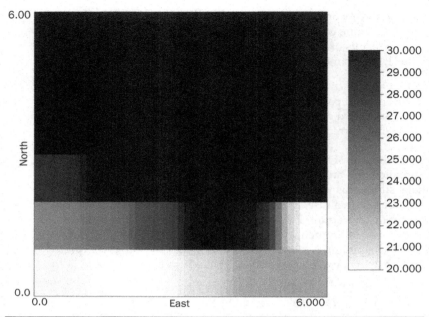

Figure 3-22   Map of E-type estimates obtained by median indicator kriging using the data configuration shown in Fig. 3-6(a) Covariance function $C(\mathbf{h})$ assumed for the application example and Fig. 3-6(b) Data configuration for the same example, and assuming the covariance model shown in Fig. 3-6(a) is representative of the variability corresponding to the median threshold. (The color version of this figure is available at www.mhprofessional.com/PRMS.)

The indicator kriging results corresponding to the variogram/ covariance model are shown in Fig. 3-23. The specification of different covariance models results in the kriged map having more gradual transition from low to high values.

The flexibility in specifying covariance models for each threshold brings with it, an increase in order-relations violations especially when the covariance models exhibit significantly different characteristics from one threshold to the next. This is observed in the above example where the average order-relations violation is 0.087 (in *cdf* units) compared to 0.043 for median indicator kriging. The maximum order-relations correction for the full indicator kriging case is 0.24 (again in *cdf* units), while that for median indicator kriging is 0.115.

3.8   Data Integration in Kriging

So far, we have only been concerned with the estimation of $Z(\mathbf{u})$ using the data $z(\mathbf{u}_\alpha)$, $\alpha = 1, \ldots, n$ that are directly related to $Z(\mathbf{u})$ (and consequently in the units of the variable being estimated). Such data is also frequently termed as "hard" data. Now, we consider the estimation of $Z(\mathbf{u})$ conditioned to both hard data and secondary data $z_k(\mathbf{u}_{\alpha_k})$,

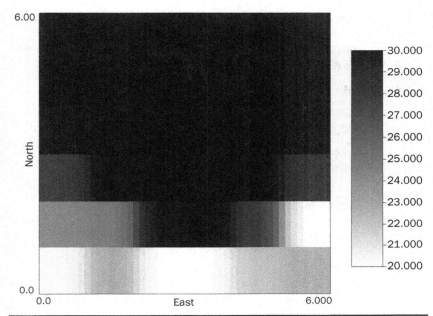

Figure 3-23   Map of thickness obtained by indicator kriging. The model uses a different covariance model for each threshold. (The color version of this figure is available at www.mhprofessional.com/PRMS.)

$\alpha_k = 1, \ldots, n_k; k = 1, \ldots, K$ that are indirectly related to the modeled phenomena. The following are some important observations regarding the secondary data:

- Variable $Z_k(\mathbf{u})$ may be imprecisely related to $Z(\mathbf{u})$ (e.g., the spread in the porosity-permeability cross-plot).

- $Z_k(\mathbf{u})$ may be at a different volume support than $Z(\mathbf{u})$ (e.g., 2-D seismic map for a reservoir layer indicative of vertically averaged porosity variations).

- The data $Z_k(\mathbf{u}_{\alpha_k})$ may be available at locations different from the estimation node $\mathbf{u}$.

We have previously talked about introducing secondary data (such as seismic two-way travel time) into the estimation of an indirectly related variable (such as layer depth) via the external drift procedure. In that case, the secondary variable is assumed linearly related to mean of the primary variable. One can argue that the disparity in support between the primary and secondary variables is implicitly accounted for by the assumed relationship between the secondary variable and the average of the primary variable. Regardless, the lack of precision in that relationship is unaccounted for in the external drift procedure. That imprecision in the relationship between the primary and the secondary variables can be accounted for in the case of multi-variate Gaussian distributions through the covariance or the correlation between the two. This precisely is the underlying premise of the co-kriging estimator.

3.8.1   Simple Co-Kriging Estimator

The co-kriging estimator is a straightforward extension of SK estimator (Wackernagel, 1994). An extended linear formulation is assumed implying that the estimate is optimal if the spatial distributions of the primary and secondary attributes are multi-Gaussian. The linear estimator is written as:

$$[Z(\mathbf{u}_o) - m_o]^* = \sum_{\alpha_o=1}^{n_o} \lambda_{\alpha_o} \cdot [Z(\mathbf{u}_{\alpha_o}) - m_{\alpha_o}]$$

$$+ \sum_{k=1}^{k} \sum_{\alpha_k=1}^{n_k} \lambda_{\alpha_k} \cdot [Z(\mathbf{u}_{\alpha_k}) - m_{\alpha_k}],$$

$$k = 1, \ldots, K \text{ secondary data}$$

The expression for the estimator can be written more compactly as:

$$[Z(\mathbf{u}_o) - m_o]^* = \sum_{k=0}^{k} \sum_{\alpha_k=1}^{n_k} \lambda_{\alpha_k} \cdot [Z(\mathbf{u}_{\alpha_k}) - m_{\alpha_k}] \tag{3-38}$$

The variable corresponding to $k = 0$ corresponds to the primary variable that is being kriged. We go about the usual process of defining an estimation error and subsequently deriving the moments of the Gaussian distribution characterizing the error distribution.

Estimation Error

$$Z(\mathbf{u}_o) - Z^*(\mathbf{u}_o) = [Z(\mathbf{u}_o) - m_o] - [Z^*(\mathbf{u}_o) - m_o]$$

$$= R(\mathbf{u}_o) - \sum_{k=0}^{K} \sum_{\alpha_k=1}^{n_k} \lambda_{\alpha_k} \cdot R(\mathbf{u}_{\alpha_k})$$

Because the RF $Z(\mathbf{u})$ is assumed stationary:

$$E\{R(\mathbf{u}_o)\} = E\{R(\mathbf{u}_{\alpha_k})\} = 0, \alpha = 1,\ldots,n_k, k = 1,\ldots,K$$

This implies that the estimator is unbiased.

Error Variance

$$\sigma^2_{SCK}(\mathbf{u}_o) = Var\left\{Z(\mathbf{u}_o) - Z^*(\mathbf{u}_o)\right\} = Var\left\{R(\mathbf{u}_o) - \sum_{k=0}^{k}\sum_{\alpha_k=1}^{n_k}\lambda_{\alpha_k} \cdot R(\mathbf{u}_{\alpha_k})\right\}$$

$$= Var\{R(\mathbf{u}_o)\} + Var\left\{\sum_{k=0}^{k}\sum_{\alpha_k=1}^{n_k}\lambda_{\alpha_k} \cdot R(\mathbf{u}_{\alpha_k})\right\}$$

$$- 2 \cdot Cov\left\{R(\mathbf{u}_o), \sum_{k=0}^{k}\sum_{\alpha_k=1}^{n_k}\lambda_{\alpha_k} \cdot R_{\alpha_k})\right\}$$

Each term in the expression for the estimation variance can be further expanded:

$$Var\left\{\sum_{k=0}^{k}\sum_{\alpha_k=1}^{n_k}\lambda_{\alpha_k} \cdot R_{\alpha_k}\right\} = \sum_{k'=0}^{k}\sum_{\alpha_{k'}=1}^{n_{k'}}\sum_{k=0}^{k}\sum_{\alpha_k=1}^{n_k}\lambda_{\alpha_k}\lambda_{\alpha_k} \cdot Cov\left\{R_{\alpha_k}, R_{\alpha_{k'}}\right\}$$

Because: $Cov\{X, aY\} = aCov\{X,Y\}$

$$Cov\left\{R(\mathbf{u}_o), \sum_{k=0}^{k}\sum_{\alpha_k=1}^{n_k}\lambda_{\alpha_k} \cdot R_{\alpha_k}\right\} = \sum_{k=0}^{k}\sum_{\alpha_k=1}^{n_k}\lambda_{\alpha_k} \cdot Cov\left\{R(\mathbf{u}_o), R_{\alpha_k}\right\}$$

Denoting: $Cov\left\{R(\mathbf{u}_o), R_{\alpha_k}\right\} = C_{R_{k0}}(u_{\alpha_k} - u_0)$ and $Cov\left\{R_{\alpha_k}, R_{\alpha_k}\right\} = C_{R_{kk'}}(u_{\alpha_k} - u_{\alpha_k})$

The term $C_{R_{k0}}$ represents the cross-covariance between the variable k and the primary variable o. When $k = 0$, we have the auto-covariance of the primary variable. Similarly, $C_{R_{kk'}}$ represents the cross-covariance between the variables k and k', unless $k = k'$, in which case, we are referring to the auto-covariance of variable k.

Substituting the above expansions, we get the following expression for the co-kriging estimation variance:

$$\sigma^2_{SCK}(u_o) = \sigma^2_{oo} + \sum_{k'=0}^{k}\sum_{\alpha_{k'}=1}^{n_{k'}}\sum_{k=0}^{k}\sum_{\alpha_k=1}^{n_k}\lambda_{\alpha_k}\lambda_{\alpha_{k'}} \cdot C_{R_{kk'}}(u_{\alpha_k} - u_{\alpha_{k'}})$$

$$- 2 \cdot \sum_{k=0}^{k}\sum_{\alpha_k=1}^{n_k}\lambda_{\alpha_k} \cdot C_{R_{k0}}(u_{\alpha_k} - u_0)$$

In order to arrive at the BEST estimate, this expression for the error variance has to be minimized. This is achieved by taking the derivative of the error variance with respect to the parameters λ_{α_k} and setting the derivative equal to 0:

$$\frac{\partial \sigma^2_{SCK}(u_o)}{\partial \lambda_{\alpha_k}} = 2\sum_{k'=0}^{k}\sum_{\alpha_{k'}=1}^{n_{k'}}\lambda_{\alpha_k} \cdot C_{R_{kk'}}(u_{\alpha_k} - u_{\alpha_{k'}}) - 2 \cdot C_{R_{k0}}(u_{\alpha_k} - u_0)$$

Setting to 0 yields:

$$\sum_{k'=0}^{k} \sum_{\alpha_{k'}=1}^{n_{k'}} \lambda_{\alpha_k} \cdot C_{R_{kk'}}(u_{\alpha_k} - u_{\alpha_{k'}}) = C_{R_{k0}}(u_{\alpha_k} - u_0) \forall \alpha_k$$
$$= 1,\ldots,n_k; k = 0,\ldots,k \qquad (3\text{-}39)$$

This is a system of $\sum_{k=0}^{K} n_k$ equations to be solved for the corresponding number of unknowns. Under stationarity, the covariance:

$$C_{R_{kk'}}\left(u_{\alpha_k} - u_{\alpha_{k'}}\right) = E\{R_k(\mathbf{u}), R_{k'}(\mathbf{u} + \mathbf{h})\}$$
$$= E\{Z_k(\mathbf{u}) - m_{k'}Z_{k'}(\mathbf{u}) - m_{k'}\} = C_{kk'}(\mathbf{h})$$

$C_{kk'}(\mathbf{h})$ is a measure of the relationship between variables k and k' at locations that are separated by a lag $\mathbf{h}$. When $k = k'$ then $C_{kk'}(\mathbf{h})$ is simply the auto-correlation of the variable k at a lag $\mathbf{h}$. $C_{k0}(\mathbf{h})$ is the cross-correlation between the primary variable and the secondary variable k separated by a lag $\mathbf{h}$. We will talk more about the inference and modeling of these cross-covariances in the next chapter.

To summarize, the simple co-kriging estimator is:

$$(Z(\mathbf{u}_o) - m_o)^* = \sum_{\alpha_o=1}^{n_o} \lambda\alpha_o \cdot \left[Z(\mathbf{u}_{\alpha_o}) - m_{\alpha_o}\right]$$
$$+ \sum_{k=1}^{k} \sum_{\alpha_k=1}^{n_k} \lambda\alpha_k \cdot \left[Z(\mathbf{u}_{\alpha_k}) - m_{\alpha_k}\right]$$

The $K + 1$ means $m_{\alpha_{k'}}$ $k = 0, \ldots, K$ are assumed to be known and stationary. The weights are obtained by solving the co-kriging system:

$$\sum_{k'=0}^{k} \sum_{\alpha_{k'}=1}^{n_k} \lambda\alpha_{k'} \cdot C_{kk'}(u_{\alpha_k} - u_{\alpha_{k'}}) = C_{k0}(u_{\alpha_k} - u_0)$$
$$\forall\, \alpha_k = 1,\ldots,n_k; \ k = 0,\ldots,K$$

Solution of this co-kriging system requires licit models for the cross-covariances $C_{kk'}(\mathbf{h})$.

3.8.2 Simplified Models for Data Integration

The co-kriging systems that was expounded earlier requires legitimate models for the auto- and cross-covariances to be available that will provide values corresponding to all lags and ensure that the kriging system is non-singular for the estimation at all locations. Furthermore, by legitimate models, we mean models that will ensure that the error variance is positive no matter the weights assigned to the data and the kriging system yield a unique set of weights.

The first simplification that can be performed is to restrict the influence of the secondary data on the estimation such that only the collocated data exerts its influence. Let us write the expression for the co-kriging estimator:

$$(Z(\mathbf{u}_o) - m_o)^* = \sum_{\alpha_o=1}^{n_o} \lambda_{\alpha_0} \cdot [Z(\mathbf{u}_{\alpha_0}) - m_{\alpha_o}] + \sum_{\alpha_T=1}^{n_T} \lambda_{\alpha_T} \cdot [T(\mathbf{u}_{\alpha_T}) - m_{\alpha_T}]$$

In the above equation, a single secondary variable $T(\mathbf{u})$ is assumed to inform the primary variable $Z(\mathbf{u})$. If we assume that for the estimation at location $\mathbf{u}_0$, it is only the collocated datum $T(\mathbf{u}_0)$ that influences the estimate, then the estimator simplifies to:

$$(Z(\mathbf{u}_o) - m_o)^* = \sum_{\alpha_o=1}^{n_o} \lambda_{\alpha_o} \cdot [Z(\mathbf{u}_{\alpha_o}) - m_{\alpha_o}] + v \cdot (T(\mathbf{u}_o) - m_T) \qquad (3\text{-}40)$$

The co-kriging system then becomes:

$$\sum_{\beta_k=1}^{n_k} \lambda_{\beta_k} \cdot C_Z(u_{\alpha_k} - u_{\beta_k}) + v \cdot C_{ZT}(u_{\alpha_k} - u_0) = C_z(u_{\alpha_k} - u_0) \; \forall \; \alpha_k = 1,\ldots,n_k$$

$$\sum_{\beta_k=1}^{n_k} \lambda_{\beta_k} \cdot C_{ZT}(u_0 - u_{\beta_k}) + v \cdot C_T(0) \qquad\qquad = C_{ZT}(0) \qquad (3\text{-}41)$$

The advantage of the collocated co-kriging assumption is evident by looking at the reduced system above. The system does not require knowledge of the secondary covariance $C_T(\mathbf{h})$. However, collocated co-kriging still requires specification of the cross-covariance between the primary and secondary variables $C_{ZT}(\mathbf{h})$ and the covariance of the primary variable $C_Z(\mathbf{h})$.

The assumption that only the secondary variable collocated with the estimation node exerts an influence on the estimation is commonly referred to as a screening assumption (i.e., the collocated secondary data screens the influence of all other secondary data) or more technically as a Markov assumption. It would be interesting if such a screening or Markov assumption can also alleviate the task of inferring and modeling the cross-covariance between the primary and secondary variables. Two such models have been put forth—Markov model I (Almeida, 1994) and Markov model II (Shmaryan, 1999).

Markov Model I
We make a Markov-type assumption that at an estimation node $\mathbf{u}$ where secondary data $T(\mathbf{u})$ is available, the collocated primary data $Z(\mathbf{u}) = z$ screens out further-away primary data $Z(\mathbf{u} + \mathbf{h}) = z'$. This can be written mathematically as:

$$E\{T(\mathbf{u}) \,|\, Z(\mathbf{u}) = z, Z(\mathbf{u} + \mathbf{h}) = z'\} = E\{T(\mathbf{u}) \,|\, Z(\mathbf{u}) = z\} \; \forall \mathbf{u}, \mathbf{h}, z, z'$$

$$(3\text{-}42)$$

Now let us look at the cross-correlation $\rho_{ZT}(\mathbf{h})$. For simplicity, let us assume that the RVs $Z(\mathbf{u})$ and $T(\mathbf{u})$ are standard variables (mean zero, unit variance). Also recognizing by the law of total probability that:

$$f(A) = \int_B f(A \mid B) \cdot f_B dB$$

In terms of expectation: $E(A) = \int_B E(A \mid B) \cdot f_B dB$

This yields the cross-correlation as:

$$\rho_{ZT}(\mathbf{h}) = E\{Z(\mathbf{u}) \cdot T(\mathbf{u} + \mathbf{h})\} = \int_{z'} \int_z E\{Z(\mathbf{u}) \cdot T(\mathbf{u} + \mathbf{h}) \mid Z(\mathbf{u})$$
$$= z, Z(\mathbf{u} + \mathbf{h}) = z'\} \cdot f_h(z, z') \, dz dz'$$
$$= \int_{z'} \int_z z \cdot E\{T(\mathbf{u} + \mathbf{h}) \mid Z(\mathbf{u}) = z, Z(\mathbf{u} + \mathbf{h}) = z'\} \cdot f_h(z, z') \, dz dz'$$

The last equality comes about because the RV $Z(\mathbf{u})$ has assumed a specific outcome $Z(\mathbf{u}) = z$. Substituting the previous screening hypothesis:

$$\rho_{ZT}(h) = \int_{z'} \int_z z \cdot E\{T(\mathbf{u} + \mathbf{h}) \mid Z(\mathbf{u} + \mathbf{h}) = z'\} \cdot f_h(z, z') dz dz'$$

Given that the co-kriging estimator is optimum when the variables are multi-Gaussian, the conditional expectation $E\{T(\mathbf{u} + \mathbf{h}) \mid Z(\mathbf{u} + \mathbf{h}) = z'\}$ can be expressed as a linear regression estimate:

$$E\{T(\mathbf{u} + \mathbf{h}) \mid Z(\mathbf{u} + \mathbf{h}) = z'\} = \rho_{ZT}(0) \cdot z'$$

Substituting in the expression for the cross-correlation at lag $\mathbf{h}$:

$$\rho_{ZT}(\mathbf{h}) = \rho_{ZT}(0) \cdot \int_{z'} \int_z z \cdot z' \cdot f_h(z, z') dz dz' = \rho_{ZT}(0) \cdot \rho_Z(\mathbf{h}) \quad (3\text{-}43)$$

The cross-correlation at lag $\mathbf{h}$ is thus expressed as a simple rescaling of the auto-correlation of the primary variable $Z(\mathbf{u})$.

Remarks
- The Markov model I (MMI) as detailed above uses the covariance or alternatively, the variogram of the primary variable in order to develop the cross-covariance between the primary and secondary variables. In most practical situations, the sampled hard or primary data may be very sparse and hence not adequate for reliably inferring the correlation. In that case, the modeled cross-covariance will also be unreliable.

- In case the support (the representative volume scale) of the primary data is much more than that of the secondary data, then the hypothesis that the collocated primary data screens the influence of the farther-away datum appears reasonable. In most modeling situation, it is the primary variable that has substantially smaller volume support (e.g., the well log data regarding porosity of the rock) compared to the secondary variable (e.g., seismic impedance data available exhaustively). The validity of the Markov assumption is questionable in those cases.

Markov Model II

In order to address some of the drawbacks of the MMI, consider the following revised statement of the Markov hypothesis:

$$E\{Z(\mathbf{u}) \mid T(\mathbf{u}) = t, T(\mathbf{u} + \mathbf{h}) = t'\} = E\{Z(\mathbf{u}) \mid T(\mathbf{u}) = t\} \ \forall \mathbf{u}, \mathbf{h}, t, t'$$

$$(3\text{-}44)$$

At an estimation node $\mathbf{u}$ where secondary data $T(\mathbf{u})$ is available, the collocated secondary data $T(\mathbf{u}) = t$ screens out further-away secondary data $T(\mathbf{u} + \mathbf{h}) = t'$.

Let us derive an expression for the cross-correlation $\rho_{ZT}(\mathbf{h})$ using this revised model:

$$
\begin{aligned}
\rho_{ZT}(\mathbf{h}) &= E\{Z(\mathbf{u}) \cdot T(\mathbf{u} + \mathbf{h})\} = \int_{t'} \int_{t} E\{Z(\mathbf{u}) \cdot T(\mathbf{u} + \mathbf{h}) \mid T(\mathbf{u}) \\
&= t, T(\mathbf{u} + \mathbf{h}) = t'\} \cdot f_h(t, t') dt dt' \\
&= \int_{t'} \int_{t} t' \cdot E\{Z(\mathbf{u}) \mid T(\mathbf{u}) = t, T(\mathbf{u} + \mathbf{h}) = t'\} \cdot f_h(t, t') dt dt' \\
&= \int_{t'} \int_{t} t' \cdot E\{Z(\mathbf{u}) \mid T(\mathbf{u}) = t\} \cdot f_h(t, t') dt dt'
\end{aligned}
$$

The last equality comes about because of the assumed Markov hypothesis. Again, because the estimation is being developed in a multi-Gaussian framework, we can assume the following linear form for the conditional expectation:

$$E\{Z(\mathbf{u}) \mid T(\mathbf{u}) = t\} = \rho_{ZT}(0) \cdot t$$

Substituting the above development for the cross-correlation results in:

$$\rho_{ZT}(\mathbf{h}) = \rho_{ZT}(0) \cdot \rho_T(\mathbf{h}) \qquad (3\text{-}45)$$

The cross-correlogram between the variables $Z(\mathbf{u})$ and $T(\mathbf{u})$ at lag $\mathbf{h}$ can thus be developed in terms of the correlogram of the secondary variable $T(\mathbf{u})$.

The above expression for the Markov hypothesis can be interpreted in a slightly different fashion also. Because the collocated secondary data screens the influence of other secondary data, we can write the following linear relationship to describe the relationship between $Z(\mathbf{u})$ and $T(\mathbf{u})$ in a Gaussian framework:

$$Z(\mathbf{u}) = \rho_{ZT}(0) \cdot T(\mathbf{u}) + R(\mathbf{u})$$

The residue $R(\mathbf{u})$ is added to introduce scatter (uncertainty) into the relationship between the two variables and is assumed to be independent of $T(\mathbf{u})$. Under such an assumption:

$$
\begin{aligned}
\rho_{ZT}(\mathbf{h}) = E\{Z(\mathbf{u}) \cdot T(\mathbf{u} + \mathbf{h})\} &= E\{(\rho_{ZT}(0) \cdot T(\mathbf{u}) + R(\mathbf{u})) \cdot T(\mathbf{u} + \mathbf{h})\} \\
&= \rho_{ZT}(0) \cdot \begin{pmatrix} E\{T(\mathbf{u}) \cdot T(\mathbf{u} + \mathbf{h})\} \\ + E\{R(\mathbf{u}) \cdot T(\mathbf{u} + \mathbf{h})\} \end{pmatrix} \\
&= \rho_{ZT}(0) \cdot \rho_T(\mathbf{h})
\end{aligned}
$$

The term $E\{R(\mathbf{u}) \cdot T(\mathbf{u} + \mathbf{h})\}$ is equal to 0 because of the assumed independence between $R(\mathbf{u})$ and $T(\mathbf{u})$. The Markov assumption interpreted in this manner results in the same expression for the cross-correlogram. In addition, the regression model with the added stochastic residue provides some other interesting insights. Consider the variance of the residue $R(\mathbf{u})$:

$$R(\mathbf{u}) = Z(\mathbf{u}) - \rho_{ZT}(0) \cdot T(\mathbf{u})$$

$$
\begin{aligned}
Var\{R(\mathbf{u})\} &= Var\{Z(\mathbf{u}) - \rho_{ZT}(0) \cdot T(\mathbf{u})\} \\
&= Var\{Z(\mathbf{u})\} - 2 \cdot \rho_{ZT}(0) \cdot Cov\{Z(\mathbf{u}), T(\mathbf{u})\} + \rho_{ZT}^2(0)Var\{T(\mathbf{u})\} \\
&= Var\{Z(\mathbf{u})\} - 2 \cdot \rho_{ZT}^2(0) + \rho_{ZT}^2(0)Var\{T(\mathbf{u})\}
\end{aligned}
$$

And because $Z(\mathbf{u})$ and $T(\mathbf{u})$ are assumed to be standardized variables:

$$Var\{R(\mathbf{u})\} = 1 - \rho_{ZT}^2(0)$$

The variance of the residue is therefore a constant regardless of the outcome of $T(\mathbf{u})$. A practical check for the validity of the Markov model is to compute the residue based on the decomposition $R(\mathbf{u}) = Z(\mathbf{u}) - \rho_{ZT}(0) \cdot T(\mathbf{u})$ for various observed $Z(\mathbf{u})$ and $T(\mathbf{u})$ data and check if the variance of $R(\mathbf{u})$ remains constant across all classes of $T(\mathbf{u})$.

Now let us take a look at the correlogram of the primary variable under such a Markov hypothesis:

$$
\begin{aligned}
\rho_Z(\mathbf{h}) &= E\{Z(\mathbf{u}) \cdot Z(\mathbf{u} + \mathbf{h})\} \\
&= E\{(\rho_{ZT}(0) \cdot T(\mathbf{u}) + R(\mathbf{u})) \cdot (\rho_{ZT}(0) \cdot T(\mathbf{u} + \mathbf{h}) + R(\mathbf{u} + \mathbf{h}))\} \\
&= \rho_{ZT}^2(0)\rho_T(\mathbf{h}) + E\{R(\mathbf{u}) \cdot R(\mathbf{u} + \mathbf{h})\}
\end{aligned}
$$

Because the residue $R(\mathbf{u})$ has a variance $Var\{R(\mathbf{u})\} = 1 - \rho^2_{ZT}(0)$ (different from 1), the expectation: $E\{R(\mathbf{u}) \cdot R(\mathbf{u} + \mathbf{h})\} = (1 - \rho^2_{ZT}(0)) \cdot \rho_R(\mathbf{h})$. This leads to:

$$\rho_Z(\mathbf{h}) = \rho^2_{ZT}(0)\rho_T(\mathbf{h}) + (1 - \rho^2_{ZT}(0)) \cdot \rho_R(\mathbf{h}) \qquad (3\text{-}46)$$

This model for the auto-correlogram of the primary variable provides another check for the Markov assumption. Under this model, the auto-correlogram for the primary can be modeled as a linear combination of the correlogram for the secondary variable and any other assumed model for the correlogram of the residual. If the correlogram for the primary cannot be fitted using such a linear combination, it may be indicative of the inappropriateness of the Markov model II (MMII) for the particular problem at hand.

3.8.3 Linear Model of Coregionalization

In the event that the previous Markov assumptions for simplifying the co-kriging process are invalid, we have to go back to the full co-kriging estimator and devise techniques for developing models for the auto- and cross-covariances required by the procedure. We are still going to think of simplifying the assumptions but not as drastic as the screening assumptions discussed previously. Let us start with a recap of how we developed covariance models for spatial phenomena exhibiting complex heterogeneities. In that case (Chap. 3), we started with a linear decomposition of the form:

$$Z(\mathbf{u}) = \sum_{k=1}^{K} \lambda_k \cdot Z_k(\mathbf{u})$$

We represent the random function $Z(\mathbf{u})$ as a linear combination of several other random functions. One can interpret this as the decomposition of a random function into a set of frequency (or in the context of spatial coordinates, various length scale) components. This resulted in the covariance model:

$$C(\mathbf{h}) = \sum_{k=1}^{K} \lambda_k \cdot C_k(\mathbf{h})$$

The preceding equation states that given some legitimate (positive definite) basic structures $C_k(\mathbf{h})$, it is possible to model the covariance (or equivalently, the variogram) describing a stationary RF $Z(\mathbf{u})$ as a positive linear combination of the known structures. This concept of nested variogram or covariance modeling is used to model the covariance structure corresponding to complex reservoir heterogeneity at multiple length scales.

In the case of co-kriging, the requirement is for the entire set of auto-covariances/variograms and cross-covariances/variograms to be jointly positive definite (i.e., they yield a unique solution to the kriging system and ensure positive error variance). In other words, we need a model $C_{ij}(\mathbf{h})$ for all combinations of i and j. Let us suppose a linear decomposition:

$$Z_i(\mathbf{u}) = \sum_{k=1}^{K} a_{ik} \cdot Y_k(\mathbf{u})$$

Assuming that the $K - RF$ $Y_k(\mathbf{u})$ are independent, zero mean and auto-covariance given by $E\{Y_k(\mathbf{u}+\mathbf{h}) \cdot Y_k(\mathbf{u})\} = C_k(\mathbf{h})$ and $C_{kk'}(\mathbf{h}) = 0, \forall k' \neq k$

Then: $C_{ij}(\mathbf{h}) = E\{Z_i(\mathbf{u}) \cdot Z_j(\mathbf{u}+\mathbf{h})\} = \sum_{m=1}^{M}\sum_{k=1}^{K} a_{ik} \cdot a_{jm} \cdot C_{km}(\mathbf{h})$

But $C_{km}(\mathbf{h}) = 0, \forall k \neq m$ and so $C_{ij}(\mathbf{h}) = \sum_{k=1}^{K} a_{ik} \cdot a_{jk} \cdot C_k(\mathbf{h})$

In the formulation above, the parameters a_{ik} could be positive, negative, or zero. The functions $C_k(\mathbf{h})$ are positive definite. Though the formulation above seems to result in a fairly straightforward approach to model the cross-covariance by simply extending the previous linear model of regionalization, it is missing something. We would like to derive a positive definite condition that would ensure that the entire coregionalization model comprising auto-covariances $C_{ii}(\mathbf{h}),C_{jj}(\mathbf{h})$, as well as all their combinations, are jointly positive definite. In order to do this, we have to generalize the previous linear combination $C_{ii}(\mathbf{h}),C_{jj}(\mathbf{h})$ and write in terms of vectors of basic RFs $Y_k(\mathbf{u})$ (Journel & Huijbregts, 1978; Goovaerts, 1997):

$$Z_i(\mathbf{u}) = \sum_{k=1}^{K}\sum_{\ell=1}^{L} a_{ik}^{\ell} \cdot Y_k^{\ell}(\mathbf{u}) \tag{3-47}$$

The L-independent random functions $Y_k^{\ell}(\mathbf{u})$ that make up $Y_k(\mathbf{u})$ are assumed to share the same covariance $E\{Y_k^{\ell}(\mathbf{u})Y_k^{\ell}(\mathbf{u}+\mathbf{h})\} = C_k(\mathbf{h})$. It is also assumed that the cross-covariance $E\{Y_k^{\ell}(\mathbf{u})Y_k^{\ell}(\mathbf{u}+\mathbf{h})\} = 0$ when $\ell \neq \ell'$. Then:

$$C_{ij}(\mathbf{h}) = \sum_{k'=1}^{K}\sum_{k=1}^{K}\sum_{\ell'=1}^{L}\sum_{\ell=1}^{L} a_{ik}^{\ell} \cdot a_{jk'}^{\ell'} \cdot E\{Y_k^{\ell}(\mathbf{u}) \cdot Y_{k'}^{\ell'}(\mathbf{u}+\mathbf{h})\}$$

and since $Y_k^{\ell}(\mathbf{u})$ and $Y_{k'}^{\ell}(\mathbf{u}+\mathbf{h})$ are orthogonal (independent) and $E\{Y_k^{\ell}(\mathbf{u}) \cdot Y_{k'}^{\ell'}(\mathbf{u}+\mathbf{h})\}$ is non-zero only for $k = k'$ and $\ell = \ell'$, the quadruple sum reduces to:

$$C_{ij}(\mathbf{h}) = \sum_{k=1}^{K}\left[\sum_{\ell=1}^{L} a_{ik}^{\ell} \cdot a_{jk}^{\ell}\right] \cdot K_k(\mathbf{h})$$

If we denote:
$$\left[\sum_{\ell=1}^{L} a_{ik}^{\ell} \cdot a_{jk}^{\ell}\right] = b_{ik}^{k}$$

Then:

$$C_{ij}(\mathbf{h}) = \sum_{k=1}^{K} b_{ij}^{k} \cdot K_{k}(\mathbf{h}) \tag{3-48}$$

The covariances $C_{ij}(\mathbf{h})$ form a $[N_v \times N_v]$ matrix with auto-covariances along the diagonal. The coefficients $b_{ik}^{k} = \left[\sum_{\ell=1}^{L} a_{ik}^{\ell} \cdot a_{jk}^{\ell}\right]$ are thus entries in a $[N_v \times N_v]$ matrix B_{ij}^{k} and there is one such matrix for each basic structure $C_k(\mathbf{h})$, $k = 1, \ldots, K$. There are thus K such $[N_v \times N_v]$ matrices. The components b_{ij}^{k} can be interpreted as the aggregated contributions of the L RFs $Y_k^{\ell}(\mathbf{u})$ toward the covariance $C_{ij}(\mathbf{h})$ observed at lag $\mathbf{h}$.

The operation $\left[\sum_{\ell=1}^{L} a_{ik}^{\ell} \cdot a_{jk}^{\ell}\right]$ is equivalent to $a^T a$ and so the coefficient b_{ij}^{k} can be interpreted as the variance contribution of the kth structure toward the covariance $C_{ij}(\mathbf{h})$. Because the basic structures $C_k(\mathbf{h})$, $k = 1, \ldots, K$ are positive definite, we require the matrices B_{ij}^{k} to be positive definite in order to ensure that the entire set of auto- and cross-covariances are positive definite. Practically, how do we ensure the matrix B_{ij}^{k} to be positive semi-definite?

From linear algebra, a $[N_v \times N_v]$ symmetric matrix is positive semi-definite if all its minor diagonal determinants are non-negative. Checking this for different dimensions of the $[N_v \times N_v]$ matrices:

$$1 \times 1 : b_{11} \geq 0 \qquad 2 \times 2 : \begin{vmatrix} b_{11} & b_{12} \\ b_{21} & b_{22} \end{vmatrix} \geq 0 \text{ implies } b_{11}b_{22} \geq b_{12}^2$$

$$3 \times 3 : \begin{vmatrix} b_{11} & b_{12} & b_{13} \\ b_{21} & b_{22} & b_{23} \\ b_{31} & b_{32} & b_{33} \end{vmatrix} \geq 0$$

$$[N_v \times N_v] : \begin{vmatrix} b_{11} & \cdots & b_{1,N_v} \\ \vdots & & \vdots \\ b_{N_v,1} & \cdots & b_{N_v,N_v} \end{vmatrix} \geq 0$$

Rigorously checking the positive definiteness of matrix with dimension $N_v > 3$ is very tedious and difficult (Goovaerts, 1994). There have been some attempts to develop an automated model-fitting scheme using various regression and other iterative techniques (Marcotte, 2012; Emery, 2010).

Remarks

- Following up on the positive definiteness of the coregionalization matrices, if a particular structure $C_k(\mathbf{h})$ is absent in an auto-covariance, then it must also be absent from all corresponding cross-covariances.

  Taking a $[2 \times 2]$ case, if $b_{ii}^k = 0$ then:

  $$\begin{vmatrix} b_{ii}^k & b_{ij}^k \\ b_{ji}^k & b_{jj}^k \end{vmatrix} = -[b_{ij}^k]^2 \geq 0 \quad \text{implies that } b_{ij}^k = 0 \forall j$$

  In other words, if the diagonal terms of the matrices $B_{ij}^k = [b_{ij}^k]$ are 0, then the corresponding rows and columns must be 0; that is, if $C_k(\mathbf{h})$ is present in the cross-covariance, it must be present in all corresponding cross-covariances.

- Some measures of joint spatial variability:

  Cross-covariance function:

  $$C_{ij}(\mathbf{h}) = E\{[Z_i(\mathbf{u}) - m_i] \cdot [Z_j(\mathbf{u} + \mathbf{h}) - m_j]\}$$
  $$\neq E\{[Z_j(\mathbf{u}) - m_j] \cdot [Z_i(\mathbf{u} + \mathbf{h}) - m_i]\}$$
  $$\neq C_{ji}(\mathbf{h})$$

  i.e., the cross-covariance is not reversible generally. In the above linear model of coregionalization (LMC), the coregionalization matrix is assumed to be symmetric; i.e., the lag effect is ignored.

  Cross-variogram function:

  $$2 \cdot \gamma_{ij}(\mathbf{h}) = E\{[Z_i(\mathbf{u}) - Z_i(\mathbf{u} + h)] \cdot [Z_j(\mathbf{u}) - Z_j(\mathbf{u} + h)]\}$$

  There is no longer a moment of inertia-based interpretation of the cross-variogram, as the 45° bisector does not have a meaning.

  $$\gamma_{ij}(\mathbf{h}) = \frac{1}{2} E\{[Z_i(\mathbf{u}) - Z_i(\mathbf{u} + h)] \cdot [Z_j(\mathbf{u}) - Z_j(\mathbf{u} + h)]\}$$
  $$= \gamma_{ij}(\mathbf{h})$$

  Note that because of the above definition, $\gamma_{ij}(\mathbf{h})$ can be negative. The cross-covariance function and cross-semivariogram are related through the expression:

  $$\gamma_{ij}(\mathbf{h}) = C_{ij}(0) - \frac{1}{2}\{C_{ij}(\mathbf{h}) + C_{ij}(-\mathbf{h})\}$$

  This implies that the sill value of a bounded cross-semivariogram tends to the cross-covariance at lag zero $C_{ij}(0)$.

Example 3-12

First we demonstrate the application of collocated co-kriging using MMI for simplifying the cross-covariance modeling. The conditioning data and the covariance model for the primary variable are the same as shown in Fig. 3-6. The same exhaustive map that was used as the external drift variable in Fig. 3-17 is used as the secondary data for performing collocated co-kriging. Recall that for collocated co-kriging, the exhaustive secondary map that yields a secondary value at each estimation node is a requirement. A correlation coefficient of 0.65 is assumed between the primary and the secondary variables; i.e., $\rho_{TZ} = \rho_{TZ}(0) = 0.65$. Knowing the covariance model for the primary variable, the cross-covariance mode is given as:

$$\rho_{ZT}(\mathbf{h}) = \rho_{ZT}(0) \cdot \rho_Z(\mathbf{h})$$

Equations (3-40) and (3-41) give the collocated co-kriging estimator and the accompanying system. For this particular example, the co-kriging system at location (1,1) is given by:

$$
\begin{bmatrix}
1.000 & 0.485 & 0.620 & 0.360 & 0.427 \\
0.485 & 1.000 & 0.620 & 0.258 & 0.161 \\
0.620 & 0.620 & 1.000 & 0.532 & 0.277 \\
0.360 & 0.258 & 0.532 & 1.000 & 0.234 \\
0.427 & 0.161 & 0.277 & 0.234 & 1.000
\end{bmatrix}
\cdot
\begin{bmatrix}
\lambda_1 \\ \lambda_1 \\ \lambda_2 \\ \lambda_3 \\ \lambda_4 \\ v
\end{bmatrix}
=
\begin{bmatrix}
0.656 \\ 0.247 \\ 0.426 \\ 0.360 \\ 0.650
\end{bmatrix}
$$

The solution to the kriging system is:

$$
\begin{bmatrix}
\lambda_1 \\ \lambda_2 \\ \lambda_3 \\ \lambda_4 \\ v
\end{bmatrix}
=
\begin{bmatrix}
0.465 \\ -0.083 \\ 0.012 \\ 0.105 \\ 0.437
\end{bmatrix}
$$

The kriging estimate is thus:

$$
Z^{*}_{MMI} = \left\{ \begin{bmatrix} 0.465 \cdot (-1.162) - 0.083 \cdot (-0.387) + 0.012 \cdot 1.162 \\ + 0.105 \cdot 0.387 \end{bmatrix} \cdot 12.91 + 25.00 \right\}
$$
$$
+ 0.437 \cdot 5.23 = 21.438
$$

The standardized data are -1.162, -0.387, 1.162, and 0.387, the standard deviation of the data is 12.91, the mean of the data is 25.00, and the collocated secondary data value is 5.23.

Performing co-kriging at all locations on the estimation grid, we can generate the map of estimates shown in Fig. 3-24. Comparing to the map of external drift in Fig. 3-17 and the SK in Fig. 3-10, the

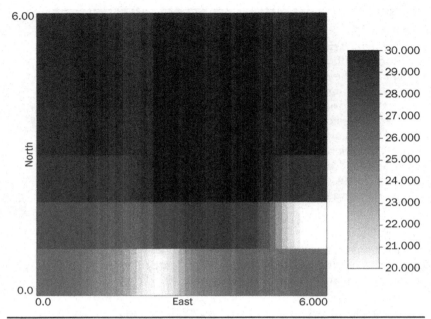

Figure 3-24   Map of estimates obtained by collocated kriging with the Markov model I assumption. (The color version of this figure is available at www .mhprofessional.com/PRMS.)

influence of the secondary variable on the estimate is apparent. The secondary data causes the co-kriging estimates to be increasing gradually in the lower part of the map.

Collocated co-kriging performed using the MMI assumption only requires the covariance model for the primary variable and the correlation coefficient between the primary and the secondary variables at lag 0 to be specified along with the mean of the primary and the secondary variables. The covariance model for the secondary variable is not needed for implementing this method.

Next we implement collocated co-kriging with the MMII. From the map of the secondary variable (Fig. 3-17), it is observed that the secondary values exhibit isotropic variability and the range of the covariance is approximately six blocks. Consequently, the secondary covariance model is written as:

$$\rho_T(\mathbf{h}) = 1 - \left[1.5 \cdot \frac{|\mathbf{h}|}{a} - 0.5 \cdot \left(\frac{|\mathbf{h}|}{a} \right)^2 \right]$$

The range of the covariance model a is assumed to be 6. In MMII, the cross-covariance between the primary and the secondary variables is modeled as:

$$\rho_{ZT}(\mathbf{h}) = \rho_{ZT}(0) \cdot \rho_T(\mathbf{h})$$

Keeping the correlation coefficient $\rho_{TZ} = \rho_{TZ}(0) = 0.65$ as in the previous case, the cross-covariance model can be easily computed. However, before proceeding with that, let us check if we can specify a model for the residual covariance $\rho_R(\mathbf{h})$ such that when that is combined with the secondary covariance model, we match the primary covariance model given by Eq. (3-46).

We follow a trial and error procedure starting with an initial guess for the range of the residual covariance model and comparing the resultant $\rho_Z(\mathbf{h})$ obtained using Eq. (3-46) with the target model in Fig. 3-6. The final converged model for the residual covariance is:

$$\rho_R(\mathbf{h}) = 1 - \left[1.5 \cdot \frac{|\mathbf{h}|}{9.91} - 0.5 \cdot \left(\frac{|\mathbf{h}|}{9.91}\right)^2\right]$$

The match between the reconstructed model for $\rho_Z(\mathbf{h})$ and the original model is shown in Fig. 3-25. The reconstructed model matches the original model closely for the short lags. There is some deviation at the larger lags but the model fit is acceptable for the most part.

The system of equations at the estimation node (1,1) is:

$$\begin{bmatrix} 1.00 & 0.49 & 0.62 & 0.36 & 0.36 \\ 0.49 & 1.00 & 0.62 & 0.26 & 0.06 \\ 0.62 & 0.62 & 1.00 & 0.53 & 0.18 \\ 0.36 & 0.26 & 0.53 & 1.00 & 0.13 \\ 0.36 & 0.06 & 0.18 & 0.13 & 1.00 \end{bmatrix} \cdot \begin{bmatrix} \lambda_1 \\ \lambda_2 \\ \lambda_3 \\ \lambda_4 \\ v \end{bmatrix} = \begin{bmatrix} 0.66 \\ 0.25 \\ 0.43 \\ 0.36 \\ 0.65 \end{bmatrix}$$

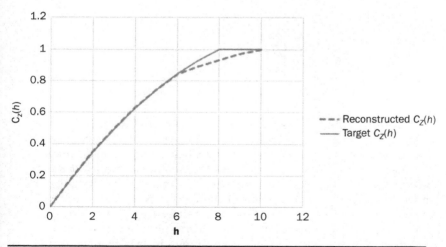

Figure 3-25 Comparison of the target model for the primary variable with the reconstructed model using the Markov II model. (The color version of this figure is available at www.mhprofessional.com/PRMS.)

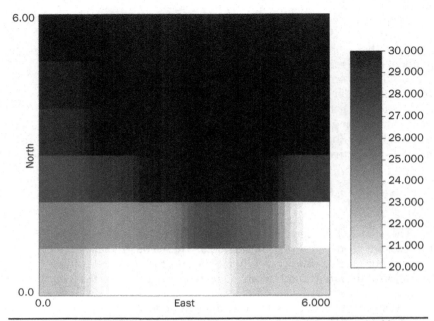

Figure 3-26 Map of estimates obtained by collocated co-kriging using the Markov model II for the cross-covariances. (The color version of this figure is available at www.mhprofessional.com/PRMS.)

The resulting solution is:

$$\begin{bmatrix} \lambda_1 \\ \lambda_2 \\ \lambda_3 \\ \lambda_4 \\ v \end{bmatrix} = \begin{bmatrix} 0.448 \\ -0.041 \\ 0.014 \\ 0.139 \\ 0.469 \end{bmatrix}$$

The map of co-kriged estimates obtained using the MMII is shown in Fig. 3-26. Comparing this map to the one in Fig. 3-23, it can be seen that the influence of the secondary data is slightly stronger in the map obtained using MMII. This is because the covariance model of the secondary variable is explicitly used to construct the cross-covariance model and thereby, control the information being borrowed from the secondary map. Though the difference between the MMI-derived map and the MMII-derived map may be very subtle, it is to be noted that the secondary data used in these examples have been generated by averaging a map of the primary variable and subjecting it to a non-linear transform. Because of this implicit assumption regarding the volume support of the secondary variable, the MMII screening hypothesis may be more appropriate for representing the influence of the secondary variable on the estimation of the primary map.

Linear Model of Coregionalization   In this example, thus far, we have seen that the following covariance model characterizes the primary variable:

$$\rho_Z(\mathbf{h}) = 1 - \left(1.5 \cdot \frac{|\mathbf{h}|}{8.0} - 0.5 \cdot \left(\frac{|\mathbf{h}|}{8.0} \right)^2 \right)$$

In addition, the secondary variable shown in Fig. 3-17 is characterized by the covariance:

$$\rho_T(\mathbf{h}) = 1 - \left(1.5 \cdot \frac{|\mathbf{h}|}{6.0} - 0.5 \cdot \left(\frac{|\mathbf{h}|}{6.0} \right)^2 \right)$$

As it stands, it is difficult to implement the LMC in this case. This is because, as we have learnt before, if a structure is absent in the primary or the secondary covariance models, it will be absent in the cross-covariance model. In this case, there are two covariance structures, one of range 6 units and the other of range 8 units. The first structure is missing from the primary covariance model, while the second is missing from the secondary model, respectively. In order to implement the LMC clearly, a compromise will have to be worked out. Let us suggest the following modification to the primary covariance model:

$$\rho_Z(\mathbf{h}) = 1 - 0.2 \cdot \left[1.5 \cdot \frac{|\mathbf{h}|}{6.0} - 0.5 \cdot \left(\frac{|\mathbf{h}|}{6.0} \right)^2 \right] - 0.8 \cdot \left[1.5 \cdot \frac{|\mathbf{h}|}{9.0} - 0.5 \cdot \left(\frac{|\mathbf{h}|}{9.0} \right)^2 \right]$$

We add a first structure and assign it a small sill contribution and in order to compensate for this short-range structure, we add a second structure of slightly longer range and assign it a larger sill contribution, in order to mimic a single structure of range 8. Figure 3-27 shows a comparison of the original primary covariance model and this revised model with two structures. Again, it is seen that the two models are very close for the short lags and exhibit some differences only closer to the range of the covariance. Similarly, the following modification is performed to the secondary covariance model:

$$\rho_T(\mathbf{h}) = 1 - 0.8 \cdot \left[1.5 \cdot \frac{|\mathbf{h}|}{6.0} - 0.5 \cdot \left(\frac{|\mathbf{h}|}{6.0} \right)^2 \right] - 0.2 \cdot \left[1.5 \cdot \frac{|\mathbf{h}|}{9.0} - 0.5 \cdot \left(\frac{|\mathbf{h}|}{9.0} \right)^2 \right]$$

Figure 3-28 reveals that the modification does capture the spatial variability of the secondary variable reasonably well. The motivation for performing these modifications to the covariance models of the primary and secondary variables is quite obvious. Now, both the structures are present in both primary and secondary covariance

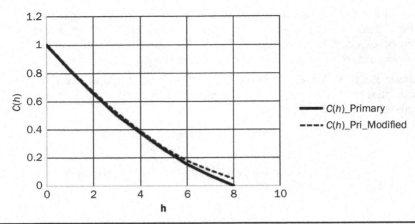

Figure 3-27 Comparison between the original covariance model for the primary variable and the modification performed in order to implement the linear model of coregionalization.

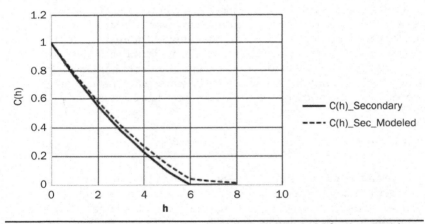

Figure 3-28 Comparison between the original covariance model for the secondary variable and the modification performed in order to implement the linear model of coregionalization.

models, albeit with different sill contributions. We are now ready to implement the LMC modeling process.

The LMC matrix for the first structure is:

$$\begin{bmatrix} 0.2 & x \\ x & 0.8 \end{bmatrix} \geq 0 \Rightarrow x^2 \leq 0.16$$

i.e., the cross-covariance model can have a first structure with a range of 6 units whose sill contribution cannot exceed $\sqrt{0.16} = 0.4$. Similarly, the LMC matrix for the second structure is:

$$\begin{bmatrix} 0.8 & x \\ x & 0.2 \end{bmatrix} \geq 0 \Rightarrow x^2 \leq 0.16$$

i.e., the second structure of range 8 units can also be present in the cross-covariance model with a maximum sill contribution of 0.4. In the MMI and MMII examples presented earlier, the correlation coefficient between the primary and secondary variables was assumed to be 0.65. Because the sill of the cross-covariance model is bound by the correlation coefficient at lag 0, we will propose the following model for the cross-covariance:

$$\rho_{ZT}(\mathbf{h}) = 0.65 - 0.325 \cdot \left(1.5 \cdot \frac{|\mathbf{h}|}{6.0} - 0.5 \cdot \left(\frac{|\mathbf{h}|}{6.0} \right)^2 \right)$$

$$- 0.325 \cdot \left(1.5 \cdot \frac{|\mathbf{h}|}{9.0} - 0.5 \cdot \left(\frac{|\mathbf{h}|}{9.0} \right)^2 \right)$$

Clearly, the sill contribution of each structure satisfies the LMC condition developed above.

The full co-kriging estimator with the LMC for the cross-covariances can account for primary and secondary data that are non-collocated. We can use the exhaustive secondary map if we so desire. However, in order to simplify the example calculations, we will only retain the secondary values at the primary data locations for the estimation. The primary and secondary data used for the estimation is presented in Table 3-3.

The co-kriging system at location (1,1) is:

$$\begin{bmatrix} 1.00 & 0.50 & 0.63 & 0.38 & 0.65 & 0.29 & 0.38 & 0.20 \\ 0.50 & 1.00 & 0.63 & 0.28 & 0.29 & 0.65 & 0.38 & 0.14 \\ 0.63 & 0.63 & 1.00 & 0.54 & 0.38 & 0.38 & 0.65 & 0.32 \\ 0.38 & 0.28 & 0.54 & 1.00 & 0.20 & 0.14 & 0.32 & 0.65 \\ 0.65 & 0.29 & 0.38 & 0.20 & 1.00 & 0.39 & 0.54 & 0.25 \\ 0.29 & 0.65 & 0.38 & 0.14 & 0.39 & 1.00 & 0.54 & 0.14 \\ 0.38 & 0.38 & 0.65 & 0.32 & 0.54 & 0.54 & 1.00 & 0.44 \\ 0.20 & 0.14 & 0.32 & 0.65 & 0.25 & 0.14 & 0.44 & 1.00 \end{bmatrix} \cdot \begin{bmatrix} \lambda_1 \\ \lambda_2 \\ \lambda_3 \\ \lambda_4 \\ v_1 \\ v_2 \\ v_3 \\ v_4 \end{bmatrix} = \begin{bmatrix} 0.664 \\ 0.266 \\ 0.439 \\ 0.375 \\ 0.405 \\ 0.129 \\ 0.247 \\ 0.203 \end{bmatrix}$$

X	Y	Primary	Secondary
3	1	10	−0.644
6	2	20	−0.811
4	3	40	1.319
2	5	30	0.454

TABLE 3-3 Location of Primary and Secondary Data Used for the Co-Kriging Example

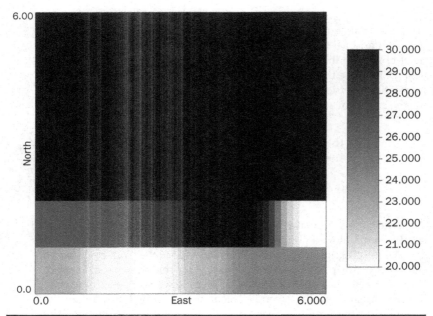

Figure 3-29 Map of co-kriged estimates obtained by employing the liner model of coregionalization for modeling the cross-covariance between the primary and secondary variables. (The color version of this figure is available at www .mhprofessional.com/PRMS.)

This yields the solution:

$$\begin{bmatrix} \lambda_1 \\ \lambda_2 \\ \lambda_3 \\ \lambda_4 \\ v_1 \\ v_2 \\ v_3 \\ v_4 \end{bmatrix} = \begin{bmatrix} 0.674 \\ -0.109 \\ 0.016 \\ 0.166 \\ -0.033 \\ -0.006 \\ 0.000 \\ -0.023 \end{bmatrix}$$

The map of co-kriged estimates is shown in Fig. 3-29. The full co-kriging estimator using the LMC does not make any data screening assumptions. Instead, the procedure allows for rigorously modeling the redundancy between the primary and the secondary data. In this case, as explained previously, the secondary variable has been obtained by applying a coarsening filter on a map of the primary variable. Consequently, there is a high degree of redundancy between the information from the hard and soft data. Furthermore, in this particular example, the conditioning data is restricted to collocated locations of primary and secondary data. Nevertheless, the influence of the secondary data on the resultant estimation of thicknesses is evident. The secondary data serves to soften the transitions from the high to low thickness values. This is evident when comparing the result in Fig. 3-29 to the map obtained by SK in Fig. 3-10 that only considered the primary data for the estimation.

3.9   References

Almeida, A. J. (1994). Joint simulation of multiple variables with a Markov-type coregionalization model. *Mathematical Geology, 26*, 565–588.

Armstrong, M. (1984). Problems with Universal Kriging. *Mathematical Geology, 16* (1), 101-108.

Beale, C. M. (2010). Regression analysis of spatial data. *Ecology Letters, 13*, 246-264.

Chiles, J. P. (1999). *Geostatistics: Modeling Spatial Uncertainty*. New York: Wiley Interscience.

Deutsch, C. V. (1998). *GSLIB: Geostatistical Software Library and User's Guide*. New York, New York, U.S.A.: Oxford University Press.

Emery, X. (2010). Iterative algorithms for fitting a linear model of coregionalization. *Computers and Geosciences, 36* (9), 1150-1160.

Goovaerts, P. (1994). Comparative performance of indicator algorithms for modeling conditional probability distribution functions. *Mathematical Geology, 26*, 389–411.

Goovaerts, P. (1997). *Geostatistics for Natural Resources Evaluation*. Oxford, U.K.: Oxford University Press.

Goovaerts, P. (1994). On a controversial method for modeling a coregionalization. *Mathematical Geology, 26*, 197-204.

Journel, A. G., & Huijbregts, C. J. (1978). *Mining Geostatistics*. London: Academic Press.

Journel, A. (1983). Nonparametric estimation of spatial distributions. *Mathematical Geology, 15*, 445-468.

Journel. A. G., & Posa, D. (1990). Characteristic Behavior and Order Relations for Indicator Variograms. *Mathematical Geology, 22* (8), 1011-1025.

Kilmartin, D. (2009). *Development of Reservoir Models using Economic Loss Functions*. Austin: University of Texas.

Lewis-Beck, M. S. (Ed.). (2004). *The SAGE Encyclopedia of Social Science Research Methods*. Thousand Oaks, California: SAGE Publications.

Longford, N. T. (2013). *Statistical decision theory*. Berlin, Heidelberg: Springer.

Luenberger, D. (1969). *Optimization by Vector Space Methods*. New York, New York: John Wiley & Sons, Inc.

Marcotte, D. (2012). Revisiting the Linear Model of Coregionalization. In H. R. Abrahamsen P. (Ed.), *Geostatistics Oslo 2012*. Dordrecht: Springer.

Mardia, K. K. (2000). *Multivariate Analysis*. London: Academic Press.

Montgomery, D. P. (2012). *Introduction to Linear Regression Anaylsis* (5th Edition ed.). John Wiley and Sons.

Olea, R. (1999). *Geostatistics for Engineers and Earth Scientists*. New York: Springer Science.

Rao, C. (1973). *Linear Statistical Inference and Its Applications*. New York: Wiley-Interscience.

Rockafellar, R. (1993). Lagrange Multipliers and Optimality. *SIAM Review, 35* (2), 183-238.

Shmaryan, L. J. (1999). Two Markov Models and Their Application. *Mathematical Geology, 31*, 965–988.

Wackernagel, H. (1994). Cokriging versus kriging in regionalized multivariate data analysis. *Geoderma, 62* (1-3).

Weisberg, S. (2005). *Applied Regression Analysis* (3rd Edition ed.). John Wiley and Sons.

Weiss, L. (1961). *Statistical decision theory*. New York: McGraw-Hill.

CHAPTER 4

Spatial Simulation

4.1 Introduction

The topic of estimation or prediction using the sparse information that might be available for a reservoir was the subject of Chap. 3. The methods, starting from linear regression to kriging to more sophisticated variants of kriging, all offer many merits. They are "best" in the sense of minimum residual or error variance, "unbiased" in the sense that they faithfully reproduce the expected value of the variable under study. Some of them also produce estimates such that the covariance between the estimated values and the available "hard" data reproduces the covariance model specified for the estimation. Kriging is also "exact" in that the kriging estimate exactly reproduces the data values. We also discussed approaches to integrate the information from many data sources within the kriging framework.

For all these merits of kriging, there are also some demerits. The one obvious demerit is that kriging (and indeed any regression technique) yields only one estimate for the variable being modeled. Assessment of uncertainty associated with the prediction or estimation is thus not possible. One may think that the combination of the kriged estimate and the estimation variance would suffice for quantifying the uncertainty associated with the estimation. But in the Gaussian context (when the linear estimator is optimal), the estimation variance is homoscedastic and is therefore not a good measure of local accuracy or uncertainty. One may then argue that kriging in a non-parametric framework (such as indicator kriging) that directly yields the estimated posterior probability would be sufficient for uncertainty assessment. It turns out that there are some other drawbacks of kriging that render it imperative so that we develop methodologies for uncertainty assessment that build on the good properties of the kriging-based estimation procedure. That is the subject of this chapter.

Note The simulation examples in this chapter are included as an Excel Workbook on this book's website, www.mhprofessional.com/PRMS. The Excel file should be opened with the macros enabled as

139

the calculations for sequential indicator simulation, including the Markov–Bayes option, are solved using macros.

4.2 Kriging—Limitations

Generalized regression (kriging) provides a minimum error variance estimate for the random function $Z(\mathbf{u})$. Assuming a linear estimator:

$$Z^*(u) = \sum_\alpha \lambda_\alpha Z(\mathbf{u}_\alpha)$$

The weights λ_α are determined such that the variance $E\{[Z(\mathbf{u}) - Z^*(\mathbf{u})]^2\}$ is minimized.

Although kriging provides viable means to interpolate and extrapolate data, it suffers from some drawbacks:

- Kriging is a minimum error variance estimator—there may be other "goodness" criteria, e.g., $E\{L[e]\}$ with $L[e] \neq e^2$ more conducive to spatial estimation

- Assessment of uncertainty is not possible since σ_E^2, the estimation variance is homoscedastic and cannot be treated as a measure of local accuracy. Indicator kriging does provide an assessment of local uncertainty, as it provides an avenue to directly estimate the local conditional probability distribution: $F^*(\mathbf{u};z_k \mid (n)) = \sum_\alpha \lambda_\alpha(z_k) \cdot I(\mathbf{u}_\alpha;z_k)$

However, kriging suffers from another major drawback:

- The kriging system:

$$\sum_\beta \lambda_\beta C_I(\mathbf{u}_\alpha - \mathbf{u}_\beta) = C_I(\mathbf{u}_\alpha - \mathbf{u}_o) \tag{4-1}$$

is a statement of the reproduction of covariance between the data and the unknown. The kriging system states that the weights are such that the weighted linear combination of the covariance between the data reproduces the covariance between the data and the unknown estimation node. *However, the kriging system does not ensure that the covariance between a pair of estimated values $Z^*(\mathbf{u})$, $Z^*(\mathbf{u}')$ will reproduce the model covariance at lag $\mathbf{h} = |\mathbf{u} - \mathbf{u}'|$.*

Indeed: $Cov\{Z^*(\mathbf{u}), Z(\mathbf{u}_\alpha)\} = E\{Z^*(\mathbf{u}) \cdot Z(\mathbf{u}_\alpha)\} - E\{Z^*(\mathbf{u})\} \cdot E\{Z(\mathbf{u}_\alpha)\}$

Under stationarity and substituting: $Z^*(\mathbf{u}) = \sum_\beta \lambda_\beta Z(\mathbf{u}_\beta)$, we get:

$$Cov\{Z^*(\mathbf{u}), Z(\mathbf{u}_\alpha)\} = \sum_\beta \lambda_\beta E\{Z(\mathbf{u}_\beta) \cdot Z(\mathbf{u}_\alpha)\} - m^2$$

$$= \sum_\beta \lambda_\beta C(\mathbf{u}_\alpha - \mathbf{u}_\beta)$$

As per the kriging system, the right hand side (RHS) is equal to $C(\mathbf{u}_\alpha - \mathbf{u})$, and so:

$$Cov\{Z^*(\mathbf{u}), Z(\mathbf{u}_\alpha)\} = C(\mathbf{u}_\alpha - \mathbf{u})$$

i.e., the covariance between the data and the estimation node is indeed equal to the model covariance. What about the covariance between any two estimation nodes?

- Covariance reproduction in kriging: Let us consider two estimated values $Z^*(\mathbf{u})$ and $Z^*(\mathbf{u}')$

$$
\begin{aligned}
Cov\{Z^*(\mathbf{u}), Z^*(\mathbf{u}')\} &= E\{Z^*(\mathbf{u}) \cdot Z^*(\mathbf{u}')\} - E\{Z^*(\mathbf{u})\} \cdot E\{Z^*(\mathbf{u}')\} \\
&= E\left\{ \sum_\beta \lambda_\beta(\mathbf{u}) Z(\mathbf{u}_\beta) \cdot \sum_\beta \lambda_\beta(\mathbf{u}') Z(\mathbf{u}_\alpha) \right\} - m^2 \\
&= \sum_\alpha \sum_\beta \lambda_\alpha(\mathbf{u}) \lambda_\beta(\mathbf{u}') E\{Z(\mathbf{u}_\beta) \cdot Z(\mathbf{u}_\alpha)\} - m^2 \\
&= \sum_\alpha \sum_\beta \lambda_\alpha(\mathbf{u}) \lambda_\beta(\mathbf{u}') Cov(\mathbf{u}_\beta - \mathbf{u}_\alpha) \\
&= \sum_\alpha \lambda_\alpha(\mathbf{u}) Cov(\mathbf{u}' - \mathbf{u}_\alpha) \\
&\neq Cov(\mathbf{u} - \mathbf{u}')
\end{aligned}
$$

That is, the covariance between two estimates does not identify the covariance model. This implies that in the scenario that we have sparse data using which we need to estimate the value of the attribute at several locations, the pattern of spatial variability emerging by kriging will not honor the prior model for spatial variability. In most modeling cases, there will be an abundance of unsampled nodes and this deficiency of kriging will imply that the covariance structure of the kriged map will not reproduce the input model for covariance. This in turn implies that the kriging-derived maps may not be geologically realistic.

- Variance of kriged estimates: What about the variability represented by the kriged maps? For simplicity, let us assume that the random function (RF) under consideration has a mean equal to zero. Then:

$$
\begin{aligned}
Var\{Z^*(\mathbf{u})\} &= E\{[Z^*(\mathbf{u})]^2\} \\
&= E\left\{ \sum_\beta \lambda_\beta(\mathbf{u}) Z(\mathbf{u}_\beta) \cdot \sum_\alpha \lambda_\alpha(\mathbf{u}) Z(\mathbf{u}) \right\} \\
&= \sum_\alpha \sum_\beta \lambda_\alpha(\mathbf{u}) \lambda_\beta(\mathbf{u}) E\{Z(\mathbf{u}_\beta) \cdot Z(\mathbf{u}_\alpha)\} \\
&= \sum_\alpha \sum_\beta \lambda_\alpha(\mathbf{u}) \lambda_\beta(\mathbf{u}) \rho(\mathbf{u}_\beta - \mathbf{u}_\alpha) \\
&= \sum_\alpha \lambda_\alpha(\mathbf{u}) \rho(\mathbf{u} - \mathbf{u}_\alpha)
\end{aligned}
$$

Recall that the minimum error (estimation) variance in kriging is:

$$\sigma_K^2 = \rho(0) - \sum_\alpha \lambda_\alpha(\mathbf{u})\rho(\mathbf{u} - \mathbf{u}_\alpha)$$

Assuming $Z(\mathbf{u})$ to have unit variance:

$$\sigma_K^2 = 1 - \sum_\alpha \lambda_\alpha(\mathbf{u})\rho(\mathbf{u} - \mathbf{u}_\alpha)$$

Substituting in the expression for the variance of the kriged estimates, we get:

$$Var\{Z^*(\mathbf{u})\} = 1 - \sigma_K^2 \qquad (4\text{-}2)$$

This implies that even though our target RF $Z(\mathbf{u})$ has unit variance, the kriged values $Z^*(\mathbf{u})$ do not reproduce that variance—in fact, the variance of the kriged map is less than the target variance by precisely the estimation variance σ_E^2. In other words, the *kriged maps are smooth and reflect less variability,* a feature seen in all our results in the earlier chapter.

4.3   Stochastic Simulation

To summarize, kriging suffers from two main drawbacks:

1. A kriged map is very smooth.
2. The texture of the kriged map is not correct (covariance structure is not reproduced).

How do we address these shortcomings of kriging while still retaining some of the good properties of kriging, such as data reproduction, unbiasedness of estimates, minimum residual variance, etc.? One strategy would be to add a residual $R(\mathbf{u})$ to the kriged estimate $Z^*(\mathbf{u})$ (Journel, 1978):

$$Z_S(\mathbf{u}) = Z^*(\mathbf{u}) + R(\mathbf{u}) \qquad (4\text{-}3)$$

In order to address the drawback of kriging $Z^*(\mathbf{u})$ that it is deterministic, we can make the residue $R(\mathbf{u})$ stochastic; in other words, the residue can assume different outcomes and consequently, $Z_S(\mathbf{u})$ are simulated values that can assume different outcomes.

Because $Z^*(\mathbf{u})$ is an unbiased estimate of $Z(\mathbf{u})$, in order for $Z_S(\mathbf{u})$ to preserve this unbiasedness property:

$$E\{Z_S(\mathbf{u})\} = E\{Z(\mathbf{u})\} = m$$

$$E\{Z^*(\mathbf{u}) + R(\mathbf{u})\} = m + E\{R(\mathbf{u})\} = m$$

This implies that:

$$E\{R(\mathbf{u})\} = 0 \qquad (4\text{-}4)$$

We also require that the simulated values reproduce the prior variability of $Z(\mathbf{u})$. This implies that we require $Var\{Z_S(\mathbf{u})\} = 1$. This, in turn, implies that we require $Var\{Z^*(\mathbf{u}) + R(\mathbf{u})\} = 1$. If we assume that $Var\{R(\mathbf{u})\} = \sigma_E^2$ and $Cov\{Z^*(\mathbf{u}), R(\mathbf{u})\} = 0$

Then: $Var\{Z^*(\mathbf{u}) + R(\mathbf{u})\} = 1 - \sigma_E^2 + 0 + \sigma_E^2 = 1$

Our conclusion is therefore that the simulated RF $Z_S(\mathbf{u}) = Z^*(\mathbf{u}) + R(\mathbf{u})$ will have the correct variance provided: $Var\{R(\mathbf{u})\} = \sigma_E^2$

and: $$Cov\{Z^*(\mathbf{u}), R(\mathbf{u})\} = 0$$

So the question is, under what condition is the residual orthogonal to the kriged estimate $Z^*(\mathbf{u})$?

Let us assume:

$$R(\mathbf{u}) = Z(\mathbf{u}) - Z^*(\mathbf{u}) \qquad (4\text{-}5)$$

That is, we equate the residue to the error of estimation. If we assume the RF $Z(\mathbf{u})$ to have zero mean, the SK estimate can then be written as:

$$Z^*(\mathbf{u}) = \sum_\alpha \lambda_\alpha Z(\mathbf{u}_\alpha)$$

with the kriging system:

$$\sum_\beta \lambda_\beta \rho(\mathbf{u}_\alpha - \mathbf{u}_\beta) = \rho(\mathbf{u}_\alpha - \mathbf{u}_o)$$

Let us look at the covariance $Cov\{Z^*(\mathbf{u}), R(\mathbf{u}')\}$ where $R(\mathbf{u}') = Z(\mathbf{u}') - Z^*(\mathbf{u}')$:

$$
\begin{aligned}
Cov\{Z^*(\mathbf{u}), R(\mathbf{u}')\} &= E\left\{\left[\sum_\alpha \lambda_\alpha(\mathbf{u})Z(\mathbf{u}_\alpha)\right] \cdot \left[Z(\mathbf{u}') - \sum_\beta \lambda_\beta(\mathbf{u}')Z(\mathbf{u}_\alpha)\right]\right\} \\
&= \sum_\alpha \lambda_\alpha(\mathbf{u})\rho(\mathbf{u}_\alpha - \mathbf{u}') - \sum_\beta\sum_\alpha \lambda_\alpha(\mathbf{u})\lambda_\beta(\mathbf{u}')\rho(\mathbf{u}_\alpha - \mathbf{u}_\beta) \\
&= \sum_\alpha \lambda_\alpha(\mathbf{u})\rho(\mathbf{u}_\alpha - \mathbf{u}') - \sum_\alpha \lambda_\alpha(\mathbf{u})\rho(\mathbf{u}_\alpha - \mathbf{u}') \\
&= 0
\end{aligned}
$$

We can therefore conclude that if we equate the residue to the estimation error, then it is orthogonal to the kriging estimate. The aforementioned development suggests the following recipe for rectifying the shortcomings of kriging:

- Perform kriging at a node.
- Obtain the kriging estimate and variance at the node.
- Assuming a mean zero and variance equal to the error variance, sample a value for the error and equate it to the residue $R(\mathbf{u})$.
- Construct the simulated value as $Z_S(\mathbf{u}) = Z^*(\mathbf{u}) + R(\mathbf{u})$.

The algorithm suggested above preserves the data exactitude property of kriging. At the data locations, the estimation variance will be zero and so the added residue will be zero. It restores the additional variance σ_E^2 to the kriged estimates and it preserves the unbiasedness property of kriging. However, the problem of lack of reproduction of the model covariance between estimated/simulated nodes remains un-addressed. We will discuss some strategies to overcome this last shortcoming of kriging.

4.3.1 Lower-Upper (LU) Simulation

Let us consider the RF $Z(\mathbf{u})$ with a target covariance function $C(\mathbf{h})$. Let us assume that the RF is defined over locations $\mathbf{u}_1, \mathbf{u}_2, \ldots, \mathbf{u}_N$. The corresponding covariance matrix is:

$$C = \begin{bmatrix} C(\mathbf{u}_1 - \mathbf{u}_1) & C(\mathbf{u}_1 - \mathbf{u}_2) & \ldots & C(\mathbf{u}_1 - \mathbf{u}_N) \\ C(\mathbf{u}_2 - \mathbf{u}_1) & C(\mathbf{u}_2 - \mathbf{u}_2) & \ldots & C(\mathbf{u}_2 - \mathbf{u}_N) \\ \vdots & \ldots & \ddots & \vdots \\ C(\mathbf{u}_N - \mathbf{u}_1) & \ldots & \ldots & C(\mathbf{u}_N - \mathbf{u}_N) \end{bmatrix}$$

Since C is positive definite, its decomposition into upper and lower triangle matrices is possible: $C = C_L \cdot C_U \cdot C_L$ is a lower triangle matrix such that:

$$[C_L]^T = C_U$$

Now let us consider a vector of uncorrelated standard normal random variables (RVs): $Y = [Y_1, Y_2, \ldots, Y_N]^T$
Let us then apply the following transformation:

$$W = C_L \cdot Y \tag{4-6}$$

The covariance of W is:

$$Cov\{W\} = E\{W \cdot W^T\} = E\{(C_L \cdot Y) \cdot (C_L \cdot Y)^T\}$$
$$= E\{C_L \cdot Y \cdot Y^T \cdot C_L^T\}$$

Because Y is standard normal $E\{Y \cdot Y^T\} = Var\{Y\} = I$

Therefore: $\qquad Cov\{W\} = CL \cdot I \cdot C_U = C$

That is, the transformation does result in $W(\mathbf{u})$ reflecting the spatial structure of the target RF $Z(\mathbf{u})$. Because multiple outcomes of Y are possible corresponding to different random draws, multiple outcomes $Z^S(\mathbf{u})$ can be obtained as (Alabert, 1987):

$$W(\mathbf{u}) = CL \cdot Y(\mathbf{u}) = Z^S(\mathbf{u}) \tag{4-7}$$

Remarks
- If Y is standard normal, $Z^S(\mathbf{u})$ are also standard normal. In practice (Alabert, 1987), Y can be sampled from any other

distribution. However, the transformation $W = CL \cdot Y$ involves product and sums of many independent RVs, and consequently, the resultant distribution will tend toward normality.

- Covariance reproduction is in an ergodic sense, over an ensemble of realizations.

- If simulation domain is large, covariance matrix C will be large and storage of LU decomposition and storage of C_L, etc. may be expensive.

- The simulation, as described above, is unconditional. It does not take into consideration the influence of the data.

In order to relate this LU simulation algorithm to our previous discussion about kriging and the influence of data, we need to talk about *conditional LU simulation*.

Let $\mathbf{u}_1, \mathbf{u}_2, \dots, \mathbf{u}_N$ be unsampled nodes on a grid (where simulations have to be performed) and $\mathbf{u}_{o_1}, \mathbf{u}_{o_2}, \dots, \mathbf{u}_{o_m}$ be locations with data. We now carefully order the $(n + m) \times (n + m)$ covariance matrix made up of covariance between all locations (both nodes with conditioning data and simulation nodes).

$$C = \begin{bmatrix} C_{11} & C_{12} \\ C_{21} & C_{22} \end{bmatrix} \tag{4-8}$$

$C_{11} = C(\mathbf{u}_{o_i} - \mathbf{u}_{o_j}), i, j = 1, \dots, m$ (data-data covariances)

$C_{12} = C(\mathbf{u}_i - \mathbf{u}_{o_j}), i = 1, \dots, n; j = 1, \dots, m$ (data-unknown covariances)

$C_{21} = C_{12}^T$

$C_{22} = C(\mathbf{u}_i - \mathbf{u}_j), i, j = 1, \dots, n$ (unknown-unknown covariances)

Again we perform LU decomposition of the matrix C:

$$C_U = \begin{bmatrix} U_{11} & U_{12} \\ 0 & U_{22} \end{bmatrix} \quad \text{and} \quad C_L = \begin{bmatrix} L_{11} & 0 \\ L_{21} & L_{22} \end{bmatrix}$$

Let us in this case use a vector: $Y = \begin{bmatrix} Y_1 \\ Y_2 \end{bmatrix}$ of dimension $(n + m) \times 1$, where both Y_1 and Y_2 are random draws from the univariate standard normal distribution. Then, the transformation:

$$C_L \cdot Y = \begin{bmatrix} L_{11} & 0 \\ L_{21} & L_{22} \end{bmatrix} \cdot \begin{bmatrix} Y_1 \\ Y_2 \end{bmatrix} = \begin{bmatrix} L_{11} Y_1 \\ L_{21} \cdot Y_1 + L_{22} \cdot Y_2 \end{bmatrix}$$

Equating this to the desired vector of simulated values $Z^S(\mathbf{u})$:

$$\begin{bmatrix} L_{11} Y_1 \\ L_{21} \cdot Y_1 + L_{22} \cdot Y_2 \end{bmatrix} = \begin{bmatrix} Z_1^s \\ Z_2^s \end{bmatrix} \tag{4-9}$$

where $Z_1^s = [Z(\mathbf{u}_{o_1}), Z(\mathbf{u}_{o_2}), \ldots, Z(\mathbf{u}_{o_m})]^T$ is the vector of data values.

Now, equating the elements on the right-hand and left-hand side of Eq. (4-9), we get:

$$L_{11}Y_1 = Z_1^S \qquad \text{and that implies} \qquad Y_1 = L_{11}^{-1} \cdot Z_1^S$$

i.e., the standard normal variates at data locations are fixed equal to a particular set of values $Y_1 = L_{11}^{-1} \cdot Z_1^S$. Then the simulated values at the uninformed nodes are given as:

$$
\begin{aligned}
Z_2^S &= L_{21} \cdot Y_1 + L_{22} \cdot Y_2 \\
&= L_{21} \cdot (L_{11}^{-1} \cdot Z_1^S) + L_{22} \cdot Y_2
\end{aligned}
\tag{4-10}
$$

This yields the simulated values at the unsampled locations.

Remarks

- The simulated values Z_2^S can be computed knowing the covariance matrix $C(\mathbf{h})$ and the random samples from the standard normal distribution Y_2.

- Dual kriging: The kriging estimate can be written in matrix form as $R^* = R_\alpha^T \cdot \lambda$, where R^* is the residue $(Z^* - m)$ at the estimation node and R_α is the residue at the data location. The weights are obtained as solution to the system $k = K \cdot \lambda$ or $\lambda = K^{-1} \cdot k$. The kriging estimate can thus be written as $R^* = R_\alpha^T \cdot (K^{-1} \cdot k)$. If we designate $R_\alpha^T \cdot K^{-1} = d^T$, then $R^* = d^T \cdot k$. This is a new interpretation of kriging—*dual kriging* where the estimate is now expressed as a linear combination of the basis (covariance) function. The accompanying system for obtaining the dual kriging weights d is now a statement of the data exactitude property of kriging, i.e., $R_\alpha^T = d^T \cdot K$; i.e., the dual kriging weights are such that the residue at the data locations is exactly reproduced.

  The expression $L_{21} \cdot (L_{11}^{-1} \cdot Z_1)$ in the LU simulation algorithm described above can be re-interpreted in the context of dual kriging. The weights in dual kriging are given as $d^T = R_\alpha^T \cdot K^{-1}$ or $d = K^{-1^T} \cdot R_\alpha = K^{-1} \cdot R_\alpha$. The expression $(L_{11}^{-1} \cdot Z_1)$ bears resemblance to the expression for d by recognizing that L_{11} is the lower triangular matrix corresponding to the data-to-data covariance matrix C_{11} or K in the notation for dual kriging. The dual kriging estimate is $R^* = d^T \cdot k$ or $R^{*^T} = k^T \cdot d$. This last expression is similar to $L_{21} \cdot (L_{11}^{-1} \cdot Z_1)$ where L_{21} is the lower triangular matrix corresponding to the data-to-unknown covariance (or k in the notation of dual kriging). Putting all these together, we can say that the simulated values Z_2 are equivalent to the kriged values corrected by a residual $R = L_{22} \cdot Y_2$.

- In the correction term $R = L_{22} \cdot Y_2$, the lower triangular matrix L_{22} corresponds to the covariance matrix between the simulation nodes themselves. Consider a small 2×2 case:

$$\begin{bmatrix} a_{11} & 0 \\ a_{21} & a_{22} \end{bmatrix} \cdot \begin{bmatrix} y_1 \\ y_2 \end{bmatrix}$$

For the simulated value at the first node, the contribution of the residual is $R_1 = a_{11} \cdot y_1$. Subsequently for the simulated value at the second node, the contribution of the residual is

$R_2 = a_{21} \cdot y_1 + a_{22} \cdot y_2 = a_{21} \cdot \dfrac{R_1}{a_{11}} + a_{22} \cdot y_2$; i.e., the simulation

at the second node utilizes the residual (or simulated value at the previous node). This suggests that in order to impart the correct spatial structure to the simulation results, a potential strategy is to utilize the simulated value at a previous location as conditioning data for the simulation at subsequent nodes. This idea is explored further in the ensuing discussion.

Example 4-1

We return to the same example described in Chap. 3, which was developed with the data configuration shown in Fig. 4-1.

Assume a covariance model with a single spherical structure of range 8 (as in the simple and ordinary kriging examples, discussed in

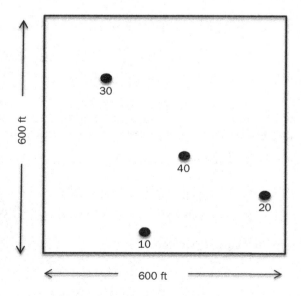

FIGURE 4-1   Data configuration assumed for the LU simulation example.

Chap. 3), the following covariance matrix between the nodes of the simulation grid is obtained:

$$C = \begin{bmatrix}
1.00 & 0.49 & 0.62 & 0.36 & 0.66 & 0.82 & 0.82 & \cdots\cdots & 0.25 & 0.22 & 0.17 \\
0.49 & 1.00 & 0.62 & 0.26 & 0.25 & 0.36 & 0.62 & \cdots\cdots & 0.32 & 0.36 & 0.38 \\
0.62 & 0.62 & 1.00 & 0.53 & 0.43 & 0.53 & 0.66 & \cdots\cdots & 0.51 & 0.49 & 0.43 \\
0.36 & 0.26 & 0.53 & 1.00 & 0.36 & 0.38 & 0.32 & \cdots\cdots & 0.62 & 0.49 & 0.36 \\
0.66 & 0.25 & 0.43 & 0.36 & 1.00 & 0.82 & 0.51 & \cdots\cdots & 0.17 & 0.12 & 0.06 \\
0.82 & 0.36 & 0.53 & 0.38 & 0.82 & 1.00 & 0.66 & \cdots\cdots & 0.22 & 0.17 & 0.12 \\
\vdots & \vdots & \vdots & \vdots & \vdots & \vdots & \vdots & \ddots & \vdots & \vdots & \vdots \\
0.25 & 0.32 & 0.51 & 0.62 & 0.17 & 0.22 & 0.26 & \cdots\cdots & 1.00 & 0.82 & 0.66 \\
0.22 & 0.36 & 0.49 & 0.49 & 0.12 & 0.17 & 0.25 & \cdots\cdots & 0.82 & 1.00 & 0.86 \\
0.17 & 0.38 & 0.43 & 0.36 & 0.06 & 0.12 & 0.22 & \cdots\cdots & 0.66 & 0.82 & 1.00
\end{bmatrix}$$

The upper-left quadrant of this matrix represents the data-data covariance matrix (C_{11}); the upper-right and the bottom-left quadrants represent the data-unknown covariance and its transpose (C_{12} and C_{21}), respectively; and the bottom-right quadrant is the unknown-unknown covariance matrix (C_{22}). Cholesky decomposition of the matrix C results in the following lower triangular matrix:

$$L = \begin{bmatrix}
1 & 0 & 0 & 0 & 0 & 0 & & 0 & 0 & 0 & 0 \\
0.49 & 0.872 & 0 & 0 & 0 & 0 & & 0 & 0 & 0 & 0 \\
0.62 & 0.363 & 0.696 & 0 & 0 & 0 & & 0 & 0 & 0 & 0 \\
0.36 & 0.096 & 0.391 & 0.842 & 0 & 0 & & 0 & 0 & 0 & 0 \\
0.656 & -0.086 & 0.0714 & 0.123 & 0.736 & 0 & & 0 & 0 & 0 & 0 \\
0.820 & -0.0484 & 0.059 & 0.073 & 0.360 & 0.432 & 0 & & 0 & 0 & 0 \\
0.820 & 0.250 & 0.082 & -0.04 & -0.014 & -0.004 & 0.506 & 0 & 0 & 0 & 0 \\
\vdots & \vdots & \vdots & \vdots & \vdots & \vdots & \vdots & \vdots & \vdots & \vdots & \vdots \\
0.247 & 0.226 & 0.392 & 0.423 & -0.069 & -0.009 & \cdots & \cdots & 0.101 & 0 & 0 \\
0.217 & 0.291 & 0.353 & 0.287 & -0.079 & \cdots\cdots & & \cdots & 0.218 & 0.454 & 0 \\
0.172 & 0.333 & 0.284 & 0.184 & -0.085 & & & 0.057 & 0.110 & 0.232 & 0.474
\end{bmatrix}$$

Let us now compute the terms that we require for LU simulation. First, computing the inverse of the top-left quadrant, we get:

$$L_{11}^{-1} = \begin{bmatrix}
1.00 & 0.00 & 0.00 & 0.00 \\
-0.56 & 1.15 & 0.00 & 0.00 \\
-0.60 & -0.60 & 1.44 & 0.00 \\
-0.09 & 0.15 & -0.67 & 1.19
\end{bmatrix}$$

Directly from the matrix L, we get:

$$L_{22} = \begin{bmatrix} 0.736 & 0 & \cdots & 0 & 0 & 0 \\ 0.360 & 0.432 & 0 & \cdots & 0 & 0 \\ -0.014 & -0.004 & 0.506 & 0 & 0 & 0 \\ \vdots & \vdots & \vdots & \vdots & \vdots & \vdots \\ -0.069 & -0.009 & \cdots & \cdots & 0.101 & 0 \\ -0.079 & \cdots\cdots & \cdots\cdots\cdots & 0.218 & 0.454 & 0 \\ -0.085 & & \cdots\cdots 0.057 & 0.110 & 0.232 & 0.474 \end{bmatrix}$$

and:

$$L_{21} = \begin{bmatrix} 0.656 & -0.086 & 0.0714 & 0.123 \\ 0.820 & -0.0484 & 0.059 & 0.073 \\ 0.820 & 0.250 & 0.082 & -0.04 \\ \vdots & \vdots & \vdots & \vdots \\ 0.247 & 0.226 & 0.392 & 0.423 \\ 0.217 & 0.291 & 0.353 & 0.287 \\ 0.172 & 0.333 & 0.284 & 0.184 \end{bmatrix}$$

Using these matrices, the quantity $L_{21} \cdot (L_{11}^{-1} \cdot Z_1)$ can be calculated. First, we standardize the data and compute:

$$\left(L_{11}^{-1} \cdot Z_1\right) = \begin{bmatrix} -1.16 \\ 0.21 \\ 2.60 \\ -0.27 \end{bmatrix}$$

The subsequent multiplication by L_{21} yields a vector of length 32×1:

$$L_{21} \cdot \left(L_{11}^{-1} \cdot Z_1\right) = \begin{bmatrix} -0.63 \\ -0.83 \\ -0.68 \\ \vdots \\ \vdots \\ 0.66 \\ 0.65 \\ 0.56 \end{bmatrix}$$

The next step is to generate a vector of random quantiles retrieved from the standard normal distribution: one value at each simulation node and multiply this vector by the matrix L_{22} in order to get:

$$L_{22} \cdot Y_2 = \begin{bmatrix} -0.56 \\ -0.31 \\ 0.15 \\ \vdots \\ \vdots \\ \vdots \\ 0.30 \\ 0.18 \\ -0.0008 \\ -0.42 \end{bmatrix}$$

Finally, the simulated values $Z^S(\mathbf{u})$ are obtained as:

$$Z^s(\mathbf{u}) = L_{21} \cdot \left(L_{11}^{-1} \cdot Z_1^s \right) + L_{22} \cdot Y_2 = \begin{bmatrix} 9.65 \\ 10.33 \\ 18.23 \\ \vdots \\ \vdots \\ 35.88 \\ 33.33 \\ 26.76 \end{bmatrix}$$

Multiple realizations of simulated values are possible by drawing different sets of random quantiles from the standard normal distribution. The map of simulated values corresponding to two such realizations is shown in Fig. 4-2. The maps exhibit some similar characteristics—lower depths in the bottom-left regions, higher depths toward the top of the map. However, they also exhibit variability especially at locations far away from the conditioning data locations. Most importantly, the simulated realizations exhibit more variability in depth than the corresponding kriged results. In fact, if the process is repeated 50 times to yield 50 realizations of the depth, the average variance calculated over the suite is 110 (in units of depth^2), while the variance of depths obtained by simple kriging is 50.77. The variance of the conditioning data is 164. The simulated map exhibits more variability, but the variance is not exactly the variance exhibited by the conditioning data and that is because of the *ergodic* fluctuations.

Remark   A stationary RF $Z(\mathbf{u}, \omega)$ is said to be ergodic if (Yaglom, 1962):

$$\lim_{V \to \infty} \frac{1}{|V|} \int_V Z(\mathbf{u}, \omega) d\mathbf{u} = m \tag{4-11}$$

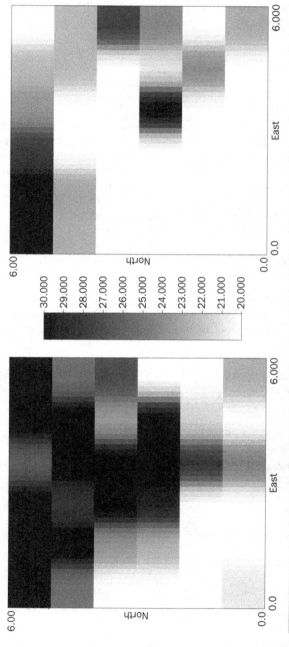

FIGURE 4-2 Two realizations of depths obtained by the LU simulation method. These realizations correspond to different random draws from the standard normal distribution. (The color version of this figure is available at www.mhprofessional.com/PRMS.)

The above ergodic property suggests that when developing a spatial model for a stationary ergodic random function, it is possible to compute the mean from a single realization (Chiles, 1999). In practice, the mean from a particular realization of the RF may be $m(\omega)$. The ergodic theorem states that if $Z(\mathbf{u}, \omega)$ is a stationary random function in space $\mathbf{u}$, and depending on the realization ω, then:

$$\lim_{V \to \infty} \frac{1}{|V|} \int_V Z(\mathbf{u}, \omega) d\mathbf{u} = m(\omega) \tag{4-12}$$

In other words, if the size of the domain is large compared to the range of the covariance or variogram, the ergodic theorem states that the realization will converge to a stationary mean $m(\omega)$. The random variable $m(\omega)$ has a mean equal to m. The implication of all these is that if the size of the reservoir domain is large compared to the variogram or covariance range and a large number of realizations of the RF $Z(\mathbf{u})$ are drawn, then we approach the stationary mean m. It has also been pointed out (Chiles, 1999) that non-ergodicity usually arises when the modeled phenomenon comprises a number of ergodic stationary functions that have been arbitrarily put together.

4.3.2   Sequential Simulation

Our objective in simulating a reservoir model that exhibits the prescribed variability (through the specified covariance model) is to sample outcomes $\{z_1^s, z_2^s, \ldots, z_N^s\}$ conditioned to the available data $\{Z(\mathbf{u}_\alpha), \alpha = 1, \ldots, n\}$. As discussed earlier, in order to reflect the uncertainty associated with our model because of the availability of sparse data, we are interested not in one set of outcomes but several outcomes: $\{z_1^{s^\ell}, z_2^{s^\ell}, \ldots, z_N^{s^\ell}\}$, $\ell = 1, L$. As we want each of these outcomes to reflect the prescribed spatial variability, our objective is to sample the outcomes from the joint probability distribution $F(\mathbf{u}_1, \mathbf{u}_2, \ldots, \mathbf{u}_N \mid (n))$ (Goovaerts, 1997).

Let us first consider the joint simulation of outcomes at pairs of locations, $\mathbf{u'}_1$ and $\mathbf{u}_2$. This entails sampling outcomes from the bivariate *cdf*:

$$F(\mathbf{u}_1, \mathbf{u}_2; z_1^\ell, z_2^\ell \mid (n)) = \text{Prob}\left\{Z(\mathbf{u}_1) \leq z_1^\ell, Z(\mathbf{u}_2) \leq z_2^\ell\right\}$$

By the definition of conditional probability: $P(A, B) = P(A \mid B) \cdot P(B)$

$$F(\mathbf{u}_1, \mathbf{u}_2; z_1^\ell, z_2^\ell \mid (n)) = F(\mathbf{u}_1; z_1^\ell \mid (n), z_2^\ell) \cdot F(\mathbf{u}_2; z_2^\ell \mid (n))$$

That is, sampling from target bivariate distribution is the same as:

- sampling from a univariate *cdf* at location $\mathbf{u}_2$ conditioned to (n) data;
- adding the outcome Z_2^ℓ to the dataset and constructing *cdf* at $\mathbf{u}_1$ conditioned to these $(n + 1)$ data; and
- sampling from that *cdf* to get outcome at $\mathbf{u}_1$ Z_1^ℓ.

The same procedure implemented with sampling at $\mathbf{u}_1$ followed by $\mathbf{u}_2$ will yield a pair Z'^ℓ_1, Z'^ℓ_2, that will also accurately represent the bivariate *cdf* albeit with a pair of outcomes different from Z^ℓ_1, Z^ℓ_2.

In summary, sampling from the correct bivariate distribution is possible by sequentially sampling univariate *cdfs* progressively conditioned to original and the simulated data. The outcomes will be dependent on the random path selected to visit the simulation nodes.

This sequential sampling approach to sample outcomes from a desired bivariate distribution can now be generalized to sample outcomes from a desired N-point *cdf*. The sequential sampling process can now be written as:

$$
\begin{aligned}
F(\mathbf{u}_1, \mathbf{u}_2, ..., \mathbf{u}_N; z^\ell_1, z^\ell_2, ..., z^\ell_N \mid (n)) &= F(\mathbf{u}_N; z^\ell_N \mid (n), z^\ell_{N-1}, ..., z^\ell_1) \\
&\times F(\mathbf{u}_1, \mathbf{u}_2, ..., \mathbf{u}_{N-1}; z^\ell_1, z^\ell_2, ..., z^\ell_{N-1} \mid (n)) \\
&= F(\mathbf{u}_N; z^\ell_N \mid (n), z^\ell_{N-1}, ..., z^\ell_1) \times F(\mathbf{u}_{N-1}; z^\ell_{N-1} \mid (n), z^\ell_{N-2}, ..., z^\ell_1) \\
&\times F(\mathbf{u}_1, \mathbf{u}_2, ..., \mathbf{u}_{N-2}; z^\ell_1, z^\ell_2, ..., z^\ell_{N-2} \mid (n)) \\
&= F(\mathbf{u}_N; z^\ell_N \mid (n + N - 1)) \times F(\mathbf{u}_{N-1}; z^\ell_{N-1} \mid (n + N - 2))... \\
&\times F(\mathbf{u}_1; z^\ell_1 \mid (n))
\end{aligned}
\tag{4-13}
$$

$F(\mathbf{u}_N; z^\ell_N \mid (n + N - 1))$ implies a conditional distribution conditioned to the (n) original data and the $(N-1)$ previously simulated values.

The decomposition above implies the following N steps:

- Model the *cdf* at the first location $\mathbf{u}_1$ conditioned to the n-original data $Z(\mathbf{u}_\alpha)$, $\alpha = 1, ..., n$.

- Draw an outcome Z^ℓ_1 from that *cdf*. Add that simulated value to the conditioning dataset for the next node.

- Model *cdf* at the *i*th node— $Z(\mathbf{u}_i)$ given the n-original data and all the $(i-1)$ previously simulated values.

- Draw an outcome Z^ℓ_1 from that *cdf*.

- Repeat the procedure by visiting the nodes in a random order until all the nodes are visited. This will yield one realization of the RF $Z(\mathbf{u})$.

A number of realization $\ell = 1, ..., L$ can be sampled by repeating the procedure following different random paths and by using different random numbers to sample from the conditional distributions.

The sequential simulation procedure requires a model for the conditional *cdfs* $F^*(\mathbf{u}; z_k \mid (n))$ at each location $\mathbf{u}$ being simulated using an updated set of conditioning data. Two avenues are available for modeling the requisite *cdfs*:

1. Parametric approach, e.g., the multi-Gaussian approach

2. Non-parametric or indicator approach

Before detailing the sequential simulation algorithm using these two modeling routes, let us formally look at the issue of reproduction of the specified covariance model by following the sequential simulation approach.

Covariance Reproduction in Sequential Simulation

Sequential simulation allows sampling realizations from the multi-variate distribution $F\{\mathbf{u}_1, \mathbf{u}_2, \ldots, \mathbf{u}_N; Z_1^\ell, Z_2^\ell, \ldots, Z_N^\ell \mid (n)\}$. Any realization sampled from the N-variate distribution will also reflect bivariate (covariance) characteristics accurately. However, in order to formally prove covariance reproduction, consider the following representation of the simulated value at any location:

$$Z_S(\mathbf{u}) = Z_k^*(\mathbf{u}) + R(\mathbf{u})$$

$Z_S(\mathbf{u})$ is obtained by randomly sampling from a distribution with mean $Z_k^*(\mathbf{u})$. The addition of the stochastic residue $R(\mathbf{u})$ mimics the effect of randomly sampling from the *ccdf* at location $\mathbf{u}$. The variability introduced due to this random sampling process is restricted by the kriging variance. As discussed previously:

$$Var\{R(\mathbf{u})\} = \sigma_K^2 \qquad E\{R(\mathbf{u})\} = 0$$

The covariance reproduction between simulated value $Z_S(\mathbf{u})$ and any data $Z(\mathbf{u}_\alpha)$ is unhampered by the addition of the residue:

$$Cov\{Z_S(\mathbf{u}), Z(\mathbf{u}_\alpha)\} = E\{Z_S(\mathbf{u}) \cdot Z(\mathbf{u}_\alpha)\} = E\{[Z_K^*(\mathbf{u}) + R(\mathbf{u})] \cdot Z(\mathbf{u}_\alpha)\}$$
$$= E\{Z_K^*(\mathbf{u}) \cdot Z(\mathbf{u}_\alpha)\} + E\{R(\mathbf{u}) \cdot Z(\mathbf{u}_\alpha)\} = C(\mathbf{u} - \mathbf{u}_\alpha)$$

$E\{R(\mathbf{u}) \cdot Z(\mathbf{u}_\alpha)\}$ is equal to zero since $R(\mathbf{u})$ is orthogonal to data, and the last equality comes about because kriging guarantees that the covariance between the data and the kriged node, $E\{Z_K^*(\mathbf{u}) \cdot Z(\mathbf{u}_\alpha)\}$, will reproduce the model covariance at lag $\mathbf{h}_{\alpha o} = \mathbf{u}_\alpha - \mathbf{u}_o$.

Now let us consider the sequential simulation at node $\mathbf{u}$ immediately followed by the simulation at $\mathbf{u}'$. According to the sequential simulation paradigm, the simulated value at node $\mathbf{u}$ is added to the conditioning dataset prior to performing simulation at node $\mathbf{u}'$. Thus:

$$Z_K^*(\mathbf{u}') = \sum_{\alpha=1}^{n} \lambda_\alpha Z(\mathbf{u}_\alpha) + \lambda_{n+1} Z_S(\mathbf{u})$$

Because kriging is guaranteed to reproduce covariance between data and unknown and $Z_S(\mathbf{u})$ is treated as an additional datum for estimation at $\mathbf{u}'$:

$$Cov\{Z_K^*(\mathbf{u}'), Z_S(\mathbf{u})\} = C(\mathbf{u}' - \mathbf{u})$$

Because $Z_S(\mathbf{u}') = Z_k^*(\mathbf{u}') + R(\mathbf{u}')$

$$Cov\{Z_S(\mathbf{u}'), Z_S(\mathbf{u})\} = E\{Z_k * (\mathbf{u}') \cdot Z_S(\mathbf{u})\}$$
$$+ E\{R(\mathbf{u}') \cdot Z_S(\mathbf{u})\} = C(\mathbf{u}' - \mathbf{u})$$

The residual $R(\mathbf{u}')$ is orthogonal to the data and $Z_S(\mathbf{u})$ are conditioning data for covariance calculated over the estimation at $\mathbf{u}'$ and that renders $E\{R(\mathbf{u}) \cdot Z^S(\mathbf{u})\} = 0$. Thus, the pair of simulated values at $\mathbf{u}$ and $\mathbf{u}'$ reproduces the model covariance.

Remarks

- As discussed previously, covariance reproduction is on the average over a large number of realizations—per theory of ergodic random functions.

- Covariance reproduction in any one realization is affected by the size of the simulated domain, variogram range (simulation over a large domain assuming a variogram model with small range will result in fairly accurate reproduction of the model covariance).

- The finite size of the simulation domain generally introduces artifacts at the boundaries. It is therefore a good strategy to simulate over a larger domain than what is required and then trim off the edges.

Sequential Gaussian Simulation

If the n-variate joint distribution $F(\mathbf{u}_1, \mathbf{u}_2; Z_1^\ell, Z_2^\ell \mid (n))$ is assumed to be multi-Gaussian, then all lower-order distributions $F^*(\mathbf{u}_i; z_k \mid (n + i - 1))$ conditioned to n-original data and $i-1$ previously simulated values are Gaussian (Mardia, 1979). Furthermore, the kriging estimator $Z^*(\mathbf{u}) = \sum_{\alpha=1}^{n+i-1} \lambda_\alpha Z(\mathbf{u}_\alpha)$ and the kriging variance

$$\sigma_{SK}^2 = C(0) - \sum_{\alpha=1}^{n} \cdot \lambda_\alpha C(\mathbf{h}_{\alpha 0})$$ satisfy all the requirements for the mean and variance of Gaussian conditional distributions:

- Mean expressed as a linear combination of data $Z(\mathbf{u}_\alpha), \alpha = 1, \ldots, (n + i - 1)$

- Mean corresponding to minimum error variance

- Estimation variance σ_K^2 being homoscedastic

Thus, in sequential Gaussian simulation, the kriged estimate $Z_{SK/OK}^*(\mathbf{u})$ and estimation variance $\sigma_{SK/OK}^2$ are treated as mean and variance of the local Gaussian *cdf* $F^*(\mathbf{u}_i; z_k \mid (n + i - 1))$. The simulated value $Z^S(\mathbf{u})$ is obtained by randomly sampling from the Gaussian *pdf*.

Implementation Notes

- All available data has to be normal score transformed prior to simulation, i.e., $Z(\mathbf{u}_\alpha), \alpha = 1, \ldots, n$ have to be transformed to $Y(\mathbf{u}_\alpha), \alpha = 1, \ldots, n$.

- The covariance has to be inferred and modeled using the transformed data $C_Y(\mathbf{h})$.

- Prior to implementation, verify that the multi-Gaussian hypothesis is valid.

  - Check univariate *cdf* and verify if it is Gaussian.

  - Check if bivariate distribution is Gaussian. In the case of bivariate Gaussian distributions, given the symmetric shape of the distribution, an expression of the indicator covariance in terms of the covariance of the Gaussian distribution can be derived (Deutsch, 1998):

$$C_I(\mathbf{h}; z) = \frac{1}{2\pi} \int_0^{C_Y(\mathbf{h})} \frac{1}{\sqrt{1-u^2}} e^{-\left(\frac{z^2}{1+u}\right)} du \qquad (4\text{-}14)$$

In order to use this expression to check if the bivariate distribution is Gaussian:

- First infer and model the covariance $C_Y(\mathbf{h})$ using the transformed data $Y(\mathbf{u}_\alpha)$, $\alpha = 1, \ldots, n$.
- Evaluate $C_I(\mathbf{h}; z)$ corresponding to a set of thresholds z using the above expression.
- For the same thresholds, evaluate the actual indicator covariance $\hat{C}_I(\mathbf{h}; z)$ using available data.
- If $\hat{C}_I(\mathbf{h}; z) \approx C_I(\mathbf{h}; z)$, phenomenon is at least bi-Gaussian
- The above procedure for checking whether a spatial distribution is bi-Gaussian is not unique. There are other checks that can be implemented including the calculation of quantile variograms and comparison of their characteristics to that for a Gaussian distribution.
  - An alternate procedure for checking bi-Gaussianity using the concept of quantile variograms is discussed in App. A.
- At the first node, perform kriging using the n conditioning data and obtain $Y^*(\mathbf{u}_o)$ and variance: $\sigma_K^2(\mathbf{u}_o)$.
- Using $Y^*(\mathbf{u}_o)$ and $\sigma_K^2(\mathbf{u})$ as the mean and variance of a Gaussian *ccdf*, draw a random value from that distribution.
- Add $Y^\ell(\mathbf{u}_o)$ to the conditioning dataset and proceed to the next node on a random path. Repeat kriging using $(n + 1)$ data.
- Visit all remaining $(N{-}1)$ nodes and obtain simulated values.
- Back-transform the simulated values to the original data space: $Y^\ell(\mathbf{u}) \rightarrow Z^\ell(\mathbf{u})$ using the original data histogram to accomplish the back transformation.
- Repeat the process L times to generate multiple realizations of the Gaussian RF $Z(\mathbf{u})$.
- Data integration is possible within this sequential Gaussian framework by replacing kriging for estimating the mean and variance of the local conditional distribution with co-kriging either using the linear model of coregionalization (LMC) or using one of the Markov models.

Example 4-2

The data in Fig. 4-1 is used to perform sequential simulation at all nodes on the 6×6 grid. The same variogram that was used for the LU simulation is used for the sequential simulation. The random order along which the nodes are visited is:

$$[20 \quad 18 \quad 4 \quad \dots \quad 5 \quad 15]$$

At the first node, the kriging system can be written as:

$$0.485 \qquad 0.318 \qquad 0.620 \qquad 0.820$$

$$\begin{bmatrix} 1.00 & 0.49 & 0.62 & 0.36 & 1 \\ 0.49 & 1.00 & 0.62 & 0.26 & 1 \\ 0.62 & 0.62 & 1.00 & 0.53 & 1 \\ 0.36 & 0.26 & 0.53 & 1.00 & 1 \\ 1 & 1 & 1 & 1 & 0 \end{bmatrix} \cdot \begin{bmatrix} \lambda_1 \\ \lambda_2 \\ \lambda_3 \\ \lambda_4 \\ \mu \end{bmatrix} = \begin{bmatrix} 0.75 \\ 0.66 \\ 0.82 \\ 0.43 \\ 1.00 \end{bmatrix}$$

The weights are calculated as:

$$\begin{bmatrix} \lambda_1 \\ \lambda_2 \\ \lambda_3 \\ \lambda_4 \\ \mu \end{bmatrix} = \begin{bmatrix} 0.149 \\ -0.041 \\ 0.206 \\ 0.686 \\ -0.018 \end{bmatrix}$$

This yields the kriged estimate $Z^*_{OK} = 0.35$ and $\sigma^2_{OK} = 0.27$. Assuming these to be the mean and variance of a Gaussian local conditional *pdf*, we get the simulated value $Y^S = 0.65$ in the standardized space and multiplying it by the standard deviation of the data and adding the mean of the data, we get $Z^S = 33.44$.

At the second simulation node ($i = 18$), the size of the kriging system increases by one:

$$\begin{bmatrix} 1.00 & 0.49 & 0.62 & 0.36 & 0.49 & 1.00 \\ 0.49 & 1.00 & 0.62 & 0.26 & 0.32 & 1.00 \\ 0.62 & 0.62 & 1.00 & 0.53 & 0.62 & 1.00 \\ 0.36 & 0.26 & 0.53 & 1.00 & 0.82 & 1.00 \\ 0.49 & 0.32 & 0.62 & 0.82 & 1.00 & 1.00 \\ 1.00 & 1.00 & 1.00 & 1.00 & 1.00 & 0.00 \end{bmatrix} \cdot \begin{bmatrix} \lambda_1 \\ \lambda_2 \\ \lambda_3 \\ \lambda_4 \\ \lambda_5 \\ \mu \end{bmatrix} = \begin{bmatrix} 0.426 \\ 0.820 \\ 0.656 \\ 0.318 \\ 0.360 \\ 1.000 \end{bmatrix}$$

This yields the weights:

$$\begin{bmatrix} \lambda_1 \\ \lambda_2 \\ \lambda_3 \\ \lambda_4 \\ \lambda_5 \\ \mu \end{bmatrix} = \begin{bmatrix} -0.053 \\ 0.724 \\ 0.267 \\ 0.082 \\ -0.019 \\ -0.058 \end{bmatrix}$$

The fifth weight is assigned to the previously simulated value $Y^S = 0.65$. The kriged estimate and the variance are 0.11 and 0.29, respectively. The sampled simulated value is $Y^S = -0.57$ at this node. The simulated value in the units of the data at this node is $Z^S = 17.59$. This procedure is repeated for all nodes along the random path. The simulated map of depths is shown in Fig. 4-3. As expected, the sequential simulation results bear resemblance to the models obtained by LUSIM. The variances of the two realizations are 88 and 46, respectively, which are different from the target of 166. Again, as explained, the reproduction of target statistics is in an ergodic sense, as an average over several realizations.

4.4 Non-Parametric Sequential Simulation

The previous discussion on sequential simulation is centered on kriging and specifically on corrections to kriging to ensure that the resultant models reproduce the target variance and covariance. Because it has already been pointed out that simple/ordinary (co-)kriging is an optimal estimate for a Gaussian RF, it is to be expected that sequential simulation with kriging as the estimation engine would result in Gaussian models. However, the sequential paradigm can be easily generalized to non-Gaussian reservoir models by recognizing that any arbitrary RF model can be retrieved as long as we have a technique for modeling local distributions conditioned to progressively updated data. One example for such an estimation technique would be indicator kriging that directly yields the local conditional distribution $(F(\mathbf{u}; z \mid (n))$ without resorting to any Gaussian hypothesis. The sequential simulation procedure using indicator kriging would consist of the following steps:

- Given (n) original data, the local conditioning distribution can be constructed as:

$$F(\mathbf{u}; z \mid (n)) = I^*(\mathbf{u} : z) - p(\mathbf{u}) = \sum_{(n)} \lambda_\alpha(\mathbf{u}; z) \cdot (I(\mathbf{u}_\alpha : z) - p(\mathbf{u}_\alpha)) \quad (4\text{-}15)$$

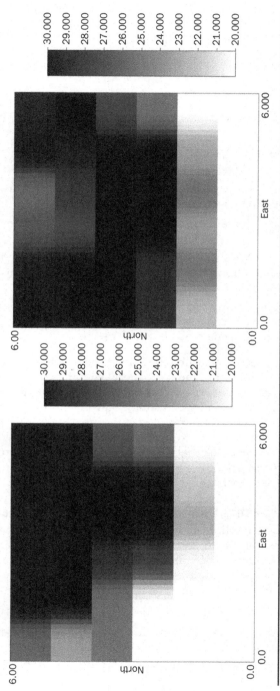

Figure 4-3 Two simulated realizations obtained by sequential Gaussian simulation. The conditioning data for the simulation is as given in Fig. 4-1. (The color version of this figure is available at www.mhprofessional.com/PRMS.)

or alternatively using the ordinary indicator kriging estimator. The simple indicator kriging implementation requires the prior probabilities $p(\mathbf{u})$ to be specified. The weights $\lambda_\alpha(\mathbf{u}; z)$ are obtained by solving the indicator kriging system:

$$\sum_\beta \lambda_\beta(z) \cdot C_I(\mathbf{h}_{\alpha\beta}; z) = C_I(\mathbf{h}_{\alpha o}; z) \qquad (4\text{-}16)$$

This requires a legitimate model for the indicator covariance $C_I(\mathbf{h}; z)$ specific to the threshold z.

- Repeat the indicator kriging procedure corresponding to the identified thresholds z.

- Perform order-relations corrections to render the local distribution legitimate.

- Perform Monte Carlo sampling in order to draw a simulated value.

- Add simulated value to the conditioning dataset and visit the next simulation node along a random path. Repeat indicator kriging at this node.

- A complete simulated map is obtained when all the simulation nodes have been visited. Multiple realizations of the map can be obtained using random number seeds.

A crucial aspect of the algorithm described above is the sampling of simulation outcomes from the kriged local conditional distribution. Indicator kriging is performed corresponding to a few thresholds, while the random number used to sample outcomes from the distribution could be any value of probability in the range [0,1]. This implies that a mechanism for inter/extrapolating the local *cdf* is needed that can allow for quantiles to be retrieved corresponding to any probability value different from the specified thresholds. Several options for interpolation/extrapolation exist, which are discussed below.

4.4.1   Interpolating within the Range of Thresholds Specified

A straightforward option is to perform linear interpolation between the two nearest thresholds specified. Another option is to interpolate according to a reference distribution specified. The reference distribution must be deemed analogous to the local conditional *cdf* estimated using indicator kriging.

Example
Let us say that the specified thresholds for indicator kriging and the corresponding *cdf* values are:

21	0.7
29	0.9

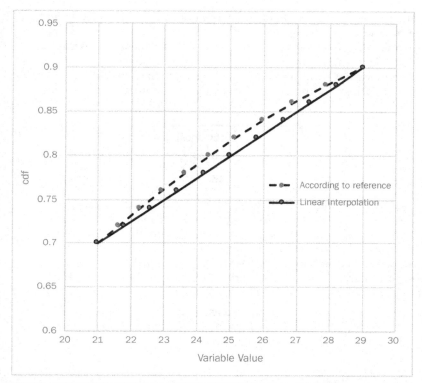

Figure 4-4   Linear interpolation between a pair of thresholds specified in indicator kriging compared to interpolation based on a reference distribution (Gaussian in this case). (The color version of this figure is available at www.mhprofessional.com/PRMS.)

Linear interpolation between the thresholds is shown in Fig. 4-4. If instead of linear interpolation, it is desired to interpolate using a reference distribution, say Gaussian, the quantiles corresponding to values of probability in the range 0.7–0.9 can be used to perform the interpolation of the target local conditional *cdf*. The corresponding interpolated values are also shown in Fig. 4-4. The interpolated *cdf* in this case exhibits a curved profile consistent with the curvature of the reference Gaussian distribution.

4.4.2 Tail Extrapolation

Several options are available for the tail extrapolation:

- Non-linear extrapolation: An expression of the form:

$$p = a \cdot z^{\omega} + b$$

can be calibrated. The power ω would have to be specified, as well as the maximum value of z (corresponding to $p = 1$). The last specified threshold and the corresponding probability value are also known. Using these, the constants a and b can be calculated. The calibrated equation can then be used for extrapolation. The characteristics of the tail corresponding to three different values of ω are shown in Fig. 4-5. These have been calculated corresponding to the following threshold and maximum values:

Cdf Value	Threshold
0.9	29
1.0	45

As can be seen, all the extrapolations done using the above non-linear expression approach the specified maximum value in different manner. When $\omega = 4$, values around the last specified threshold value are less likely, while the values around the specified maximum (45) are more likely. On the other hand, when $\omega = -4$, values around the last specified threshold value (29) are more likely.

- Extrapolation based on a reference distribution: Supposing for example the reference distribution exhibits the characteristics of a Weibull distribution given by:

$$f_X(x) = \begin{cases} e^{-(-x)^v} & x < 0 \\ 1 & x \geq 0 \end{cases} \text{ for } v > 0$$

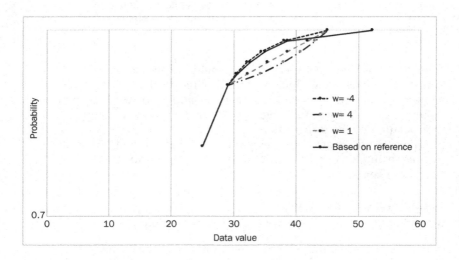

Figure 4-5 Various options for tail extrapolation. The extrapolation based on a reference distribution is based on a reference Weibull distribution. (The color version of this figure is available at www.mhprofessional.com/PRMS.)

Corresponding to a parameter value $v = 3$, the quantiles from the Weibull distribution are:

Cdf Values	Quantiles
0.9	−0.472
0.92	−0.437
0.94	−0.396
0.96	−0.344
0.98	−0.272
1	0.000

The quantiles from the target distribution can then be obtained by using these tabulated values. For example, the quantile corresponding to $p = 0.92$ can be obtained as: $z_{0.92} = \dfrac{(-0.437 - (-0.472))}{(0.92 - 0.9)} \cdot 0.92 +$ $29 = 30.61$. The extrapolated *cdf* obtained in this manner is also shown in Fig. 4-5. It can be seen that the tail extrapolated in this manner need not be bound at the previously specified maximum value.

There are other techniques for performing tail extrapolation, such as the extrapolation based on computation of the confidence limits associated with extrapolation of tail values (Scholz, 1995). These confidence limits are derived based on the resemblance to an extreme value distribution, such as the Weibull distribution.

Now we present an example for sequential indicator simulation.

Example 4-3

We demonstrate indicator simulation using the same conditioning data as used in Fig. 4-1. Four thresholds are specified in order to classify the depths. These thresholds are 15, 25, 35, and 45 in-depth units. The prior *cdf* values corresponding to these four thresholds are (0.20, 0.52, 0.70, and 0.87). The covariance model characterizing the spatial variability corresponding to the four thresholds is specified to be:

$$\rho_I(\mathbf{h}; 15) = 1 - \left[1.5 \cdot \frac{|\mathbf{h}|}{12.0} - 0.5 \cdot \left(\frac{|\mathbf{h}|}{12.0} \right)^3 \right]$$

$$\rho_I(\mathbf{h}; 25) = 1 - \left[1.5 \cdot \frac{|\mathbf{h}|}{8.0} - 0.5 \cdot \left(\frac{|\mathbf{h}|}{8.0} \right)^3 \right]$$

$$\rho_I(\mathbf{h}; 35) = 1 - \left[1.5 \cdot \frac{|\mathbf{h}|}{4.0} - 0.5 \cdot \left(\frac{|\mathbf{h}|}{4.0} \right)^3 \right]$$

$$\rho_{I}(\mathbf{h}; 45) = 1 - \left(1.5 \cdot \frac{|\mathbf{h}|}{4.0} - 0.5 \cdot \left(\frac{|\mathbf{h}|}{4.0}\right)^{3}\right)$$

The random order in which the simulation nodes are visited is:

[34   32   2   10   6   30   27]

Indicator kriging is performed corresponding to each threshold. The kriging system at the first simulation node corresponding to the threshold 15 is:

$$\begin{bmatrix} 1.000 & 0.653 & 0.411 & 0.505 \\ 0.653 & 1.000 & 0.724 & 0.724 \\ 0.411 & 0.724 & 1.000 & 0.614 \\ 0.505 & 0.724 & 0.614 & 1.000 \end{bmatrix} \cdot \begin{bmatrix} \lambda_1 \\ \lambda_2 \\ \lambda_3 \\ \lambda_4 \end{bmatrix} = \begin{bmatrix} 0.724 \\ 0.653 \\ 0.467 \\ 0.401 \end{bmatrix}$$

and this yields the conditional probability: $I^*(\mathbf{u}_1; 15) = F^*(\mathbf{u}_1; 15 \mid (n)) = -0.1445$. The negative *cdf* value indicates an order-relations violation that has to be corrected once the *cdf* corresponding to the remaining thresholds have been estimated.

The kriging system corresponding to the threshold 25 is different because of the change in the covariance model. The new system is:

$$\begin{bmatrix} 1.00 & 0.49 & 0.18 & 0.30 \\ 0.49 & 1.00 & 0.59 & 0.59 \\ 0.18 & 0.59 & 1.00 & 0.44 \\ 0.30 & 0.59 & 0.44 & 1.00 \end{bmatrix} \cdot \begin{bmatrix} \lambda_1 \\ \lambda_2 \\ \lambda_3 \\ \lambda_4 \end{bmatrix} = \begin{bmatrix} 0.592 \\ 0.464 \\ 0.249 \\ 0.173 \end{bmatrix}$$

This yields the conditional probability value: $F^*(\mathbf{u}_1; 25 \mid (n)) = 0.0675$. Repeating kriging for the remaining thresholds yields:

	$z = 15$	$z = 25$	$z = 35$	$z = 45$
$F^*(\mathbf{u} : z \mid (n))$	−0.1445	0.0675	0.7172	0.9059

After performing the order-relations correction detailed in Chap. 3, we get the *cdf*:

	$z = 15$	$z = 25$	$z = 35$	$z = 45$
$F^*(\mathbf{u} : z \mid (n))$	0.0000	0.0675	0.7172	0.9059

Monte Carlo sampling is performed using the updated *cdf* by implementing linear interpolation at the tails. The sampled value is 25.77.

The simulated value is added to the conditioning dataset when kriging at the next node along the random path. The kriging system at node 32 corresponding to the first threshold (15) is:

$$\begin{bmatrix} 1.00 & 0.72 & 0.65 & 0.50 & 0.41 \\ 0.72 & 1.00 & 0.63 & 0.40 & 0.47 \\ 0.65 & 0.63 & 1.00 & 0.72 & 0.72 \\ 0.50 & 0.40 & 0.72 & 0.61 & 0.61 \\ 0.41 & 0.47 & 0.72 & 0.61 & 1.00 \end{bmatrix} \cdot \begin{bmatrix} \lambda_1 \\ \lambda_2 \\ \lambda_3 \\ \lambda_4 \\ \lambda_5 \end{bmatrix} = \begin{bmatrix} 0.88 \\ 0.75 \\ 0.56 \\ 0.40 \\ 0.35 \end{bmatrix}$$

This yields the weights:

$$\begin{bmatrix} \lambda_1 \\ \lambda_2 \\ \lambda_3 \\ \lambda_4 \\ \lambda_5 \end{bmatrix} = \begin{bmatrix} 0.73 \\ 0.28 \\ -0.05 \\ -0.02 \\ -0.04 \end{bmatrix}$$

The estimated local conditional *cdf* value corresponding to this threshold is $F^*(\mathbf{u}_1; 15 \mid (n)) = -0.0029$. Repeating the kriging corresponding to the remaining thresholds, the following local conditional *cdf* is obtained:

	$z = 15$	$z = 25$	$z = 35$	$z = 45$
$F^*(\mathbf{u} : z \mid (n))$	−0.0029	0.0232	0.9977	0.9646

The local conditional *cdf* after order-relations correction is:

	$z = 15$	$z = 25$	$z = 35$	$z = 45$
$F^*(\mathbf{u} : z \mid (n))$	0.0000	0.0232	0.9811	0.9811

The sampled value from this local conditional *cdf* $Z^S(\mathbf{u}) = 27.79$ is added to the conditioning dataset and the simulation then proceeds to the next node. Two simulated realizations obtained by following this procedure are shown in Fig. 4-6.

In comparison to the previous simulation results obtained using the Gaussian and LUSIM methods, the simulation map indicates more abrupt transitions from high (red) values to low (blue) values. In the Gaussian-based models, that transition is more gradual (through green or yellow). This is a characteristic of indicator-based simulation methods.

4.4.3 Data Integration within the Indicator Framework

We discussed the properties of the indicator transform in Chap. 3. Corresponding to a continuous variable $Z(\mathbf{u})$, the indicator transform is given by:

$$I(\mathbf{u}) = \begin{cases} 1 & \text{if } Z(\mathbf{u}) \\ 0 & \text{otherwise} \end{cases}$$

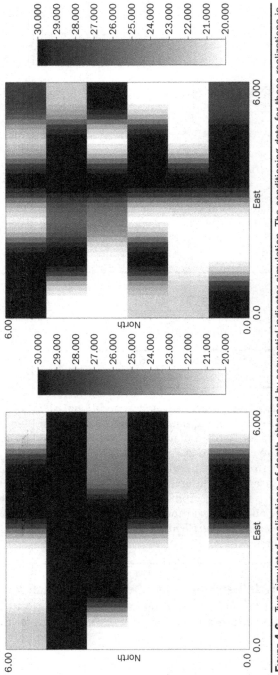

Figure 4-6 Two simulated realizations of depth obtained by sequential indicator simulation. The conditioning data for these realizations is as shown in Fig. 4-1. (The color version of this figure is available at www.mhprofessional.com/PRMS.)

166

The indicator estimate:

$$I^* - p(u_0; z) = \sum_\alpha \lambda_\alpha \cdot (I(u_\alpha; z) - p(u_\alpha; z)) \qquad (4\text{-}17)$$

yields the conditional probability $F(\mathbf{u}_0; z \mid (n))$. The weights in the indicator kriging expression are obtained by solving the system:

$$\sum_\beta \lambda_\beta C_I(\mathbf{h}_{\alpha\beta}; z) = C_I(\mathbf{h}_{\alpha 0}; z) \qquad (4\text{-}18)$$

In order to integrate secondary data (such as seismic impedance) into this indicator kriging framework, one strategy would be to quantify the information in the secondary data in the form of a soft probability. That is, we quantify:

$$y(\mathbf{u}; \mathbf{t}) = \mathbf{Prob}\{Z(\mathbf{u}) \le z \mid t(\mathbf{u})\}$$

One of the big advantages of working with such soft probability is that all data is now encoded in the form of probabilities. Hard data $I(u_\alpha; z)$ that assumes binary value of either 0 or 1 implies a step function (that in turn corresponds to a Dirac *pdf*). The modeled conditional probability $F(\mathbf{u}_0; z \mid (n))$ is thus modeled as a linear combination of probabilities.

The extended indicator (co-)kriging estimator is:

$$I^*(\mathbf{u}_0; z) - F_Z(\mathbf{u}_0; z) = \sum_\alpha \lambda_\alpha \cdot (I(\mathbf{u}_\alpha; z) - F_Z(\mathbf{u}_\alpha; z))$$
$$+ \sum_{\alpha_s} v_{\alpha_s} \cdot (y(\mathbf{u}_{\alpha_s}; z) - E\{y(\mathbf{u}_{\alpha_s}; z)\} \qquad (4\text{-}19)$$

with the accompanying system:

$$\sum_\beta \lambda_\beta C_I(\mathbf{h}_{\alpha\beta}) + \sum_{\alpha_s} v_{\alpha_s} C_{IY}(\mathbf{h}_{\alpha\alpha_s}) = C_I(\mathbf{h}_{\alpha 0}), \alpha = 1,\dots,n$$
$$\sum_\alpha \lambda_\alpha C_{IY}(\mathbf{h}_{\alpha\alpha_s}) + \sum_{\beta_s} v_{\beta_s} C_Y(\mathbf{h}_{\alpha_s\beta_s}) = C_{IY}(\mathbf{h}_{\alpha_s 0}), \alpha_s = 1,\dots,n_s \qquad (4\text{-}20)$$

or in matrix form:

$$
\begin{bmatrix}
C_{I_1 I_1} & C_{I_1 I_2} & \cdots & C_{I_1 I_n} & C_{I_1 Y_1} & C_{I_1 Y_2} & \cdots & C_{I_1 Y_{n_s}} \\
C_{I_2 I_1} & C_{I_2 I_2} & \cdots & C_{I_2 I_n} & C_{I_2 Y_1} & C_{I_2 Y_2} & \cdots & C_{I_2 Y_{n_s}} \\
\vdots & \vdots & \cdots & \vdots & \vdots & \vdots & \cdots & \vdots \\
C_{I_n I_1} & C_{I_n I_2} & \cdots & C_{I_n I_n} & C_{I_n Y_1} & C_{I_n Y_2} & \cdots & C_{I_n Y_{n_s}} \\
C_{Y_1 I_1} & C_{Y_1 I_2} & \cdots & C_{Y_1 I_n} & C_{Y_1 Y_1} & C_{Y_1 Y_2} & \cdots & C_{Y_1 Y_{n_s}} \\
C_{Y_2 I_1} & C_{Y_2 I_2} & \cdots & C_{Y_2 I_n} & C_{Y_2 Y_1} & C_{Y_2 Y_2} & \cdots & C_{Y_2 Y_{n_s}} \\
\vdots & \vdots & \cdots & \vdots & \vdots & \vdots & \cdots & \vdots \\
C_{Y_{n_s} I_1} & C_{Y_{n_s} I_2} & \cdots & C_{Y_{n_s} I_n} & C_{Y_{n_s} Y_1} & C_{Y_{n_s} Y_2} & \cdots & C_{Y_{n_s} Y_{n_s}}
\end{bmatrix}
\cdot
\begin{bmatrix}
\lambda_1 \\ \lambda_2 \\ \vdots \\ \lambda_n \\ v_1 \\ v_2 \\ \vdots \\ v_{n_s}
\end{bmatrix}
=
\begin{bmatrix}
C_{I_1 I_0} \\ C_{I_2 I_0} \\ \vdots \\ C_{I_n I_0} \\ C_{Y_1 I_0} \\ C_{Y_2 I_0} \\ \vdots \\ C_{Y_{n_s} I_0}
\end{bmatrix}
\qquad (4\text{-}21)
$$

Solution of the indicator kriging system and the subsequent indicator estimate is contingent in the availability of:

- The prior *cdf* of $Z(\mathbf{u})$: $F_Z(\mathbf{u}_0; z)$. Note $E\{y(\mathbf{u}_{\alpha_s}; z)\} = E\{\text{Prob}\{Z(\mathbf{u}_{\alpha_s}) \le z \mid t(\mathbf{u}_{\alpha_s})\}\}$ is the same as $\text{Prob}\{Z(\mathbf{u}_{\alpha_s}) \le z\} = F_Z(z)$.

- Models for the indicator covariance $C_I(\mathbf{h})$, the soft data covariance $C_Y(\mathbf{h})$ and the primary soft cross-covariance $C_{IY}(\mathbf{h})$. These models have to be jointly positive definite for each threshold z.

Remarks

- The co-indicator kriging also suffers from order-relations violations; i.e., the estimated posterior *cdf* may behave illegitimately especially at locations far away from conditioning data.

- The specification of the soft secondary variable $y(\mathbf{u}; t)$ requires us to work with the scatterplot between the collocated primary and secondary data. Rather than summarizing the joint relationship between the primary and secondary variables using a correlation coefficient (as in co-kriging), indicator co-kriging strives to exploit the joint relationship (or the scatterplot) fully by calculating conditional probabilities corresponding to several thresholds.

- Co-indicator kriging can be used as the estimation engine within a sequential simulation framework. In that case, the local conditional distributions $F(\mathbf{u}; z \mid (n), (n_s))$ are estimated using co-indicator kriging.

Example 4-4

We will assume the same data configuration as shown in Fig. 4-1. Similar to the indicator kriging example discussed in Chap. 3, we will assume the thresholds (15.0, 25.0, 35.0, 45.0) for the primary variable. Note that these thresholds are difficult to justify based on the four data values available in this example. Nevertheless, we will assume these thresholds in order to illustrate the workings of the indicator kriging procedure. In order to refresh our memory regarding the LMC and in order to demonstrate the flexibility of indicator kriging to use different covariance models for each threshold, we will assume the following indicator covariance models:

$$\rho_1(\mathbf{h}; 15) = 1 - \left(1.5 \cdot \frac{|\mathbf{h}|}{12.0} - 0.5 \cdot \left(\frac{|\mathbf{h}|}{12.0}\right)^3\right)$$

$$\rho_1(\mathbf{h}; 25) = 1 - \left(1.5 \cdot \frac{|\mathbf{h}|}{8.0} - 0.5 \cdot \left(\frac{|\mathbf{h}|}{8.0}\right)^3\right)$$

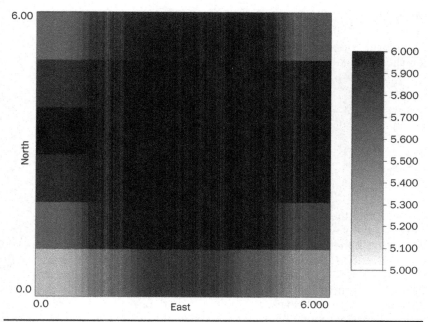

Figure 4-7 The map of secondary values used for the numerical example. (The color version of this figure is available at www.mhprofessional.com/PRMS.)

$$\rho_1\left(\mathbf{h}; 35\right) = 1 - \left(1.5 \cdot \frac{|\mathbf{h}|}{4.0} - 0.5 \cdot \left(\frac{|\mathbf{h}|}{4.0}\right)^3\right)$$

$$\rho_1\left(\mathbf{h}; 45\right) = 1 - \left(1.5 \cdot \frac{|\mathbf{h}|}{4.0} - 0.5 \cdot \left(\frac{|\mathbf{h}|}{4.0}\right)^3\right)$$

In addition, the secondary variable shown in Fig. 4-7 is assumed to be characterized by the covariance:

$$\rho_T\left(\mathbf{h}\right) = 1 - \left(1.5 \cdot \frac{|\mathbf{h}|}{6.0} - 0.5 \cdot \left(\frac{|\mathbf{h}|}{6.0}\right)^3\right)$$

Note that strictly speaking, the indicator co-kriging procedure requires the covariance model for the soft probabilities $(Y(\mathbf{u}; \mathbf{t}))$. However, because the inference of these soft probabilities is based on calibration using very sparse data available for this example, the covariance model for the secondary data will be directly used as the model for the soft data covariance.

Linear Model of Coregionalization

In order to ensure the positive definiteness of the requisite auto and cross-covariances, an LMC was employed. In order to model the first threshold, the following decomposition is assumed:

$$\rho_I(\mathbf{h};15) = 0.1 \cdot \left[1 - \left[1.5 \cdot \frac{|\mathbf{h}|}{5.8} - 0.5 \cdot \left(\frac{|\mathbf{h}|}{5.8}\right)^3\right]\right] + 0.9 \cdot \left[1 - \left[1.5 \cdot \frac{|\mathbf{h}|}{12.5} - 0.5 \cdot \left(\frac{|\mathbf{h}|}{12.5}\right)^3\right]\right]$$

$$\rho_T(\mathbf{h}) = 0.9 \cdot \left[1 - \left[1.5 \cdot \frac{|\mathbf{h}|}{5.8} - 0.5 \cdot \left(\frac{|\mathbf{h}|}{5.8}\right)^3\right]\right] + 0.9 \cdot \left[1 - \left[1.5 \cdot \frac{|\mathbf{h}|}{12.5} - 0.5 \cdot \left(\frac{|\mathbf{h}|}{12.5}\right)^3\right]\right]$$

Note that these models for the primary variable (depth) and the soft probability represent a compromise, so as to make the LMC work. Recall that in the LMC framework, any structure missing in either the primary or the secondary covariance model would be missing from the cross-covariance model also. There are no common structures in the original primary and secondary (soft) covariance models and so it would be impossible to model the cross-covariance within the LMC framework. Consequently, we compromise and model the primary and secondary covariance using two structures—one with a range of 5.8 units and the other with a range of 12.5 units. The extent of this compromise is displayed visually in Fig. 4-8.

Thus, according to the LMC, the cross-covariance would have a structure with range 5.8 units with a sill contribution: $sill \leq \sqrt{0.1 \cdot 0.9}$ and another structure with a range of 12.5 units with a limit on $sill \leq \sqrt{0.1 \cdot 0.9}$. Assuming that the correlation coefficient between the primary and secondary variables is 0.6 (i.e., $\rho_{IY}(0) = 0.6$) which will be consistent with the assumption in the LMC example presented in Chap. 3, the LMC for the cross-covariance is written as:

$$\rho_{IY}(\mathbf{h};15) = 0.3 \cdot \left[1 - \left[1.5 \cdot \frac{|\mathbf{h}|}{5.8} - 0.5 \cdot \left(\frac{|\mathbf{h}|}{5.8}\right)^3\right]\right] + 0.3 \cdot \left[1 - \left[1.5 \cdot \frac{|\mathbf{h}|}{12.5} - 0.5 \cdot \left(\frac{|\mathbf{h}|}{12.5}\right)^3\right]\right]$$

Note that the sill contribution of each structure in the above model satisfies the LMC condition ($sill \leq \sqrt{0.1 \cdot 0.9}$) stated above.

LMCs for the other thresholds are indicated below:

$$\rho_I(\mathbf{h}; 25) = 0.2 \cdot \left[1 - \left[1.5 \cdot \frac{|\mathbf{h}|}{6} - 0.5 \cdot \left(\frac{|\mathbf{h}|}{6}\right)^3\right]\right] + 0.8 \cdot \left[1 - \left[1.5 \cdot \frac{|\mathbf{h}|}{9} - 0.5 \cdot \left(\frac{|\mathbf{h}|}{9}\right)^3\right]\right]$$

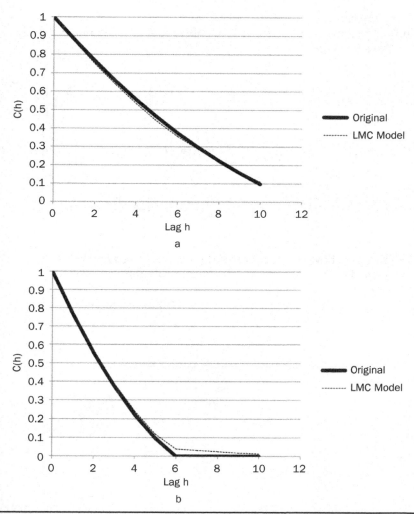

FIGURE 4-8 a) Comparison of the original covariance model for depth and the LMC at a depth threshold of 15. **b)** Comparison of the original covariance model for the secondary data and the LMC for the same data.

$$\rho_I(\mathbf{h}) = 0.8 \cdot \left[1 - \left(1.5 \cdot \frac{|\mathbf{h}|}{6} - 0.5 \cdot \left(\frac{|\mathbf{h}|}{6}\right)^3\right)\right] + 0.2 \cdot \left[1 - \left(1.5 \cdot \frac{|\mathbf{h}|}{9} - 0.5 \cdot \left(\frac{|\mathbf{h}|}{9}\right)^3\right)\right]$$

$$\rho_{IY}(\mathbf{h}; 25) = 0.3 \cdot \left[1 - \left(1.5 \cdot \frac{|\mathbf{h}|}{6} - 0.5 \cdot \left(\frac{|\mathbf{h}|}{6}\right)^3\right)\right] + 0.3 \cdot \left[1 - \left(1.5 \cdot \frac{|\mathbf{h}|}{9} - 0.5 \cdot \left(\frac{|\mathbf{h}|}{9}\right)^3\right)\right]$$

The thresholds *Depth* = 35 and 45 have the same primary variogram models and hence the LMCs are also the same. The LMCs for these thresholds are:

$$\rho_I(\mathbf{h};35) = 0.9 \cdot \left[1 - \left(1.5 \cdot \frac{|\mathbf{h}|}{3.9} - 0.5 \cdot \left(\frac{|\mathbf{h}|}{3.9}\right)^3\right)\right] + 0.1 \cdot \left[1 - \left(1.5 \cdot \frac{|\mathbf{h}|}{6.1} - 0.5 \cdot \left(\frac{|\mathbf{h}|}{6.1}\right)^3\right)\right]$$

$$\rho_T(\mathbf{h}) = 0.1 \cdot \left[1 - \left(1.5 \cdot \frac{|\mathbf{h}|}{3.9} - 0.5 \cdot \left(\frac{|\mathbf{h}|}{3.9}\right)^3\right)\right] + 0.9 \cdot \left[1 - \left(1.5 \cdot \frac{|\mathbf{h}|}{6.1} - 0.5 \cdot \left(\frac{|\mathbf{h}|}{6.1}\right)^3\right)\right]$$

$$\rho_{IY}(\mathbf{h};35) = 0.3 \cdot \left[1 - \left(1.5 \cdot \frac{|\mathbf{h}|}{3.9} - 0.5 \cdot \left(\frac{|\mathbf{h}|}{3.9}\right)^3\right)\right] + 0.3 \cdot \left[1 - \left(1.5 \cdot \frac{|\mathbf{h}|}{6.1} - 0.5 \cdot \left(\frac{|\mathbf{h}|}{6.1}\right)^3\right)\right]$$

Figure 4-9 shows the location of the soft data used for indicator co-kriging. The next task would be to calibrate the soft probabilities from a scatterplot of collocated primary (depth) and secondary data. However, in this case, it is deliberately assumed that the

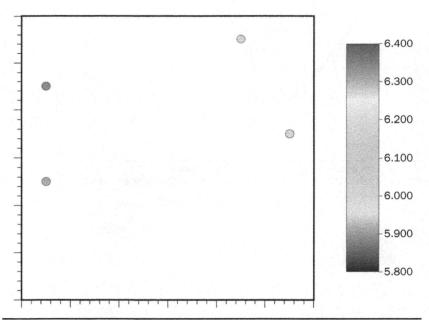

Figure 4-9   Location of soft data used for co-indicator kriging. (The color version of this figure is available at www.mhprofessional.com/PRMS.)

secondary data is at locations different from the primary data and so it is not possible to prepare the calibration plot. Instead, we generate a calibration plot by simulating pairs of variables from a bivariate Gaussian distribution with a correlation coefficient of 0.6 between the variables. The procedure for simulating these pairs of primary and secondary values is by randomly sampling primary data values from a normal distribution with a mean equal to the mean of the primary data 25 and a standard deviation of 12.9. These randomly sampled values are then transformed to a standard normal distribution denoted as U. Another set of standard normal values is randomly sampled and that is denoted as V. Then a new random variable is generated as:

$$Z = U + \rho \cdot V$$

with ρ being the desired correlation of 0.6. The generated Z values are then transformed to a normal distribution with a mean of 5.9875 and standard deviation of 0.1258 representing the secondary data. The resultant cross-plot between the primary and secondary values is shown in Fig. 4-10. The calibration plot then yields the soft probabilities, as shown in Table 4-1 corresponding to the different secondary data values. These probabilities are computed by counting the number of data in each bin.

Using these values of the soft probabilities, and the LMC model, the following indicator co-kriging system corresponding to the

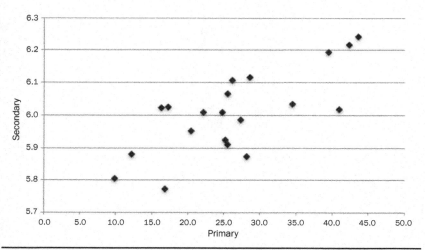

Figure 4-10   Calibration plot between the primary and secondary variables.

	Secondary	5.88	5.91	6.00	6.16
Primary					
15		0.176	0.000	0.000	0.000
25		0.941	0.864	0.270	0.000
35		1.000	1.000	0.838	0.405
45		1.000	1.000	1.000	0.919

TABLE 4-1   Calibrated Values of Soft Probability $y(u; t) = $ **Prob**$\{Z(u) £ z \mid t(u)\}$ Corresponding to Thresholds of the Primary Variable and for Given Values of the Secondary Variable at the Conditioning Data Locations

threshold, $Depth = 15$ for estimation at node (1,2) (which is the first node visited along a random path) is written as:

$$
\begin{bmatrix}
1.00 & 0.59 & 0.71 & 0.48 & 0.18 & 0.38 & 0.36 & 0.19 \\
0.59 & 1.00 & 0.71 & 0.39 & 0.17 & 0.11 & 0.23 & 0.49 \\
0.71 & 0.71 & 1.00 & 0.63 & 0.30 & 0.13 & 0.28 & 0.36 \\
0.48 & 0.39 & 0.63 & 1.00 & 0.12 & 0.19 & 0.27 & 0.27 \\
0.18 & 0.38 & 0.36 & 0.19 & 1.00 & 0.06 & 0.06 & 0.48 \\
0.17 & 0.11 & 0.23 & 0.49 & 0.06 & 1.00 & 0.53 & 0.15 \\
0.30 & 0.13 & 0.28 & 0.36 & 0.06 & 0.53 & 1.00 & 0.07 \\
0.12 & 0.19 & 0.27 & 0.27 & 0.48 & 0.15 & 0.07 & 1.00
\end{bmatrix}
\cdot
\begin{bmatrix}
\lambda_1 \\ \lambda_2 \\ \lambda_3 \\ \lambda_4 \\ v_1 \\ v_2 \\ v_3 \\ v_4
\end{bmatrix}
=
\begin{bmatrix}
0.71 \\ 0.39 \\ 0.59 \\ 0.59 \\ 0.12 \\ 0.28 \\ 0.49 \\ 0.11
\end{bmatrix}
$$

This results in the following solution for the weights:

$$
\begin{bmatrix}
\lambda_1 \\ \lambda_2 \\ \lambda_3 \\ \lambda_4 \\ v_1 \\ v_2 \\ v_3 \\ v_4
\end{bmatrix}
=
\begin{bmatrix}
0.5410 \\ -0.1241 \\ 0.0571 \\ 0.2964 \\ -0.0066 \\ -0.0928 \\ 0.2724 \\ -0.0265
\end{bmatrix}
$$

The posterior probability $I^*(u_o; 15)$ can be obtained as:

$$
\begin{aligned}
I^*(\mathbf{u}_o;15) = {} & 0.541 \cdot 1 - 0.124 \cdot 0 + 0.057 \cdot 0 + 0.296 \cdot 0 - 0.007 \cdot 0.17 \\
& - 0.093 \cdot 0.33 + 0.272 \cdot 0.08 - 0.027 \cdot 0.0 \\
& + (1 - (0.54 - 0.12 + 0.06 + 0.30) - (-0.01 - 0.09 + 0.27 \\
& - 0.03)) \cdot 0.2 \\
= {} & 0.55
\end{aligned}
$$

The prior probability $F_Z(15)$ is assumed to be 0.2 in the above equation. The kriging system is developed for the other thresholds and the indicator estimates are obtained. The following posterior *cdf* is obtained:

$$I^*(\mathbf{u}_o) = \begin{bmatrix} 0.55 \\ 0.72 \\ 0.81 \\ 0.935 \end{bmatrix}$$

The result above seems to indicate that there are no order-relations violations at the first simulation node. A value is sampled from the *cdf* by performing linear interpolation between the thresholds and from first *cdf* value at a threshold of 15 to a *cdf* value of 0 at a minimum value of 5. Similarly, a linear extrapolation is done at the upper tail to a maximum value of 65. The resultant simulated value randomly sampled from the *cdf* is 31.61. This value is added to the dataset and the indicator kriging is again performed at the next node along the random path. The resultant map of simulated values is shown in Fig. 4-11. Comparing this map to the indicator simulation results shown in Fig. 4-6, the conditioning influence of the soft data is apparent. In addition, the indicator simulation map also exhibits abrupt transitions from the blue regions to the red regions without much transition into the intermediate colors as would be expected for indicator simulation. The extent of the region with higher depths is restricted according to the information in the secondary data. This is especially noticeable in the southern regions of the figure with shallower depths.

The indicator co-simulation procedure discussed in this section entails the calibration of soft probabilities on the basis of the available secondary information, the inference and modeling of the requisite auto and cross-covariances using the LMC. A faster approach to data integration within the indicator framework is available using the Markov–Bayes approach that is discussed next.

4.4.4 Markov–Bayes Approach

As noted earlier, full indicator co-kriging requires models for the auto-covariance of the primary variables and the soft probabilities, as well as the cross-covariance between the primary and soft variables. As was evident from the example presented, the inference of these covariances is especially challenging in the indicator framework because a different covariance model may be assumed for each threshold. A simplification is possible by making a screening assumption as we did in Chap. 3 when we talked about the Markov models.

Let us make the following Markov assumption that the closest primary data screen the influence of all other primary and soft data. This can be written as (Zhu, 1993):

$$P(I(\mathbf{u} + \mathbf{h}) \mid I(\mathbf{u}) = i, Y(\mathbf{u};z) = y) = P(I(\mathbf{u} + \mathbf{h}) \mid I(\mathbf{u}) = i) \qquad (4\text{-}22)$$
$$P(Y(\mathbf{u}) \mid I(\mathbf{u}) = i, Y(\mathbf{u} + \mathbf{h};z) = y') = P(I(\mathbf{u}) \mid I(\mathbf{u}) = i)$$
$$P(I(\mathbf{u}) \mid I(\mathbf{u}) = i, I(\mathbf{u} + \mathbf{h};z) = j) = P(Y(\mathbf{u}) \mid I(\mathbf{u}) = i)$$

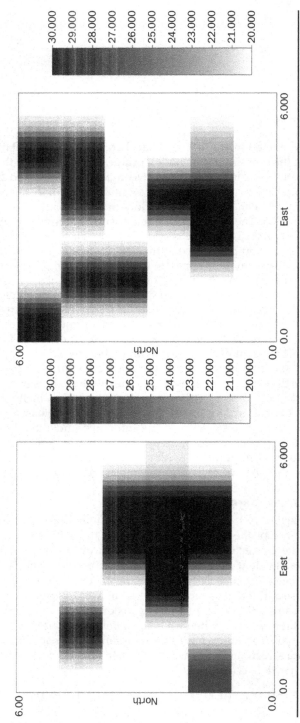

Figure 4-11 Map of simulated values obtained by indicator co-simulation using the primary and secondary data shown in Fig. 4-1. (The color version of this figure is available at www.mhprofessional.com/PRMS.)

The implication of this Markov assumption is that for the estimation of an indicator outcome at a location, only relationship between the collocated primary and soft variables needs to be considered. This leads to the following simplified estimator (Yao, 2002):

$$
\begin{aligned}
I^*(\mathbf{u}_0; z) - F_Z(\mathbf{u}_0; z) &= \sum_\alpha \lambda_\alpha \cdot (I(\mathbf{u}_\alpha; z) - F_Z(\mathbf{u}_\alpha; z)) \\
&\quad + v \cdot (y(\mathbf{u}_0; z) - E\{y(\mathbf{u}_0; z)\})
\end{aligned}
\tag{4-23}
$$

with the accompanying system:

$$
\begin{aligned}
\sum_\beta \lambda_\beta C_I(\mathbf{h}_{\alpha\beta}) + v \cdot C_{IY}(\mathbf{h}_{\alpha 0}) &= C_I(\mathbf{h}_{\alpha 0}), \quad \alpha = 1, \ldots, n \\
\sum_\alpha \lambda_\alpha C_{IY}(\mathbf{h}_{\alpha 0}) + v_{\beta_s} \cdot C_Y(0) &= C_{IY}(0)
\end{aligned}
\tag{4-24}
$$

It is evident from the previous equation that the auto-covariance of the soft variable is no longer needed, thanks to the Markov assumption. A positive definite model for the cross-covariance between the primary and soft variables is needed.

Note
In order to perform estimation using the Markov assumption detailed above, the secondary (soft) data must be available at all estimation locations. The Markov–Bayes model is therefore only feasible when the secondary data is available exhaustively over the whole domain.

In order to develop the model for the cross-covariances, we define some additional parameters. Let us define the means:

$$
\begin{aligned}
m^{(1)} &= E\{Y(\mathbf{u}) \mid I(\mathbf{u}; z) = 1\} \\
m^{(0)} &= E\{Y(\mathbf{u}) \mid I(\mathbf{u}; z) = 0\}
\end{aligned}
\tag{4-25}
$$

These means can be calculated from the calibration plot between the primary and the secondary (soft) variables. Using these parameters, it can be shown (see App. B) that the cross-covariance:

$$
C_{IY}(\mathbf{h}; z) = C_I(\mathbf{h}; z) \cdot B(z)
\tag{4-26}
$$

The parameter $B(z)$ is defined as:

$$
B(z) = m^{(1)} - m^{(0)}
\tag{4-27}
$$

Even though the auto-covariance of the secondary variable is not needed within this Markov–Bayes procedure, it can be shown that:

$$C_Y(\mathbf{h}) = C_I(\mathbf{h}; z) \cdot B^2(z) \qquad (4\text{-}28)$$

Because the secondary data is required to be known exhaustively, the covariance of the secondary variable can most likely be inferred from the available data. By comparing the inferred secondary covariance with the model based on the screening assumption, the validity of the Markov–Bayes approach for integrating secondary data into the indicator kriging process can be assessed.

Example 4-5

The data configuration in Fig. 4-1 is used for the example. As discussed previously, four thresholds for depth (15.0, 25.0, 35.0, 45.0) are assumed. The auto-covariance model for the primary variable corresponding to the first threshold is assumed to have a range of 12 units, for the second threshold 8 units, while the last two thresholds have a range of 4 units.

The calibration plot between the primary and secondary variables is used to compute the $B(z)$. The corresponding primary–secondary data is shown in Table 4-2. This data has been obtained by sampling from a bivariate Gaussian distribution as discussed earlier.

The corresponding table of soft probabilities $y(\mathbf{u}; t) = \mathbf{Prob}\{Z(\mathbf{u}) \leq z \mid t(\mathbf{u})\}$ shown in Table 4-1 can be calculated. The calibration table can also be used to compute $m^{(1)} = E\{Y(\mathbf{u}) \mid I(\mathbf{u}; z) = 1\}$ and $m^{(0)} = E\{Y(\mathbf{u}) \mid I(\mathbf{u}; z) = 0\}$. For example, considering the primary threshold of 15, the average of the secondary values corresponding to $I(\mathbf{u}; z) = 1$, i.e., $Depth \leq 15$ is 5.742. The corresponding value of y is (using the calibration data as in Table 4-2) 0.382 and consequently $m^{(1)} = 0.382$. The mean of the secondary values corresponding to $I(\mathbf{u}; z) = 0$, i.e., $Depth > 15$ is 6.043. Similarly, translating these secondary values to soft probability values (using the calibration data in Table 4-2) results in $m^{(0)} = 0.0$. This leads to $B(z) = 0.382$. Repeating this for the other thresholds yields:

$B(z)$	0.382	0.132	0.595	0.04

Using the Markov–Bayes hypothesis and the corresponding reduced form for the auto- and cross-covariances, kriging can be performed. For example, at the first simulation node (1,2), the following is the kriging system corresponding to the threshold $Depth = 15$:

$$
\begin{bmatrix}
1.00 & 0.59 & 0.71 & 0.48 & 0.27 \\
0.59 & 1.00 & 0.71 & 0.39 & 0.15 \\
0.71 & 0.71 & 1.00 & 0.63 & 0.22 \\
0.71 & 0.39 & 0.63 & 1.00 & 0.22 \\
0.27 & 0.15 & 0.22 & 0.22 & 1.00
\end{bmatrix}
\cdot
\begin{bmatrix}
\lambda_1 \\ \lambda_2 \\ \lambda_3 \\ \lambda_4 \\ v_1
\end{bmatrix}
=
\begin{bmatrix}
0.706 \\ 0.392 \\ 0.592 \\ 0.592 \\ 0.380
\end{bmatrix}
$$

X (Depth)	Y (Secondary)
39.435	6.194
40.980	6.019
27.346	5.986
9.869	5.806
43.566	6.242
24.804	6.010
17.314	6.026
34.497	6.035
16.891	5.775
42.283	6.217
20.458	5.951
26.140	6.107
25.148	5.926
12.229	5.881
16.317	6.023
25.523	5.911
22.109	6.009
25.474	6.065
28.162	5.875
28.567	6.115

TABLE 4-2   Corresponding Values of Primary and Secondary Variables

The weights are:

$$\begin{bmatrix} \lambda_1 \\ \lambda_2 \\ \lambda_3 \\ \lambda_4 \\ v_1 \end{bmatrix} = \begin{bmatrix} 0.554 \\ -0.143 \\ 0.207 \\ 0.084 \\ 0.187 \end{bmatrix}$$

The collocated value of the secondary data at (1,2) is 5.62 (based on the secondary map shown in Fig. 4-7) and that corresponds to a soft probability $y(\mathbf{u}; \mathbf{t}) = 0.0$ (obtained by extrapolating the value in Table 4-1). The posterior probability $I^*(\mathbf{u}_o; 15)$ is thus:

$$\begin{aligned} I^*(\mathbf{u}_o; 15) &= 0.554 \cdot 1 - 0.143 \cdot 0 + 0.207 \cdot 0 + 0.084 \cdot 0 + 0.187 \cdot 0 \\ &\quad + (1 - (0.554 - 0.143 + 0.207 + 0.084 + 0.187)) \cdot 0.2 \\ &= 0.58 \end{aligned}$$

Repeating the kriging for the other thresholds yields the following result for the posterior probability:

$$I^*(\mathbf{u}_o) = \begin{bmatrix} 0.58 \\ 0.59 \\ 0.91 \\ 0.87 \end{bmatrix}$$

As can be seen above, there is an order-relation violation corresponding to the simulation at this node. The correction procedure discussed earlier is implemented and the corrected conditional *cdf* is:

$$I^*(\mathbf{u}_o) = \begin{bmatrix} 0.58 \\ 0.59 \\ 0.89 \\ 0.89 \end{bmatrix}$$

The simulated value at (1,2) can be obtained by randomly sampling from the conditional *cdf* above. The simulation procedure can be repeated at other nodes in the grid and multiple realizations of the simulated depth map can be generated. Two such realizations are shown in Fig. 4-12. Comparing this figure to Fig. 4-11, the influence of collocated values of the soft probabilities used in the Markov–Bayes procedure is to smear the values of the estimates; i.e., the map of depths now looks more diffused.

Next, we discuss an alternate framework for data integration within the indicator framework without resorting to a Markov assumption or a linear model of co-regionalization.

4.5 Data Integration Using the Permanence of Ratio Hypothesis

The data integration methods that we have discussed so far are based either on a linear combination of data from different sources or a linear combination of the information from different sources calibrated in the form of probabilities. Let us assume that A is the rock type or property being modeled, B is the prior "hard" information (generally in the same units as A) available to model A, and C is the secondary information such as seismic data that may be indirectly related to the attribute being modeled. The objective has been to either derive the conditional expectation $E\{A \mid B, C\}$ (as in co-kriging/co-simulation) or to synthesize the conditional probability $P(A \mid B, C)$ (as in indicator co-kriging). In this section, we will discuss a general procedure for synthesizing the conditional probability distribution $P(A \mid B, C)$ from the conditional probabilities $P(A \mid B)$ and $P(A \mid C)$

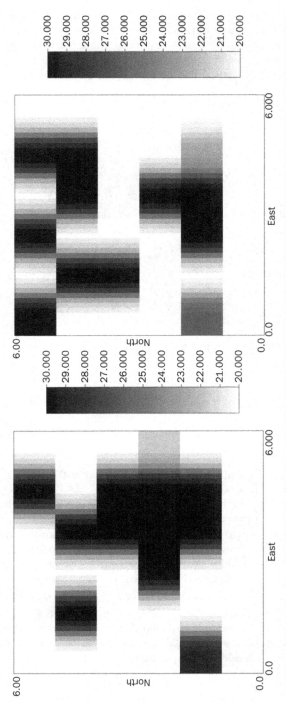

Figure 4-12 Two simulated realizations obtained by performing indicator co-simulation employing the Markov–Bayes assumption. (The color version of this figure is available at www.mhprofessional.com/PRMS.)

using the permanence of ratio hypothesis proposed by Journel (2002). The permanence of ratio hypothesis is predicated on the calibration of information from each source of data and then merging the information accounting for the redundancy between the data.

The joint probability of $P(A \mid B, C)$ is by definition:

$$P(A \mid B, C) = \frac{P(A, B, C)}{P(B, C)} \tag{4-29}$$

The joint probability is given as:

$$P(A, B, C) = P(A) \cdot P(B \mid A) \cdot P(C \mid A, B) = P(A) \cdot$$
$$P(C \mid A) \cdot P(B \mid A, C) \tag{4-30}$$

If we make an assumption that the two data sources are independent, i.e.,

$$P(C \mid A, B) = P(C \mid A); \quad P(B \mid C) = P(B) \tag{4-31}$$

then simplifies to:

$$P(A \mid B, C) = P(A) \cdot \frac{P(B \mid A)}{P(B)} \cdot \frac{P(C \mid A)}{P(C)} \tag{4-32}$$

which can be rearranged as:

$$P(A \mid B, C) = P(A \mid B) \cdot \frac{P(A \mid C)}{P(A)} \tag{4-33}$$

The simplicity of Eq. (4-33) is due to the assumption of full independence of data events and leads to some inconsistencies as described below:

- $P(A \mid B, C)$ is not guaranteed to be between 0 and 1. This will happen if the conditional probabilities $P(A \mid B)$ and $P(A \mid C)$ are evaluated independently and they are inconsisent with the assumption of independence between B and C. Consider the case where $P(A \mid B) = 0.6$, $P(A \mid C) = 0.4$, and $P(A) = 0.2$. These are all legitimate probabilities and do not presume any dependence between the events B and C. However, Eq. (4-33) leads to $P(A \mid B, C) = 1.2$, and that cannot happen.

- If C is fully informative of A, i.e, $P(A \mid C) = 1$ or 0 from Eq. (4-33), then $P(A \mid B, C) \neq P(A \mid C)$. This means that by the assumption of full independence between data events, the resultant data integration scheme implies uncertainty even if one of the data sources is rendering the prediction deterministic.

Now, if we assume conditional independence between event B and event C; i.e., we assume that in the presence of a common conditiong event (such as the existence of a reservoir property field A), we can write (Dawid, 1979):

$$P(C \mid A, B) = P(C \mid A) \tag{4-34}$$

$$P(B \mid A, C) = P(B \mid A) \tag{4-35}$$

Plugging these in equation will lead to the following:

$$P(A \mid B, C) = \frac{P(A)P(B \mid A)P(C \mid A)}{P(B, C)} \tag{4-36}$$

Now consider the complement event $\overline{A}$, such that $P(A) + P(\overline{A}) = 1$ Doing the same developments for $P(\overline{A} \mid B, C)$ will yield:

$$P(\overline{A} \mid B, C) = \frac{P(\overline{A})P(B \mid \overline{A})P(C \mid \overline{A})}{P(B, C)} \tag{4-37}$$

Thus:

$$\frac{P(A \mid B, C)}{P(\overline{A} \mid B, C)} = \frac{P(A)P(B \mid A)P(C \mid A)}{P(\overline{A})P(B \mid \overline{A})P(C \mid \overline{A})} \tag{4-38}$$

Recall that:

$$P(B \mid A) = \frac{P(A \mid B)P(B)}{P(A)} \tag{4-39}$$

Hence the above equation can be written as:

$$\frac{P(A \mid B, C)}{P(\overline{A} \mid B, C)} = \frac{\left(\dfrac{P(A \mid B)}{P(\overline{A} \mid B)}\right)\left(\dfrac{P(A \mid C)}{P(\overline{A} \mid C)}\right)}{\dfrac{P(A)}{P(\overline{A})}} \tag{4-40}$$

Next we define the following odds ratios:

$$x = \frac{1 - P(A \mid B, C)}{P(A \mid B, C)}, a = \frac{1 - P(A)}{P(A)}, b = \frac{1 - P(A \mid B)}{P(A \mid B)},$$
$$c = \frac{1 - P(A \mid C)}{P(A \mid C)} \tag{4-41}$$

We can write Eq. (4-41) as:

$$\frac{x}{a} = \frac{b}{a} \cdot \frac{c}{a} \tag{4-42}$$

The odds ratios measure the relative distance to the occurrence of A given each data source, e.g., B. If A is guaranteed not to occur given the data, then the relative distance to occurrence of A is infinity; while if A is guaranteed to occur because of the occurrence of the data source, then the relative distance is zero. As the odds ratio x is in terms of the required conditional probability $P(A \mid B, C)$, the permanence of ratio hypothesis provides an approach to merge the conditional probabilities from individual sources of information in order to obtain $P(A \mid B, C)$.

Remarks

- Since all the odd ratios in Eq. (4-41) are positive, x remains positive, thus:

$$x \geq 0 \ implies \ that \ 0 \leq P(A \mid B, C) = \frac{1}{1 + x} \leq 1 \qquad (4\text{-}43)$$

- When one source of information guarantees the occurrence of an event A, say for example:

$$P(A \mid B) = 1$$

then

$$b = \frac{1 - P(A \mid B)}{P(A \mid B)} = 0 \qquad (4\text{-}44)$$

Plugging Eq. (4-44) into Eq. (4-42) leads to:

$$x = b \cdot \frac{c}{a} = 0 \qquad (4\text{-}45)$$

That in turn implies that:

$$P(A \mid B, C) = \frac{1}{1 + x} = 1 \qquad (4\text{-}46)$$

Therefore, if one of the data sources guarantees occurrence of a simulation event A, integrating a secondary source of information will not introduce any uncertainty in the occurrence of that event and the merged probability also guarantees the occurrence of that event. On the other hand, in case a source of information guarantees that the simulation event does not occur, then:

$$P(A \mid B) = 0$$

Therefore

$$b = \frac{1 - P(A \mid B)}{P(A \mid B)} = \infty \qquad (4\text{-}47)$$

This leads to:

$$x = b \cdot \frac{c}{a} = \infty \tag{4-48}$$

This in turns implies:

$$P(A \mid B, C) = \frac{1}{x+1} = 0 \tag{4-49}$$

Therefore, if one of the data sources guarantees that the simulation event does not occur, integrating a secondary source of information will not alter the probability of the event occurring and the merged probability also guarantees that the event does not occur.

4.5.1 The Tau Model

It should be noted that the above discussion of the permanence of ratio model assumes conditional independence between events B and C given the event A. One way to reintroduce dependence between the data events is to set the contribution $\frac{c}{a}$ to a power which is itself a function of both B and C

$$\frac{x}{b} = \left(\frac{c}{a}\right)^{\tau(B,C)} \tag{4-50}$$

When $\tau = 0$ the data source C is completely ignored, for $\tau > 1$ the influence of C is increased, and if $\tau < 1$ the influence of C is decreased.

As pointed out by Journel (2002), data dependency is different from data redundancy. The elementary data sources B, C could be related to each other and that is indicative of their dependency. However, the concept of data redundancy comes from how much information inferred about event A from data source B overlaps with the information coming from data source C. Traditional data integration techniques address the concept of data redundancy through measures such as data-to-data covariance and mutual information (McEliece, 2002). In order to account for the dependency between data, Krishnan (2008) presented the tau model. In fact, that model is a tautology, whereby the exact expression for the *tau* parameter can be derived starting from the definition of the conditional probability $P(A \mid B, C)$. The odds ratio of A given B and C can be expressed as:

$$\frac{P(A \mid B, C)}{P(\bar{A} \mid B, C)} = \frac{P(A) \cdot P(B \mid A) \cdot P(C \mid A, B)}{P(\bar{A}) \cdot P(B \mid \bar{A}) \cdot P(C \mid \bar{A}, B)} \tag{4-51}$$

By applying Bayes rule, we get:

$$P(B \mid A) = \frac{P(A \mid B)P(B)}{P(A)}, \quad P(B \mid \bar{A}) = \frac{P(\bar{A} \mid B)P(B)}{P(\bar{A})} \tag{4-52}$$

Implementing Eq. (4-52) in Eq. (4-51) and using the definition of the odds ratio b, we get:

$$\frac{x}{b} = \frac{P(C \mid \bar{A}, B)}{P(C \mid A, B)} \tag{4-53}$$

Equating Eq. (4-53) to Eq. (4-50), we get:

$$\tau = \frac{\log\left(\dfrac{P(C \mid \bar{A}, B)}{P(C \mid A, B)}\right)}{\log\left(\dfrac{C}{a}\right)} \tag{4-54}$$

The advantages of the permanence of ratio hypothesis over other data integration approaches is detailed in Journel (2002). However, it is evident that a robust procedure for ccalculating the tau weight is crucial for implemeznting the permanence of ratio hypothesis. In several implementation examples (Caers & Srinivasan, 2002; Caers & Srinivasan, 2003; Kashib & Srinivasan, 2006; Srinivasan & Sen, 2010), the tau weight has been arbitrarily assumed to be unity. According to Eq. (4-54), setting τ equal to 1 is equivalent to setting the ratios $\dfrac{P(C \mid \bar{A}, B)}{P(C \mid A, B)}$ and $\dfrac{P(C \mid \bar{A})}{P(C \mid A)}$ to be equal to each other. One possibility for accomplishing this is to set:

$$P(C \mid A, B) = P(C \mid A) \ \text{ and } \ P(C \mid \bar{A}, B) = P(C \mid \bar{A}) \tag{4-55}$$

This means that the datum C is independent of the previous data B given A, i.e., conditional independence. However, in general, equality of these ratios does not necessarily require conditional independence but only that: $\dfrac{P(C \mid A, B)}{P(C \mid A)} = \dfrac{P(C \mid \bar{A}, B)}{P(C \mid \bar{A})} = \text{someconstant } r$. The value $r = 1$ arises from conditional independence. Any other value of $r \in [0, \infty]$ would also yield tau to be unity (Krishnan, Boucher & Journel, 2004). The interpretation then is that the relative information in data source C toward the discrimination of event A remains invariant regardless of the event B. This is in fact the fundamental hypothesis underlying this data integration approach and the key emphasis is on the term "relative information" represented by the odds ratio.

Krishnan et al. (2004) proposed the following relationship between the tau parameter and correlation coefficient between data sources:

$$\tau = 1 - \rho_{B,C|A}^2 \tag{4-56}$$

where $\rho_{B,C|A}^2$ is conditional correlation of B and C given A. Intuitively the greater the correlation coefficient, the less informative is data source C, and its weight should decrease. However, the main drawback of Eq. (4-56) is that the tau weight is always less than 1, which means that the sensitivity of datum C to change in A is always diminished by the data source B.

Another Approach for Calculating Tau Weights

A robust method for calculating the τ parameters accounting for redundancy between data sources is presented in Naraghi & Srinivasan (2015). The derivation is presented in the indicator framework where A, the property being estimated, is a categorical attribute, such as rock type. The probability distribution of data B and C is assumed to exhibit multimodal characteristics consistent with the presence of multiple categories of event A. Such multimodal distributions can be modeled employing a mixture of Gaussian hypothesis. The tau weights can then be calculated using the mixture of Gaussian assumption as follows:

$$\tau_i = \frac{\log\left(\frac{P(C \mid \bar{A}_1, B)}{P(C \mid A_i, B)}\right)}{\log\left(\frac{P(C \mid \bar{A}_i)}{P(C \mid A_i)}\right)} \tag{4-57}$$

$$P(C \mid \bar{A}_l, B) = \frac{\sum_{j \neq i} P(A_j) P(B, C \mid A_j)}{\sum_{j \neq i} P(A_j) P(B \mid A_j)} \tag{4-58}$$

$$P(C \mid A_i, B) = \frac{P(B, C \mid A_i)}{P(B \mid A_i)} \tag{4-59}$$

$$P(C \mid \bar{A}_l) = \frac{\sum_{j \neq i} P(A_j) P(C \mid A_j)}{1 - P(A_j)} \tag{4-60}$$

Keeping with the multi-Gaussian assumption made above, the likelihood of data B given A can be written as:

$$P(B \mid A_i) \propto (2\pi)^{-\frac{N_B}{2}} \left|\Sigma_{B_i}\right|^{-\frac{1}{2}} exp\left(\frac{-1}{2}\left(B - \mu_{B_i}\right)^t \Sigma_{B_i}^{-1}\left(B - \mu_{B_i}\right)\right) \tag{4-61}$$

$$P(C \mid A_i) \propto (2\pi)^{-\frac{N_C}{2}} \left|\Sigma_{C_i}\right|^{-\frac{1}{2}} exp\left(\frac{-1}{2}\left(C - \mu_{C_i}\right)^t \Sigma_{C_i}^{-1}\left(C - \mu_{C_i}\right)\right) \tag{4-62}$$

$$P(B, C \mid A_i) \propto (2\pi)^{-\frac{N_X}{2}} \left|\Sigma_{X_i}\right|^{-\frac{1}{2}} exp\left(\frac{-1}{2}\left(X - \mu_{X_i}\right)^t \Sigma_{X_i}^{-1}\left(X - \mu_{X_i}\right)\right) \tag{4-63}$$

where A_j is A belonging to category i. B is deemed the primary data source, N_B is the number of samples of primary data, and Σ_{B_i} and μ_{B_i} are the covariance matrix and mean values, respectively, of primary data within class i. C is deemed the secondary data source, N_C is the size of vector C, and Σ_{C_i} and μ_{C_i} are the covariance matrix and mean values, respectively, of secondary data within class i. N_X is the size of vector X, Σ_{X_i} and μ_{X_i} are the covariance matrix and mean values, respectively, of vector X within class i. By calculating these parameters using the available data, we can calculate the tau weights, merge the conditional probabilities, and find $P(A_i \mid B, C)$ for all categories.

It should be noted that Eq. (4-57) for calculating tau is data dependent and requires the actual outcomes of B and C data event. However, the outcome of one or both the data events may not be known at every location. In order to overcome this problem, the expected value of tau can be computed:

$$\bar{\tau} = \iint \tau(b, c) f_{BC}(b, c) db\, dc \tag{4-64}$$

The joint *pdf* of variables B and C is $f_{BC}(b, c)$. $\bar{\tau}$ can be computed by taking various realizations of the data events B and C and computing tau corresponding to each realization and finally computing the average over all values of tau. It can be proved that the average value of tau can be related to the mutual information between the data sources and discrimination density expectation and is guaranteed to be positive (Krishnan, 2008).

Example 4-6

In order to demonstrate the process for assimilating data using the permanence of ratios hypothesis, we go back to the data configuration in Fig. 4-1. The secondary data for this example is as shown in Fig. 4-7. Because the primary data available for this exercise is extremely sparse (only 4 values), we will rely on a calibration plot between the primary and secondary variables that was generated using the procedure described earlier (under the Markov–Bayes section, Fig. 4-10). The calibration plot was used to compute the conditional probability $P(A \mid C)$ and these values are tabulated in Table 4-3.

For this example, the correlation coefficient between the primary and secondary variables is assumed to be $\rho_{XY} = 0.6$ (consistent with the previous examples discussed in Chap. 3 and in this chapter) which yields the estimate for tau to be $\tau_{XY} = 1 - \rho_{XY}^2 = 0.64$. In the formulation for tau presented in Eq. (4-56), the conditional correlation is required. The calibration plot is computed based on the data for the same reservoir and so the correlation coefficient ρ_{XY} is assumed to be the conditional correlation. The procedure using the mixture of Gaussian hypothesis is rather more involved and hence not shown here.

In order to model the conditional probability derived using the primary (depth) data, fifty realizations of the spatial depth

Secondary Value	<=15	<=25	<=35	<=45
5.229	1.000	1.000	1.000	1.000
5.594	0.940	1.000	1.000	1.000
5.850	0.500	1.000	1.000	1.000
5.656	0.500	1.000	1.000	1.000
5.404	0.920	1.000	1.000	1.000
5.624	1.000	1.000	1.000	1.000
5.982	0.860	1.000	1.000	1.000
6.226	0.375	0.875	1.000	1.000
:	:	:	:	:

TABLE 4-3 Tabulated Values for $P(A \mid C)$ Corresponding to Four Thresholds Computed Using the Calibration Plot in Fig. 4-10 and the Secondary Data in Fig. 4-7

distribution were generated using the sequential indicator simulation method discussed earlier. Thus, at each simulation node, 50 outcomes are available corresponding to the number of realizations generated. The conditional *pdf* or the *cdf* at each location can be generated at each location corresponding to the four thresholds indicated in Table 4-3. The resultant values of $P(A \mid B)$ for the first few nodes of the simulation grid are shown in Table 4-4.

The prior probabilities corresponding to the four depth thresholds are (0.20, 0.52, 0.70, 0.87) consistent with the previous examples presented. The *pdf* values corresponding to the *cdf* values for $P(A)$, $P(A \mid B)$, and $P(A \mid C)$ can be calculated by differencing the *cdf* values between adjacent thresholds. Using the *pdf* values, the odds ratios a, b and c can be calculated for each threshold. For example, corresponding to the first threshold:

$$a = \frac{1 - 0.2}{0.2} = 4.0$$

$$b = \frac{1 - 0.66}{0.66} = 0.515$$

$$c = \frac{1 - 1.0}{1.0} = 0.0$$

This results in: $K = \left(\dfrac{c}{a}\right)^{\tau} \cdot b = \left(\dfrac{0}{4}\right)^{0.64} \cdot 0.515 = 0.0$

<=15	<=25	<=35	<=45
0.660	0.660	0.820	0.920
0.740	0.740	0.840	0.980
1.000	1.000	1.000	1.000
0.700	0.980	0.980	0.980
0.420	0.960	0.960	0.980
0.220	0.920	0.920	0.940
0.520	0.520	0.780	0.900
0.540	0.540	0.760	1.000
:	:	:	:

TABLE 4-4   Tabulated Value of P(A | B) Computed Using a Suite of 50 Simulated Realization of Depth Conditioned to the Primary Data

The required conditional probability $P(A \mid B, C)$ is then:

$$P(A \mid B, C) = \frac{1}{1 + K} = 1.0$$

This suggests that the value of the secondary variable at the simulation location is such that the simulated value is guaranteed to be within the first interval. Similar calculations can be performed for the other thresholds. This will yield the *pdf* values and by aggregating the values up to a threshold, we can compute the *cdf* values. The calculated *cdfs* at the first few simulation nodes are tabulated in Table 4-5.

The table reveals that there is no uncertainty associated with the outcomes at the first and the third simulation nodes. In order to generate a spatial map of depth, we can employ Monte Carlo sampling that will require us to specify how we want to interpolate values between the thresholds and how we want to extrapolate at the tails. For this example, we will stick to linear inter/extrapolation; however, any of the interpolation/extrapolation schemes discussed earlier can be used. Two realizations obtained by Monte Carlo sampling are shown in Fig. 4-13. It can be seen that the results are broadly consistent with the results obtained using other data integration approaches (indicator co-kriging, Markov–Bayes, etc.). However, the influence of the secondary data is more pronounced in the result shown in Fig. 4-13. The simulated map looks more diffused. In addition, the big advantage using the permanence of ratio approach is that the residual uncertainty associated with the estimation of the simulated value at a node ($P(A \mid B, C)$) is explicitly available.

<=15	<=25	<=35	<=45
1.0000	1.0000	1.0000	1.0000
0.9757	0.9757	0.9759	0.9762
1.0000	1.0000	1.0000	1.0000
0.6874	1.0000	1.0000	1.0000
0.7582	1.0000	1.0000	1.0000
0.9976	1.0000	1.0000	1.0000
0.8937	0.8937	0.8943	0.8945
0.6727	0.6728	0.8492	0.8498
0.4202	0.4370	0.6818	0.9896
:	:	:	:

TABLE 4-5 Calculated Values for the Conditional *cdf* Values $P(A \mid B, C)$ at the First Few Simulation Nodes

4.5.2 The Nu Model

Equation (4-50) is the fundamental equation underlying the tau model discussed in the earlier section. Taking the natural log of both sides, we can write:

$$\ln\left(\frac{x}{a}\right) = \ln\left(\frac{b}{a}\right) + \tau(B,C) \cdot \ln\left(\frac{c}{a}\right) \qquad (4\text{-}65)$$

That is, the tau model can be interpreted as an additive model for data integration of the form:

$$\ln\left(\frac{x}{a}\right) = \sum_{k=1}^{K} \tau_k \cdot \ln\left(\frac{x_k}{a}\right) \qquad (4\text{-}66)$$

x_k are the odds ratios for the K data events. Polyakova & Journel (2007) instead present a multiplicative model for data integration. Starting with the definition of a conditional probability:

$$P(A \mid D_1, D_2, ..., D_n) = \frac{P(A \mid D_1, D_2, ..., D_n)}{P(D_1, D_2, ..., D_n)} = \frac{P(D_n \mid \bar{D}_{n-1}, A) \cdot P(\bar{D}_n, A)}{P(D_1, D_2, ..., D_n)}$$

$$= P(A)\frac{P(D_n \mid \bar{D}_{n-1}, A) \cdot P(D_{n-1} \mid \bar{D}_{n-2}, A)...P(D_1 \mid A)}{P(D_1, D_2, ..., D_n)} \qquad (4\text{-}67)$$

$\bar{D}_{i-1}$ represents all data events other than the event D_i. As before, we can define the odds ratio:

$$x = \frac{P(\bar{A} \mid D_1, D_2, ..., D_n)}{P(A \mid D_1, D_2, ..., D_n)} = \frac{P(\bar{A})}{P(A)}\prod_K\frac{P(D_i \mid D_{i-1}, \bar{A})}{P(D_i \mid D_{i-1}, A)} \qquad (4\text{-}68)$$

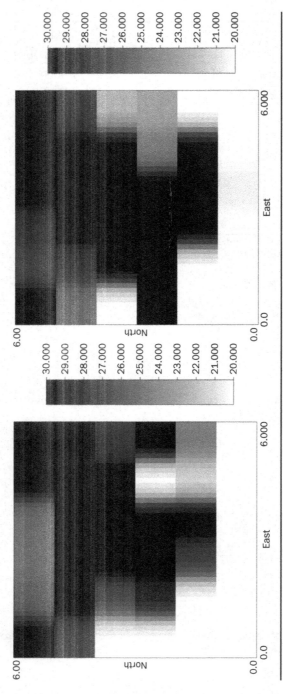

Figure 4-13 Two realizations obtained using the permanence of ratio hypothesis to combine information from the primary and secondary data. (The color version of this figure is available at www.mhprofessional.com/PRMS.)

Define the *Nu* parameters:

$$\eta_i = \frac{\dfrac{P\left(D_i \mid D_{i-1}, \bar{A}\right)}{P\left(D_i \mid D_{i-1}, A\right)}}{\dfrac{P\left(D_i \mid \bar{A}\right)}{P\left(D_i \mid A\right)}} \qquad (4\text{-}69)$$

and recognizing that:

$$\frac{P\left(D_i \mid \bar{A}\right)}{P\left(D_i \mid A\right)} = \frac{P\left(\bar{A} \mid D_i\right)}{P\left(A \mid D_i\right)} \cdot \frac{P(A)}{P(\bar{A})} = \frac{x_i}{x_o}$$

The parameter $x_o = \dfrac{P(\bar{A})}{P(A)}$ is analogous to the odds ratio a in our earlier discussions on the tau model. Substituting Eq. (4-68), we get the *Nu* model as:

$$\frac{x}{x_o} = \prod_K \frac{P\left(D_i \mid D_{i-1}, \bar{A}\right)}{P\left(D_i \mid D_{i-1}, A\right)} = \prod_K \eta_i \cdot \left(\frac{x_i}{x_o}\right) \qquad (4\text{-}70)$$

Equation (4-70) reveals that the *Nu* model is in fact a multiplicative model. The η parameters have been derived on the basis of Eq. (4-67) and depend on the order in which the data is assimilated, i.e., on the definition of D_i and D_{i-1}. The application of this *Nu* model on a synthetic example is also presented in Polyakova & Journel (2007).

4.6 References

Alabert, F. (1987). The practice of fast conditional simulations through the LU decomposition of the covariance matrix. *Mathematical Geology*, 19, 369-386.

Caers, J., & Srinivasan, S. (2003). Combining geological information with seismic and production data. In M. Nikravesh, F. Aminzadeh, & L. A. Zadeh (Eds.), *Soft Computing and Intelligent Data Analysis in Oil Exploration, Developments in Petroleum Science* (Vol. 51, pp. 499-525). Elsevier.

Caers, J., & Srinivasan, S. (2002). Statistical Pattern Recognition and Geostatistical Data Integration. In P. Wong, F. Aminzadeh, & M. Nikravesh (Eds.), *Soft Computing for Reservoir Characterization and Modeling. Studies in Fuzziness and Soft Computing* (Vol. 80). Physica, Heidelberg.

Chiles, J. P. (1999). *Geostatistics: Modeling Spatial Uncertainty.* New York: Wiley Interscience.

Dawid, A. P. (1979). Conditional Independence in Statistical Theory. *Journal of the Royal Statistical Society: Series B (Methodological)*, 41 (1), 1-15.

Deutsch, C. V. (1998). *GSLIB: Geostatistical Software Library and User's Guide* (2nd ed.). New York, U.S.A.: Oxford University Press.

Goovaerts, P. (1997). *Geostatistics for Natural Resources Evaluation.* Oxford, U.K.: Oxford University Press.

Journel, A. G. (1978). *Mining Geostatistics.* London: Academic Press.

Journel, A. G. (2002). Combining Knowledge from Diverse Sources: An Alternative to Traditional Data Independence Hypotheses. *Mathematical Geology*, 34, 573-596.

Journel, A. (1983). Nonparametric estimation of spatial distributions. *Mathematical Geology, 15*, 445-468.

Kashib, T., & Srinivasan, S. (2006). A probabilistic approach to integrating dynamic data in reservoir models. *Journal of Petroleum Science and Engineering, 50* (3-4), 241-257.

Krishnan, S. (2008). The Tau Model for Data Redundancy and Information Combination in Earth Sciences: Theory and Application. *Mathematical Geosciences, 40* (6), 705-727.

Krishnan, S., Boucher, A., & Journel, A. G. (2004). Evaluating information redundancy through the tau model. Geostatistics Banff.

Mardia, K. K. (1979). *Multivariate Analysis*. London, UK: Academy Press.

McEliece, R. (2002). *The Theory of information and coding: a mathematical framework for communication*. Cambridge University Press.

Naraghi, M. E., & Srinivasan, S. (2015). Integration of Seismic and Well Data to Characterize Facies Variation in a Carbonate Reservoir - The Tau Model Revisited. *Petroleum Geostatistics 2015*. European Association of Geoscientists & Engineers.

Polyakova, E. I., & Journel, A. G. (2007). The Nu Expression for Probabilistic Data Integration. *Mathematical Geology, 39*, 715-733.

Scholz, F. T. (1995). *Nonparametric Tail Extrapolation, Simulation Results*. Information & Support Services. Seattle: Boeing Corporation.

Srinivasan, S., & Sen, M. (2010). Mapping of diagenesis in a carbonate reservoir in the Gulf of Mexico by a stochastic data integration technique. *SEG Technical Program Expanded Abstracts*. Society of Exploration Geophysicists.

Yaglom, A. (1962). *An Introduction to Mathematical Stationary Random Functions*. Englewood Cliffs, NJ: Prentice-Hall.

Yao, T. (2002). Integrating Seismic Data for Lithofacies Modeling: A Comparison of Sequential Indicator Simulation Algorithms. *Mathematical Geology, 34*, 387-403.

Zhu, H. J. (1993). Formatting and Integrating Soft Data: Stochastic Imaging via the Markov-Bayes Algorithm. In S. A., *Geostatistics Tróia' 92*. Dordrecht, NL: Springer.

CHAPTER 5

Geostatistical Simulation Constrained to Higher-Order Statistics

As seen in the previous chapters, when indicator-based methods were discussed, the conditional probability of the outcome of a random variable (RV) can be conveniently expressed in the form of a conditional expectation of the indicator RV. By computing that expectation threshold by threshold, the entire conditional *cdf* can be computed. In indicator kriging, this conditional expectation is calculated by linearly combining the available data that have been indicator coded for the selected thresholds. This unique property of indictor RVs to directly yield the conditional probability when linearly combined, together with the insensitivity of indicators to any non-linear transformation, allows indicator geostatistics to break the shackles of traditional multi-Gaussian and linear geostatistics (kriging and Gaussian simulation). Although the indicator-based approaches presented thus far possess these important advantages, they do suffer from a limitation. Spatial continuity is represented in these algorithms in the form of the indicator covariance function. The classical indicator covariance, though it does have an interesting interpretation as a transition probability, still represents spatial variability over pairs of locations. It may not be possible to represent spatial features that exhibit complex spatial patterns using two-point or covariance statistics. The connectivity of complex curvilinear objects cannot be represented using a covariance computed on the bases of linear lag templates. We need to seek ways to introduce higher-order or multiple-point (mp) statistics within the indicator-based approaches.

In this chapter, the conditional expectation of an indicator RV is developed in terms of an indicator basis function. The indicator basis function yields the conditional probability at an estimation or simulation location accounting for multiple-point interactions between all RVs that characterize the domain. In order to ascertain the weights in this indicator basis function, statistics such as proportions and multiple-point indicator covariance functions are required. Once the required proportions and multiple-point indicator covariances functions are inferred, the weights in the indicator basis functions can be evaluated. This interpretation of the conditional expectation therefore becomes the basis of many algorithms for stochastic simulation accounting for multiple-point or higher-order statistics that describe the complex spatial relationships exhibited by several natural systems. These commonly implemented techniques come under the broad classification of single normal equation simulation where the probability of a simulation event conditioned to a specific multiple-point data event at the conditioning locations is inferred by scanning a training image (TI). Some of those techniques are reviewed in this chapter. In addition, an extended normal simulation (ENESIM) algorithm is presented that utilizes the powerful feature of the full indicator basis expansion. A proposal for sequential simulation using this paradigm is presented and illustrated through some simple examples.

5.1   Indicator Basis Function

Consider the indicator RV:

$$I(\mathbf{u}) = \begin{cases} 1, & \text{if } A(\mathbf{u}) = a \\ 0, & \text{otherwise} \end{cases}$$

As noted previously, the indicator RV has the important property that:

$$\begin{aligned} E\{I(\mathbf{u})\} &= 1 \times \text{Prob}(A(\mathbf{u}) = a) + 0 \times \text{Prob}(A(\mathbf{u}) \neq a) \\ &= \text{Prob}(A(\mathbf{u}) = a) \end{aligned}$$

That is, the expectation of the indicator directly yields the proportion (or equivalently, the probability) corresponding to the outcome a. Note that if the indicator RV is defined for a continuous variable by specifying a threshold a, the conditional expectation yields the conditional cumulative probability corresponding to the threshold a. This important result can be extended to the conditional expectation of the indicator given n data. In that case:

$$E\{I(\mathbf{u}) \mid (n)\} = \text{Prob}(A(\mathbf{u}) = a \mid (n))$$

In the case of continuous variables:

$$E\{I(\mathbf{u}) \mid (n)\} = \text{Prob}(A(\mathbf{u}) \leq a \mid (n))$$

Now consider a function $\varphi(n)$ defined on the basis of n indicator random variables I_α, $\alpha = 1, \ldots, n$ (Journel & Huijbregts, 1978):

$$\varphi(n) = a + \sum_{j_1=1}^{n} b_{j_1}^{(1)} \cdot I\left(\mathbf{u}_{j_1}\right) + \sum_{j_1=1}^{n}\sum_{j_2=1}^{n} b_{j_1 j_2}^{(2)} \cdot I\left(\mathbf{u}_{j_1}\right)I\left(\mathbf{u}_{j_2}\right) + \cdots$$
$$+ \sum_{j_1=1}^{n}\sum_{j_2=1}^{n} \cdots \sum_{j_n=1}^{n} b_{j_1,j_2,\ldots,j_n}^{(n)} \prod_{i=1,n} I\left(\mathbf{u}_{j_i}\right) \tag{5-1}$$

If the above expansion is truncated at the single-point term $I\left(\mathbf{u}_{j_1}\right), j_1 = 1, \ldots, n$, then $\varphi(n)$ represents a function in n dimensional space defined by the basis vectors $I\left(\mathbf{u}_{j_1}\right)$ with the projection along each basis vector $I\left(\mathbf{u}_{j_1}\right)$ given by $b_{j_1}^{(1)}$. There are $1 + \binom{n}{1} + \binom{n}{2} + \cdots + \binom{n}{n} 2^n$ terms in the above expansion. This implies that $\varphi(n)$ can be viewed as a hyper-surface in 2^n dimensional space defined by the basis vectors in Eq. (5-1) with the projection (resolution) along each basis vector direction given by the coefficients $b_{j_1,j_2,\ldots,j_n}^{(n)}$.

Given the n indicator RVs, the projection/expansion above is the most complete possible. It uses all possible combinations of the n indicator RVs. Furthermore, the indicator $I(\mathbf{u})$ is immune to non-linear transformations. Higher-order basis functions, e.g., $I^{\omega_1}\left(\mathbf{u}_{j_1}\right)$, $I^{\omega_1}\left(\mathbf{u}_{j_1}\right)I^{\omega_2}\left(\mathbf{u}_{j_2}\right)$, $I^{\omega_1}\left(\mathbf{u}_{j_1}\right)I^{\omega_2}\left(\mathbf{u}_{j_2}\right)I^{\omega_3}\left(\mathbf{u}_{j_3}\right)$, etc., revert to the base quantities $I\left(\mathbf{u}_{j_1}\right)$, $I\left(\mathbf{u}_{j_1}\right)I\left(\mathbf{u}_{j_2}\right)$, $I\left(\mathbf{u}_{j_1}\right)I\left(\mathbf{u}_{j_2}\right)I\left(\mathbf{u}_{j_3}\right)$ regardless of the exponent ω_1, ω_2 or ω_3.

Remarks

- Since we are dealing with binary indicator RVs, there are two outcomes for each $I\left(\mathbf{u}_{j_1}\right)$ and since there are n RVs, there are 2^n outcomes possible for the function $\varphi(n)$.

- There are a total of $1 + \binom{n}{1} + \binom{n}{2} + \cdots + \binom{n}{n} = 2^n$ terms on the right-hand side of the equation. Solving the $2^n \times 2^n$ system of equation corresponding to all combinations of outcomes of the RVs can yield the coefficients $b_{j_1,j_2,\ldots,j_n}^{(n)}$. However, it is evident that the system will quickly become intractable when the number of nodes that make up the domain exceeds 25 or 30.

Given n random variables, $I(\mathbf{u}_\alpha)$, $\alpha = 1, \ldots, n$, the most general estimator I^* of the indicator RV at a location $\mathbf{u}_0$ is a measureable function $f(I(\mathbf{u}_1), I(\mathbf{u}_2), \ldots, I(\mathbf{u}_n))$ of the available data. In other words, I^* is itself an RV. The function $\varphi(n)$ can be interpreted as the vector space consisting of all measurable functions of n data. Indeed, the basis functions in the expansion $\prod_{i=1,n} I\left(\mathbf{u}_{j_i}\right)$ are themselves RVs. In particular, a linear combination of the type $\sum_{j_1=1}^{n} b_{j_1}^{(1)} \cdot I\left(\mathbf{u}_{j_1}\right)$ is contained within the space defined by $\varphi(n)$ (obtained by truncating the expansion at the first-order term). Thus $\varphi(n)$ is the widest possible space where a search for an estimator I^* can be conducted.

The estimator I^* can be viewed as the projection of the "true" random variable $I(\mathbf{u})$ to the space defined by $\varphi(n)$, or equivalently, the projection to the space defined by the basis $L_n : \left\{1, I_{j_1}, I_{j_1} I_{j_2}, I_{j_1} I_{j_2} I_{j_3}, ..., I_{j_1} I_{j_2} I_{j_3} ... I_{j_n}\right\}$. Because $\varphi(n)$ is the widest possible space where a search for an estimator I^* can be conducted, the projection of the unknown I^* to $\varphi(n)$ will be the best possible estimator given the n data. Recognizing the characteristic property of conditional expectation $E\{I(\mathbf{u}) \mid (n)\}$ that it is the best estimator given n data and denoting the projection of I^* on to $\varphi(n)$ as $E_n I$, we can state that:

$$I^* = E\left\{I(\mathbf{u}) \mid I(\mathbf{u}_1), I(\mathbf{u}_2), ..., I(\mathbf{u})_n\right\} = E_n I \tag{5-2}$$

Remark Suppose D is an observed data event $D = 1$. For example, D could be a four-point data event $I_{j_1} \cdot \left(1 - I_{I_2}\right) \cdot I_{j_3} \cdot I_{j_4}$, i.e., the outcomes of the indicator random variables $I_{j_k}, k = 1, ..., 4$ are such that the product as defined above is 1. Also suppose that the basis L_n is truncated at L_1:$\{1, D\}$. In this case, the estimate I^* can be written as:

$$I^* = E\{I_o \mid D\} = a + b \cdot D$$

We still need a system to solve for the two coefficients a and b. In general, for the full expansion, we need a system of equations to solve for the 2^n coefficients $a, b_{j_1, j_2}, ..., b_{j_1, j_2, ..., j_n}$.

5.1.1 Establishing the Basis Function—Projection Theorem

The indicator expansion stated in Eq. (5-1) represents a function $\varphi(n)$ that is defined in terms of 2^n basis vectors. This implies that the notion of a Cartesian space that is restricted to two-dimensional planes and three-dimensional objects has to be extended to a Hilbert space H that generalizes the notion of Euclidean space to higher dimensions. Just like the Euclidean space, H is a vector space in which the operation of inner product of two vectors is defined that allows lengths and angles to be measured or defined.

Let H be a Hilbert space, then the target random variable $I_o \in H$. Now consider the projection of I_o to a subspace M of H. The projection theorem (Luenberger, 1969) states that corresponding to any $I_o \in H$ there exists a unique vector $I_o^* \in M$ such that:

$$\left\|\left(I_o - I_o^*\right)\right\| < \left\|\left(I_o - m\right)\right\| \forall m \in M$$

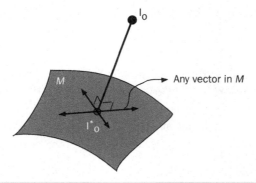

FIGURE 5-1 Conceptual drawing showing the projection of a random variable I_o to a subspace M and the existence of a unique projection I_o^* such that $(I_o - I_o^*)$ is orthogonal to all vectors in the subspace M. (The color version of this figure is available at www.mhprofessional.com/PRMS.)

m are the basis vectors defining M. Furthermore, a necessary and sufficient condition that $I_o^* \in M$ be a unique minimizing vector is that $(I_o - I_o^*)$ be orthogonal to all vectors belonging to M, as depicted in Fig. 5-1.

The projection theorem thus confirms the existence of a unique best estimator I_o^* and furthermore provides the condition that has to be satisfied in order to arrive at that unique estimator.

Remarks

- Recall that in kriging, the best estimate is obtained as the one minimizing the error variance. The variance of indicators is difficult to interpret and hence the projection theorem provides an effective alternative where we simply have to ensure that the error $(I_o - I_o^*)$ is orthogonal to the basis of the indicator expansion $\varphi(n)$.

- The orthogonality condition will yield a system of equations similar to the kriging or normal system that can be solved to obtain the coefficients $b_{j_1, j_2, \ldots, j_n}^{(n)}$. However, the solution of the full $2^n \times 2^n$ is likely to be infeasible for any reasonably sized reservoir model comprising n gridblocks.

Consider the reduced basis $L_k : \left\{ 1, I_{j_1}, I_{j_1} I_{j_2}, I_{j_1} I_{j_2} I_{j_3}, \ldots, I_{j_1} I_{j_2} \cdots I_{j_k} \right\}$. The projection $E\{I(\mathbf{u}) \mid (n)\}$ on this reduced basis can be written as:

$$I_k^* = E_k \big(I(u) \mid (n) \big) = a + \sum_{j_1=1}^{n} b_{j_1}^{(1)} \cdot I\big(\mathbf{u}_{j_1}\big) + \sum_{j_1=1}^{n} \sum_{j_2=1}^{n} b_{j_1,j_2}^{(2)} \cdot I\big(\mathbf{u}_{j_1}\big) I\big(\mathbf{u}_{j_2}\big)$$

$$+ \cdots + \sum_{j_1=1}^{n} \sum_{j_2=1}^{n} \cdots \sum_{j_n=1}^{n} b_{j_1, j_2, \ldots, j_k}^{(n)} \prod_{i=1,k} I\big(\mathbf{u}_{j_i}\big)$$

$$(5\text{-}3)$$

The above expansion has n_k terms:

$$n_k = 1 + \binom{n}{1} + \binom{n}{2} + \cdots + \binom{n}{k}$$

Let us try and simplify the expression for the expansion in Eq. (5-3) by using the following notation for the basis vectors:

$V_1 = \{1\}$, the zeroth order term

$V_\ell = [I_i], i = 1, ..., n$, the single-point terms, $\ell = i + 1$

$V_\ell = [I_i I_j], i = 1, ..., n, j = 1, ..., n$, the two-point terms,

$$\ell = (j - 1) \cdot n + i + \binom{n}{1} + 1$$

$\vdots$

$$V_\ell = \left[\prod_{m=1}^{k} I_{m_{i_m}} \right], i_m = 1, ..., n, \text{ the k-point terms,}$$

$$\ell = n^{m-1} + i_m + \sum_{m=1}^{k-1} \binom{n}{m} + 1$$

In this notation, the projection I_k^* can be expressed compactly as:

$$I_{o_k}^* = \sum_{\ell=1}^{n_k} \lambda_\ell \cdot V_\ell$$

In the notation of the projection theorem, I_k^* is a vector in L_k and it is a unique, minimum norm projection if and only if $(I - I_k^*)$ is orthogonal to all vectors $V_\ell, \ell = 1, ..., n_k$ defining the space L_k. The orthogonality condition of the projection theorem amounts to:

$$\left\langle (I - I_k^*), V_\ell \right\rangle = 0 \Leftrightarrow \left\langle I \cdot V_\ell \right\rangle = \left\langle I_k^* \cdot V_\ell \right\rangle$$

or in terms of expectations:

$$E\{I \cdot V_\ell\} = E\{I_k^* \cdot V_\ell\}, \forall \ell = 1, ..., n_k \tag{5-4}$$

This yields a system of n_k normal equations. In order to understand this system of equations, let us start with some simple examples.

Example 5-1

Let us apply the normal equation for the first basis function $V_1 = \{1\}$. Then the left-hand side of Eq. (5-4) is:

$$E\{I \cdot 1\} = E\{I\}$$

The right-hand side of the same equation is:

$$E\{I_k^* \cdot 1\} = E\left\{ \sum_{\ell=1}^{n_k} \lambda \cdot V_\ell \cdot 1 \right\} = \sum_{\ell=1}^{n_k} \lambda_\ell \cdot E\{V_\ell\}$$

Equating the left-hand and right-hand sides, we get:

$$\sum_{\ell=1}^{n_k} \lambda_\ell \cdot E\{V_\ell\} = E\{I\}$$

That is, the bases are such that the linear weighting of the expected value of the bases recovers back the expectation of the indicator random variable I, which would be a statement of unbiasedness.

Example 5-2

Now look at the system corresponding to the single-point terms, $V_\ell = I_j, j = 1,\ldots, n$. The left-hand side of Eq. (5-4) in this case works out to be $E\{I \cdot I_j\}$. The right-hand side of the same equation is:

$$E\{I_k^* \cdot I_j\} = E\left\{\sum_{\ell=1}^{n_k} \lambda_\ell V_\ell \cdot I_j\right\} = \sum_{j=1}^{n_k} \lambda_\ell \cdot E\{V_\ell I_j\}$$

Equating the left-hand and right-hand sides results in:

$$E\{I \cdot I_j\} = \sum_{\ell=1}^{n_k} \lambda_\ell E\{V_\ell \cdot I_j\}, j = 1,\ldots, n$$

This equation suggests that the normal system results in weights λ_ℓ such that the weighted linear combination of bases (V_ℓ) to data (I_j) covariances identifies the data (I_j) to unknown (I) covariance. This is often referred to as the covariance reproduction property of kriging.

Example 5-3

Now look at the orthogonality condition for the two-point bases, $V_\ell = I_{j_1} \cdot I_{j_2}; j_1, j_2 = 1,\ldots, n$.

The right-hand side of Eq. (5-4) is $E\{I_k^* \cdot I_{j_1} I_{j_1}\} = \sum_{\ell=1}^{n_k} \lambda_\ell E\{V_\ell I_{j_1} I_{j_1}\}$.

The left-hand side is $E\{I \cdot V_\ell\} = E\{I \cdot I_{j_1} I_{j_2}\}$. Equating the right- and left-hand sides, we get:

$$E\{I \cdot I_{j_1} I_{j_2}\} = \sum_{\ell=1}^{n_k} \lambda_\ell E\{V_\ell \cdot I_{j_1} I_{j_2}\}, j_1, j_2 = 1,\ldots, n$$

that is, a statement of reproduction of the three-point indicator covariances. Note that the above equation requires knowledge of up to the four-point indicator covariances if the indicator basis expansion is truncated at the second-order term. As a generalization of the above, the orthogonality condition can be represented in the form:

$$E\{I \cdot V_{\ell'}\} = \sum_{\ell=1}^{n_k} \lambda_\ell \cdot E\{V_\ell \cdot V_{\ell'}\}, \ell' = 1,...,n_k$$

$$E\{I \cdot V_\ell\} = E\{I_k^* \cdot V_\ell\}, \forall \ell = 1,...,n_k \qquad (5\text{-}5)$$

If the expansion is truncated at the same order as ℓ', then the expansion will require knowledge of $2 \times \text{Order}(\ell')$ indicator covariances.

5.1.2   Single Extended Normal Equation

The normal system and the number of equations in the normal system detailed above scale according to the total number of realizations (2^N) of the N binary indicator RV data. If our interest is to obtain the estimate I_o^* corresponding to a specific realization of the indicator RVs, say $D = \{i_1 \cdot i_2, ..., i_n\} = 1$, then we can write the indicator basis expansion in terms of the specific data event D as (Guardiano and Srivastava, 1993):

$$I_o^* = \varphi(D) = \lambda_o + \lambda_1 \cdot D \qquad (5\text{-}6)$$

Applying the orthogonality condition to the basis $\{1\}$:

$$E\{I_o \cdot 1\} = E\{I_o^* \cdot 1\} = E\{\lambda_o + \lambda_1 D \cdot 1\} = \lambda_o + \lambda_1 \cdot E\{D\}$$

This yields in: $\lambda_o = p_o - \lambda_1 \cdot E\{D\}$
Substituting Eq. (5-6), we get:

$$I_o^* - p_o = \lambda_1 \cdot (D - E\{D\}) \qquad (5\text{-}7)$$

Now only the weight λ_1 assigned to the pattern of data D needs to be solved. Applying the orthogonality condition to the basis D:

$$E\{I_o \cdot D\} = E\{I_o^* \cdot D\} = E\{P_o + \lambda_1 \cdot (D - E\{D\}) \cdot D\}$$
$$= P_o \cdot E\{D\} + \lambda_1 \cdot E\{D^2\} - \lambda_1 \cdot [E\{D\}]^2$$

This yields:

$$\lambda_1 = \frac{E\{I_o \cdot D\} - P_o \cdot E\{D\}}{Var\{D\}}$$

Because $p_o = E\{I_o\}$, this implies that:

$$\lambda_1 = \frac{E\{I_o \cdot D\} - E\{I_o\} \cdot E\{D\}}{Var\{D\}}$$

Because D is a binary (0 or 1) event, $Var\{D\} = E\{D\} \cdot (1 - E\{D\})$.

Specifically, the conditioning data $D = 1$ (i.e., D is the existence of a specific pattern) and substituting this in Eq. (5-7), we get:

$$I_o^* - p_o = \lambda_1 \cdot (1 - E\{D\})$$

Substituting the expression for the weight λ_1, we get:

$$I_0^* = \frac{E\{I_o D\}}{E\{D\}} = E(I_o \mid D) \tag{5-8}$$

indicating that the estimate for I_0^* is identically the conditional expectation of I_o given D; i.e., I_0^* is the best estimate possible given the data event D.

Remarks

- The weight λ_1 in this case is data-dependent (specific to the event D).
- The indicator estimate $I_0^* = \dfrac{E\{I_o D\}}{E\{D\}}$ entails computing the joint probability of the data event D, as well as the indicator outcome I_o (binary) at the simulation node. The denominator is the probability of the data event D. Both the probability in the numerator and the probability in the denominator can be obtained by scanning an image and counting the number of occurrences of spatial patterns. In the numerator, the pattern is dictated by the join occurrence of both an outcome of I_o and D. In the denominator, it is only the occurrence of D. However, in order to get these counts, an exhaustive image is necessary that exhibits the spatial patterns deemed representative of the target reservoir. That image is referred to as the TI.
- Considering a three-point data template $N = 3$, with the data event $i_1 = 1, i_2 = 0$, and $i_3 = 1$, the data event $D = 1$, corresponds to $D = i_1 \cdot (1 - i_2) \cdot i_3 = 1$.

5.1.3 Single Normal Equation Simulation

Guardiano and Srivastava (1993), and later Caers (1998) and Strebelle (2002), proposed the idea of borrowing conditional probabilities directly from a TI, allowing the use of higher-order or mp-statistics to reproduce geological structures and patterns. The main components of such multiple-point simulation algorithms based on the single extended normal equation derived earlier include the generation or acquisition of TIs, optimal definition of scanning templates to capture the proper conditioning information from the TI, computational schemes to scan, store, and reproduce complex spatial features, and a simulation algorithm for imposing the scanned statistics on a reservoir model, while at the same time honoring all the available data.

Training Image

Training image (Journel and Zhang, 2006) is a conceptual description of the random process that captures the essential pattern of spatial variability characterizing the subsurface deposit. TIs have great impact on reproduced patterns and are analogous to the set of face-images used, for example, in a face recognition algorithm.

Scanning Template

Scanning template is an interface for retrieving mp-statistics from TIs. The reliability of the final reproduced patterns is highly sensitive to the template shape and size (Huang and Srinivasan, 2012).

Scanning Technique

The process of scanning the TIs using a spatial template and keeping track and storing the frequency of different observed patterns in an efficient way has received considerable research focus in the recent years. The search tree, a TI scanning technique (Strebelle, 2002) motivated widespread application of the multiple-point statistics approach. In the search tree approach, the scanned statistics are recorded in a dynamic data structure that renders it convenient to store nested information, such as number of outcomes of data events (joint probabilities).

Pattern Reproduction

Several methods for reproducing patterns in subsurface geologic models by incorporating mp-statistics or patterns extracted from TIs have been proposed. Qiu & Kelkar (1995) used simulated annealing to honor mp-statistics borrowed from a TI of categorical facies data. Strebelle (2000) developed an algorithm based on the single extended normal equation. That algorithm sequentially simulates one node at a time conditioned to the hard data. Conditional probabilities at each step are built based on the frequencies of different patterns extracted from TIs.

In the subsequent sections, details are provided regarding each of these crucial aspects of single extended normal equation-based implementations.

Training Image Selection

The common practice is to import the required statistics for the single extended normal equation from a TI, which is a numerical representation of the spatial law that describes the geological structures and patterns to be reproduced in the target reservoir. TIs embody reservoir characteristics that are typical for a particular depositional environment and are generally based on outcrops and in some cases, aerial photographs. A valid TI would be one that is sufficiently large such that it yields all required mp-statistics. Process-based modeling for TI generation could result in a library of TI with realistic geological features. Training models for deepwater systems (Pyrcz, Catuneanu and Deutsch, 2005) and for fluvial environments (Pyrcz, Boisvert and Deutsch, 2008) have been developed.

In most single extended normal [multiple point statistics (MPS)] implementations, a gridded TI is required for inferring the statistics. However, in cases such as modeling karst networks (Erzeybek, Srinivasan and Janson, 2012) or natural fracture networks (Liu and Srinivasan, 2004), the TI comprises surveys reporting XYZ coordinates of feature central line. Therefore, it is not practical to perform gridding and apply a rigid gridded template to calibrate statistics using such "point" data surveys. In order to address this issue, Erzeybek (2012) presents an MPS algorithm that works on a non-gridded basis to characterize cave and paleokarst networks using "point-set" information. The algorithm basically consists of MPS analysis using flexible spatial templates (Fig. 5-2) and subsequent pattern simulation using the calibrated statistics. Instead of a rigid template with a configuration of data nodes surrounding a central node, the template is specified with angle and lag tolerances that can capture the complex

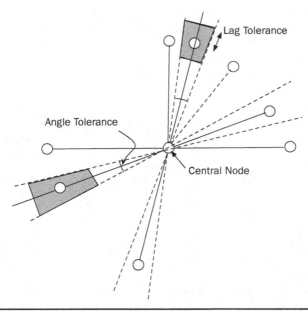

FIGURE 5-2 A flexible eight-node template. Points falling within the specified lag and angle tolerances are pooled together in order to compute the multiple-point pattern statistics. (The color version of this figure is available at www.mhprofessional.com/PRMS.)

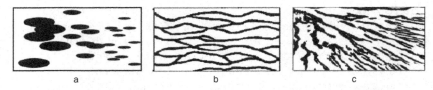

a   b   c

FIGURE 5-3 Three possible candidates for training images: **a)** Elliptical shale lenses exhibiting a spatial trend, **b)** fluvial channel system, and **c)** deltaic-type reservoir exhibiting a progress orientation change in tributary channel orientations. Only **b)** exhibits spatial stationarity and is therefore suitable as a training image.

patterns exhibited by cave connections or fracture/fault junctions specified using point-set data.

An important challenge associated with TI selection lies in the fact that the construction and use of TIs are restricted by the principles of stationarity and ergodicity. In other words, the frequency of patterns in various parts of the image should be same and the TI should not show any progressive trend in the orientation of objects, etc. This means that simple drawings cannot be used as a TI if they do not follow these principles. Considering the example drawn from Caers and Zhang's research (2004), Fig. 5-3 shows three possible candidates for a TI: a set of elliptical shapes, a fluvial channel reservoir and a fan-like deposit. According to Caers & Zhang (2004), under the principle of stationarity, only TI 2 can be used as a TI. Training image 1 does not follow the principle of stationarity because the patterns are changing over the TI. Training image 3 exhibits a systematic trend in the orientation of objects.

Scanning Template

An interface or a scanning template is needed to search and extract pattern replicates from the TI. A typical search template T_J is defined by J vectors radiating from a central node $\mathbf{u}_0$ (Strebelle and Journel, 2001), i.e., $(T(\mathbf{u}_0 + \mathbf{h}_j), j = 1, \ldots, J$. For a typical central node $\mathbf{u}'_0$ of the template T_J, a pattern from TI can be extracted as:

$$pattern\left\{\mathbf{u}'_0\right\} = \left\{ti(\mathbf{u}'_0); ti(\mathbf{u}'_0 + \mathbf{h}_j), j = 1, \ldots, J\right\}$$

Here, $ti(\mathbf{u}'_0 + h_j)$ is the TI value at grid node $\mathbf{u}'_0 + h_j$. All these training patterns extracted from TIs using the template are stored in a search tree for fast access during simulation. The template size and shape affect the speed and effectiveness of the algorithm for pattern reproduction.

The higher-order statistics to which the simulated models will be constrained are fully characterized by the spatial template. A conventional template could be characterized by seven parameters:

1. Number of nodes in the search template
2. Three parameters defining the template dimensions in 3D
3. Three angles defining the template orientation—azimuth, dip, and rake

Depending on the type of geological features present in the TI and the continuity of these features, the template definition should be flexible enough to follow the geometry of the different objects. The impact of template size and shape on final reproduced patterns is shown in Fig. 5-4. Based on these results, it is evident that models generated using the two sub-optimal templates are incompatible

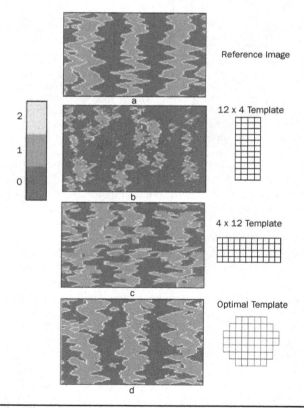

Figure 5-4 Effect of the multiple-point data template on multiple-point simulation results: **a)** Reference image, **b)** image simulated using a sub-optimal template, **c)** image simulated using sub-optimal 4 × 12 template, and **d)** image simulated using optimal spatial template. (The color version of this figure is available at www.mhprofessional.com/PRMS.)

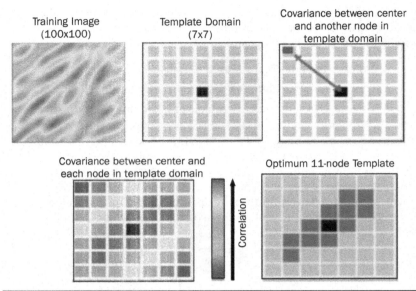

Training Image (100x100)

Template Domain (7x7)

Covariance between center and another node in template domain

Covariance between center and each node in template domain

Correlation

Optimum 11-node Template

FIGURE 5-5 Schematic for optimal template selection procedure. From top-left to bottom-right: the training image (TI), template frame which translates on TI, covariance calculated between values observed at central node and a node in the template domain. The covariance map between each node and center node is computed. Optimal template is finalized by defining a cutoff to retain high covariance nodes in the covariance map. (The color version of this figure is available at www.mhprofessional.com/PRMS.)

with the reference. In contrast, an "optimal" template defined as one consistent with the pattern of features observed in the geologic system results in a more representative model, as also seen in Fig. 5-4.

A fast and robust algorithm to derive optimal spatial templates is presented in Eskandari (2008). Given a user-specified template grid, the algorithm searches the TI with two-point templates using the central node in the template grid and any other node in the grid. These two-point spatial correlations are ranked and finally a restricted set of template nodes are selected based on the ranking. The details of the calculation procedure are given below and illustrated in Fig. 5-5:

- Algorithm starts with defining a 3D cubic template grid in rectangular coordinates. The numbers of cells in the three coordinate directions are user specified. In the next steps, the final template shape and size would be determined by removing some of the cells from the specified grid. The size of the template grid is generally guided by the available *cpu*, complexity of the geological image, etc.

- The TI is scanned with the grid defined in the previous step. During the scanning process, each TI value observed at each template grid node is stored in an array corresponding to that specific node.

- The covariance between the array corresponding to the central node and the array at any node in the template grid is calculated. This procedure is repeated for all the nodes in the grid. Final outcome in this step is the covariance map that shows the linear correlation between central node in the template grid and the rest of the grid nodes.

- The cells in the template grid are ranked from the highest to the lowest covariance values (with respect to the central node).

  - The final configuration of the template grid is based on the ranking in the previous step and the user specification for the maximum size of the template. The selected nodes are nodes with the higher covariance values. For example, if the total number of nodes in the template grid is 40 and the user stipulates the final template to be made up of a maximum of 23 nodes, the nodes with highest 23 covariance values are retained and the rest are removed from the grid.

Figure 5-6 depicts cases in which optimal templates result in accurate pattern reproduction. In this figure, TIs are shown in the first row, the corresponding optimal template determined by implementing the algorithm detailed above is shown in the second row, and the corresponding simulated images obtained using the optimal templates are shown in the third row.

Scanning Strategy

In order to perform mp-statistics-based simulation, the probability of drawing an outcome for the simulation event given the configuration of data on the spatial template is essential. That probability is stored in a multiple-point histogram. Such a histogram summarizes the frequency of observed patterns in the TIs. Figure 5-7 depicts the procedure for constructing the mp histogram from a typical TI. The scanning and extraction of mp-statistics as depicted in Fig. 5-7 are not a trivial task. For most cases, where the average size of geological objects in the TIs and consequently the templates is big, the scanning technique and the mechanism for storage of the statistics need to be designed such that the scanning is *cpu* efficient.

In order to render the task of scanning, storing, and subsequently retrieving mp pattern frequencies more efficiently, a tree concept that helps keep track of pattern occurrences in an efficient manner can be devised. A scheme could be to store each nine digits of the pattern in a different array and then compare patterns by comparing the first array comprising the first nine digits, if they are equal, then comparing the second array, and so on. Further efficiencies to this task are possible by implementing the procedure in a distributed computing environment. The task of comparing and sorting a particular array

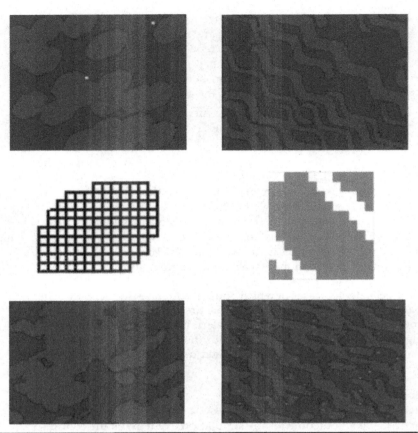

FIGURE 5-6 Cases demonstrating the optimal template computed using the algorithm. The top row depicts the training images, the middle row depicts the optimal template computed using the algorithm, and the bottom row depicts the corresponding simulated images. (The color version of this figure is available at www.mhprofessional.com/PRMS.)

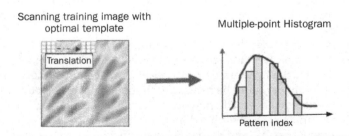

FIGURE 5-7 The procedure for building a multiple-point histogram. (The color version of this figure is available at www.mhprofessional.com/PRMS.)

for several observed patterns, e.g., the first array of nine digits, can be assigned to the first processor and the subsequent arrays to other different processors.

As an example, suppose a pattern is coded as 124563968914(pat1) and the second one is coded 458921573512(pat2). These integer strings correspond to a twelve-point spatial template. In the first pattern, for example, the indicator code1 is observed in the first spatial node of the template, whereas an indicator code 4 is observed at the same node for the second pattern. The first nine digits of the patterns can be written in the form of 124563968 (array 1, pattern 1) and 458921573 (array 1, pattern 2) and the rest of pattern are in the form of 914 (array 2, pattern 1) and 512 (array 2, pattern 2). For comparing these two patterns, we can first compare the first arrays; in case they are equal, we compare the second array. In this example, arrays 1 are different for the two patterns and furthermore pattern 2 is sorted as a higher pattern index than pattern 1. In the event, the integers in the first array were identical for the two patterns, and the sorting of pattern indices would be based on the second array. In this example, the size of the first array is specified to be nine digits. This is to enable the arrays to be stored as a four-byte integer.

Strebelle (2002) presents a variation of this idea where the generated pattern statistics are stored in a search tree. In order to understand this, let us consider a scenario where the search template in Fig. 5-8 is used to scan the TI shown in the same Fig. 5-8. Table 5-1 shows the resulting pattern statistics generated.

Note that the counts in the table correspond only to the occurrences when the entire template is contained within the TI. Strebelle (2002) and Zhang et al. (2012) provide full details of how a search tree is constructed and how pattern statistics can be retrieved from the search tree. The first branch of the search tree distinguishes between the occurrences of two data events (yellow or white) in the central node of the template. Looking at Table 5-1, there are

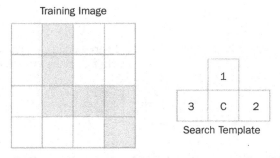

FIGURE 5-8 A 4 × 4 training image (left) and a search template (right). (The color version of this figure is available at www.mhprofessional.com/PRMS.)

Unique Pattern	Frequency of Occurrence of Pattern	Frequency of Occurrence of Event in the Central Node of the Template	
		C	C
[1 / 3 2]	2	1	1
[1 / 3 2]	0	0	0
[1 / 3 2]	2	1	1
[1 / 3 2]	0	0	0
[1 / 3 2]	0	0	0
[1 / 3 2]	1	1	0
[1 / 3 2]	0	0	0
[1 / 3 2]	1	0	1

TABLE 5-1 Pattern Statistics Contained in Training Image Shown in Fig. 5-8

three occurrences of yellow (gray) and three occurrences of white. The first branch of the search tree can therefore be denoted as:

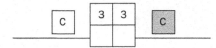

Next, we can look at the second level of the search tree: when the central node is white, the first node of the template can be either white or yellow. Using the results tabulated in Table 5-1, the frequency of data events at template node 1 corresponding to the occurrence of white in the central node can be denoted as:

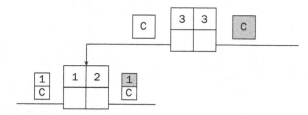

Completing this level corresponding to the central node being yellow, we get:

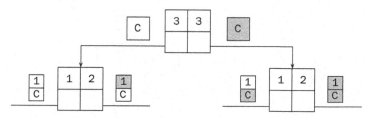

Continuing in this manner, the entire search tree shown in Fig. 5-9 can be constructed.

The search tree is efficient for storing pattern statistics and facilitates retrieval of these statistics during stochastic simulation. However, the search tree is also very memory-intensive and hence not very attractive when using a large TI to retrieve the pattern statistics. Straubharr et al. (2011) proposed the replacement of the dynamic search tree with a list for storage of multiple-point pattern statistics This list-based alternative has a low memory requirement and allows for easy parallelization (Straubharr et al., 2013). However, MPS simulations based on a list structure are generally slower than equivalent MPS simulations that use a dynamic search tree. This is because branches of a search tree provide shortcuts that speed up the retrieval of the mp-statistics. Lists do not have branches and therefore do not benefit from the shortcuts made possible by the presence of branches in a data structure. Retrieving conditional statistics from a list requires scanning all elements of the list to find a pattern that matches the conditioning data event. The process of scanning the entire list is slow and memory-intensive.

Scanning and Storing in a Hash Table

As noted previously, storage of pattern statistics in a dynamic search tree has a large memory requirement especially for large complex TIs. Storage of pattern statistics in a list mitigates the memory issue associated with storage in a search tree. However, list-based storage of pattern statistics leads to slower MPS simulations because retrieval of statistics from a list is slow. This is because all elements in the list have to be scanned to look up the pattern corresponding to a conditioning data event.

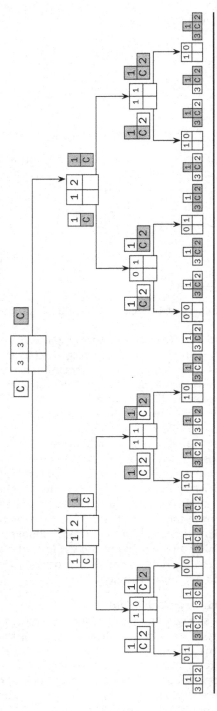

Figure 5-9 Search tree corresponding to the training image and the search table template in Fig. 5-8. The pattern along a branch of the search tree is shown and the corresponding frequency of occurrence is shown as [x y]. The frequency of pattern occurrence is calculated using the outcomes presented in Table 5-1. (The color version of this figure is available at www.mhprofessional.com/PRMS.)

Chukwudiuto (2018) demonstrates a hash table based approach for storing the multiple point statistics. A hash table has the potential to mitigate the memory issue associated with search trees and the slow retrieval of conditioning statistics associated with lists. A hash table is a data structure that stores data in an associative form. This means that data stored in a hash table come as a key-value pair. In this approach keys are mapped to values for highly efficient lookup (McDowell, 2015). The implementation of a hash table consists of an underlying array and a hash function. Each data stored in the array is contained in a unique array index. This implies that access or retrieval of any data will be very fast if we know the array index of the data. When data (data value and its corresponding key) is to be inserted into a hash table, the hash function maps the key to an integer, which indicates the index in the array where the data will be stored. When searching for a data value contained in a hash table, the hash function uses the key that corresponds to the data value to compute the array index, which points to the location where the data value will be found. The implication is that unlike lists, retrieving conditional statistics from a hash table will not require a scan of all elements of the underlying array in the hash table. The use of hash table for storage of TI pattern statistics provides a fast and direct access to the conditional statistics needed for MPS simulation. When storing pattern statistics in a hash table, the configuration of neighborhood data events is used as the key, while the facies outcome at the central node is used as the data value.

To demonstrate how pattern statistics are stored in a hash table, consider the list of all patterns presented in Table 5-1 corresponding to an imaginary TI.

As discussed earlier, a hash table consists of an array and a hash function. An example of a hash function is:

$$\text{hash function} = \sum_{i=1}^{n} \left\{ 2^{\text{data value}(i)} \cdot 31^{n-1} \right\} \bmod N \qquad (5\text{-}9)$$

The size of the template is denoted as n. The hash table array is assumed to be of size N. In the example detailed below, the size of the array is assumed to be 9 (to cover the number of patterns presented in Table 5-2). We need each element of the array to be directly accessed with a unique integer called an index. For an N-sized array, each integer in the set of consecutive integers ranging from 0 to $N-1$ inclusive can be used to directly access a unique array element. The prime number 31 in the function is used so that the resultant outcomes of the hash function follow a fairy even probability distribution. This is the prime number used in the implementation of the hash-code function in Java. As such, the quantity within the summation can yield rise to impossibly large numbers. We would like the generated values to fall within the range 0 to $N-1$. This is accomplished by applying the modulo operator mod N (i.e., the remainder of the integer division by N). The application of this hash function to the entries in Table 5-2 yields Table 5-3.

Pattern	Frequency of Data Event at Center Node of Search Template	
	C	C
1 2 3	2	4
1 2 **3**	0	7
1 **2** 3	2	9
1 **2 3**	5	1
1 2 3	5	7
1 2 **3**	3	3
1 2 3	1	6
1 2 3	8	0

Table 5-2 Patterns and Assumed Statistics for Demonstrating Pattern Scanning and Retrieval Using Hash Functions

Pattern	String	Hash Function Value
1 2 3	0 0 0	$\{2^0 \cdot 31^2 + 2^0 \cdot 31^1 + 2^0 \cdot 31^0\} \bmod 9 = 3$
1 2 **3**	0 0 1	$\{2^0 \cdot 31^2 + 2^0 \cdot 31^1 + 2^1 \cdot 31^0\} \bmod 9 = 4$
1 **2** 3	0 1 0	$\{2^0 \cdot 31^2 + 2^1 \cdot 31^1 + 2^0 \cdot 31^0\} \bmod 9 = 7$
1 **2 3**	0 1 1	$\{2^0 \cdot 31^2 + 2^1 \cdot 31^1 + 2^1 \cdot 31^0\} \bmod 9 = 8$
1 2 3	1 0 0	$\{2^1 \cdot 31^2 + 2^0 \cdot 31^1 + 2^0 \cdot 31^0\} \bmod 9 = 1$
1 2 **3**	1 0 1	$\{2^0 \cdot 31^2 + 2^0 \cdot 31^1 + 2^1 \cdot 31^0\} \bmod 9 = 2$
1 2 3	1 1 0	$\{2^1 \cdot 31^2 + 2^1 \cdot 31^1 + 2^0 \cdot 31^0\} \bmod 9 = 5$
1 2 3	1 1 1	$\{2^1 \cdot 31^2 + 2^1 \cdot 31^1 + 2^1 \cdot 31^0\} \bmod 9 = 6$

Table 5-3 Storage of Patterns and Their Corresponding Statistics in a Hash Table

As the search template scans the TI, a pattern (data events) captured by the search template is converted to integer value using the hash function shown in Table 5-3. This integer value is used as an index integer to directly access the element of the array, where the statistics of the captured pattern is to be stored or updated. For example, consider the array index 1 that corresponds to the pattern | 1 | 2 | 3 | has a frequency of occurrence of 5 with the value in the central node being 0 and a frequency of 7 with the value in the central node being 1. The array index 1 therefore corresponds to the (key, value) pair: ["1 0 0," [5,7]]. Table 5-4 shows how the patterns and their corresponding statistics for the array of the hash table are stored.

It is important to note that for a search template with m neighborhood nodes and a TI with n data event types, the total number of possible patterns that can be captured by the search template and stored in the hash table is n^m. Hence if the size of the hash table's underlying array is less than n^m, a scenario where the hash function generates the same index integer value for 2 or more different patterns may occur. This scenario is called a hash collision. Corman et al. (2009) provides more details regarding strategies to overcome hash collision.

Retrieval of Statistics during Simulation

Consider the simulation on a small grid, as shown in Fig. 5-10. The statistics of the data events observed on the spatial template are as shown in Table 5-2. Consider the simulation at two nodes marked x and y. Figure 5-11 shows the template centered on node x and the corresponding pattern observed in the neighborhood nodes.

Since all the neighborhood nodes of the search template capture a valid data event on the simulation grid, the hash function can be applied to retrieve the statistics corresponding to the observed pattern. Application of the hash function generates the index integer value of 5 for the pattern shown in Fig. 5-11. The element at index 5 of the underlying hash table array is accessed. From Table 5-4, we can see that with the index value of 5, we have the pattern statistics of 1 that the central node is white and 6 that the central node is yellow. In terms of probabilities:

$$\text{Prob}\{\text{Central Node} = \text{white} \mid 110\} = \frac{1}{2}$$

$$\text{Prob}\{\text{Central Node} = \text{yellow} \mid 110\} = \frac{6}{7}$$

Array Index	Key-Value Pair Data at Array Index [Key, Value] $\Rightarrow$ [String, Array [Integer]]
1	["1 0 0," [5,7]]
2	["1 0 1," [3,3]]
3	["0 0 0," [2,4]]
4	["0 0 1," [0,7]]
5	["1 1 0," [1,6]]
6	["1 1 1," [8,0]]
7	["0 1 0," [2,9]]
8	["0 1 1," [5,1]]

TABLE 5-4 Storage of Patterns and Their Corresponding Statistics in a Hash Table

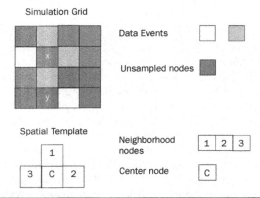

Figure 5-10 A 4 × 4 simulation grid and spatial template used for simulation. (The color version of this figure is available at www.mhprofessional.com/PRMS.)

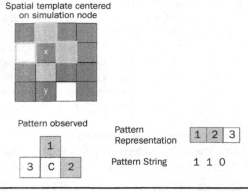

Figure 5-11 Spatial template centered on node x and the corresponding neighborhood pattern. (The color version of this figure is available at www.mhprofessional.com/PRMS.)

In the second case, for simulation at node y, not all neighborhood nodes of the search template contain a valid data event.

As seen in Fig. 5-12, the neighborhood node at position 3 coincides with an unsimulated node on the simulation grid. This means that unlike the previous scenario where one unique pattern is searched for in the hash table, this scenario involves a search in the hash table for multiple patterns whose data event at the first and second positions match the data event at the first and second positions of the incomplete pattern captured in Fig. 5-12. This search is done by visiting all elements of the hash table array. The statistics of all patterns that meet the search criteria are summed across all possibilities of events at search node 3 in order to compute the statistics for the incomplete pattern. Table 5-5 shows all patterns in the hash table's array that

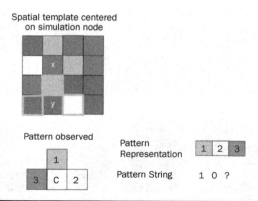

Pattern observed

Pattern
Representation

Pattern String   1   0   ?

FIGURE 5-12   Spatial template centered on node y and the corresponding neighborhood pattern. (The color version of this figure is available at www.mhprofessional.com/PRMS.)

Pattern	String	Hash Function Value	Key-Value Pair
1 2 3	1 0 0	$\{2^1 \cdot 31^2 + 2^0 \cdot 31^1 + 2^0 \cdot 31^0\}$ mod9 = 1	["1 0 0," [5,7]]
1 2 3	1 0 1	$\{2^0 \cdot 31^2 + 2^0 \cdot 31^1 + 2^1 \cdot 31^0\}$ mod9 = 2	["1 0 1," [3,3]]

TABLE 5-5   Patterns that Meet Search Criteria and Their Corresponding Statistics

meet the search criteria. The statistics for the incomplete pattern are also shown in Table 5-5.

Based on the statistics indicated, the probability that the center node is white is:

$$\text{Prob}\{\text{Central Node} = \text{white} \mid 10?\} = \frac{5+3}{(5+7+3+3)} = \frac{4}{9}$$

while the probability that the center node is yellow is:

$$\text{Prob}\{\text{Central Node} = \text{yellow} \mid 10?\} = \frac{7+3}{(5+7+3+3)} = \frac{5}{9}$$

In essence, it is important to note that retrieving pattern statistics from a hash table during simulation is very fast when the pattern captured from the simulation grid is entirely composed of valid data events. For the case where the pattern captured from the simulation grid is incomplete, the process of retrieving statistics from the hash table requires the statistics of partially filled templates to be aggregated.

The hash table-based simulation was implemented on *gpus* (Chukwudiuto, 2018) and the results are shown in Fig. 5-13. For comparison, slices through a realization obtained using a traditional

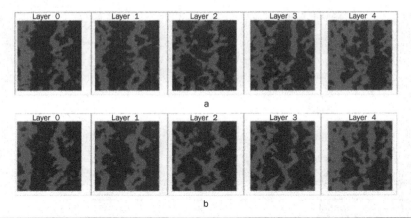

Figure 5-13 Simulated realizations: **a)** Snesim with hash table for storing and retrieving mp statistics. Simulation implemented on a GPU. **b)** Traditional snesim using search tree. (The color version of this figure is available at www.mhprofessional.com/PRMS.)

implementation of *snesim* is also shown in the figure. As can be seen, the connectivity of channel features is represented accurately.

The *gpu* implementation with a hash table is also computationally efficient, as shown in Fig. 5-14. The details of the desktop computer used for this comparison are also presented by Chukwudiuto (2018).

Growthsim
As discussed previously, multiple-point geostatistics aim at developing probability distribution functions that characterize frequency of occurrence of a simulation event (single- or multiple-point) in a spatial template conditioned to the pattern exhibited by the data in the neighborhood of that location. The traditional *mp statistics*-based workflow is based on the calibration of the probability of an outcome in the central node of the spatial template conditioned on the observed data pattern in the template. The mp simulation framework is extended to permit the inference and simulation of a spatial pattern (multiple points) conditioned to the multiple-point pattern exhibited by the data in the spatial template (Eskandari and Srinivasan, 2010; Huang and Srinivasan, 2012). This subtle and yet important shift in the *mp paradigm* results in a much more efficient, growth-based methodology to simulate models exhibiting complex geologic structures.

In *growthsim*, instead of building the simulated image node by node, entire patterns are simulated conditioned to a pattern observed on the simulation template. The algorithm is therefore growth-based, with the patterns grown from seed nodes where data is initially available. As the simulation proceeds, the set of nodes from where the patterns can be grown expands. At any step in the simulation algorithm, a list of simulatable nodes is created. These simulatable nodes are such that when the spatial template is centered at that location, there is at least one conditioning data located on the template. A node is

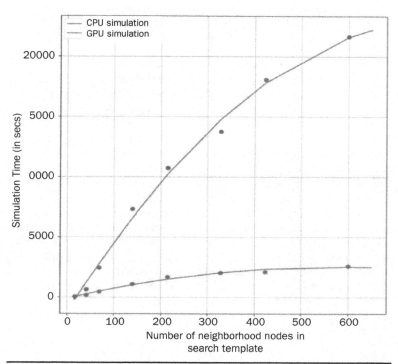

Figure 5-14 Computation time as a function of template size for snesim implemented on *cpu* and for the implementation with hash table on *gpus*. (The color version of this figure is available at www.mhprofessional.com/ PRMS.)

selected at random from this list of simulatable nodes, and becomes the site from which patterns are grown. The different patterns that are possible based on the conditioning pattern observed on the spatial template are retrieved as discussed previously with the help of a search tree or a hash table.

As discussed in the section developing the single extended normal simulation, the probability of a simulation event at the central node of the spatial template conditioned to the pattern observed on the remaining nodes of the template can be obtained as:

$$I_o^* = \frac{E(I_o D)}{E(D)} = E(I_o \mid D) = \text{Prob}\{I_o = k \mid D\}$$

I_o is the indicator variable at the simulation node in the original scheme that can assume one of K outcomes and D is the conditioning pattern. In *growthsim*, I_o is treated as a random function defined by the indicator outcomes at all unsimulated nodes in the spatial template. As before, the conditional probability is computed for various patterns of occurrence in the unsimulated nodes. The hash table or search tree is used for aggregating the statistics over various combinations of partially filled spatial template. Once the conditional probability is obtained, the simulated pattern is derived by sampling from that probability distribution function.

In order to demonstrate simulation within the *growthsim* framework, consider the simulation at the node *y* shown in Fig. 5-12. Data events are observed at nodes 1 and 2, while node 3 and the center node are as yet unsimulated. In the base algorithm, as discussed earlier, the probabilities corresponding to various patterns in Table 5-4 that match the pattern in data nodes 1 and 2 are aggregated in order to get the statistics for the two events possible at the center node. In the *growthsim* algorithm, the calculation is extended to include all combination of patterns that are possible at both the center node and node 3 in the spatial template. The possible patterns are the same as shown in Table 5-5.

There are a total of 18 occurrences corresponding to the four patterns exhibited in the four nodes of the spatial template. These patterns are shown in Table 5-6 where the last element in the pattern is the outcome at the central node of the template. The statistics for occurrence of the pattern are the same tabulated in Table 5-5 except that the occurrences corresponding to white and yellow in the central node are written as two separate entries in Table 5-5.

Based on the occurrences tabulated in Table 5-6, the following conditional probabilities can be computed:

$$\text{Prob}\{\text{Central Node} = \text{white}, \text{Node 3} = \text{white} \mid 10?\}$$
$$= \frac{5}{(5+7+3+3)} = \frac{5}{18}$$

$$\text{Prob}\{\text{Central Node} = \text{yellow}, \text{Node 3} = \text{white} \mid 10?\}$$
$$= \frac{7}{(5+7+3+3)} = \frac{7}{18}$$

$$\text{Prob}\{\text{Central Node} = \text{white}, \text{Node 3} = \text{yellow} \mid 10?\}$$
$$= \frac{3}{(5+7+3+3)} = \frac{1}{6}$$

$$\text{Prob}\{\text{Central Node} = \text{yellow}, \text{Node 3} = \text{yellow} \mid 10?\}$$
$$= \frac{3}{(5+7+3+3)} = \frac{1}{6}$$

Pattern	Occurrences
1 0 0 0	5
1 0 0 1	7
1 0 1 0	3
1 0 1 1	3

TABLE 5-6 Patterns Possible in All Four Nodes of the Template Corresponding to the Data Event in Fig. 5-12

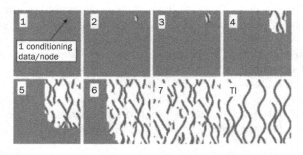

Figure 5-15 The progress of the *growthsim* simulation algorithm. (The color version of this figure is available at www.mhprofessional.com/PRMS.)

A particular outcome for the central node, as well as node 3, can be obtained by drawing from these conditional probabilities. This will result in the growth of a pattern at both the central node and node 3 of the spatial template.

Figure 5-15 shows the progressive growth of patterns during *growthsim* starting from a conditioning node (Huang and Srinivasan, 2012).

Direct Sampling

Single pixel and pattern-growth class of MPS methods still suffer from several shortcomings. Both class of methods require memory for storage of all patterns scanned from the TI. Furthermore, the traditional *snesim* methods can be used for simulation of categorical properties but are computationally inefficient for the simulations of continuous properties (Huang and Srinivasan, 2012). To mitigate these issues, Mariethoz, Renard, and Straubhaar (2010) proposed the direct sampling-based multiple-point simulation technique. The underlying principle on which this method is based is that constructing the conditional probability of the data event from the TI and then randomly drawing a sample from it is equivalent to directly sampling the first event in a TI that exhibits the data pattern in the neighboring data nodes. If the direct sampling process is continued many times, the probability distribution corresponding to the patterns observed in the simulated images would equal the distribution inferred from a TI.

During sequential simulation using direct sampling, each unsimulated node **u** on the simulation grid is randomly visited and its local conditioning data event $T(\mathbf{u})$ is recorded. The data event $T(\mathbf{u})$ may be obtained by spiraling away from the simulation node until a user-specified number of neighboring data have been identified. The TI is then directly and randomly sampled until an appropriate pattern that closely matches the data event $T(\mathbf{u})$ is found. For each pattern randomly sampled from the TI, a distance (mismatch) function is used to calculate the difference between the data event $T(\mathbf{u})$ observed in the simulation grid and the pattern sampled from the TI

(Meerschmann et al., 2013). If the distance (mismatch) is lower than a given threshold, the random sampling process on the TI is stopped and the value at the central node of the sampled pattern is assigned to the simulation grid at the current node location $\mathbf{u}$. The advantage of direct sampling is that it has a low memory requirement as it does not require storage of patterns found in the TI, and because there is no overhead associated with storage of statistics and subsequently recall of the statistics, it can be structured to handle both continuous and categorical variables.

Remarks

- The direct sampling algorithm described above does not use a spatial template formally defined for the simulation. Instead, the template is implicitly defined when the user-specified number of neighborhood data is searched and located. In this framework, it is likely that the simulation starts with neighborhood data that are spread farther (i.e., coarser template) and as the simulation gets filled, the template becomes tighter.

- Shannon and Weaver (1948) in their seminal work state, "When the source for the statistics is assumed to be ergodic, averages along a sequence can be identified with averages over an ensemble of possible sequences (the probability of a discrepancy being zero). For example, the relative frequency of the letter A in a particular infinite sequence will be, with probability one, equal to its relative frequency in the ensemble of sequences." This assurance that the simulation of patterns on a first observed basis will eventually approach the probability of the pattern in an infinite sequence is the basis for the direct sampling algorithm. The algorithm is predicated on the ergodicity of the TI, but that is a requirement for the other TI-based algorithms as well.

5.1.4 Returning to the Full Indicator Basis Function

All implementations of multiple-point simulation algorithms till date have been based on the single extended normal equation. However, the full indicator basis function provides some intriguing insights into the statistics of spatial variability and the quantification of uncertainty. In order to understand the characteristics of the full expansion, consider the estimation of a central node given four surrounding data locations (see following table). Consider a random binary media such that when $I(\mathbf{u}_3) = 1$, $I(\mathbf{u}_0) = 1$ always. Thus, for example, a possible pattern observed in the spatial template could be 11110, i.e., the outcome in the central node $I(\mathbf{u}_0) = 1$ when the outcome in the third conditioning node is $I(\mathbf{u}_3) = 1$. That is, the only outcome possible in the central node given that the outcome in the third node is 1. On the other hand, when the outcome in the third node is 0, for example, in 11100, then two outcomes in the central node are possible: 11100 and 11000. The list of patterns that can be observed is tabulated below. Note that normally, given only binary outcomes and five nodes in the template, a total of 32 independent patterns are theoretically possible. In this case, because of the rule stipulated, only 25 independent patterns are possible.

11111	11111
11110	11110
11101	11001
10111	10111
01111	01111
11100	11000
10110	10110
01110	01110
10101	10001
01101	01001
00111	00111
10100	10000
01100	01000
00101	00001
00110	00110
00100	00000

The indicator basis expansion in this case can be written as:

$$
\begin{aligned}
I^*(\mathbf{u}_3) &= E\{I(\mathbf{u}_0) \mid [I_1, I_2, I_3, I_4]\} \\
&= a_0 + a_1 I_1 + a_2 I_2 + a_3 I_3 + a_4 I_4 + a_5 I_1 I_2 + a_6 I_1 I_3 + a_7 I_1 I_4 \\
&\quad + a_8 I_2 I_3 + a_9 I_2 I_4 + a_{10} I_3 I_4 \\
&= a_{11} I_1 I_2 I_3 + a_{12} I_1 I_2 I_4 + a_{13} I_1 I_3 I_4 + a_{14} I_2 I_3 I_4 + a_{15} I_1 I_2 I_3 I_4
\end{aligned}
$$

The sixteen weights in the expansion can be obtained by taking the dot product of the error $(I - I^*)$ with each basis vector and setting it to zero. Thus:

$$
\begin{aligned}
\langle (I - I^*), \mathbf{1} \rangle = 0 &\Rightarrow E\{I(\mathbf{u}_0)\} = E\{I^*(\mathbf{u}_0)\} \\
&\Rightarrow P_o = a_0 + a_1 p_1 + a_2 p_2 + a_3 p_3 + a_4 p_4 + a_5 E\{I_1 I_2\} \\
&\quad + a_6 E\{I_1 I_3\} \\
&\quad + a_7 E\{I_1 I_4\} + a_8 E\{I_2 I_3\} + a_9 E\{I_2 I_4\} + a_{10} E\{I_3 I_4\} \\
&\quad + a_{11} E\{I_1 I_2 I_3\} + a_{12} E\{I_1 I_2 I_4\} + a_{13} E\{I_1 I_3 I_4\} \\
&\quad + a_{14} E\{I_2 I_3 I_4\} + a_{15} E\{I_1 I_2 I_3 I_4\}
\end{aligned}
$$

and

$$\langle (I - I^*), I_1 \rangle = 0 \Rightarrow E\{I(\mathbf{u}_0)I_1\} = E\{I^*(\mathbf{u}_0)I_1\}$$
$$\Rightarrow E\{I_oI_1\} = a_0p_1 + a_1p_1 + a_2E\{I_1I_2\} + a_3E\{I_1I_3\}$$
$$+ a_4E\{I_1I_4\} + a_5E\{I_1I_2\}$$
$$+ a_6E\{I_1I_3\} + a_7E\{I_1I_4\} + a_8E\{I_1I_2I_3\}$$
$$+ a_9E\{I_1I_2I_4\} + a_{10}E\{I_1I_3I_4\}$$
$$+ a_{11}E\{I_1I_2I_3\} + a_{12}E\{I_1I_2I_4\} + a_{13}E\{I_1I_3I_4\}$$
$$+ a_{14}E\{I_1I_2I_3I_4\} + a_{15}E\{I_1I_2I_3I_4\}$$

and so for the remaining dot products.

From the above table of outcomes, it can be seen that:

$$P_o = \text{prob}\{\text{central node} = 1\} = \frac{3}{4}$$

$$P_1 = \text{prob}\{I(\mathbf{u}_1) = 1\} = \frac{1}{2}, \text{ and similarly } P_2 = P_3 = P_4 = \frac{1}{2}$$

$$E\{I_1I_2\} = \text{prob}\{I(\mathbf{u}_1) = 1, I(\mathbf{u}_2) = 1\} = \frac{8}{32} = E\{I_1I_3\}$$
$$= E\{I_1I_4\} = E\{I_2I_3\} = E\{I_2I_3\} = E\{I_2I_4\}$$
$$= E\{I_3I_4\}$$

$$E\{I_1I_2I_3\} = \text{Prob}\{I(\mathbf{u}_1) = 1, I(\mathbf{u}_2) = 1, I(\mathbf{u}_3) = 1\}$$
$$= \frac{4}{32} = E\{I_1I_2I_4\} = E\{I_1I_3I_4\} = E\{I_2I_3I_4\}$$

$$E\{I_1I_2I_3I_4\} = \text{Prob}\{I(\mathbf{u}_1) = 1, I(\mathbf{u}_2) = 1, I(\mathbf{u}_3) = 1, I(\mathbf{u}_4) = 1\}$$
$$= \frac{2}{32}$$

These capture all the interactions between the data themselves. Look at the interaction between the data and the unknown that is required on the left-hand side of the system of equations:

$$E\{I_1I_o\} = \text{Prob}\{I(\mathbf{u}_1) = 1, I(\mathbf{u}_0) = 1\} = \frac{3}{8} = E\{I_2I_o\} = E\{I_4I_o\}$$

$$E\{I_3I_o\} = \frac{1}{2}$$

$$E\{I_1I_2I_o\} = \text{Prob}\{I(\mathbf{u}_1) = 1, I(\mathbf{u}_2) = 1, I(\mathbf{u}_0) = 1\} = \frac{3}{16}$$
$$= E\{I_1I_4I_o\} = E\{I_2I_4I_o\}$$

$$E\{I_1I_3I_o\} = \frac{1}{4} = E\{I_3I_4I_o\} = E\{I_2I_4I_o\}$$

$$E\{I_1I_2I_3I_o\} = \frac{1}{8} = E\{I_1I_3I_4I_o\} = E\{I_2I_3I_4I_o\}$$

$$E\{I_1I_2I_4I_o\} = \frac{3}{32}$$

and finally, $E\{I_1I_2I_3I_4I_o\} = \frac{1}{16}$

Solving the matrix system yields the following weights:

a_0	0.5
a_1	0
a_2	0
a_3	0.5
a_4	0
a_5	0
a_6	0
a_7	0
a_8	0
a_9	0
a_10	0
a_11	0
a_12	0
a_13	0
a_14	0
a_15	0

These weights imply that $I^*(\mathbf{u}_0) = \text{Prob}\{I(\mathbf{u}_0) \mid (n)\} = 1$ for any combination of outcomes in the four conditioning nodes where $I(\mathbf{u}_3) = 1$. Thus, for example when the combination of outcomes in the conditioning nodes is 1010, the corresponding conditional probability that $I(\mathbf{u}_0) = 1$ is 1.0. On the other hand, the conditional probability is 0.5 for all combinations where $I(\mathbf{u}_3) = 0$. These results agree with our prior expectation given the relationship between the outcomes at nodes 0 and 3.

Truncated Expansion
The previous example is ideal for studying the influence of multiple-point indicator function on the computation of conditional probability at the estimation node. The original example where the outcomes at the conditioning nodes and the estimation node are sampled independent to each other was repeated but this time truncating the indicator expansion at the two-point covariances. The truncated expansion is written as:

$$I^{**}(\mathbf{u}_0) = a_o + a_1I_1 + a_2I_2 + a_3I_3 + a_4I_4 + a_5I_1I_2 + a_6I_1I_3 \\ + a_7I_1I_4 + a_8I_2I_3 + a_9I_2I_4 + a_{10}I_3I_4$$

The 11 weights a_1 in the above expansion are obtained as before by solving the projection equations. The weights obtained are as before: $a_o = 0.5$, $a_3 = 0.5$, $a_1 = a_2 = a_4 = \ldots = 0.0$.

This implies that the central node is guaranteed to have an outcome if the outcome in the third conditioning node is 1. The probability drops to 0.5 if the outcome at the third node is 0.

In the second case, the exhaustive set of outcomes was changed such that $I(\mathbf{u}_0) = 1$ whenever $I_1 = I_2 = I_3 = 1$ and for all other combination of data events, the probability that $I(\mathbf{u}_0) = 1$ is 0.5. The revised list of data events is tabulated below. The shaded cells are the ones changed because of the new condition.

11111	11111
11110	11110
11101	11001
10111	10011
01111	01011
11100	11000
10110	10010
01110	01010
10101	10001
01101	01001
00111	00011
10100	10000
01100	01000
00101	00001
00110	00010
00100	00000

This results in a slight change in the calculated values of p_0 and the covariances between the data and $\mathbf{u}_0$.

p_o	0.5625
p_10	0.3125
p_20	0.3125
p_30	0.3125
p_40	0.28125
p_120	0.1875
p_130	0.1875
p_140	0.15625
p_230	0.1875
p_240	0.15625
p_340	0.15625

p_1230	0.125
p_1240	0.09375
p_1340	0.09375
p_2340	0.09375
p_12340	0.0625

When the system of equations for the full indicator basis function is solved, the following weights result:

a_0	0.5
a_1	0
a_2	0
a_3	0
a_4	0
a_5	0
a_6	0
a_7	0
a_8	0
a_9	0
a_10	0
a_11	0.5
a_12	0
a_13	0
a_14	0
a_15	0

The weight a_{11} multiplies the terms $I_1 I_2 I_3$ and this implies that when $I_1 = I_2 = I_3 = 1$ the conditional probability $I^*(\mathbf{u}_0) = E\{I(\mathbf{u}_0) \mid (n) = 1.0$. For all other combination of outcomes in the data nodes, the conditional probability $I^*(\mathbf{u}_0) = 0.5$. For this same example when the indicator basis function is truncated at the two-point covariance term, the corresponding weights (a_0 through a_{10}) are:

a_0	0.56
a_1	−0.1
a_2	−0.1
a_3	−0.1
a_4	0
a_5	0.25
a_6	0.25
a_7	0
a_8	0.25
a_9	0
a_10	0

In this case when $I_1 = I_2 = I_3 = 1$, the conditional probability $I^*(\mathbf{u}_0) = 0.94 \neq 1.0$ as it should be. When $I_1 = 1, I_2 = 0, I_3 = 1, I_4 = 1$, then $I^*(\mathbf{u}_0) = 0.56$. The slight departure from the true value of 0.5 is a consequence of ignoring the higher-order interactions between the conditioning locations.

A More General Example

In this example, the outcomes at the four conditioning nodes are assigned following a Monte Carlo sampling scheme. For nodes 1, 2, and 3:

$$I(\mathbf{u}) = \begin{cases} 1, & \text{if } p < 0.5 \\ 0 & \text{otherwise} \end{cases}$$

where $p \rightarrow U[0, 1]$. The outcome at the central $(\mathbf{u}_0)$ is also assigned an outcome following the same criterion. At the fourth node:

$$I(\mathbf{u}) = \begin{cases} 1, & \text{if } I(\mathbf{u}_0) = 1, \text{ and } p < 0.75 \text{ or if } I(\mathbf{u}_0) = 0, \text{ and } p < 0.5 \\ 0 & \text{otherwise} \end{cases}$$

This condition helps ensure that the outcome in the central node is somewhat correlated to the outcome at the fourth conditioning node, but that correlation is not perfect.

A thousand realizations of outcomes at the five (4 conditioning nodes + 1 central node) nodes were sampled. Based on these realizations, the calculated moments are summarized in the table below.

p_0	0.516		
p_1	0.489	p_10	0.256
p_2	0.486	p_20	0.244
p_3	0.516	p_30	0.273
p_4	0.638	p_40	0.388
p_12	0.226	p_120	0.115
p_13	0.239	p_130	0.125
p_14	0.311	p_140	0.194
p_23	0.266	p_230	0.146
p_24	0.313	p_240	0.189
p_34	0.321	p_340	0.196
p_123	0.118	p_1230	0.063
p_124	0.146	p_1240	0.09
p_134	0.146	p_1340	0.089
p_234	0.167	p_2340	0.111
p_1234	0.071	p_12340	0.047

The condition stipulated on the outcome at the fourth node manifests itself in the form of higher values for $p_4 = E\{I(\mathbf{u}_4)\}$, higher values of $p_{40} = E\{I_4I_0\}$, and three-point covariances, such as $p_{140} = E\{I_1I_4I_0\}$, p_{240}, p_{340}, etc. The corresponding weights are:

Weights
0.341
−0.015
−0.073
0.099
0.301
0.019
0.009
0.061
−0.001
−0.076
−0.196
−0.039
0.014
−0.041
0.273
−0.015

As expected, the fourth datum gets significantly higher weight than all others. Similarly, the data combinations I_1I_4, I_2I_4, I_3I_4, $I_1I_2I_4$, $I_1I_3I_4$, $I_2I_3I_4$ get higher weights than other two- and three-point combinations.

If the particular data event in the surrounding nodes is $I_1 = 1$, $I_2 = 1$, $I_3 = 1$, $I_4 = 1$, then $I^*(\mathbf{u}_0) = \text{Prob}\{I(\mathbf{u}_0) = 1 \mid (n)\} = 0.66$, more than the nominal value of 0.5 because of the correlation between the outcomes at $\mathbf{u}_4$ and that at $\mathbf{u}_0$. If on the other hand, the data event is $I_1 = 1$, $I_2 = 1$, $I_3 = 1$, $I_4 = 0$, then the conditional probability $I^*(\mathbf{u}_0) = \text{prob}\{I(\mathbf{u}_0) \mid (n)\} = 0.29$, considerably less than 0.5 because conditioning data $I_4 = 0$. When $I_1 = 1$, $I_2 = 1$, $I_3 = 1$, $I_4 = 0$, $I^*(\mathbf{u}_0) = \text{Prob}\{I(\mathbf{u}_0) = 1 \mid (n)\} = 0.308$. The slight difference in conditional probability for this case compared to the previous one (both for $I_4 = 0$) may be because of small non-ergodic fluctuations in the inferred mp-statistics that influence the calculation of weights.

Simulation Using the Indicator Basis Function

The indicator basis function yields the conditional probability at location $\mathbf{u}_0$ accounting for all multiple-point interactions between the RVs at conditioning data locations. Once the required proportions and multiple-point indicator covariance functions are inferred, the weights assigned to the terms in the indicator basis functions can be evaluated. It must be emphasized that once the weights have been computed, they can be applied to whatever combination of outcomes observed at the conditioning data locations. Thus, though there is the initial computational expense of inferring the indicator covariances and solving the (possibly large) system of equations in order to compute the weights, the subsequent calculation of conditional probability at an estimation/simulation node is simply an algebraic calculation for any combination of outcomes at the conditioning locations. This is in contrast to the single normal equation simulation in which an extended system of equations does not have to be solved, but instead at every step of the simulation, the probability of a simulation event conditioned to the particular data event at the conditioning locations has to be inferred by scanning a TI.

Implementing the full-extended normal equation within the sequential simulation framework requires one further consideration. During the process of sequential simulation, nodes within a domain are visited along a random path. The spatial template that was used for inferring the pattern statistics is centered on the visited node and the conditional probability of a data event at the central node conditioned to the data or previously simulated nodes within the template has to be computed. In single extended normal equation simulations, that conditional probability is obtained by scanning the TI and recording the frequency of the conditioning data event. As mentioned previously, the required conditional probability can also be computed knowing the weights in the extended normal equation. However, the extended normal equations would be solved taking into consideration all the nodes of the spatial template used for the simulation. The presence of a partially informed spatial template during sequential simulation therefore presents a problem. In the following section, it is demonstrated that the conditional probability for the simulation event given a partially informed data template can be computed using the weights for the full-extended normal equation and after some manipulations.

Conditional Probabilities Given a Partially Filled Data Template

Consider a simple spatial template, as shown in Fig. 5-16.

The dark nodes of the spatial template represent nodes that are informed by data. That data could either be the original data used for conditioning the model or values that have been already simulated. These dark nodes are collectively referred to as the set B in the ensuing discussion. The light nodes do not have any data values associated with them. These are referred to as the set C. Figure 5-16 therefore shows a partially filled template and the objective is to evaluate the conditional probability for a simulation event in the central node (white). The simulation event is referred to as A.

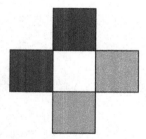

FIGURE 5-16 Illustration of a typical spatial template with some informed nodes (dark) and some uniformed nodes (light). (The color version of this figure is available at www.mhprofessional.com/PRMS.)

FIGURE 5-17 Spatial template with informed nodes **B** and uninformed nodes **C** clearly marked. (The color version of this figure is available at www.mhprofessional.com/PRMS.)

The full spatial template is represented as $B \cup C$ in Fig. 5-17. Denoting the data event in the informed nodes a: $D_B = \prod_{i \in B} I(\mathbf{u}_i)$ and similarly $D_C = \prod_{i \in C} I(\mathbf{u}_i)$ where the subscript i refers to the node index. The complementary event $D^C C = 1 - D_C$ can also be defined. Thus if $D_C = 1$ then $D^C{}_C = 0$. Using set theory notation, the marginal probability of the data event D_B can be written as:

$$\text{prob}\{D_B\} = (D_B \cap D_C) \cup \left(D_B \cap D_C^C\right)$$

Similarly,

$$\text{prob}\{D_A \cap D_B\} = \text{prob}(D_A \cap D_B \cap D_C) + \text{prob}\left(D_A \cap D_B \cap D_C^C\right)$$

This implies that:

$$\text{prob}(D_A \mid D_B)$$
$$= \frac{\text{prob}\{D_A \cap D_B\}}{\text{prob}\{D_B\}} = \frac{\text{prob}\{D_A \cap D_B \cap D_C\} + \text{prob}\{D_A \cap D_B \cap D_C^C\}}{\text{prob}\{D_B\}}$$

Recognizing that:

$$\text{prob}\{D_A \cap D_B \cap D_C\} = \text{prob}(D_A \mid D_B \cap D_C) \cdot \text{prob}(D_B \cap D_C)$$
$$= \text{prob}(D_A \mid D_B \cap D_C) \cdot \text{prob}(D_C \mid D_B) \cdot \text{prob}(D_B)$$

and similarly for $\text{Prob}\{D_A \cap D_B \cap D_C^c\}$ and because $\text{prob}\{D_C^C \mid D_B\} = 1 - \text{prob}\{D_C \mid D_B\}$

$$
\begin{aligned}
&\text{prob}(D_A \mid D_B) \\
&= \frac{P(D_A \mid D_B \cap D_C) \cdot P(D_C \mid D_B) \cdot P(D_B) + P\left(D_A \mid D_B \cap D_C^C\right) \cdot (1 - P(D_C \mid D_B) \cdot P(D_B))}{P\{D_B\}} \\
&= P(D_A \mid D_B, D_C) \cdot P(D_C \mid D_B) + P\left(D_A \mid D_B, D_C^C\right) \cdot (1 - P(D_C \mid D_B)) \\
&= P\left(D_A \mid D_B, D_C^C\right) + P(D_C \mid D_B) \cdot \left(P(D_A \mid D_B, D_C) - P\left(D_A \mid D_B, D_C^C\right)\right)
\end{aligned}
$$

The conditional probability $P(D_A \mid D_B, D_C)$ corresponds to the probability of the event in the central node given data on a full spatial template. Recall that the full spatial template has been separated into the nodes that are informed (B) and nodes that do not have any data (C). D_C corresponds to the data event $D_C = \prod_{i \in C} I(\mathbf{u}_i) = 1$. Similarly, $P\left(D_A \mid D_B, D_C^C\right)$ also corresponds to the probability given a full template except, now the combination of events in the uninformed nodes is such that $\prod_{i \in C} I(\mathbf{u}_i) = 0$.

In order to demonstrate the application of the above formula, a simple numerical example is presented and discussed next.

Single Data Example

Consider the case of simulation on an elementary two-point template.

The location $\mathbf{u}_1$ is the data node and $\mathbf{u}_0$ is the simulation node. In this opening example, we assume a random media where the prior probability (p) is $\frac{1}{2}$ for the outcome to be 0 or 1 at each node of the template. We are interested in the posterior probability $P\left(D(\mathbf{u}_0) \mid p, D(\mathbf{u}_1)\right)$.

Now consider a simulation situation where there is no data at location $\mathbf{u}_1$. In that case, the posterior probability reduces to $P\left(D(\mathbf{u}_0) \mid p\right)$. Using the previously developed expression:

$$
\begin{aligned}
P\left(D(\mathbf{u}_0) \mid p\right) = {}& P\left(D(\mathbf{u}_0) \mid p, D^C(\mathbf{u}_1)\right) + P\left(D(\mathbf{u}_1) \mid P\right) \cdot \\
& \left(P\left(D(\mathbf{u}_0) \mid p, D(\mathbf{u}_1) - P(D(\mathbf{u}_0))\right) \mid p, D^C(\mathbf{u}_1)\right)
\end{aligned}
$$

Since the outcomes at the nodes are uncorrelated:

$$
P\left(D(\mathbf{u}_0) \mid p, D(\mathbf{u}_1)\right) = P\left(D(\mathbf{u}_0) \mid p, D^C(\mathbf{u}_1)\right) = \frac{1}{2}
$$

Consequently, the posterior probability $P(D(\mathbf{u}_0) \mid p) = \dfrac{1}{2}$ as expected.

Now consider the same example but the outcomes are correlated such that if the outcome at the data node $\mathbf{u}_1$ is 1, then the probability that the outcome at the simulation node $\mathbf{u}_0$ is 1 increases to 0.85. If on the other hand, the outcome at $\mathbf{u}_1$ is 0, then the probability of any outcome (0 or 1) at the simulation node remains the prior probability $p = \dfrac{1}{2}$. Performing this as a Monte Carlo simulation exercise and drawing 1000 outcomes of $Z(\mathbf{u}_1)$ and $Z(\mathbf{u}_0)$ following the rule outlined the above results in the following:

$P\big(D(\mathbf{u}_0) \mid p, D^C(\mathbf{u}_1)\big) = P\big(Z(\mathbf{u}_0) \mid p, Z(\mathbf{u}_1) = 0\big)$	0.468	Close to the prior probability of 0.5
$P\big(D(\mathbf{u}_0) \mid p, D(\mathbf{u}_1)\big) = P\big(Z(\mathbf{u}_0) \mid p, Z(\mathbf{u}_1) = 1\big)$	0.861	Close to the higher probability of 0.85 assumed
$P\big(D(\mathbf{u}_1) \mid p\big) = P\big(Z(\mathbf{u}_1) = 1 \mid p\big)$	0.5025	Close to the prior probability of 0.5 for any outcome at $\mathbf{u}_1$

Substituting the above quantities in the equation for the probability $P\big(D(\mathbf{u}_0) \mid p\big)$

$$P\big(D(\mathbf{u}_0) \mid p\big) = P\big(D(\mathbf{u}_0) \mid p, D^C(\mathbf{u}_1)\big) + p\big(D(\mathbf{u}_1) \mid p\big)$$
$$\big(P\big(D(\mathbf{u}_0) \mid p, D(\mathbf{u}_1)\big) - P\big(D(\mathbf{u}_0) \mid p, D^C(\mathbf{u}_1)\big)\big)$$

$$P\big(D(\mathbf{u}_0) \mid p\big) = 0.468 + 0.5025 \cdot (0.861 - 0.468) = 0.665$$

Going back to the set of simulated outcomes, $P\big(D(\mathbf{u}_0 \mid p) = 0.665$ by counting the number of 1s simulated at the location of the simulation node over the total set of 1000 outcomes.

The fundamental theory underlying the use of higher-order statistics to constrain geostatistical simulation is presented. In this chapter, the conditional probability of a simulated outcome at a spatial location is expressed in the form of an indicator basis expansion that includes higher-order statistics describing the multiple-point variability in the reservoir. Current implementations of mp-statistics-based algorithms are based on a truncation of the indicator basis expansion considering only the observed data pattern. Some details of the resultant single extended normal equation simulation methods are presented. The full-extended normal equation is again revisited and the chapter ends with an example for the implementation of the full indicator expansion. With the advent of faster computers and parallel computations, it is hoped that research will yield practical implementation of spatial simulation using the full indicator expansion.

5.2 References

Caers, J C, and T Zhang. 2004. "Multiple-point geostatistics: a quantitative vehicle for integrating geologic analogs into multiple reservoir models." In *Integration of outcrop and modern analogs in reservoir modeling*, by G M Grammer, P M Harris and G P Eberly, 383-394. AAPG.

Caers, J. 1998. "Stochastic Reservoir Simulation Using Neural Networks Trained on Outcrop Data." *Proceedings of SPE Annual Technical Conference and Exhibition.* New Orleans: Society of Petroleum Engineers. 321-337.

Chukwudiuto, M U. 2018. *Parallelization stratgey for multiple point statistics based simulation on GPUs.* Thesis, Energy and Mineral Engineering, The Pennsylvania State University, University Park: The Pennsylvania State University.

Corman, T H, C E Leiserson, R L Rivest, and C Stein. 2009. *Introduction to algorithms.* Cambridge, MA: MIT Press.

Erzeybek, S, S Srinivasan, and X Janson. 2012. "Multiple-Point Statistics in a Nongridded Domain: Application to Karst/Fracture Network Modeling." Edited by P Abrahamsen, R Hauge and O Kolbjørnsen. *Geostatistics Oslo.* Springer.

Eskandari, K, and S Srinivasan. 2010. "Reservoir modeling of complex geological systems - a multiple-point perspective." *Journal of Canadian Petroleum Technology* 49: 59-69.

Guardiano, F B, and R M Srivastava. 1993. "Multivariate Geostatistics: Beyond Bivariate Moments." Edited by A Soares. *Geostatistics Tróia '9.* Dordrecht: Springer.

Huang, Y, and S Srinivasan. 2012. "GrowthSim - Efficient Conditional Simulation of Spatial Patterns Using a Pattern-Growth Algorithm." Edited by P Abrahamsen, R Hauge and O Kolbjørnsen. *Proceedings of the Ninth International Geostatistics Congress:.* Oslo: Springer-Verlag. 209-220.

Journel, A G, and C J Huijbregts. 1978. *Mining Geostatistics.* London: Academic Press.

Journel, A G, and T Zhang. 2006. "The Necessity of a Multiple-Point Prior Model." *Mathematical Geology* (Springer) 38 (5): 591-610.

Liu, X, and S Srinivasan. 2004. "Geological characterization of naturally fractured reservoirs using multiple point geostatistics." Edited by O Leuangthong and C V Deutsch. *Geostatistics Banff.* Springer. 75-84.

Luenberger, D G. 1969. *Optimization by Vector Space Methods.* New York: John Wiley and Sons.

Mariethoz, G, P Renard, and J Straubhaar. 2010. "The direct sampling method to perform multiple-point geostatistical simulations." *Water Resources Research* 46 (11).

McDowell, G L. 2015. *Cracking the Coding Interview: 189 Programming Questions and Solutions.* Palo Alto, CA: CareerCup.

Meerschmann, E, G Pirot, G Mariethoz, J Straubhaar, M Meirvenne, and P Renard. 2013. "A practical guide to performing multiple-point simulations with the Direct sampling algorithm." *Computers and Geosciences* 52: 307-324.

Pyrcz, M J, J Boisvert, and C V Deutsch. 2008. "A Library of Training Images for Fluvial and Deepwater Reservoirs and Associated Code." *Computers and Geosciences* (Elsevier) 34 (5).

Pyrcz, M J, O Catuneanu, and C V Deutsch. 2005. "Stochastic Surface-based Modeling of Turbidite Lobes." *American Association of Petroleum Geologists Bulletin* (AAPG) 89 (2): 177-191.

Qiu, W V, and M Kelkar. 1995. "Simulation of Geological Models Using Multipoint Histogram." *SPE Annual Technical Confefence and Exhibition*. Dallas: Society of Petroleum Engineers. 771-781.

Shannon, C, and W Weaver. 1948. *The Mathematical Theory of Communication*. Champaign: University of Illinois Press.

Straubharr, J, A Walgenwitz, and P Renard. 2013. "Parallel multiple-point statistics algorithm based on list and treet structures." *Mathematical Geosciences* (Springer) 45 (2): 131-147.

Straubharr, J, P Renard, G Mariethoz, R Froideveaux, and O Besson. 2011. "An improved parallel multiple-point algorithm using a list approach." *Mathematical Geosciences* (Springer) 43 (3): 305-328.

Strebelle, S. 2002. "Conditional Simulation of Complex Geological Structures Using Multiple-Point Statistics." *Mathematical Geology* (Springer) 34: 1-21.

Strebelle, S, and A G Journel. 2001. "Reservoir Modeling Using Multiple-Point Statistics." *Proccedings of the Annual Technical Conference and Exhibition*. New Orleans: Society of Petroleum Engineers.

Zhang, T, S I Pedersen, C Knudby, and D McCormick. 2012. "Memory-efficient categorical multi-point statistics algorithms based on compact search trees." Mathematical Geosciences (Springer) 44 (7): 863-879.

CHAPTER 6

Numerical Schemes for Flow Simulation

This chapter will present the basic mass conservation and momentum balance equations for multi-phase flow, relevant auxiliary equations, and the corresponding initial and boundary conditions required for solving these equations. The black-oil model is used for phase behavior description. In addition, as isothermal condition is assumed, the conservation of energy is omitted. Numerical approximation and solution techniques for these differential equations using the control volume finite difference (CVFD) schemes are discussed. The discussion will be centered on presenting the implementation of these solution schemes and robust solvers for obtaining solutions to the linear system of equations. Explicit, semi-implicit, and fully implicit schemes for solving the flow equations will be presented.

6.1 Governing Equations

6.1.1 Conservation of Mass

Following the nomenclature in Lake (2010), the conservation of mass for component i for an arbitrary control volume (CV) can be formulated as:

$$\int_V \frac{\partial W_i}{\partial t} dV = - \int_A (\vec{N}_i \cdot \vec{n}) dA + \int_V R_i dV \qquad (6\text{-}1)$$

for $i = 1, ..., N_c$ (total number of components)

Rate of appearance and disappearance of component i into CV	Net rate of component i into CV	Rate of accumulation of component i in CV

W_i is the mass of i per bulk volume, $\vec{N}_i$ is the flux or the rate of mass transport of i across the superficial area (perpendicular to flow), and

R_i is the rate of mass of i "reacted" per bulk volume; these quantities can be defined as:

$$W_i = \phi\sum_{j=1}^{N_p} \rho_j S_j \varpi_{ij} + (1 - \phi)\rho_s \varpi_{is} \qquad (6\text{-}2)$$

$$\vec{N}_i = \sum_{j=1}^{N_p}\left(\rho_j\varpi_{ij}\vec{u}_j - \phi S_j\rho_j\vec{\vec{D}}_{ij}\cdot\nabla\varpi_{ij}\right) \qquad (6\text{-}3)$$

ϕ is porosity; p_j and p_s refer to the density of phase j and solid, respectively, S_j is the saturation of phase j, and N_p is the total number of phases. w_{ij} and w_{is} refer to the mass (mole) fraction of component i in phase j or solid, respectively. $\vec{\vec{D}}_{ij}$ is the dispersion tensor. It is assumed that mass can enter or leave the CV in any direction. The normal vector $(\vec{n})$ is directing outward. The CV should be simply connected, though its surface can be piecewise continuous. Detailed discussion on the subject of material balance can be found in the reference by Bird et al. (2007).

6.1.2 Conservation of Momentum

Conservation of momentum in a porous medium is represented by Darcy's law (Bear, 1972). For multi-directional multi-phase flow, the superficial velocity of phase j ($\vec{u}_j$) is described as:

$$\vec{u}_j = \vec{\vec{\lambda}}_j\cdot(\nabla P_j + \rho_j\vec{g}) \qquad (6\text{-}4)$$

$$\vec{\vec{\lambda}}_j = \vec{\vec{k}}\lambda_{rj} = \vec{\vec{k}}\frac{k_{rj}(S_j)}{\mu_j(\varpi_{ij},\vec{u}_j)} \qquad (6\text{-}5)$$

$\vec{\vec{k}}$ = permeability tensor; $\vec{g}$ is the gravitational vector along the vertical direction (positive upwards); P_j, μ_j, k_{rj} are the pressure, viscosity, and relative permeability, respectively, for phase j; and λ_j is the phase mobility. Darcy's law is an approximation to the full conservation of momentum as the Reynolds number approaches 0:

$$\frac{\partial(\rho_j\vec{v}_j)}{\partial t} = -(\nabla\cdot\rho_j\vec{v}_j)\vec{v}_j - \nabla\cdot\vec{\vec{\tau}}_j - \nabla p_j - \rho_j\vec{g} \qquad (6\text{-}6)$$

At steady state and low velocity, the above equation can be reduced to:

$$\nabla\cdot\vec{\vec{\tau}}_j = -\nabla p_j - \rho_j\vec{g} \qquad (6\text{-}7)$$

$\vec{\vec{\tau}}$ is the shear stress tensor and $\vec{V}$ is the interstitial velocity. Equation (6-4) is obtained using $\nabla\cdot\vec{\vec{\tau}}_j = \vec{\vec{\lambda}}_j^{-1}\cdot\vec{u}_j$.

6.2 Single-Phase Flow

6.2.1 Simulation Equations

We begin with the formulation for isothermal immiscible flow (but only one phase is mobile). Assuming zero rock interaction ($\omega_{is} = 0$) and complete immiscibility ($\omega_{ij} = 1$ for $i = j$ and $= 0$ for $i \neq j$), the conservation of mass can be simplified:

$$\int_V \frac{\partial(\phi\rho_j S_j)}{\partial t} dV = -\int_A (\rho_j \vec{u}_j \cdot \vec{n}) dA + \int_V R_i dV \tag{6-8}$$

To illustrate the meaning of the above equation, let us consider a system consisting of two immiscible phases: aqueous ($j = 1$) and oleic ($j = 2$). There are two components: water ($i = 1$) and oil ($i = 2$). The immiscibility condition implies that water exists in the aqueous phase, while the oil exists in the oleic phase. If only the oleic phase is flowing, $\vec{u}1 = 0$ and $\vec{u}_2 \neq 0$, one can also assume the source term corresponding to any production or injection of the immobile phase ($j = 1$) to be 0.

$$\int_V \frac{\partial(\phi\rho_1 S_1)}{\partial t} dV = 0 \tag{6-9}$$

$$\int_V \frac{\partial(\phi\rho_2 S_2)}{\partial t} dV = -\int_A (\rho_2 \vec{u}_2 \cdot \vec{n}) dA + \int_V R_2 dV \tag{6-10}$$

Considering a CV with constant properties within a volume of V, in which an average pressure P can be defined, Eq. (6-9) becomes:

$$V\frac{\partial(\phi\rho_1 S_1)}{\partial t} = 0$$

$$V\left[\phi\rho_1\frac{\partial(S_1)}{\partial t} + \phi S_1\frac{\partial(\rho_1)}{\partial t} + \rho_1 S_1\frac{\partial\phi}{\partial t}\right] = 0$$

$$V\left[-\phi\rho_1\frac{\partial(S_2)}{\partial t} + \phi S_1\frac{\partial(\rho_1)}{\partial t} + \rho_1 S_1\frac{\partial\phi}{\partial t}\right] = 0 \tag{6-11}$$

$$V\left[-\phi\rho_1\frac{\partial(S_2)}{\partial t} + \phi S_1\rho_1\frac{1}{\rho_1}\frac{\partial\rho_1}{\partial P}\frac{\partial P}{\partial t} + \rho_1 S_1\phi\frac{1}{\phi}\frac{\partial\phi}{\partial P}\frac{\partial P}{\partial t}\right] = 0$$

$$V\left[-\phi\frac{\partial(S_2)}{\partial t} + \phi(S_1 c_1 + S_1 c_f)\frac{\partial P}{\partial t}\right] = 0$$

Fluid and rock compressibilities are denoted by c_j and c_f, respectively. If we set $P_1 = P_2 = P$, one can show that Eq. (6-10) becomes:

$$V\left[-\phi\rho_2\frac{\partial(S_1)}{\partial t} + \phi\rho_2(S_2 c_2 + S_2 c_f)\frac{\partial P}{\partial t}\right] = -\int_A (\rho_2 \vec{u}_2 \cdot \vec{n}) dA + \int_V R_2 dV \tag{6-12}$$

Multiply Eq. (6-11) by r_2 and add to Eq. (6-12):

$$V\left[\phi\rho_2(S_1c_1 + S_1c_2 + c_f)\frac{\partial P}{\partial t}\right] = -\int_A (\rho_2\vec{u}_2 \cdot \vec{n})dA + \int_V R_2 dV$$

$$V\phi c_t\frac{\partial P}{\partial t} + \frac{1}{\rho_2}\int_A (\rho_2\vec{u}_2 \cdot \vec{n})dA = \frac{1}{\rho_2}\int_V R_2 dV$$

$$V\phi c_t\frac{\partial P}{\partial t} + B_2\int_A\left(\frac{1}{B_2}\vec{u}_2 \cdot \vec{n}\right)dA = \frac{B_2}{\rho_2^0}\int_V R_2 dV \tag{6-13}$$

The quantity $S_1c_1 + S_2c_2 + c_f$ is often referred to as the total compressibility, c_t. The superscript "0" refers to some reference (e.g., standard) conditions. The above equation is the conservation of mass for a grid cell (i.e., control volume V) with constant properties. Since only one phase is flowing, the subscript "2" will be omitted in the subsequent discussion to simplify the notation. To further formulate each term of Eq. (6-13), a specific grid configuration must be defined. We will consider a rectangular grid cell, as shown in Fig. 6-1. The x and y axes are oriented in the horizontal plane. Since the phase and component subscripts have been dropped, the variables i, j, and k are used from now on to refer to the grid indices.

If we further assume that there is no flow along the z direction (i.e., no flow coming into and out of the top and bottom of the CV), only flow in 2D needs to be considered (see Example 6-2 for an extension to 3D that also incorporates gravitational force along the vertical direction). In that case, the accumulation term can be expressed:

$$V\phi c_t\frac{\partial P}{\partial t} = (\Delta x\Delta y\Delta z\phi c_t)_{ij}^m\frac{\partial P_{ij}}{\partial t} \tag{6-14}$$

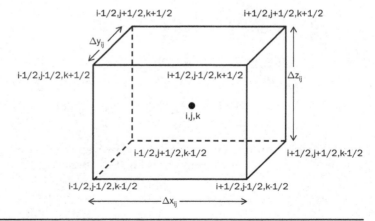

FIGURE 6-1   Representation of the control volume in 3D.

The surface integral can be expressed explicitly for each face:

$$\int_V \left(\frac{1}{B}\vec{u}\cdot\vec{n}\right)dA = \sum_{i=1}^{\#faces}\int_{A_i}\left(\frac{1}{B}\vec{u}\cdot\vec{n}\right)dA \tag{6-15}$$

$$\int_{A_i}\left(\frac{1}{B}\vec{u}\cdot\vec{n}\right)dA = \left(\frac{1}{B}u_x\Delta y\Delta z\right)^m_{i+1/2,j} - \left(\frac{1}{B}u_x\Delta y\Delta z\right)^m_{i-1/2,j}$$
$$+ \left(\frac{1}{B}u_y\Delta x\Delta z\right)^m_{i,j+1/2} - \left(\frac{1}{B}u_y\Delta x\Delta z\right)^m_{i,j-1/2} \tag{6-16}$$

There is a negative sign in front of the second and fourth terms on the right-hand side because the normal vector is pointing in the opposite direction of the x and y axes. The superscript m refers to the time level at which the term is evaluated. The formulation of the time derivative and m will be explained later in the discussion of time stepping. The velocity terms, u_x and u_y, are evaluated using Darcy's law. Here is an example for the first flux term across the interface between cells (i, j) and $(i + 1, j)$, assuming all off-diagonal elements in the permeability tensor are 0:

$$\left(\frac{1}{B}u_x\Delta y\Delta z\right)^m_{i+1/2,j} = -\left(\Delta y\Delta z\frac{k_x}{\mu B}\frac{\partial P}{\partial x}\right)^m_{i+1/2,j} \tag{6-17}$$

Formulation of the source term will be explained later in the discussion of well models. For the moment, it will be expressed as Q_{ij}.

$$\frac{B}{\rho_2^0}\int_V R_2 dV = Q_{ij}^m \tag{6-18}$$

6.2.2 External Boundary Conditions

A number of boundary conditions can be prescribed to represent most scenarios encountered in subsurface flow. Neumann (second-type) boundary refers to a condition where the derivative of the solution is specified. This is particularly useful for prescribing a flux boundary defined by a constant flow rate or constant pressure gradient (Islam et al., 2010). The flux term in Eq. (6-15) corresponding to the boundary face can be expressed as:

$$\int_{A_i}\left(\frac{1}{B}\vec{u}\cdot\vec{n}\right)dA = -\vec{\lambda}_j\cdot\nabla P_j\Big|^m_{boundary} \tag{6-19}$$

Specifying a constant flow rate or pressure gradient can be achieved by setting the right-hand side of Eq. (6-19) with a constant. A no-flow boundary (e.g., next to a sealing fault) would be a special case where the constant is equal to 0.

Another option is the Dirichlet (first-kind) boundary, which refers to a condition where the solution is specified. In this case, a constant pressure ($P_{boundary}$) is imposed at the boundary. This can be implemented by formulating the pressure gradient in Eq. (6-19) in terms of $P_{boundary}$ and the unknown pressure of the boundary grid cell.

6.2.3   Initialization

Initial conditions are needed to define the solution at $t = 0$. The system is considered to be static (i.e., no fluid movement) at $t < 0$. In other words, $= 0$ and $w_{ij} = $ constant. Equation (6-3) can be simplified as:

$$0 = \phi \sum_{j=1}^{N_p} (p_j \varpi_{ij} \vec{u}_j) \tag{6-20}$$

The only way the above equation can be satisfied is when $\vec{u}_j = 0$. Furthermore, if $\vec{k} \neq 0$, the following conditions would apply:

$$\frac{\partial P_j}{\partial x} + \rho_j g_x = 0 \tag{6-21a}$$

$$\frac{\partial P_j}{\partial y} + \rho_j g_y = 0 \tag{6-21b}$$

$$\frac{\partial P_j}{\partial z} + \rho_j g_z = 0 \tag{6-21c}$$

If the x and y axes are oriented in the horizontal plane, the above equations can be further simplified.

$$\frac{\partial P_j}{\partial x} = 0 \tag{6-22a}$$

$$\frac{\partial P_j}{\partial y} = 0 \tag{6-22b}$$

$$\frac{\partial P_j}{\partial z} + \rho_j g = 0 \tag{6-22c}$$

The set of relationships in Eq. (6-22) describes the initial pressure distribution.

6.2.4   Well Models

Well models are often the most significant part of a simulation. It defines the inner boundary where mass may enter and leave the simulation domain. The physical dimension of a well is typically much smaller than the grid cell where it is located; therefore, well models must include the sub-scale physics occurring below the scale of the grid cell. Due to the intricacy in well operations, well models should be flexible such that complex schedules, constraints, and completions configurations can be integrated.

Well models are used to define and compute the source term Q_{ij}^m in cell (i, j). For example:

$$Q_{ij}^m = \sum_{w=1}^{N_w} Q_w^m \delta_{i_w, i} \delta_{j_w, j} \qquad (6\text{-}23)$$

where $w = $ index of a well ($w = 1, \ldots, N_w$). Q_w^m refers to the rate of well w at time level m. The location of well w is denoted by (i_w, j_w). Two well locator functions, $\delta_{i_w, i}$ and $\delta_{j_w, j}$, are introduced:

$$\delta_{i_w, i} = \begin{cases} 1 & i = i_w \\ 0 & i \neq i_w \end{cases}; \quad \delta_{j_w, j} = \begin{cases} 1 & j = j_w \\ 0 & j \neq j_w \end{cases} \qquad (6\text{-}24)$$

Q_w^m can be formulated using the concept of productivity index (Economides et al., 1994), which allows us to relate the pressure of the well cell to the flow rate from a well that is located somewhere inside the well cell:

$$J_w^m = \frac{Q_w^m}{\left(P_{wf, w}^m - P_{i_w, j_w}^m\right)} \qquad (6\text{-}25)$$

$P_{wf, w}^m$ and J_w^m are the bottom-hole pressure and productivity index of well w, respectively, at time level m. P_{i_w, j_w}^m is the pressure of the well cell. The sign convention is that $Q_w^m > 0$ for an injector and $Q_w^m < 0$ for a producer. If a rate constraint is applied (i.e., Q_w^m is specified), the well model can be used to express $P_{wf, w}^m$; if a pressure constraint is applied (i.e., $P_{wf, w}^m$ is specified), the well model can be used to express Q_w^m.

J_w^m generally depends on the well and fluid properties. For the flow of an incompressible fluid, it can be considered a constant. Different formulations of the productivity index are available in the literature. They differ primarily in the way the sub-scale physics are represented. For example, the diffusivity equation can be used to describe radial flow from the well cell into the wellbore at pseudo steady-state. In particular, if the pressure at the edge of the well cell is considered to be P_{i_w, j_w}^m, the following relationship of J_w^m applies (Walsh and Lake, 2003):

$$J_w^m = \frac{2\pi k_{i_w, j_w} h_{i_w, j_w}}{\mu_j \left[\dfrac{1}{2} \ln\left(\dfrac{4 A_{i_w, j_w}}{\gamma C_{A_w} r_{w_w}^2}\right) + \dfrac{1}{4} + s_w\right]} \qquad (6\text{-}26)$$

If the average pressure of the well cell is taken to be P_{i_w, j_w}^m, then the equation should be modified slightly as:

$$J_w^m = \frac{2\pi k_{i_w, j_w} h_{i_w, j_w}}{\mu_j \left[\dfrac{1}{2} \ln\left(\dfrac{4 A_{i_w, j_w}}{\gamma C_{A_w} r_{w_w}^2}\right) + s_w\right]} \qquad (6\text{-}27)$$

where γ = Euler's constant; A_{i_w}, k_{i_w,j_w}, and h_{i_w,j_w} refer to the area, permeability, perforated thickness of the well cell located at (i_w, j_w), respectively; and s_w and r_{w_w} refer to the skin and well radius (r_w) of well w, respectively. C_{A_w} is the Dietz shape factor, which depends on the shape of and position of the well within the cell (Dietz, 1965). For a circular well located in the middle of a rectangular grid cell, C_{A_w} is approximately 31.62.

Equation (6-27) can also be written in terms of r_o, which is adjusted to match the analytical solution (Peaceman, 1978):

$$J_w^m = \frac{2\pi k_{i_w,j_w} h_{i_w,j_w}}{\mu_j \left[\ln \left(\dfrac{r_o}{r_{w_w}} \right) + s_w \right]} \tag{6-28}$$

The idea is to find an equivalent distance (r_o) from the well for which the pressure calculated by the simulator is the same as that obtained analytically: $P(r_o) = P_{i_w, j_w}^m$. For example, if $\Delta x = \Delta y$ in 2D, it can be shown that $r_o = e^{-\pi/2}\Delta x$, and the constant $e^{-\pi/2} = 0.2078$ is referred to as the "Peaceman Correction."

If the well spans or penetrates multiple grid cells, the well constraint can be split among numerous cells accordingly. There will be a source term in each cell where the well penetrates. This source term can be represented by a partial well rate of Q_{i_w,j_w}^m, and Q_w^m is the summation of Q_{i_w,j_w}^m over all grid cells penetrated by the well:

$$Q_w^m = \sum Q_{i_w,j_w}^m \tag{6-29}$$

Equation (6-25) is modified to represent Q_{i_w,j_w}^m, where J_{i_w,j_w}^m is the partial productivity index that may vary among different cells along the well trajectory:

$$Q_{i_w,j_w}^m = J_{i_w,j_w}^m \left(P_{wf,i_w,j_w}^m - P_{i_w,j_w}^m \right) \tag{6-30}$$

Remarks

- Equations (6-29) and (6-30) can be re-arranged to obtain Q_{i_w,j_w}^m from a fixed Q_w^m. For example, if a well penetrates two blocks (1 and 2), the total well rate according to Eqs. (6-29) and (6-30) is:

$$Q_w^m = J_1^m \left(P_{wf,1}^m - P_1^m \right) + J_2^m \left(P_{wf,2}^m - P_2^m \right) \tag{6-31}$$

- P_{wf,i_w,j_w}^m may vary along the well trajectory as well. Relationships that describe flow and friction loss in tubing and wellbore can be used to obtain P_{wf} in different cells. For instance, $P_{wf,2}^m$ can be described in terms of $P_{wf,1}^m$ using one of these flows in wellbore relationships.

$$Q_w^m = J_1^m P_{wf,1}^m + J_2^m f\left(P_{wf,1}^m\right) - J_1^m P_1^m - J_2^m P_2^m \qquad (6\text{-}32)$$

The above equation can be re-arranged to express $P_{wf,1}^m$ in terms of Q_w^m, such that it can be substituted back into Eq. (6-30). This offers the flexibility of formulating Q_{i_w,j_w}^m when either $P_{wf,1}^m$ or Q_w^m is defined.

6.2.5 Linearization

The solution of pressure using Eq. (6-13) requires linearization, since pressure is a function of location and time. In this section, the treatment of spatial and temporal (time) non-linearities is addressed.

Spatial Linearization

To linearize the flux terms at the interface between neighboring cells, let us consider a mass balance across an interface. By definition, an interface has a 0 volume, implying that no mass can be accumulated at the interface. Consider an example, the following condition applies to the flux at the interface of $i + 1/2$:

$$\left(\frac{\Delta y \Delta z k_x}{\mu B}\right)_{ij}^m \left[\frac{P_{ij} - P_{i+1/2,j}}{\frac{\Delta x}{2}}\right]_{ij}^m$$

$$= \left(\frac{\Delta y \Delta z k_x}{\mu B}\right)_{i+1,j}^m \left(\frac{P_{i+1/2,j} - P_{i+1,j}}{\frac{\Delta x}{2}}\right)_{i+1,j}^m \qquad (6\text{-}33)$$

$$\frac{1}{(\mu B)_{ij}^m}(\Delta y \Delta z k_x)_{ij}^m \left[\frac{P_{ij} - P_{i+1/2,j}}{\frac{\Delta x}{2}}\right]_{ij}^m$$

$$= \frac{1}{(\mu B)_{i+1,j}^m}(\Delta y \Delta z k_x)_{i+1,j}^m \left[\frac{P_{i+1/2,j} - P_{i+1,j}}{\frac{\Delta x}{2}}\right]_{i+1,j}^m \qquad (6\text{-}34)$$

The Taylor series expansion at $P_{ij}|_x$ is used to approximate the spatial derivatives in the above equations:

$$P_{ij}\big|_{x+\Delta x/2} \approx P_{ij}\big|_x + \frac{\Delta x}{2}\frac{\partial P_{ij}}{\partial x}\bigg|_x + \frac{\left(\frac{\Delta x}{2}\right)^2}{2}\frac{\partial^2 P_{ij}}{\partial x^2}\bigg|_x + \ldots\frac{\left(\frac{\Delta x}{2}\right)^n}{n!}\frac{\partial^n P_{ij}}{\partial x^n}\bigg|_x \qquad (6\text{-}35)$$

where n is a finite number. If Δx is small, only the first-order term is retained.

$$\frac{\partial P_{ij}}{\partial x} \approx \frac{P_{ij}\big|_{x+\Delta x/2} - P_{ij}\big|_x}{\Delta x / 2} = \frac{P_{i+1/2,j} - P_{ij}}{\Delta x / 2} \tag{6-36}$$

Equation (6-34) can be re-arranged to obtain an expression for $P_{i+1/2,j}$ in terms of P_{ij} and $P_{i+1,j}$. Substituting the obtained expression back into the right-hand side of Eq. (6-34) would yield the following:

$$\frac{1}{(\mu B)_{ij}^m} (\Delta y \Delta z k_x)_{ij}^m \left(\frac{P_{ij} - P_{i+1/2,j}}{\frac{\Delta x}{2}} \right)_{ij}^m$$

$$= \frac{1}{\mu B} \times 2 \left[\left(\left(\frac{\Delta y \Delta z k_x}{\Delta x} \right)_{ij}^m \right)^{-1} + \left(\left(\frac{\Delta y \Delta z k_x}{\Delta x} \right)_{i+1,j}^m \right)^{-1} \right]^{-1} \left(P_{ij}^m - P_{i+1,j}^m \right)$$

$$= T_{x,i+1/2,j}^m \left(P_{ij}^m - P_{i+1,j}^m \right) \tag{6-37}$$

T is the transmissibility, and the subscript x denotes that it is evaluated along the x direction. T consists of two parts. One part is composed of fluid properties, $1/\overline{\mu B}$, which needs to be evaluated dynamically as a function of P. The other part is composed of static cell dimensions and permeability.

The dynamic portion can be computed in a number of ways. Let us denote $\dfrac{1}{\mu(P) \times B(P)} = f(P)$. With single-point upstream weighting, $f(P_{i+1,j}) = f(P_{ij})$ if cell (i, j) is upstream to block $(i+1, j)$, or $f(P_{i+1/2,j}) = f(P_{i+1,j})$ if cell $(i+1, j)$ is upstream to block (i, j). With average pressure weighting, $f(P_{i+1/2,j}) = f(\overline{P})$.

The static portion in this case is merely a harmonic average over the two neighboring cells. This particular formulation is referred to as the two-point flux approximation, as it is based on a mass balance across the interface involving only the two immediate neighboring cells. The harmonic averaging tends to suppress or smooth out the large contrast in properties among the neighboring cells. It is interesting to note that setting the permeability outside the domain to be 0 would result in T at the boundary being 0 (the harmonic average between a zero and a non-zero value equals 0). Therefore, the no-flow boundary is a natural consequence of the harmonic averaging implemented in the transmissibility calculation.

Finally, assembling all the terms, Eq. (6-13) becomes:

$$
\left(\Delta x \Delta y \Delta z \varphi c_t\right)_{ij}^m \frac{\partial P_{ij}}{\partial t} + T_{x,i+1/2,j}^m \left(P_{ij}^m - P_{i+1,j}^m\right) + T_{x,i-1/2,j}^m \left(P_{ij}^m - P_{i-1,j}^m\right)
$$
$$
+ T_{y,i,j+1/2}^m \left(P_{ij}^m - P_{i,j+1}^m\right) + T_{y,i,j-1/2}^m \left(P_{ij}^m - P_{i,j-1}^m\right) = Q_{ij}^m \qquad (6\text{-}38)
$$

A few comments should be highlighted. The unit corresponding to each term is reservoir volume divided by time $[L^3/t]$. The equation does not contain information at a resolution or scale smaller than the grid cell.

Example 6-1

Derive the conservation of mass, where fluxes are expressed in terms of transmissibility, for the case where permeability is a full tensor.

Solution

If permeability is a full tensor,

$$
\vec{\vec{k}} = \begin{bmatrix} k_x & k_{xy} \\ k_{xy} & k_y \end{bmatrix}
$$

Therefore, velocity in each direction becomes

$$
u_x = -\left(\frac{k_x}{\mu}\frac{\partial P}{\partial x} + \frac{k_{xy}}{\mu}\frac{\partial P}{\partial y}\right)
$$

$$
u_y = -\left(\frac{k_{xy}}{\mu}\frac{\partial P}{\partial x} + \frac{k_y}{\mu}\frac{\partial P}{\partial y}\right)
$$

The conservation of mass is formulated as follows:

$$
\left(\Delta x \Delta y \Delta z \phi c_t\right)_{ij}^m \frac{\partial P_{ij}}{\partial t} +
$$
$$
T_{x,i+1/2,j}^m \left(P_{ij}^m - P_{i+1,j}^m\right) + T_{xy,i,j+1/2}^m \left(P_{ij}^m - P_{i,j+1}^m\right)
$$
$$
+ T_{x,i-1/2,j}^m \left(P_{ij}^m - P_{i-1,j}^m\right) + T_{xy,i,j-1/2}^m \left(P_{ij}^m - P_{i,j-1}^m\right)
$$
$$
+ T_{y,i,j+1/2}^m \left(P_{ij}^m - P_{i,j+1}^m\right) + T_{xy,i+1/2,j}^m \left(P_{ij}^m - P_{i+1,j}^m\right)
$$
$$
+ T_{y,i,j-1/2}^m \left(P_{ij}^m - P_{i,j-1}^m\right) + T_{xy,i-1/2,j}^m \left(P_{ij}^m - P_{i-1,j}^m\right) = Q_{ij}^m
$$

The transmissibility T_{xy} is again evaluated as a harmonic average between two neighboring cells. For example:

$$
T_{xy,i+1/2,j}^m = \frac{1}{\mu B} \times 2 \left\{ \left[\left(\frac{\Delta y \Delta z k_{xy}}{\Delta x}\right)_{ij}^m\right]^{-1} + \left[\left(\frac{\Delta y \Delta z k_{xy}}{\Delta x}\right)_{i+1,j}^m\right]^{-1} \right\}
$$

Example 6-2

Extend the formulation in Eq. (6-38) for a rectangular grid cell in 3D.

Solution

Darcy's law [Eq. (6-4)] is substituted into each of the flux terms in Eq. (6-16). For example:

$$\left(\frac{1}{B}u_x\Delta y\Delta z\right)^m_{i+1/2,j} = -\left(\Delta y\Delta z\frac{k_x}{\mu B}\left(\frac{\partial P}{\partial x} + \rho g_x\right)\right)^m_{i+1/2,j}$$

$$g_x = -g\frac{\partial D}{\partial x}$$

D represents the depth below a certain datum (e.g., top of the domain). The same scheme for formulating the transmissibility is adopted. For example, the above flux term becomes:

$$\left(\frac{1}{B}u_x\Delta y\Delta z\right)^m_{i+1/2,j} = T^m_{x,i+1/2,j}\left[\left(P^m_{ij} - \rho^m g D_{ij}\right) - \left(P^m_{i+1,j} - \rho^m g D_{i+1,j}\right)\right]$$

$$(6\text{-}39)$$

It should be obvious that if the x axis is oriented along the horizontal plane, then $g_x = 0$ and $D_{ij} = D_{i+1,j}$. Including the flux terms entering and exiting the CV from the top and bottom faces, Eq. (6-38) is now extended to include flow in 3D:

$$\left(\Delta x\Delta y\Delta z\phi c_t\right)^m_{ijk}\frac{\partial P_{ijk}}{\partial t} + T^m_{x,i+1/2,jk}\left(P^m_{ijk} - P^m_{i+1,jk} - \gamma^m D_{ijk} + \gamma^m D_{i+1,jk}\right)$$

$$+ T^m_{x,i-1/2,jk}\left(P^m_{ijk} - P^m_{i-1,jk} - \gamma^m D_{ijk} + \gamma^m D_{i-1,jk}\right)$$

$$+ T^m_{y,i,j+1/2,k}\left(P^m_{ijk} - P^m_{i,j+1,k} - \gamma^m D_{ijk} + \gamma^m D_{i,j+1,k}\right)$$

$$+ T^m_{y,i,j-1/2,k}\left(P^m_{ijk} - P^m_{i,j-1,k} - \gamma^m D_{ijk} + \gamma^m D_{i,j-1,k}\right)$$

$$+ T^m_{z,ij,k+1/2}\left(P^m_{ijk} - P^m_{ij,k+1} - \gamma^m D_{ijk} + \gamma^m D_{ij,k+1}\right)$$

$$+ T^m_{z,ij,k-1/2}\left(P^m_{ijk} - P^m_{ij,k-1} - \gamma^m D_{ijk} + \gamma^m D_{ij,k-1}\right) = Q^m_{ijk}$$

$$(6\text{-}40)$$

where $\gamma = \rho g$. An additional subscript k is introduced to denote the dimension along the z axis. It should be noted that even in 2D, it is possible for D to vary across the entire domain.

Time Linearization

Linearization in time refers to the time-stepping scheme to compute the time-dependent variables. It is needed to advance the solution of the unknowns to the next time level. To derive an expression for the time derivative, let us represent $P_{ij}|_{t+\Delta t}$ as a Taylor series expansion at $P_{ij}|_t$:

$$P_{ij}\Big|_{t+\Delta t} \approx P_{ij}\Big|_t + \Delta t\frac{\partial P_{ij}}{\partial t}\Big|_t + \frac{(\Delta t)^2}{2}\frac{\partial^2 P_{ij}}{\partial t^2}\Big|_t + \dots \frac{(\Delta t)^n}{n!}\frac{\partial^n P_{ij}}{\partial t^n}\Big|_t \quad (6\text{-}41)$$

where n is a finite number. If Δt is small, only the first-order term is retained.

$$\frac{\partial P_{ij}}{\partial t} \approx \frac{P_{ij}\big|_{t+\Delta t} - P_{ij}\big|_{t}}{\Delta t} = \frac{P_{ij}^{n+1} - P_{ij}^{n}}{\Delta t} \qquad (6\text{-}42)$$

A new superscript is introduced here: n refers to the old or current time level, while $n + 1$ refers to the new or next time level (the level where P is unknown and to be calculated). Substituting the Taylor series approximation into Eq. (6-38) would yield the following:

$$\begin{aligned}
(\Delta x \Delta y \Delta z \phi c_t)_{ij}^m &\frac{P_{ij}^{n+1} - P_{ij}^n}{\Delta t} + T_{x,i+1/2,j}^m\left(P_{ij}^m - P_{i+1,j}^m\right) \\
&+ T_{x,i-1/2,j}^m\left(P_{ij}^m - P_{i-1,j}^m\right) + T_{y,i,j+1/2}^m\left(P_{ij}^m - P_{i,j+1}^m\right) \qquad (6\text{-}43) \\
&+ T_{y,i,j-1/2}^m\left(P_{ij}^m - P_{i,j-1}^m\right) = Q_{ij}^m
\end{aligned}$$

For the other terms, the formulation depends on the choice of m for the pressure-dependent variables:

$$P^m = \theta P^{n+1} + (1 - \theta)P^n \qquad (6\text{-}44)$$

In the above equation, $0 \leq \theta \leq 1$ is a temporal weighting factor. A scheme is deemed explicit if $\theta = 0$ or $m = n$, while it is considered implicit if $\theta = 1$ or $m = n + 1$.

Case 1—Explicit ($\theta = 0$) Equation (6-43) can be expressed as follows after evaluating all terms at the n level:

$$\begin{aligned}
(\Delta x \Delta y \Delta z \phi c_t)_{ij}^n &\frac{P_{ij}^{n+1} - P_{ij}^n}{\Delta t} + T_{x,i+1/2,j}^n\left(P_{ij}^n - P_{i+1,j}^n\right) \\
&+ T_{x,i-1/2,j}^n\left(P_{ij}^n - P_{i-1,j}^n\right) + T_{y,i,j+1/2}^n\left(P_{ij}^n - P_{i,j+1}^n\right) \qquad (6\text{-}45) \\
&+ T_{y,i,j-1/2}^n\left(P_{ij}^n - P_{i,j-1}^n\right) = Q_{ij}^n
\end{aligned}$$

Since the solution of P_{ij}^{n+1} does not depend on P^{n+1} at other locations, the algebraic expression presented in Eq. (6-45) can be used to compute P_{ij}^{n+1} at each grid cell independently.

Case 2—Implicit ($\theta = 1$) After evaluating all terms at the $n + 1$ level, Eq. (6-43) can be expressed as follows:

$$\begin{aligned}
(\Delta x \Delta y \Delta z \phi c_t)_{ij}^{n+1} &\frac{P_{ij}^{n+1} - P_{ij}^n}{\Delta t} + T_{x,i+1/2,j}^{n+1}\left(P_{ij}^{n+1} - P_{i+1,j}^{n+1}\right) \\
&+ T_{x,i-1/2,j}^{n+1}\left(P_{ij}^{n+1} - P_{i-1,j}^{n+1}\right) + T_{y,i,j+1/2}^{n+1}\left(P_{ij}^{n+1} - P_{i,j+1}^{n+1}\right) \qquad (6\text{-}46) \\
&+ T_{y,i,j-1/2}^{n+1}\left(P_{ij}^{n+1} - P_{i,j-1}^{n+1}\right) = Q_{ij}^{n+1}
\end{aligned}$$

In contrast to the explicit case, the implicit solution of P_{ij}^{n+1} depends on P^{n+1} at other locations (which are also unknown). Therefore, the unknown pressure values at all locations must be solved simultaneously, and Eq. (6-46) describes a non-linear system of equations for the solution of $P_{ij}^{n+1} \in i, j$.

6.2.6 Solution Methods

Specific solution methods are presented to demonstrate the assembling of the relevant set of equations, boundary conditions, well models, and initialization into a complete system of equations. The governing equation describing mass and momentum balance for each cell derived in Eq. (6-43) is re-arranged as:

$$
\begin{aligned}
(\Delta x \Delta y \Delta z \phi c_t)_{ij}^m \left(P_{ij}^{n+1} - P_{ij}^n \right) & \\
- \Delta t \left[T_{x,i+1/2,j}^m \left(P_{i+1,j}^m - P_{ij}^m \right) - T_{x,i-1/2,j}^m \left(P_{ij}^m - P_{i-1,j}^m \right) \right] & \quad (6\text{-}47) \\
- \Delta t \left[T_{y,i,j+1/2}^m \left(P_{i,j+1}^m - P_{ij}^m \right) - T_{y,i,j-1/2}^m \left(P_{ij}^m - P_{i,j-1}^m \right) \right] &= \Delta t Q_{ij}^m
\end{aligned}
$$

The difference form can also be introduced to simplify the notation:

$$
\begin{aligned}
(\Delta x \Delta y \Delta z \phi c_t)_{ij}^m \left(P_{ij}^{n+1} - P_{ij}^n \right) - \Delta t \left(\Delta_x T_x \Delta_x P \right)^m & \\
- \Delta t \left(\Delta_y T_y^m \Delta_y P \right)^m &= \Delta t Q_{ij}^m
\end{aligned}
\quad (6\text{-}48)
$$

For a total of N grid cells, Eq. (6-43) or (6-48) represents a system of equations with N unknowns of P_{ij}^{n+1}. For cells located at the boundary (i.e., if at least one face is aligned with the domain edge), the flux corresponding to that particular edge should be modified, as discussed previously. For example, if the cell face $(i - 1/2, j)$ is located at a no-flow boundary, $T_{x,i-1/2,j}^m$ is set to be 0 accordingly.

Explicit ($\theta = 0$) For an explicit scheme for time-stepping, Eq. (6-45) can be used:

$$
\begin{aligned}
(\Delta x \Delta y \Delta z \phi c_t)_{ij}^n P_{ij}^{n+1} & \\
- \Delta t \left[T_{x,i+1/2,j}^n \left(P_{i+1,j}^n - P_{ij}^n \right) - T_{x,i-1/2,j}^n \left(P_{ij}^n - P_{i-1,j}^n \right) \right] & \\
- \Delta t \left[T_{y,i,j+1/2}^n \left(P_{i,j+1}^n - P_{ij}^n \right) - T_{y,i,j-1/2}^n \left(P_{ij}^n - P_{i,j-1}^n \right) \right] & \quad (6\text{-}49) \\
= (\Delta x \Delta y \Delta z \phi c_t)_{ij}^n P_{ij}^n + \Delta t Q_{ij}^n
\end{aligned}
$$

Implicit ($\theta = 1$) To solve the fully implicit form of Eq. (6-46), it entails solving a non-linear system of equations in terms of $P_{ij}^{n+1} \in i, j$. Solution techniques, such as successive substitution and Newton–Raphson, can be adopted. The idea is to solve the system of equations by minimizing the residual function $R_{p,ij} \in i, j$:

$$
\begin{aligned}
R_{p,ij} &= \left(\Delta x \Delta y \Delta z \phi c_t\right)_{ij}^{n+1} \frac{P_{ij}^{n+1} - P_{ij}^n}{\Delta t} + T_{x,i+1/2,j}^{n+1}\left(P_{ij}^{n+1} - P_{i+1,j}^{n+1}\right) \\
&+ T_{x,i-1/2,j}^{n+1}\left(P_{ij}^{n+1} - P_{i-1,j}^{n+1}\right) + T_{y,i,j+1/2}^{n+1}\left(P_{ij}^{n+1} - P_{i,j+1}^{n+1}\right) \\
&+ T_{y,i,j-1/2}^{n+1}\left(P_{ij}^{n+1} - P_{i,j-1}^{n+1}\right) - Q_{ij}^{n+1}
\end{aligned}
\tag{6-50}
$$

Successive substitution involves solving a linearized version of the system of equations iteratively until the residual is close to 0. Newton–Raphson involves approximating the residual function with the Taylor's series and deriving a recursive formula to compute the unknowns iteratively. Successive substitution is easy to implement, though it can be quite slow. Newton–Raphson, on the other hand, requires fewer iterations, but it involves computing derivatives of $R_{p,ij}$ with respect to $P_{ij}^{n+1} \in i, j$.

Equation (6-46) can be linearized if T and c_t are evaluated at level n, instead of $n + 1$.

$$
\begin{aligned}
&\left[\Delta t\left(T_{x,i+1/2,j}^n + T_{x,i-1/2,j}^n + T_{y,i,j+1/2}^n + T_{y,i,j-1/2}^n\right) + \left(\Delta x \Delta y \Delta z \phi c_t\right)_{ij}^n\right] \\
&P_{ij}^{n+1} - \Delta t T_{x,i+1/2,j}^n P_{i+1,j}^{n+1} - \Delta t T_{x,i-1/2,j}^n P_{i-1,j}^{n+1} - \Delta t T_{y,i,j+1/2}^n P_{i,j+1}^{n+1} \\
&- \Delta t T_{y,i,j-1/2}^n P_{i,j-1}^{n+1} = \left(\Delta x \Delta y \Delta z \phi c_t\right)_{ij}^n P_{ij}^n + \Delta t Q_{ij}^{n+1}
\end{aligned}
\tag{6-51}
$$

The above linearized equation can be used in a successive substitution scheme, where n and $n + 1$ are replaced by l and $l + 1$, with l being the iteration counter. In fact, a spectrum of implicitness is possible depending on how each of these terms is evaluated. The term Q_{ij}^{n+1} should be expressed using the well model discussed earlier. Re-arranging Eq. (6-25) for $m = n + 1$ gives the following:

$$
Q_w^{n+1} = J_w^{n+1}\left(P_{wf,w}^{n+1} - P_{i_w,j_w}^{n+1}\right)
\tag{6-52}
$$

Two possibilities exist: (1) if the well is rate-constrained with known Q_w^{n+1}, its value can be substituted directly into Eq. (6-51) together with Eq. (6-23) and (2) if the well is pressure-constrained with known $P_{iw,jw}^{n+1}$, the substitution of Eqs. (6-52) and (6-23) into Eq. (6-51) becomes:

$$
\begin{aligned}
&\left[\Delta t\left(T_{x,i+1/2,j}^n + T_{x,i-1/2,j}^n + T_{y,i,j+1/2}^n + T_{y,i,j-1/2}^n + \boxed{J_w^{n+1}}\right) + \left(\Delta x \Delta y \Delta z \phi c_t\right)_{ij}^n\right] \\
&P_{ij}^{n+1} - \Delta t T_{x,i+1/2,j}^n P_{i+1,j}^{n+1} - \Delta t T_{x,i-1/2,j}^n P_{i-1,j}^{n+1} - \Delta t T_{y,i,j+1/2}^n P_{i,j+1}^{n+1} \\
&- \Delta t T_{y,i,j-1/2}^n P_{i,j-1}^{n+1} = \left(\Delta x \Delta y \Delta z \phi c_t\right)_{ij}^n P_{ij}^n + \boxed{\Delta t J_w^{n+1} P_{wf,w}^{n+1}}
\end{aligned}
\tag{6-53}
$$

This represents a linear system of equations for the solution of $P_{ij}^{n+1} \in i, j$, which can be expressed as:

$$\mathbf{T} \cdot \mathbf{P}^{n+1} = \mathbf{B}. \tag{6-54}$$

$\mathbf{B}$ is the forcing function that contains the "old" pressure of $\mathbf{P}^n$ and the source term; $\mathbf{P}^{n+1}$ is the unknown pressure at the "new" time level; and $\mathbf{T}$ is the transmissibility matrix. This matrix is slightly misnamed as it contains more than the terms such as T_x or T_y. It is sparse and diagonally dominant. It is a banded matrix, and its bandwidth depending on the stencil used to evaluate the flux terms. For example, a five-point stencil involving (i, j), $(i + 1, j)$, $(i - 1, j)$, $(i, j + 1)$, and $(i, j - 1)$ can be used for 2D, while a seven-point stencil involving (i, j, k), $(i + 1, j, k)$, $(i - 1, j, k)$, $(i, j + 1, k)$, $(i, j - 1, k)$, $(i, j, k + 1)$, and $(i, j, k - 1)$ can be used for 3D. It is symmetric because the flux is same for both sides of the interface. The solution to the above equation requires a matrix solver. Once the pressure is computed, the velocity across a particular interface can be obtained in a post-processing step; for instance, the velocity between the blocks (i, j) and $(i + 1, j)$ can be computed as:

$$u_x|_{i+1/2,j} = \frac{T_{x,i+1/2,j}\left(P_{ij}^{n+1} - P_{i+1,j}^{n+1}\right)}{\Delta y_{ij}\Delta z_{ij}} \tag{6-55}$$

Example 6-3
Assemble $\mathbf{T}$ and $\mathbf{B}$ for this system with $N_x = 3$ and $N_y = 2$.

Solution
We can choose to cycle the x or y axis first. If the x axis is selected, we have $i = 1, 2, 3$ and $j = 1, 2$ (Fig. 6-2).

1,2	2,2	3,2
1,1	2,1	3,1

FIGURE 6-2   Cell ordering for a
3 × 2 grid.

According to Eq. (6-51), the system of equations becomes:

$$
\begin{bmatrix}
T_{11} & -\Delta t T_{x\frac{3}{2},1} & 0 & -\Delta t T_{y1,\frac{3}{2}} & 0 & 0 \\
-\Delta t T_{x\frac{3}{2},1} & T_{21} & -\Delta t T_{x\frac{5}{2},1} & 0 & -\Delta t T_{y2,\frac{3}{2}} & 0 \\
0 & -\Delta t T_{x\frac{5}{2},1} & T_{31} & 0 & 0 & -\Delta t T_{y3,\frac{3}{2}} \\
-\Delta t T_{y1,\frac{3}{2}} & 0 & 0 & T_{12} & -\Delta t T_{x\frac{3}{2},2} & 0 \\
0 & -\Delta t T_{y2,\frac{3}{2}} & 0 & -\Delta t T_{x\frac{3}{2},2} & T_{22} & -\Delta t T_{x\frac{5}{2},2} \\
0 & 0 & -\Delta t T_{y3,\frac{3}{2}} & 0 & -\Delta t T_{x\frac{5}{2},2} & T_{32}
\end{bmatrix}
$$

$$
\begin{bmatrix} P_{11}^{n+1} \\ P_{21}^{n+1} \\ P_{31}^{n+1} \\ P_{12}^{n+1} \\ P_{22}^{n+1} \\ P_{32}^{n+1} \end{bmatrix} = \begin{bmatrix} C_{t11}V_{p11}P_{11}^{n} + \Delta t {Q_{11}}^{n+1} \\ C_{t21}V_{p21}P_{21}^{n} + \Delta t {Q_{21}}^{n+1} \\ C_{t31}V_{p31}P_{31}^{n} + \Delta t {Q_{31}}^{n+1} \\ C_{t12}V_{p12}P_{12}^{n} + \Delta t {Q_{12}}^{n+1} \\ C_{t22}V_{p22}P_{22}^{n} + \Delta t {Q_{22}}^{n+1} \\ C_{t32}V_{p32}P_{32}^{n} + \Delta t {Q_{32}}^{n+1} \end{bmatrix}
$$

where

$$
T_{11=} \left[\Delta t \left(T_{x,3/2,1}^{n} + T_{y,1,3/2}^{n} \right) + \left(\Delta x \Delta y \Delta z \phi c_{t} \right)_{1,1}^{n} \right]
$$

$$
T_{21=} \left[\Delta t \left(T_{x,5/2,1}^{n} + T_{x,3/2,1}^{n} + T_{y,2,3/2}^{n} \right) + \left(\Delta x \Delta y \Delta z \phi c_{t} \right)_{2,1}^{n} \right]
$$

$$
T_{31=} \left[\Delta t \left(T_{x,5/2,1}^{n} + T_{y,3,3/2}^{n} \right) + \left(\Delta x \Delta y \Delta z \phi c_{t} \right)_{3,1}^{n} \right]
$$

$$
T_{12=} \left[\Delta t \left(T_{x,3/2,2}^{n} + T_{y,1,3/2}^{n} \right) + \left(\Delta x \Delta y \Delta z \phi c_{t} \right)_{1,2}^{n} \right]
$$

$$
T_{22=} \left[\Delta t \left(T_{x,5/2,2}^{n} + T_{x,3/2,2}^{n} + T_{y,2,3/2}^{n} \right) + \left(\Delta x \Delta y \Delta z \phi c_{t} \right)_{2,2}^{n} \right]
$$

$$
T_{32=} \left[\Delta t \left(T_{x,5/2,2}^{n} + T_{y,3,3/2}^{n} \right) + \left(\Delta x \Delta y \Delta z \phi c_{t} \right)_{3,2}^{n} \right]
$$

An explicit scheme is assumed. It is possible to choose a different cycling scheme such that the bandwidth of the matrix is smaller. The readers should repeat this example by cycling along the y direction first.

Example 6-4

Repeat Example 6-3 where cell (2,2) is outside of the boundary domain (Fig. 6-3).

Solution

Once again, if the x axis is selected, we have $i = 1, 2, 3$ and $j = 1, 2$, and since cell (2,2) is inactive, there are only five active cells in the computational domain.

1,2	2,2	3,2
1,1	2,1	3,1

Shaded cells are part of the domain of interest
Clear cells are outside of the domain of interest

FIGURE 6-3 Cell ordering for a 3 × 2 grid for Example 6-4 using Approach #1.

There are various ways for handling a domain consisting of inactive cells.

Approach #1—Both active and inactive cells are retained in the computational domain. Inactive cells are treated as dummy cells in the linear system. This can be achieved by assigning an arbitrarily large number for P_{22}, for example, $P_{22}{}^{n+1} = 9999$.

According to Eq. (6-51), the system of equations becomes:

$$
\begin{bmatrix}
T_{11} & -\Delta t T_{x3/2,1} & 0 & -\Delta t T_{y1,3/2} & 0 & 0 \\
-\Delta t T_{x3/2,1} & T_{21} & -\Delta t T_{x5/2,1} & 0 & 0 & 0 \\
0 & -\Delta t T_{x5/2,1} & T_{31} & 0 & 0 & -\Delta t T_{y3,3/2} \\
-\Delta t T_{y1,3/2} & 0 & 0 & T_{12} & 0 & 0 \\
0 & 0 & 0 & 0 & 1 & 0 \\
0 & 0 & -\Delta t T_{y3,3/2} & 0 & 0 & T_{32}
\end{bmatrix}
$$

$$
\begin{bmatrix}
P_{11}^{n+1} \\
P_{21}^{n+1} \\
P_{31}^{n+1} \\
P_{12}^{n+1} \\
P_{22}^{n+1} \\
P_{32}^{n+1}
\end{bmatrix}
=
\begin{bmatrix}
C_{t11}V_{p11}P_{11}^{n} + \Delta t Q_{11}{}^{n+1} \\
C_{t21}V_{p21}P_{21}^{n} + \Delta t Q_{21}{}^{n+1} \\
C_{t31}V_{p31}P_{31}^{n} + \Delta t Q_{31}{}^{n+1} \\
C_{t12}V_{p12}P_{12}^{n} + \Delta t Q_{12}{}^{n+1} \\
9999 \\
C_{t32}V_{p32}P_{32}^{n} + \Delta t Q_{32}{}^{n+1}
\end{bmatrix}
$$

where

$$T_{11=}\left[\Delta t\left(T_{x,3/2,1}^{n} + T_{y,1,3/2}^{n}\right) + \left(\Delta x \Delta y \Delta z \phi c_t\right)_{1,1}^{n}\right]$$

$$T_{21=}\left[\Delta t\left(T_{x,5/2,1}^{n} + T_{x,3/2,1}^{n}\right) + \left(\Delta x \Delta y \Delta z \phi c_t\right)_{2,1}^{n}\right]$$

$$T_{31=}\left[\Delta t\left(T_{x,5/2,1}^{n} + T_{y,3,3/2}^{n}\right) + \left(\Delta x \Delta y \Delta z \phi c_t\right)_{3,1}^{n}\right]$$

$$T_{12=}\left[\Delta t\left(T_{y,1,3/2}^{n}\right) + \left(\Delta x \Delta y \Delta z \phi c_t\right)_{1,2}^{n}\right]$$

$$T_{32=}\left[\Delta t\left(T_{y,3,3/2}^{n}\right) + \left(\Delta x \Delta y \Delta z \phi c_t\right)_{3,2}^{n}\right]$$

It is clear from the above formulation that the pressure at cell $(2,2)$ does not affect the solution in the computational domain. The disadvantage, though, is that the matrix equation to be solved remains large. If the inactive cells constitute a significant portion of the domain, the efficiency of this approach would be compromised.

Approach #2—Inactive cells are not included in the computational domain. A "new" indexing system is created to keep track of only the active cells (Fig. 6-4):

$$(1,1) \to 1 \quad (2,1) \to 2 \quad (3,1) \to 3 \quad (1,2) \to 4 \quad (3,2) \to 5$$

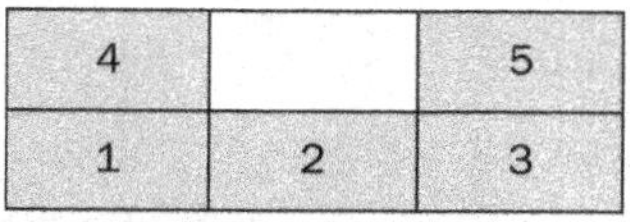

Shaded cells are part of the domain of interest
Clear cells are outside of the domain of interest

Figure 6-4 Cell ordering for a 3 × 2 grid for
Example 6-4 using Approach #2.

$$
\begin{bmatrix}
T_{11} & -\Delta t T_{x\frac{3}{2},1} & 0 & -\Delta t T_{y1,\frac{3}{2}} & 0 \\
-\Delta t T_{x\frac{3}{2},1} & T_{21} & -\Delta t T_{x\frac{5}{2},1} & 0 & 0 \\
0 & -\Delta t T_{x\frac{5}{2},1} & T_{31} & 0 & -\Delta t T_{y3,\frac{3}{2}} \\
-\Delta t T_{y1,\frac{3}{2}} & 0 & 0 & T_{12} & 0 \\
0 & 0 & -\Delta t T_{y3,\frac{3}{2}} & 0 & T_{32}
\end{bmatrix}
$$

$$
\begin{bmatrix}
P_1^{n+1} \\
P_2^{n+1} \\
P_3^{n+1} \\
P_4^{n+1} \\
P_5^{n+1}
\end{bmatrix}
=
\begin{bmatrix}
C_{t1}V_{p1}P_1^n + \Delta t Q_1^{n+1} \\
C_{t2}V_{p2}P_2^n + \Delta t Q_2^{n+1} \\
C_{t3}V_{p3}P_3^n + \Delta t Q_3^{n+1} \\
C_{t4}V_{p4}P_4^n + \Delta t Q_4^{n+1} \\
C_{t5}V_{p5}P_5^n + \Delta t Q_5^{n+1}
\end{bmatrix}
$$

where

$$
T_{11=}\left[\Delta t\left(T_{x,3/2,1}^n + T_{y,1,3/2}^n\right) + \left(\Delta x \Delta y \Delta z \phi c_t\right)_{1,1}^n\right]
$$

$$
T_{21=}\left[\Delta t\left(T_{x,5/2,1}^n + T_{x,3/2,1}^n\right) + \left(\Delta x \Delta y \Delta z \phi c_t\right)_{2,1}^n\right]
$$

$$
T_{31=}\left[\Delta t\left(T_{x,5/2,1}^n + T_{y,3,3/2}^n\right) + \left(\Delta x \Delta y \Delta z \phi c_t\right)_{3,1}^n\right]
$$

$$
T_{12=}\left[\Delta t\left(T_{y,1,3/2}^n\right) + \left(\Delta x \Delta y \Delta z \phi c_t\right)_{1,2}^n\right]
$$

$$
T_{32=}\left[\Delta t\left(T_{y,3,3/2}^n\right) + \left(\Delta x \Delta y \Delta z \phi c_t\right)_{3,2}^n\right]
$$

In this approach, the number of unknowns equal to the number of active cells and the dimension of the matrix equation is reduced. Provided one can keep track of the two indexing systems, the computational efficiency of Approach #2 is superior to Approach #1. More detail can be found in Ertekin et al.'s work (2001).

Example 6-5

Calculate the pressure distribution at 0.1 days for the following 1D reservoir with three grid cells. There is a no-flow outer boundary. A pressure-constrained well is producing at 55 m³/day and $J = 3.0 \times 10^{-8}$ m³/Pa·s is located in Cell 2.

Initial reservoir pressure $= 12$ MPa
$ct = 3 \times 10^{-9}$ Pa^{-1}
$\mu = 0.001$ Pa·s
$\Delta x = \Delta y = \Delta z = 10$ m

	Cell 1	Cell 2	Cell 3
k (m²) =	2×10^{-13}	1×10^{-14}	4×10^{-14}
$\phi =$	0.2	0.15	0.08

Solution

The following system of equations can be derived for the implicit scheme:

$$
\begin{bmatrix}
\Delta t\left(T_{3/2}\right) + C_{t1}V_{p1} & -\Delta t T_{3/2} & 0 \\
-\Delta t T_{3/2} & \Delta t\left(T_{3/2} + T_{5/2}\right) + C_{t2}V_{p2} & -\Delta t T_{5/2} \\
0 & -\Delta t T_{5/2} & \Delta t\left(T_{5/2}\right) + C_{t3}V_{p3}
\end{bmatrix}
$$

$$
\begin{bmatrix}
P_1^{n+1} \\
P_2^{n+1} \\
P_3^{n+1}
\end{bmatrix}
=
\begin{bmatrix}
C_{t1}V_{p1}P_1^{n} \\
C_{t2}V_{p2}P_2^{n} + \Delta t Q_2^{n+1} \\
C_{t3}V_{p3}P_3^{n+1}
\end{bmatrix}
$$

The calculation steps and solutions are provided in the EXCEL® spreadsheet "Example 6-5.xlsx," which can be found on this book's website, www.mhprofessional.com/PRMS.

Example 6-6

Modify Example 6-5 such that the pressure at the left boundary of Cell 1 is fixed at 12 MPa.

Solution

In the previous example, $T_{1/2} = T_{7/2} = 0$, corresponding to the no-flow boundary. To account for the constant pressure boundary, Eq. (6-19) can be used to incorporate a P_{boundary} of 12 MPa. The flux term corresponding to flow into the left boundary is $T_{1/2}(P_{\text{boundary}} - P_1)$, where

$$
T_{1/2} = \frac{\Delta y_1 \Delta z_1 k_1}{\mu \dfrac{\Delta x_1}{2}}
$$

$$
\begin{bmatrix}
\Delta t\left(T_{1/2} + T_{3/2}\right) + C_{t1}V_{p1} & -\Delta t T_{3/2} & 0 \\
-\Delta t T_{3/2} & \Delta t\left(T_{3/2} + T_{5/2}\right) + C_{t2}V_{p2} & -\Delta t T_{5/2} \\
0 & -\Delta t T_{5/2} & \Delta t\left(T_{5/2}\right) + C_{t3}V_{p3}
\end{bmatrix}
$$

$$
\begin{bmatrix} P_1^{n+1} \\ P_2^{n+1} \\ P_3^{n+1} \end{bmatrix} =
\begin{bmatrix}
C_{t1}V_{p1}P_1^n + \Delta t T_{1/2}P_{boundary} \\
C_{t2}V_{p2}P_2^n + \Delta t Q_2^{n+1} \\
C_{t3}V_{p3}P_3^{n+1}
\end{bmatrix}
$$

The calculation steps and solutions are provided in the EXCEL spreadsheet "Example 6-6.xlsx," which can be found on this book's website.

6.3 Multi-Phase Flow

6.3.1 Simulation Equations

The flow of three phases is considered oleic, aqueous, and gaseous. Once again, isothermal conditions and zero rock interaction ($w_{is} = 0$) are assumed. A black-oil model, which is a simplified fluid model of pressure-volume-temperature (PVT) description where hydrocarbons are lumped into two pseudo-components at surface conditions, a heavy component called "oil" and a light component called "gas," is used to model the oleic and gaseous phases (McCain, 1990). This assumption is valid when changes in compositions in the three phases can be neglected (e.g., no chemical or gas injection).

The three immiscible phases are aqueous ($j = 1$), oleic ($j = 2$), and gaseous ($j = 3$). There are three components of water ($i = 1$), oil ($i = 2$), and gas ($i = 3$). The water component exists only in the aqueous phase; the hydrocarbons components may exist in both the oleic and gaseous phases.

$$
\varpi_{1j} = \begin{cases} 1; j = 1 \\ 0; j = 2,3 \end{cases} \quad
\varpi_{1j} = \begin{cases} 0; j = 1 \\ > 0; j = 2,3 \end{cases} \quad
\varpi_{3j} = \begin{cases} 0; j = 1 \\ > 0; j = 2,3 \end{cases} \quad (6\text{-}56)
$$

If diffusion is neglected, the conservation of mass becomes:

$$
\int_V \frac{\partial\left(\phi \sum_{j=1}^{3} \rho_j S_j \varpi_{ij}\right)}{\partial t} dV = -\int_A \left(\sum_{j=1}^{3} \rho_j \varpi_{ij} \vec{u}_j \cdot \vec{n}\right) dA + \int_V R_i dV \quad \text{for } i = 1,\ldots,
$$

Nc (total number of components). $\qquad\qquad (6\text{-}57)$

Introducing the immiscibility condition, Eq. (6-57) can be expressed for each component.

Water: $\displaystyle\int_V \frac{\partial(\phi\rho_1 S_1)}{\partial t}dV = -\int_A (\rho_1\vec{u}_1 \cdot \vec{n})dA + \int_V R_1 dV$ (6-58a)

Oil: $\displaystyle\int_V \frac{\partial\left[\phi(\rho_2 S_2\varpi_{22} + \rho_3 S_3\varpi_{23})\right]}{\partial t}dV$

$\displaystyle = -\int_A (\rho_2\varpi_{22}\vec{u}_2 \cdot \vec{n} + \rho_3\varpi_{23}\vec{u}_3 \cdot \vec{n})dA + \int_V R_2 dV$ (6-58b)

Gas: $\displaystyle\int_V \frac{\partial\left[\phi(\rho_2 S_2\varpi_{32} + \rho_3 S_3\varpi_{33})\right]}{\partial t}dV$

$\displaystyle = -\int_A (\rho_2\varpi_{32}\vec{u}_2 \cdot \vec{n} + \rho_3\varpi_{33}\vec{u}_3 \cdot \vec{n})dA + \int_V R_3 dV$ (6-58c)

To utilize the black-oil model formation, it is necessary to replace ρ_j and w_{ij} with R_s, R_v, B_g, and B_o (solubility ratios and formation volume factors). This can be achieved if we divide each equation by ρ_1^0, ρ_2^0, and ρ_3^0, respectively, where the superscript "0" refers to the standard conditions:

Water: $\displaystyle\int_V \frac{\partial\left(\frac{\phi\rho_1}{\rho_1^0}S_1\right)}{\partial t}dV = -\int_A \left(\frac{\rho_1}{\rho_1^0}\vec{u}_1 \cdot \vec{n}\right)dA + \frac{1}{\rho_1^0}\int_V R_1 dV$ (6-59a)

Oil: $\displaystyle\int_V \frac{\partial\left[\phi\left(\frac{\rho_2\varpi_{22}}{\rho_2^0}S_2 + \frac{\rho_3\varpi_{23}}{\rho_2^0}S_3\right)\right]}{\partial t}dV$ (6-59b)

$\displaystyle = -\int_A \left(\frac{\rho_2\varpi_{22}}{\rho_2^0}\vec{u}_2 \cdot \vec{n} + \frac{\rho_3\varpi_{23}}{\rho_2^0}\vec{u}_3 \cdot \vec{n}\right)dA + \frac{1}{\rho_2^0}\int_V R_2 dV$

Gas: $\displaystyle\int_V \frac{\partial\left[\phi\left(\frac{\rho_2\varpi_{32}}{\rho_3^0}S_2 + \frac{\rho_3\varpi_{33}}{\rho_3^0}S_3\right)\right]}{\partial t}dV$ (6-59c)

$\displaystyle = -\int_A \left(\frac{\rho_2\varpi_{32}}{\rho_3^0}\vec{u}_2 \cdot \vec{n} + \frac{\rho_3\varpi_{33}}{\rho_3^0}\vec{u}_3 \cdot \vec{n}\right)dA + \frac{1}{\rho_3^0}\int_V R_3 dV$

Recognizing that $B_1 = \dfrac{\rho_1^0}{\rho_1}$, $B_2 = \dfrac{\rho_2^0}{\rho_2\varpi_{22}}$, $B_3 = \dfrac{\rho_3^0}{\rho_3\varpi_{33}}$, $R_v = \dfrac{\varpi_{23}\rho_3^0}{\varpi_{33}\rho_2^0}$ and $R_s = \dfrac{\varpi_{32}\rho_2^0}{\varpi_{22}\rho_3^0}$, Eq. (6-59) becomes:

Water: $\displaystyle\int_V \frac{\partial\left(\frac{\varphi}{B_1}S_1\right)}{\partial t}dV = -\int_A \left(\frac{1}{B_1}\vec{u}_1 \cdot \vec{n}\right)dA + \frac{1}{\rho_1^0}\int_V R_1 dV$ (6-60a)

$$(6\text{-}60b)$$

Oil: $$\int_V \frac{\partial\left[\varphi\left(\dfrac{S_2}{B_2} + \dfrac{R_v}{B_3}S_3\right)\right]}{\partial t} dV$$

$$= -\int_A \left(\frac{1}{B_2}\vec{u}_2 \cdot \vec{n} + \frac{R_v}{B_3}\vec{u}_3 \cdot \vec{n}\right) dA + \frac{1}{\rho_2^0}\int_V R_2 dV$$

$$(6\text{-}60c)$$

Gas: $$\int_V \frac{\partial\left[\varphi\left(\dfrac{R_s}{B_2}S_2 + \dfrac{S_3}{B_3}\right)\right]}{\partial t} dV$$

$$= -\int_A \left(\frac{R_s}{B_2}\vec{u}_2 \cdot \vec{n} + \frac{1}{B_3}\vec{u}_3 \cdot \vec{n}\right) dA + \frac{1}{\rho_3^0}\int_V R_3 dV$$

6.3.2   External Boundary Conditions

Similar to the single-phase case, a number of boundary conditions can be prescribed for multi-phase flow. The boundary flux terms in Eq. (6-60) can be formulated in terms of $-\vec{\vec{\lambda}}_j \cdot \nabla P_j\big|_{\text{boundary}}$. Neumann or flux boundary is implemented by replacing the flux term with a constant (e.g., no flow). Dirichlet boundary can be implemented by formulating the flux term in terms of the known pressure and saturation at the boundary. This boundary condition is commonly used to model a fluid contact (e.g., water–oil contact). Due to limited geologic control outside the reservoir domain $\vec{\vec{\lambda}}_{j}$, can be computed using properties evaluated at the boundary cell and neglecting the capillarity effect at the boundary (Islam et al., 2010). Upstream weighting (as described in subsequent sections) is needed as flow can be directed either into or out of the reservoir domain.

6.3.3   Initialization

Again, the system is considered to be static (i.e., no fluid movement) at $t < 0$. Eq. (6-22c) can be revised as:

$$\frac{\partial P_j}{\partial z} + \rho_j g = 0 \quad j \in 1, 2, 3 \tag{6-61}$$

Subtracting the equations corresponding to $j = 1$ and 2, as well as $j = 2$ and 3, would give:

$$\frac{\partial P_{c12}}{\partial z} + (\rho_2 - \rho_1)g = 0 \tag{6-62a}$$

$$\frac{\partial P_{c23}}{\partial z} + (\rho_3 - \rho_2)g = 0 \tag{6-62b}$$

Equation (6-61) can be used to initialize P_1 (in a fashion similar to the single-phase case). Based on the relationships between capillary pressure and saturations, Eqs. (6-62a) and (6-62b) can be used to initialize S_1 and S_2. The rest of the unknowns can be determined as:

$$P_2 = P_1 + P_{cow}(S_1) \tag{6-63a}$$

$$P_3 = P_2 + P_{cgo}(S_2) \tag{6-63b}$$

$$S_3 = 1 + S_1 = S_2 \tag{6-63c}$$

6.3.4   Well Models

The source term in Eq. (6-60) can be expressed as:

$$\frac{1}{\rho_1^0}\int_V R_1 dV = Q_{1ij}^m; \frac{1}{\rho_2^0}\int_V R_2 dV = Q_{2ij}^m; \frac{1}{\rho_3^0}\int_V R_3 dV = Q_{3ij}^m \tag{6-64}$$

Following the same notation as in Sec. 6.2.4:

$$Q_{1ij}^m = \sum_{w=1}^{N_w} Q_{1w}^m \delta_{i_w,i}\delta_{j_w,j}; Q_{2ij}^m = \sum_{w=1}^{N_w} Q_{2w}^m \delta_{i_w,i}\delta_{j_w,j}; Q_{3ij}^m = \sum_{w=1}^{N_w} Q_{3w}^m \delta_{i_w,i}\delta_{j_w,j} \tag{6-65}$$

Q_{1w}^m, Q_{2w}^m, and Q_{3w}^m can also be formulated using the concept of productivity index:

$$Q_{1,w}^m = \tilde{J}_w^m \frac{k_{r1}\left(S_{1,i_w,j_w}^m\right)}{\mu_1\left(P_{1,i_w,j_w}^m\right)B_1\left(P_{1,i_w,j_w}^m\right)}\left(P_{wf,w}^m - P_{1,i_w,j_w}^m\right) \tag{6-66a}$$

$$Q_{2,w}^m = \tilde{J}_w^m \frac{k_{r2}\left(S_{2,i_w,j_w}^m\right)}{\mu_1\left(P_{2,i_w,j_w}^m\right)B_2\left(P_{2,i_w,j_w}^m\right)}\left(P_{wf,w}^m - P_{2,i_w,j_w}^m\right) \tag{6-66b}$$

$$Q_{3,w}^m = \tilde{J}_w^m \frac{k_{r3}\left(S_{3,i_w,j_w}^m\right)}{\mu_1\left(P_{3,i_w,j_w}^m\right)B_3\left(P_{3,i_w,j_w}^m\right)}\left(P_{wf,w}^m - P_{3,i_w,j_w}^m\right) \tag{6-66c}$$

The variable $\tilde{J}_w^m$ denotes the productivity index for multi-phase flow. Its formulation is essentially identical to those of J_w^m presented in Sec. 6.2.4, except that viscosity (which is pressure-dependent) is now taken out of the formulation of $\tilde{J}_w^m$. Therefore, $\tilde{J}_w^m$ is a more general formulation, which is strictly a function of static properties and does not vary with pressure and saturation.

If the well is pressure constrained (i.e., $P_{wf,w}^m$ is specified), expressions in Eq. (6-66) are substituted directly into the source terms in Eq. (6-60). Similar to the single-phase case, after the pressure and saturation of the well cell are solved, Eq. (6-66) can be used to compute the well rate for each phase.

If the well is rate constrained, a few options are available. If the well rate of an individual phase, say $Q^m_{1,w}$, is specified, Eq. (6-66a) can be re-arranged to obtain an expression for $P^m_{wf,w}$, which can be substituted into Eqs. (6-66b) and (6-66c) to express $Q^m_{2,w}$ and $Q^m_{3,w}$ in terms of pressure and saturation of the well cell. Another possibility is for the total well rate, $Q^m_{total,w} = Q^m_{1,w} + Q^m_{2,w} + Q^m_{3,w}$, to be specified. Summation of Eqs. (6-66a), (6-66b), and (6-66c) would yield $P^m_{wf,w}$ as a function of $Q^m_{total,w}$ and well cell's pressure and saturation. Once again, this expression for $P^m_{wf,w}$, though slightly complicated, can be substituted into Eqs. (6-66a), (6-66b), and (6-66c) to express $Q^m_{1,w}$, $Q^m_{2,w}$, and $Q^m_{3,w}$ in terms of pressure and saturation of the well cell.

6.3.5   Linearization and Solution Methods

Water–Oil Flow

The governing equations of the black-oil model can be reduced to any of the two-phase flow models. For instance, if only the oleic and aqueous phases are flowing, the governing equations become:

Water:
$$\int_V \frac{\partial\left(\phi \frac{S_1}{B_1}\right)}{\partial t} dV = -\int_A \left(\frac{1}{B_1} \vec{u}_1 \cdot \vec{n}\right) dA + \frac{1}{\rho^0_1} \int_V R_1 dV \qquad \text{(6-67a)}$$

Oil:
$$\int_V \frac{\partial\left(\phi \frac{S_2}{B_2}\right)}{\partial t} dV = -\int_A \left(\frac{1}{B_2} \vec{u}_2 \cdot \vec{n}\right) dA + \frac{1}{\rho^0_2} \int_V R_2 dV \qquad \text{(6-67b)}$$

It should be noted that although this set of equations represents the mass balance of the water and oil components, the unit of each equation is, in fact, the standard volume of phase j per time. Complete immiscibility exists in the absence of gas (i.e., water component exists only in the aqueous phase, while oil component exists only in the oleic phase). It is possible to solve this set of equations in a coupled fashion, which is the simultaneous solution (SS) method. However, it is more efficient to add the two equations together to create the "pressure equation," which will be solved by coupling with Eq. (6-67a) (the water saturation equation). This is often referred to as IMPES, which stands for *Implicit Pressure, Explicit Saturation*. First, we expand the accumulation term for each phase (Ertekin et al., 2001):

$$
\begin{aligned}
\frac{\partial\left(\phi \frac{S_j}{B_j}\right)}{\partial t} &= \frac{\phi}{B_j} \frac{\partial S_j}{\partial t} + \frac{S_j}{B_j} \frac{\partial \phi}{\partial t} - \frac{\phi S_j}{B_j^2} \frac{\partial B_j}{\partial t} \\
&= \frac{\phi}{B_j} \frac{\partial S_j}{\partial t} + \frac{\phi S_j}{B_j} c_f \frac{\partial P}{\partial t} - \frac{\phi S_j}{B_j}\left(\frac{1}{B_j} \frac{\partial B_j}{\partial P}\right) \frac{\partial P}{\partial t} \\
&= \frac{\phi}{B_j} \frac{\partial S_j}{\partial t} + \frac{\phi S_j}{B_j}\left(c_f - \frac{1}{B_j} \frac{\partial B_j}{\partial P}\right) \frac{\partial P}{\partial t}
\end{aligned}
\qquad \text{(6-68)}
$$

It is assumed that $P_j = P$. Similar to the treatment in Sec. 6.2.1, consider a CV with constant properties within a volume of V. Substituting Eq. (6-68) into Eq. (6-67) and re-arranging would yield:

$$V\left[\phi\frac{\partial S_1}{\partial t} + \phi S_1\left(c_f - \frac{1}{B_1}\frac{\partial B_1}{\partial P}\right)\frac{\partial P}{\partial t}\right] + B_1\int_A\left(\frac{1}{B_1}\vec{u}_1 \cdot \vec{n}\right)dA = \frac{B_1}{\rho_1^0}\int_V R_1 dV$$
(6-69a)

$$V\left[\phi\frac{\partial S_2}{\partial t} + \phi S_2\left(c_f - \frac{1}{B_2}\frac{\partial B_2}{\partial P}\right)\frac{\partial P}{\partial t}\right] + B_2\int_A\left(\frac{1}{B_2}\vec{u}_2 \cdot \vec{n}\right)dA = \frac{B_2}{\rho_2^0}\int_V R_2 dV$$
(6-69b)

Adding the two equations and recognizing that $c_t = c_f - S_1\left(\frac{1}{B_1}\frac{\partial B_1}{\partial P}\right)$ $- S_2\left(\frac{1}{B_2}\frac{\partial B_2}{\partial P}\right)(S_1 + S_2) = 1$ and $\left(\frac{\partial S_1}{\partial t} + \frac{\partial S_2}{\partial t}\right) = 0$, we obtain the "pressure equation":

$$V\phi c_t\frac{\partial P}{\partial t} + B_1\int_A\left(\frac{1}{B_1}\vec{u}_1 \cdot \vec{n}\right)dA + B_2\int_A\left(\frac{1}{B_2}\vec{u}_2 \cdot \vec{n}\right)dA$$

$$= \frac{B_1}{\rho_1^0}\int_V R_1 dV + \frac{B_2}{\rho_2^0}\int_V R_2 dV$$
(6-70a)

The pressure P in Eq. (6-70) can be treated as P_1. The solution of pressure (P) and saturation (S_1) requires the coupling of Eqs. (6-70) and (6-69a), which is the "water saturation equation." The pressure equation depends on saturation through terms such as c_t and $\vec{u}_1$, which are the functions of S_j, while the water saturation equation depends on pressure through terms such as B_j, which is a function of P. After solving for S_1, S_2 is obtained as $(1 - S_1)$ and $P_2 = P_1 + P_c(S_1)$.

The two equations are analogous to Eq. (6-13). We will, again, consider a rectangular grid cell, as shown in Fig. 6-1. The accumulation is expressed according to Eq. (6-14). The flux terms are formulated in a fashion similar to that for the single-phase case; for example, the flux of Phase 1 is formulated using Darcy's law:

$$\int_A\left(\frac{1}{B_1}\vec{u}_1 \cdot \vec{n}\right)dA = \left(\frac{u_{x1}\Delta y\Delta z}{B_1}\right)^m_{i+1/2,j} = -\left(\Delta y\Delta z\frac{k_x k_{r1}}{B_1\mu_1}\frac{\partial P_1}{\partial x}\right)^m_{i+1/2,j}$$

$$= -\left(\Delta y\Delta z k_x\lambda_{r1}\frac{\partial P_1}{\partial x}\right)^m_{i+1/2,j}$$
(6-71)

We have introduced the relative mobility of Phase 1: $\lambda_{r1} = \frac{k_{r1}}{B_1\mu_1}$.

Analogous flux terms can be formulated for other faces and Phase 2.

As a result, Eq. (6-37) can be expressed for the flux of Phase 1 as follows:

$$\left(\Delta y \Delta z k_x \lambda_{r1}\right)_{ij}^m \left(\frac{P_{1ij} - P_{1i+1/2,j}}{\frac{\Delta x}{2}} \right)_{ij}^m = T_{x,i+1/2,j} \lambda_{r1,i+1/2,j}^m \left(P_{1ij}^m - P_{1i+1,j}^m \right) \quad (6\text{-}72)$$

In terms of notation, only the static portion is included in the transmissibility (T) formulation. The dynamic, pressure-dependent properties are lumped into λ_r^m. Comparing Eq. (6-72) to Eq. (6-37), an additional term $\lambda_{r1,i+1/2,j}^m$ is present. Stability analysis reveals that this term should be evaluated according to upstream weighting, as opposed to the harmonic averaging for transmissibility. A one-point upstream weighting scheme can be implemented by setting $\lambda_{r1,i+1/2,j}^m$ to be λ_{r1}^m of the cell along the upstream direction; the upstream direction refers to which the flow is coming from and can be identified by inspecting the $(\nabla P_1 + \rho_1 \vec{g})$ between the two neighboring blocks. More sophisticated multi-point upstream weighting schemes are available, where $\lambda_{r1,i+1/2,j}^m$ is a weighted average of λ_{r1}^m corresponding to multiple cells along the upstream direction.

To incorporate gravity, Eq. (6-71) can be modified following Example 6-2:

$$\left(\frac{u_{x1}\Delta y \Delta z}{B_1} \right)_{i+1/2,j}^m = T_{x,i+1/2,j} \lambda_{r1,i+1/2,j}^m$$

$$\left[\left(P_{1ij}^m - \gamma_{1ij}{}^m D_{ij} \right) - \left(P_{1i+1,j}^m - \gamma_{1ij}{}^m D_{i+1,j} \right) \right]$$

The source term can also be expressed in a fashion similar to Eq. (6-18); for example, the source term for Phase 1 becomes:

$$\frac{B_1{}^m}{\rho_1^0} \int_V R_1^m dV = B_{1ij}{}^m Q_{1ij}^m \quad (6\text{-}73)$$

Assembling all the accumulation, flux, and source terms, the difference form of the pressure equation can be written using the simplified notation introduced in Eq. (6-48):

$$\left(\Delta x \Delta y \Delta z \phi c_t \right)_{ij}^m \left(P_{ij}^{n+1} - P_{ij}^n \right)$$

$$-\Delta t B_{1ij}{}^m \left(\Delta_x T_x \lambda_{r1} (\Delta_x P_1 - \gamma_1 \Delta_x D) \right)^m - \Delta t B_{1ij}{}^m \left(\Delta_x T_y \lambda_{r1} (\Delta_y P_1 - \gamma_1 \Delta_y D) \right)^m$$

$$-\Delta t B_{2ij}{}^m \left(\Delta_x T_x \lambda_{r2} (\Delta_x P_2 - \gamma_2 \Delta_x D) \right)^m - \Delta t B_{2ij}{}^m \left(\Delta_y T_x \lambda_{r2} (\Delta_y P_2 - \gamma_2 \Delta_y D) \right)^m$$

$$= \Delta t B_{1ij}{}^m Q_{1ij}^m + \Delta t B_{2ij}{}^m Q_{2ij}^m \quad (6\text{-}74)$$

To better illustrate how the flux terms are related to gradients in pressure, elevation, and capillary pressure, we should expand each term in Eq. (6-74) to identify its components:

$$
\begin{aligned}
&\left(\Delta x \Delta y \Delta z \phi c_t\right)_{ij}^m \left(P_{ij}^{n+1} - P_{ij}^n\right) \\
&-\Delta t B_{1ij}{}^m \left(\Delta_x T_x \lambda_{r1}^m \Delta_x P_1{}^m - \Delta_x T_x \lambda_{r1}^m \gamma_{1ij}{}^m \Delta_x D\right) \\
&-\Delta t B_{1ij}{}^m \left(\Delta_y T_y \lambda_{r1}^m \Delta_y P_1{}^m - \Delta_y T_y \lambda_{r1}^m \gamma_{1ij}{}^m \Delta_y D\right) \\
&-\Delta t B_{2ij}{}^m \left(\Delta_x T_x \lambda_{r2}^m \Delta_x P_1{}^m + \Delta_x T_x \lambda_{r2}^m \Delta_x P_c{}^m - \Delta_x T_x \lambda_{r2}^m \gamma_{2ij}{}^m \Delta_x D\right) \\
&-\Delta t B_{2ij}{}^m \left(\Delta_y T_y \lambda_{r2}^m \Delta_y P_1{}^m + \Delta_y T_y \lambda_{r2}^m \Delta_y P_c{}^m - \Delta_y T_y \lambda_{r2}^m \gamma_{2ij}{}^m \Delta_y D\right) \\
&= \Delta t B_{1ij}{}^m Q_{1ij}^m + \Delta t B_{2ij}{}^m Q_{2ij}^m
\end{aligned} \tag{6-75}
$$

Note that $P_2 = P_1 + P_c$. The difference form of the water saturation Eq. (6-69a) can be written using the simplified notation introduced in Eq. (6-48).

$$
\begin{aligned}
&\left(\Delta x \Delta y \Delta z \phi\right)_{ij}^m \left(S_{1ij}^{n+1} - S_{1ij}^n\right) + \left(\Delta x \Delta y \Delta z \phi\right)_{ij}^m S_{1ij}^m \left(c_{fij}^m + c_{1ij}^m\right)\left(P_{ij}^{n+1} - P_{ij}^n\right) \\
&-\Delta t B_{1ij}{}^m \Delta_x \left[\lambda_{r1}^m T_x \Delta_x P_1{}^m\right] - \Delta t B_{1ij}{}^m \Delta_y \left[\lambda_{r1}^m T_y \Delta_y P_1{}^m\right] \\
&+\Delta t B_{1ij}{}^m \Delta_x \left[\lambda_{r1}^m \gamma_{1ij}{}^m T_x \Delta_x D\right] + \Delta t B_{1ij}{}^m \Delta_y \left[\lambda_{r1}^m \gamma_{1ij}{}^m T_y \Delta_y D\right] \\
&= \Delta t B_{1ij}{}^m Q_{1ij}^m
\end{aligned} \tag{6-75}
$$

The idea of IMPES is to solve the pressure Eq. (6-75) and the water Eq. (6-76) sequentially: first, solve for P^{n+1} using Eq. (6-75) with all pressure-dependent variables evaluated at level $n+1$ (i.e., implicitly), while evaluating all saturation-dependent terms at level n (i.e., explicitly); next, then solve for S_1^{n+1} with Eq. (6-76) with all terms evaluated at level $n+1$.

The IMPES formulation of Eqs. (6-75) and (6-76) entails solving two non-linear systems of equations in terms of P_{ij}^{n+1} and $S_{1ij}^{n+1} \in i, j$ sequentially. First, Eq. (6-75) can be solved by minimizing the residual function of $R_{p,ij}(\mathbf{P}^{n+1}, \mathbf{S}_1^n) \in i, j$, in a fashion similar to Eq. (6-50). The solution of $\mathbf{P}^{n+1}$ is then substituted into Eq. (6-76) to solve for   by minimizing the residual function of $R_{1,ij}(\mathbf{P}^{n+1}, \mathbf{S}_1^{n+1}) \in i, j$. The IMPES scheme is easy to code and can be more accurate in terms of truncation error. However, it is only conditionally stable. In fact, small time steps are required in regions with small grid cells (e.g., near the wells).

The SS method is a more stable alternative to the IMPES method; the method involves solving Eqs. (6-67a) and (6-67b) simultaneously in a coupled fashion. This can be achieved in a number of ways. One option is to minimize $R_{p,ij}(\mathbf{P}^{l+1}, \mathbf{S}_1^l)$, where l is the iteration count, to solve for $\mathbf{P}^{l+1}$, which is then used to minimize $R_{1,ij}(\mathbf{P}^{l+1}, \mathbf{S}_1^{l+1}) \in i, j$ for S_1^{l+1}. This process is repeated until $\|\mathbf{P}^{l+1} - \mathbf{P}^l\|$ and $\|\mathbf{S}^{l+1} - \mathbf{S}_1^l\|$ are less than a pre-defined tolerance. This is equivalent to solving a number of IMPES solutions within a time step. Another option is to collectively minimize a total residual function consisting of both $R_{p,ij}(\mathbf{P}^{n+1}, \mathbf{S}_1^{n+1})$ and $R_{1,ij}(\mathbf{P}^{n+1}, \mathbf{S}_1^{n+1})$ using the Newton–Raphson technique for every time step.

Example 6-7

Consider the displacement of oil by water in a 1D domain consisting of five grid blocks. Calculate the pressure and saturation distribution at 0.5 days. There is a no-flow outer boundary. A producer ($P_{wf} = 1$ MPa) is located at Block 1, while a water injector ($Q_1 = 500$ m^3/day) is located at Block 5. $\bar{J}_w$ is constant and equal to 7×10^{-10} m^3.

$\Delta x = \Delta y = \Delta z = 50$ m
$k = 1 \times 10^{-12}$ m^2 (isotropic permeability)
$\phi = 0.15$

$\mu_1 = 1$ mPa·s; $m_2 = 2$ mPa·s
$c_1 = c_2 = 2 \times 10^{-10}$ Pa^{-1}; $cf = 1 \times 10^{-8}$ Pa^{-1}
$B_1 = B_2 = 1$

Relative permeability model:

$$k_{r1} = k_{r1}^o \left(\frac{S_1 - S_{1r}}{1 - S_{1r} - S_{2r}} \right)^{n_1} \qquad k_{r2} = k_{r2}^o \left(\frac{1 - S_1 - S_{2r}}{1 - S_{1r} - S_{2r}} \right)^{n_2}$$

$S_{1r} = 0.2$	$S_{2r} = 0.25$
$k_{r1}^o = 0.4$	$k_{r2}^o = 0.7$
$n_1 = 2$	$n_2 = 2$

Capillary pressure model: $P_c = 10000e^{7(1 - S_{2r} - S_1)}$
Initial conditions: $S_1 = 0.25, P = 3$ MPa

Solution

For a 1D domain, the gravity term, as well as the flux in the y direction, can be omitted. Therefore, Eq. (6-75) becomes:

$$\left(V_p c_t \right)_i^m \left(P_i^{n+1} - P_i^n \right) - \Delta t B_{1i}^{\ m} \left(\Delta_x T_x \lambda_{r1}^m \Delta_x P_1^m \right)$$
$$- \Delta t B_{2i}^{\ m} \left(\Delta_x T_x \lambda_{r2}^m \Delta_x P_1^m + \Delta_x T_x \lambda_{r2}^m \Delta_x P_c^m \right)$$
$$= \Delta t B_{1i}^{\ m} Q_{1i}^m + \Delta t B_{2i}^{\ m} Q_{2i}^m$$

where $V_p = \Delta x \Delta y \Delta z \phi$. For an IMPES formulation, it can be re-written as:

$$V_{pi} c_{ti}^{n+1} \left(P_i^{n+1} - P_i^n \right)$$
$$+ \Delta t \left(B_{1i}^{\ n+1} \lambda_{r1,i+1/2}^n + B_{2i}^{\ n+1} \lambda_{r2,i+1/2}^n \right) T_{i+1/2} \left(P_i^{n+1} - P_{i+1}^{n+1} \right)$$
$$+ \Delta t \left(B_{1i}^{\ n+1} \lambda_{r1,i-1/2}^n + B_{2i}^{\ n+1} \lambda_{r2,i-1/2}^n \right) T_{i-1/2} \left(P_i^{n+1} - P_{i-1}^{n+1} \right)$$
$$+ \Delta t B_{2i}^{\ n+1} \lambda_{r2,i+1/2}^n T_{i+1/2} \left(P_{ci}^n - P_{ci+1}^n \right) + \Delta t B_{2i}^{\ n+1} \lambda_{r2,i-1/2}^n T_{i-1/2}$$
$$\left(P_{ci}^n - P_{ci-1}^n \right) = \Delta t B_{1i}^{\ n+1} Q_{1i}^{n+1} + \Delta t B_{2i}^{\ n+1} Q_{2i}^{n+1}$$

All pressure-dependent variables are evaluated at time level $n + 1$, while all saturation-dependent variables are evaluated at time level n. For the relative mobility, it should be evaluated as:

$$\lambda_{rj}{}^n = \frac{k_{rj}^n}{B_j^{n+1}\mu_j^{n+1}}$$

Similar treatment is considered for the source terms:

$$Q_{j,w}^{n+1} = \tilde{J}_w^{n+1}\lambda_{rj,i_w,j_w}^n \left(P_{wf,w}^{n+1} - P_{j,i_w,j_w}^{n+1}\right)$$

This yields a non-linear system of equations, which can be solved using techniques discussed earlier. In this example, let us further simply this calculation by performing only one iteration in a successive substitution scheme:

$$V_{pi}c_{ti}{}^n\left(P_i^{n+1} - P_i^n\right)$$
$$+\Delta t\left(B_{1i}{}^n\lambda_{r1,i+1/2}{}^n + B_{2i}{}^n\lambda_{r2,i+1/2}{}^n\right)T_{i+1/2}\left(P_i^{n+1} - P_{i+1}{}^{n+1}\right)$$
$$+\Delta t\left(B_{1i}{}^n\lambda_{r1,i-1/2}{}^n + B_{2i}{}^n\lambda_{r2,i-1/2}{}^n\right)T_{i-1/2}\left(P_i^{n+1} - P_{i-1}{}^{n+1}\right)$$
$$+\Delta t B_{2i}{}^n\lambda_{r2,i+1/2}{}^n T_{i+1/2}\left(P_{ci}{}^n - P_{ci+1}{}^n\right) + \Delta t B_{2i}{}^n\lambda_{r2,i-1/2}{}^n T_{i-1/2}$$
$$\left(P_{ci}{}^n - P_{ci-1}{}^n\right) = \Delta t B_{1i}{}^n Q_{1i}{}^{n+1} + \Delta t B_{2i}{}^n Q_{2i}{}^{n+1}$$

After obtaining P^{n+1}, Eq. (6-76) is used to compute the saturation at time level $n + 1$:

$$V_{pi}\left(S_{1i}^{n+1} - S_{1i}^n\right) + V_{pi}S_{1i}^{n+1}\left(c_{fi}^{n+1} + c_{1i}^{n+1}\right)\left(P_i^{n+1} - P_i^n\right)$$
$$- \Delta t B_{1i}{}^{n+1}\Delta_x\left[\lambda_{r1}^{n+1}T_x\Delta_x P_1^{n+1}\right] = \Delta t B_{1i}{}^{n+1}Q_{1i}^{n+1}$$

Expanding the flux term yields:

$$V_{pi}\left(S_{1i}^{n+1} - S_{1i}^n\right) + V_{pi}S_{1i}{}^{n+1}\left(c_{fi}{}^{n+1} + c_{1i}{}^{n+1}\right)\left(P_i^{n+1} - P_i^n\right) + \Delta t B_{1i}{}^{n+1}$$
$$\left(T_{i+1/2}\lambda_{r1,i+1/2}{}^{n+1}\left(P_i^{n+1} - P_{i+1}{}^{n+1}\right) + T_{i-1/2}\lambda_{r1,i-1/2}{}^{n+1}\left(P_i^{n+1} - P_{i-1}{}^{n+1}\right)\right)$$
$$= \Delta t B_{1i}{}^{n+1}Q_{1i}{}^{n+1}$$

An iterative scheme is needed to solve the above equation (as all variables in λ_{rj}^{n+1} must be evaluated at time level $n + 1$). Once again, let us further simply this calculation by performing only one iteration:

$$V_{pi}\left(S_{1i}^{n+1} - S_{1i}^n\right) + V_{pi}S_{1i}{}^n\left(c_{fi}{}^{n+1} + c_{1i}{}^{n+1}\right)\left(P_i^{n+1} - P_i^n\right) + \Delta t B_{1i}{}^{n+1}$$
$$\left(T_{i+1/2}\lambda_{r1,i+1/2}{}^n\left(P_i^{n+1} - P_{i+1}{}^{n+1}\right) + T_{i-1/2}\lambda_{r1,i-1/2}{}^n\left(P_i^{n+1} - P_{i-1}{}^{n+1}\right)\right)$$
$$= \Delta t B_{1i}{}^{n+1}Q_{1i}{}^{n+1}$$

In the above equation, all pressure-dependent variables are evaluated at time level $n + 1$, while all saturation-dependent variables are evaluated at time level n. The calculation steps and solutions for the IMPES scheme are provided at www.mhprofessional.com/PRMS in the EXCEL spreadsheet "Example 6-7.xlsx."

Oil–Water–Gas Flow

The entire set of governing Eqs. (6-60) for the black-oil model can be solved to model three-phase flow. One more primary unknown (e.g., S_3) is needed to solve the non-linear system of three equations. It is common to select P_2, S_1, and S_3 as the primary unknowns, and the secondary unknowns are obtained as using $S_2 = 1 - S_1 - S_3$, $P_1 = P_2 - P_c(S_1)$, and $P_3 = P_2 + P_c(S_2)$.

To solve for these three primary unknowns, similar linearization and solution schemes can be adopted. In the IMPES scheme, the "pressure equation" is obtained by expanding the accumulation term, as shown in Eq. (6-68), and combining all equations in Eq. (6-60). The method involves minimizing the residual function of $R_{p,ij}(\mathbf{P}^{n+1}, \mathbf{S}_1{}^n, \mathbf{S}_3{}^n) \in i, j$; the solution of $\mathbf{P}^{n+1}$ is then used to solve for $\mathbf{S}_3{}^{n+1}$ by minimizing the residual function of $R_{1,ij}(\mathbf{P}^{n+1}, \mathbf{S}_1{}^{n+1}, \mathbf{S}_3{}^{n+1}) \in i, j$; finally, the solutions of $\mathbf{P}^{n+1}$ and $\mathbf{S}_1{}^{n+1}$ are used to solve for $\mathbf{S}_3{}^{n+1}$ by minimizing the residual function of $R_{3,ij}(\mathbf{P}^{n+1}, \mathbf{S}_1{}^{n+1}, \mathbf{S}_3{}^{n+1}) \in i, j$. The SS method can also be used, for instance, by collectively minimizing a total residual function consisting of all three residual functions of $R_{p,ij}(\mathbf{P}^{n+1}, \mathbf{S}_1{}^{n+1}, \mathbf{S}_3{}^{n+1})$, $R_{1,ij}(\mathbf{P}^{n+1}, \mathbf{S}_1{}^{n+1}, \mathbf{S}_3{}^{n+1})$, and $R_{3,ij}(\mathbf{P}^{n+1}, \mathbf{S}_1{}^{n+1}, \mathbf{S}_3{}^{n+1})$ using the Newton–Raphson technique.

This formulation of $P-S_1-S_3$ is applicable when both hydrocarbon phases exist during the entire duration of the simulation. This assumption works well for the primary depletion case (i.e., reservoir is initially saturated below the bubble point and pressure continues to decline); as in this case, the PVT table with fixed bubble point provides an accurate description of the phase behavior. However, this assumption may not be valid under many circumstances; for example, if the reservoir is initially under-saturated, the gas phase will appear only after the pressure has dropped below the bubble point; the gas phase may also disappear if the reservoir is later pressurized from water injection. For an under-saturated fluid, the actual bubble-point pressure is higher than the one in the PVT table. As explained by Thomas et al. (1976), this scenario should be considered as one with a variable bubble-point pressure.

To handle the (dis)appearance of a gas phase or commonly known as "gas-phase discontinuity" in simulation literature, the two most common approaches are the variable substitution method and the pseudo-solution gas-oil ratio (GOR) method (Ertekin et al., 2001). In the first approach, the idea is to check certain conditions in each grid cell and determine which constraint equation and unknowns are to be used. The method switches between the $(P-S_1-S_3)$ formulation and the $(P-S_1-R_s)$ formulation for each iteration:

- If S_g is more than 0, the oil phase is saturated with gas, and R_s is the one obtained from the PVT table. In this case, the $(P-S_1-S_3)$ formulation is applicable, and the constraint of $S_2 = 1 - S_1 - S_3$ can be used.
- If $S_g = 0$, the oil phase is under-saturated with gas, and Rs is less than the true solution-gas ratio, which is unknown. As a result, the $(P-S_1-S_s)$ formulation is applicable, and the constraint as $S_2 = 1 - S_1$ can be used.

Wang (2007) presented a slightly different variable-switching scheme that is based on the two pseudo-component formulations by Shank and Vestal (1989). The variable substitution scheme is slightly complicated to implement. A simpler approach is to formulate a pseudo-solution GOR:

$$\widetilde{R}_s = R_s \psi \tag{6-77}$$

where

$$\psi = \begin{cases} 1 & S_3 \geq \varepsilon_g \\ \dfrac{S_3}{S_3 + \varepsilon_g} & \text{otherwise} \end{cases} \tag{6-78}$$

This method uses only the $(P-S_1-S_3)$ formulation all the time by setting a small amount of non-condensable gas in every grid cell. The tolerance ε_g is used to control the (dis)appearance of the gas phase. As a result, the gas phase may appear or disappear without actually achieving it.

6.4 Finite Element Formulation

The finite element (FE) method is another technique for finding approximate solutions to partial differential equations. There are typically three steps involved in the approach: (1) obtain the variational form of the governing equation; (2) define a basis function for the solution space; and (3) solve the resultant system of equations. Both CVFD and FE methods generate a system of equations to be solved at each time level; the main difference, however, lies in the approximation technique used in deriving this system of equations. The purpose of this section is to provide an overview of the theory and terminology involved in the FE method. The focus is on the IMPES scheme. For further detail on the subject, readers should refer to the relevant references (Chen, 2005; Chen et al., 2006; Ewing, 1983; Zienkiewicz et al., 2013).

The surface integrals in the pressure [Eq. (6-70)] can be expressed as volume integrals by invoking the *divergence theorem*:

$$V\phi c_t \frac{\partial P}{\partial t} + B_1 \int_V \nabla \cdot \left(\frac{1}{B_1}\vec{u}_1\right) dV + B_2 \int_V \nabla \cdot \left(\frac{1}{B_2}\vec{u}_2\right) dV$$

$$= \frac{B_1}{\rho_1^0} \int_V R_1 dV + \frac{B_2}{\rho_2^0} \int_V R_2 dV \qquad (6\text{-}79)$$

The support volume for Eq. (6-79) is that of the control volume of V. It implies that the equation is valid on an average sense over V. A stronger form of the equation can be attained with a support volume of a single point:

$$\phi c_t \frac{\partial P}{\partial t} + \nabla \cdot (\vec{u}_1 + \vec{u}_2) = \frac{B_1}{\rho_1^0} R_1 + \frac{B_2}{\rho_2^0} R_2 \qquad (6\text{-}80)$$

Incorporating Darcy's law, Eq. (6-80) takes the following general form:

$$c(P)\frac{\partial P}{\partial t} - \nabla \cdot \left(\vec{a}(P)\nabla P\right) = f(P) \qquad (6\text{-}81)$$

For illustrative purpose, let us consider the simple case of steady-state flow in 1D, with :

$$-\frac{\partial^2 P}{\partial x^2} = f(P), 0 \leq x < 1, P(0) = P(0) = 0 \qquad (6\text{-}82)$$

The equation is multiplied by a test function, $v \in V$, and integrated over $(0, 1)$:

$$-\left(\frac{\partial^2 P}{\partial x^2}, v\right) = (f, v) \qquad (6\text{-}83)$$

where

$$(v, w) = \int_0^1 v(x)w(x)dx \qquad (6\text{-}84)$$

Integrating by-parts would yield:

$$-\int_{d\Omega} v\frac{\partial P}{\partial \vec{n}}ds + \left(\frac{\partial P}{\partial x}, \frac{\partial v}{\partial x}\right) = (f, v) \qquad (6\text{-}85)$$

The first term disappears because of the boundary conditions; the equation is simplified as:

$$\left(\frac{\partial P}{\partial x}, \frac{\partial v}{\partial x}\right) = (f, v), \qquad v \in V \qquad (6\text{-}86)$$

Equation (6-86) is the *Galerkin variational* or weak form of Eq. (6-82). In order to find an approximate solution to Eq. (6-86), the domain is sub-divided into a series of elements with size $h_i = x_i - x_{i-1}$, where $i = 2, \ldots, M + 1$. The goal is to find $P_h \in V_h$, such that $\left(\dfrac{\partial P_h}{\partial x}, \dfrac{\partial v}{\partial x} \right) = (f, v)$, $\forall v \in V_h$ (Chen et al., 2006) (v is continuous on the domain and equal to the prescribed function value at the boundary). The next step is to introduce a basis function for the solution space $\varphi_i \in V_h$, $i = 1, \ldots, M$,

$$\varphi_i\left(x_j\right) = \begin{cases} 1, i = j \\ 0, i \neq j \end{cases} \tag{6-87}$$

φ_i is piecewise continuous function, with a value of 1 at the node and 0 everywhere else. The choice of the basis function is an important aspect of the FE method. It affects the approximation accuracy and depends on the problem and the solution space. A typical choice is the piecewise linear function.

The functions of v and P_h can now be expressed as:

$$v(x) = \sum_{i=1}^{M} v_i \varphi_i (x) \tag{6-88}$$

$$P_h(x) = \sum_{i=1}^{M} P_i \varphi_i (x) \tag{6-89}$$

where $v_i = v(x_i)$ and $P_i = P_h(x_i)$. For each $j = 1, \ldots, M$, $\left(\dfrac{\partial P_h}{\partial x}, \dfrac{\partial \varphi_j}{\partial x} \right) = \left(f, \varphi_j \right)$. Therefore, we obtain the following:

$$\sum_{i=1}^{M} \left(\frac{\partial \varphi_i}{\partial x}, \frac{\partial \varphi_j}{\partial x} \right) P_i = \left(f, \varphi_j \right), j = 1, \ldots, M \tag{6-90}$$

This is a system of equations with M unknowns P_i:

$$\mathbf{AP} = \mathbf{b}, \tag{6-91}$$

where $\mathbf{A}$ consisting of $a_{ij} = \left(\dfrac{\partial \varphi_i}{\partial x}, \dfrac{\partial \varphi_j}{\partial x} \right)$ is the stiffness matrix, and $\mathbf{b}$ consisting of $b_j = (f, \varphi_j)$ is the source vector. For a uniformly spaced grid, $\mathbf{A}$ is a sparse and banded matrix (just like in the case of CVFD), since by definition $\left(\dfrac{\partial \varphi_i}{\partial x}, \dfrac{\partial \varphi_j}{\partial x} \right) = 0$ if $|i-j| \geq 2$. Numerical integration or quadrature rule can be used to compute the integrals.

The formulation can be extended for a multi-dimensional problem. Equation (6-86) can be expressed as:

$$a(P, v) = (f, v) \tag{6-92}$$

where

$$a(P,v) = \int_\Omega \nabla P \cdot \nabla v d\mathbf{x} \tag{6-93}$$

$$(f,v) = \int_\Omega f v d\mathbf{x} \tag{6-94}$$

The following basis function can be used to obtain a system of equations, similar to that in Eq. (6-91):

$$\varphi_i(\mathbf{x}_j) = \begin{cases} 1, i = j \\ 0, i \neq j \end{cases} \tag{6-95}$$

Dirichlet boundary can be imposed by explicitly specifying the function values at the boundary nodes and including only the interior nodes in the matrix, such that V is 0 at the boundary. Neumann boundary, that is, $\dfrac{\partial P}{\partial \bar{n}} = g$ on $\partial \Omega$, is imposed through the first term in Eq. (6-85). As a result, Eq. (6-92) is modified as:

$$a(P, \nu) = (f, \nu) + (g, \nu)_{\partial \Omega} \tag{6-96}$$

where

$$(g,v)_{\partial \Omega} = \int_{\partial \Omega} v g d s \tag{6-97}$$

The goal is to find $P_h \in V_h$, such that $a(P, \nu) = (f, \nu) + (g, \nu)_{\partial \Omega}$, $\forall \nu \in V_h$ (Chen et al., 2006).

Finally, the variational form of the general governing Eq. (6-81) with no-flow boundary on $\partial \Omega$ is obtained as:

$$\left(c(P_h) \frac{\partial P_h}{\partial t}, v \right) + \left(\left(\bar{\bar{a}}(P_h) \nabla P_h \right), \nabla v \right) = \left(f(P_h), v \right) \tag{6-98}$$

After incorporating the basis function, Eq. (6-98) can also be cast in the matrix form:

$$\mathbf{C}(\mathbf{P}) \frac{\partial \mathbf{P}}{\partial t} + \mathbf{A}(\mathbf{P})\mathbf{P} = \mathbf{b}(\mathbf{P}) \tag{6-99}$$

Linearization and solution schemes described in Sec. 6.3.5 can be adopted. For instance, a fully implicit form of Eq. (6-99) becomes:

$$\mathbf{C}\left(\mathbf{P}^{n+1}\right) \frac{\mathbf{P}^{n+1} - \mathbf{P}^n}{\Delta t} + \mathbf{A}\left(\mathbf{P}^{n+1}\right)\mathbf{P}^{n+1} = \mathbf{b}\left(\mathbf{P}^{n+1}\right) \tag{6-100}$$

The formulation presented thus far can be considered as the *standard* FE method. In comparison to the CVFD method, which is locally conservative (over the CV), the standard FE method is globally conservative. More advanced FE formulation has been developed to address this issue, as well as to improve the accuracy of the method. Some of these techniques and various meshing schemes will be presented in the Chap. 7. However, it is appropriate to discuss one additional scheme, which is based on *discontinuous* FEs, here in this chapter, as it is a widely adopted approach for solving the saturation equation. First, by invoking the divergence theorem, a strong form of the saturation Eq. (6-69a) defined over a point support becomes:

$$\phi \frac{\partial S_1}{\partial t} + \phi S_1 \left(c_f - \frac{1}{B_1} \frac{\partial B_1}{\partial P} \right) + \nabla \cdot \vec{u}_1 = \frac{B_1}{\rho_1^0} R_1 \qquad (6\text{-}101)$$

Incorporating Darcy's law, Eq. (6-101) takes the following general form (the phase subscript has been dropped for notational convenience):

$$\frac{\partial S}{\partial t} + RS + \vec{b} \cdot \nabla S = f(S) \qquad (6\text{-}102)$$

There is an important difference between the pressure Eq. (6-80) and the saturation Eq. (6-101). Equation (6-80) is a second-order partial differential equation. Its solution is generally smooth (e.g., pressure does not vary abruptly in space); hence, selecting functions for the FE spaces that are continuous across the boundary between elements is appropriate. However, the saturation equation is a first-order partial differential equation. Its solution often involves a discontinuity at the front [e.g., the Buckley–Leverett solution (Buckley and Leverett, 1942)], particularly if the transport is an advection-dominated (hyperbolic) problem. The discontinuous Galerkin (DG) method was originally developed by Reed and Hill (1973), and it is also locally conservative (Chen et al., 2006).

For each element K, its boundary ∂K is divided into two parts: ∂K_+ where $\vec{b} \cdot \vec{n} > 0$ at the outflow (downstream) and ∂K_- where $\vec{b} \cdot \vec{n} < 0$ at the inflow (upstream). As an example, considering Eq. (6-102) at steady state, the DG method seeks a solution of S_h that satisfies the following for each element:

$$\left(RS_h + \vec{b} \cdot \nabla S_h, v \right)_K - \int_{\partial K_-} S_{h+} v_+ \vec{b} \cdot \vec{n} dl + \int_{\partial K_-} S_{h-} v_+ \vec{b} \cdot \vec{n} dl = (f, v)_K \qquad (6\text{-}103)$$

where $v_-(x) = \lim_{\varepsilon \to 0^-} v\left(x + \varepsilon \vec{b}\right)$ and $v_+(x) = \lim_{\varepsilon \to 0^+} v\left(x + \varepsilon \vec{b}\right)$. This is essentially the standard FE method, but with a flux boundary condition being weakly imposed. Using the notation introduced earlier, the method aims to find $S_h \in V_h$, such that $a(S_h, v) = (f, v)$, $\forall v \in V_h$, where

$$a(S_h, v) = \sum_K a_K(S_h, v) \qquad (6\text{-}104)$$

$$a_K(S_h, v) = \left(RS_h + \vec{b} \cdot \nabla S_h, v\right)_K - \int_{\partial K_-} [S_{h+} - S_{h-}] v_+ \vec{b} \cdot \vec{n} dl \quad (6\text{-}105)$$

Example 6-8

Develop the DG method for a special case of Eq. (6-105) in 1D:

$$a_K(S_h, v) = (S_h + \nabla S_h, v)_K - \int_{\partial K_-} [S_{h+} - S_{h-}] v_+ \cdot \vec{n} dl = (f, v)_K \quad (6\text{-}106)$$

Solution

Set to be piecewise constant within the element interval and simplify:

$$S_h h_i + [S_i - S_{i-1}] = \int_K f dx \quad (6\text{-}107)$$

The DG method essentially offers a way to impose the upstream weighting scheme implemented in the CVFD method. It is important to note that solution of S_h at node i depends only on other nodes in the upstream direction. As a result, the computation is local (Chen et al., 2006), as S_{h-} is expected to be known. It means that to solve for S_h, one should start from the most upstream element and progress toward downstream until the entire domain is visited. This is similar to imposing a convective boundary sequentially along the direction of flow. Other improved formulation, such as stabilized DG, has been developed to enhance its stability. Readers should refer to the previously cited references for further information.

6.5   Solution of Linear System of Equations

All numerical techniques presented in this chapter would result in a system of equations that is either linear (e.g., explicit) or non-linear (e.g., implicit). Even a non-linear system of equations can be linearized. Therefore, solution of the matrix equation of the following form is an important component of all reservoir simulation:

$$\mathbf{Ax = d.} \quad (6\text{-}108)$$

There are two main classes of matrix solvers: (1) direct and (2) iterative. Direct solvers can produce exact solutions (although round-off errors may accumulate over each step) and include techniques such as Gaussian elimination, Gauss–Jordan reduction, and LU (lower-upper) factorization (Kreyszig, 2011). These techniques are general and capable of handling full matrix (no non-zero elements), but they are often computationally intensive. Given that matrices encountered in reservoir simulation are typically sparse and banded, modifications to these direct solution methods are often recommended for reservoir simulation computations. If the structure of $\mathbf{A}$ is known *a priori*, the storage and operation of the zero entries can be avoided. Efficient band-solution techniques can also be applied.

The bandwidth of $\mathbf{A}$ depends on the ordering of the grid cells. Recall Example 6-3: if the cells are ordered such that the x axis is cycled first, the upper and lower bandwidths of the associated matrix are 3; however, if the cells are ordered such that the y axis is cycled first, the upper and the lower bandwidths of the associated matrix are 2. Both schemes are based on the natural ordering of the grid (Ertekin et al., 2001). Other classical ordering schemes, such as lexicographical, rotational lexicographical, checkerboard, and zebra-line, are described in Hackbusch (2013). Ordering schemes that are commonly adopted in reservoir simulation include D-4, D-2, and Cyclic-2 (Ertekin et al., 2001). The objective is to create a special matrix structure that can be subdivided into four sub-matrices, whose computations are more efficient.

Considering the accumulating round-off errors and computational costs, iterative methods are often employed as matrix solvers. These methods involve moving an initial guess of the unknown solution vector progressively toward the true (exact) solution over multiple iterations (Kreyszig, 2011). The efficiency of a method depends on its convergence behavior. Basic techniques, such as Jacobi, Gauss–Seidel, and successive over-relaxation, can be used, although they are not commonly implemented in commercial reservoir simulation packages nowadays. The more popular techniques are those classified as Krylov sub-space methods (Saad, 2003). The idea is to search for a good approximate solution from the Krylov sub-space. In fact, all Krylov sub-space algorithms are related to the choice of this sub-space. The most basic one is the conjugate gradient (CG) method for the following sub-space:

$$\text{span}\{\mathbf{r}^{(0)}, \mathbf{Ar}^{(0)}, \mathbf{A}^2\mathbf{r}^{(0)}, \ldots, \mathbf{A1}^{k-2}\mathbf{r}^{(0)}\} \tag{6-109}$$

It was first introduced by Hestenes and Stiefel (1952) for solving a $n \times n$ linear system. The idea is to minimize the functional:

$$F(\mathbf{x}) = \frac{1}{2}\left(\mathbf{r} \cdot \mathbf{A}^{-1} \cdot \mathbf{r}\right) \tag{6-110}$$

where $\mathbf{r} = \mathbf{d} - \mathbf{Ax}$ is the residual vector. This can be achieved by setting $\nabla F(\mathbf{x}) = -\mathbf{r} = 0$. Given $\mathbf{x}^{(0)}$, CG successively seeks the solution vector of the form:

$$\mathbf{x}^{(k+1)} = \mathbf{x}^{(k)} + a_k\mathbf{p}^{(k)} \tag{6-111}$$

where $k =$ iteration counter ($\leq n$), $a_k =$ scalar variable representing the step size, and $\mathbf{p}^{(k)} =$ vector representing the search direction. Set $\mathbf{r}^{(0)} = \mathbf{d} - \mathbf{Ax}^{(0)}$ and $\mathbf{p}^{(0)} = \mathbf{r}^{(0)}$, and the CG algorithm can be used to compute a_k, $\mathbf{x}^{(k+1)}$, and $\mathbf{p}^{(k+1)}$ iteratively:

$$a_k = \frac{\mathbf{r}^{(k)} \cdot \mathbf{r}^{(k)}}{\mathbf{p}^{(k)} \cdot \mathbf{Ap}^{(k)}}$$

$$\mathbf{x}^{(k+1)} = \mathbf{x}^{(k)} + a_k\mathbf{p}^{(k)}$$

$$\mathbf{r}^{(k+1)} = \mathbf{r}^{(k)} - a_k \mathbf{A} \mathbf{p}^{(k)}$$

$$\mathbf{p}^{(k+1)} = \mathbf{r}^{(k+1)} + b_{k+1} \mathbf{p}^{(k)}$$

$$\text{and } b_{k+1} = \frac{\mathbf{r}^{(k+1)} \cdot \mathbf{r}^{(k+1)}}{\mathbf{r}^{(k)} \cdot \mathbf{r}^{(k)}} \tag{6-112}$$

The CG method provides an exact solution in n steps (in the absence of round-off errors). It can be shown that for an arbitrary initial estimate, it converges to the exact solution in n_e iterations, where n_e ($\leq n$) = number of distinct eigenvalues of $\mathbf{A}$. If the conditional number of $\mathbf{A}$ is one, the CG method would converge after only one iteration ($n_e = 1$). The original coefficient matrix is unaltered during the computations; hence, the sparse, banded structure is preserved, and efficient coding can reduce the computational and storage requirement (Ertekin et al., 2001).

The above formulation assumes that $\mathbf{A}$ is symmetric and positive definite. For non-symmetric $\mathbf{A}$, modification to Eq. (6-112) is needed. Set $\mathbf{r}^{(0)} = \mathbf{d} - \mathbf{A}\mathbf{x}^{(0)}$, $\mathbf{p}^{(0)} = \mathbf{A}^T\mathbf{r}^{(0)}$, and $\mathbf{h}^{(0)} = \mathbf{p}^{(0)}$, the CG algorithm can be used to compute a_k, $\mathbf{x}^{(k+1)}$, and $\mathbf{p}^{(k+1)}$ iteratively:

$$a_k = \frac{\mathbf{h}^{(k)} \cdot \mathbf{h}^{(k)}}{\mathbf{A} \mathbf{p}^{(k)} \cdot \mathbf{A} \mathbf{p}^{(k)}}$$

$$\mathbf{x}^{(k+1)} = \mathbf{x}^{(k)} + a_k \mathbf{p}^{(k)}$$

$$\mathbf{r}^{(k+1)} = \mathbf{r}^{(k)} - a_k \mathbf{A} \mathbf{p}^{(k)}$$

$$\mathbf{h}^{(k+1)} = \mathbf{A}^T \mathbf{r}^{(k+1)}$$

$$\mathbf{p}^{(k+1)} = \mathbf{h}^{(k+1)} + b_{k+1} \mathbf{p}^{(k)}$$

$$\text{and } b_{k+1} = \frac{\mathbf{h}^{(k+1)} \cdot \mathbf{h}^{(k+1)}}{\mathbf{h}^{(k)} \cdot \mathbf{h}^{(k)}} \tag{6-113}$$

In this modified formulation, matrix $\mathbf{A}$ is multiplied by $\mathbf{A}^T$. This is similar to using $\mathbf{A}^T$ as a preconditioning matrix to create a symmetric matrix from the original $\mathbf{A}$. By doing so, the conditioning number becomes the square of $\mathbf{A}$. If the conditioning number of $\mathbf{A}$ is large, this method may be undesirable. An alternative formulation, which is the generalized conjugate gradient-like (CGL) method, is often preferred. It uses the conditioning number of the original A matrix by defining a different objective function $F(\mathbf{x}) = \mathbf{r} \cdot \mathbf{r}$. A popular class of generalized CGL algorithms commonly used in reservoir simulation is called the minimum-residual methods (Ertekin et al., 2001). One of such techniques is the Orthomin method (Vinsome, 1976). Set $\mathbf{r}^{(0)} = \mathbf{d} - \mathbf{A}\mathbf{x}^{(0)}$, $\mathbf{p}^{(0)} = \mathbf{r}^{(0)}$, and compute a_k, $\mathbf{x}^{(k+1)}$ and $\mathbf{p}^{(k+1)}$ iteratively:

$$a_k = \frac{\mathbf{r}^{(k)} \cdot \mathbf{A} \mathbf{p}^{(k)}}{\mathbf{A} \mathbf{p}^{(k)} \cdot \mathbf{A} \mathbf{p}^{(k)}}$$

$$\mathbf{x}^{(k+1)} = \mathbf{x}^{(k)} + a_k \mathbf{p}^{(k)}$$

$$\mathbf{r}^{(k+1)} = \mathbf{r}^{(k)} - a_k \mathbf{A}\mathbf{p}^{(k)}$$

$$\mathbf{p}^{(k+1)} = \mathbf{r}^{(k+1)} + \sum_{j=0}^{k} b_k^{(j)} \mathbf{p}^{(j)}$$

$$\text{and } b_k^{(j)} = -\frac{\mathbf{A}\mathbf{r}^{(k+1)} \cdot \mathbf{A}\mathbf{p}^{(j)}}{\mathbf{A}\mathbf{p}^{(j)} \cdot \mathbf{A}\mathbf{p}^{(j)}} \tag{6-114}$$

Other CGL techniques are commonly adopted in reservoir simulation, and they include the generalized minimal residual method (GMRES) (Saad and Schultz, 1986), the generalized conjugate residual method (GCR) (Eisenstat et al., 1983; Young and Jea, 1980), for which the Orthomin method is a truncated version, and the bi-conjugate gradient stabilized method (BiCGSTAB) (van der Vorst, 1992), which is a stabilized version of the bi-conjugate method (BCG) that spans the Krylov space of:

$$\text{span}\{\mathbf{r}^{(0)} = \mathbf{A}^T\mathbf{r}^{(0)}, (\mathbf{A}^T)^2\mathbf{r}^{(0)}, \ldots, (\mathbf{A}^T)^{k+1}\mathbf{r}^{(0)} \tag{6-115}$$

More detail regarding the mathematical formulation and theory can be found in Saad's book (2003). It is important to note that naïve application of these Krylov sub-space methods without preconditioning would generally fail, if the conditioning number of $\mathbf{A}$ is high. The objective of matrix preconditioning is to modify the coefficient matrix ($\mathbf{A}$) and reduce its conditioning number. Left preconditioning involves multiplying both sides by a matrix $\mathbf{M}$:

$$\mathbf{M}\mathbf{A}\mathbf{x} = \mathbf{M}\mathbf{d} \tag{6-116}$$

$$\tilde{\mathbf{A}}\mathbf{x} = \tilde{\mathbf{d}} \tag{6-117}$$

Therefore, the residual is modified:

$$\tilde{\mathbf{r}} = \tilde{\mathbf{d}} - \tilde{\mathbf{A}}\mathbf{x} = \mathbf{M}\mathbf{r} \tag{6-118}$$

Right preconditioning, on the other hand, utilizes the same residual:

$$\mathbf{A}m^{-1}\mathbf{M}\mathbf{x} = \mathbf{d} \tag{6-119}$$

$$\tilde{\mathbf{A}}\mathbf{M}\mathbf{x} = \mathbf{d} \tag{6-120}$$

$$\tilde{\mathbf{r}} = \mathbf{d} - \tilde{\mathbf{A}}\mathbf{M}\mathbf{x} = \mathbf{d} - \mathbf{A}\mathbf{x} = \mathbf{r} \tag{6-121}$$

The choice of $\mathbf{M}$ will be discussed later. The previously presented CG method can be modified if the right preconditioning is applied: Set $\mathbf{r}^{(0)} = \mathbf{d} - \mathbf{A}\mathbf{x}^{(0)}$, $\mathbf{h}^{(0)} = \mathbf{M}^{-1}\mathbf{r}^{(0)}$, and $\mathbf{p}^{(0)} = \mathbf{h}^{(0)}$ and compute a_k, $\mathbf{x}^{(k+1)}$, and $\mathbf{p}^{(k+1)}$ iteratively:

$$a_k = \frac{\mathbf{r}^{(k)} \cdot \mathbf{h}^{(k)}}{\mathbf{p}^{(k)} \cdot \mathbf{A}\mathbf{p}^{(k)}}$$

$$\mathbf{x}^{(k+1)} = \mathbf{x}^{(k)} + a_k \mathbf{p}^{(k)}$$

$$\mathbf{r}^{(k+1)} = \mathbf{r}^{(k)} - a_k \mathbf{A} \mathbf{p}^{(k)}$$
$$\mathbf{h}^{(k+1)} = \mathbf{M}^{-1} \mathbf{r}^{(k+1)}$$
$$\mathbf{p}^{(k+1)} = \mathbf{h}^{(k+1)} - b_{k+1}{}^{(k)}$$

$$\text{and } b_{k+1} = \frac{\mathbf{r}^{(k+1)} \cdot \mathbf{h}^{(k+1)}}{\mathbf{r}^{(k)} \cdot \mathbf{h}^{(k)}} \tag{6-122}$$

Similarly, the previously Orthomin (K) method can be modified with right preconditioning: set $\mathbf{r}^{(0)} = \mathbf{d} - \mathbf{A}\mathbf{x}^{(0)}$, $\mathbf{p}^{(0)} = \mathbf{M}^{-1}\mathbf{r}^{(0)}$, and compute a_k, $\mathbf{x}^{(k+1)}$, and $\mathbf{p}^{(k+1)}$ iteratively:

$$a_k = \frac{\mathbf{r}^{(k)} \cdot \mathbf{A}\mathbf{p}^{(k)}}{\mathbf{A}\mathbf{p}^{(k)} \cdot \mathbf{A}\mathbf{p}^{(k)}}$$
$$\mathbf{x}^{(k+1)} = \mathbf{x}^{(k)} + a_k \mathbf{p}^{(k)}$$
$$\mathbf{r}^{(k+1)} = \mathbf{r}^{(k)} - a_k \mathbf{A}\mathbf{p}^{(k)}$$

If $k + 1 < K$ (a certain pre-defined number of iterations before "restarting"):

$$\mathbf{h}^{(k+1)} = \mathbf{M}^{-1}\mathbf{r}^{(k+1)}$$

$$\mathbf{p}^{(k+1)} = \mathbf{h}^{(k+1)} + \sum_{j=0}^{k} b_k{}^{(j)} \mathbf{p}^{(j)}$$

$$\text{and } b_k^{(j)} = -\frac{\mathbf{A}\mathbf{h}^{(k+1)} \cdot \mathbf{A}\mathbf{p}^{(j)}}{\mathbf{A}\mathbf{p}^{(j)} \cdot \mathbf{A}\mathbf{p}^{(j)}} \tag{6-122}$$

If $k + 1 = K$, restart the procedure by setting $k = 0$, $\mathbf{r}^{(0)} = \mathbf{r}^{(k)}$, and $\mathbf{x}^{(0)} = \mathbf{x}^{(k)}$ before continuing.

The choice of the preconditioner $\mathbf{M}$ is not arbitrary. One desirable property is that its inverse should be easy to compute (in comparison to $\mathbf{A}^{-1}$). Another property, which may be less obvious, is that it should be similar to $\mathbf{A}$. If $\mathbf{M} = \mathbf{I}$, it would be easy to invert, but it would not help to improve the conditioning number of the modified matrix; however if $\mathbf{M} = \mathbf{A}$, its inverse would be difficult to compute, but the conditioning number of the modified matrix would become 1 (Ertekin et al., 2001). Therefore, a compromise between these two properties is often required.

One option is based on Jacobi preconditioning, where $\mathbf{M}$ contains the diagonal elements of $\mathbf{A}$ (a_{ii}, $i = 1, 2, \ldots, n$). The inverse of a diagonal matrix is easy to compute (whose elements $= 1/a_{ii}$). Another option is based on the incomplete Cholesky factorization of the original matrix $\mathbf{A}$. For example, the incomplete LU (ILU) factorization scheme is formulated from $\mathbf{A} = \mathbf{LU} - \mathbf{R}$, where $\mathbf{L}$ and $\mathbf{U}$ contain only parts of the matrices derived from the LU decomposition of $\mathbf{A}$. A general ILU factorization scheme involves dropping some entries (i.e., setting them to 0 following a certain zero pattern). As an example, ILU(0) involves choosing a zero pattern to be the same as the original $\mathbf{A}$ matrix (i.e., same zero entries and the degree of sparsity); it has no fill-in. It is easy to compute, but its inaccuracy may cause the Krylov sub-space algorithm difficulty in converging. The accuracy can be improved with different zero patterns and involving additional fill-in: ILU(l). The alternative formulation requires dropping entries based on their magnitude instead of their locations, and it is referred to as ILU with threshold (ILUT). Detail of these ILU schemes can be found in Saad's book (2003).

Another challenge for matrix preconditioning in reservoir simulation is that the linear system may consist of sub-systems exhibiting vastly different characteristics. For example, the matrix generated upon linearization with a fully implicit black-oil model would consist of both elliptic and hyperbolic parts, corresponding to the pressure and saturation variables, respectively. It seems reasonable to precondition each subsystem separately. Preconditioning is more important for the pressure variable, which often causes difficulty in converging. ILU techniques would work well for the hyperbolic part, while other approaches such as the algebraic multi-grid method (Stüben, 2001) are more suited for the elliptic part (Sebastian et al., 2014). One widely adopted approach for separate preconditioning is the constrained pressure residual method (Wallis et al., 1985); factorization is employed to weaken the coupling between the pressure and saturation variables; a two-step scheme is followed within a given iteration to compute an approximate solution to the pressure variable in the first step, which is then used to solve for the saturation variables in the second step; a different preconditioner is applied at each step. If this decoupling is deemed to be too restrictive, certain combinative schemes have been proposed, where the preconditioner in the second step would account for, to a certain degree, the interaction between the two variables (Behie and Vinsome, 1982). The development of robust preconditioning and matrix solution schemes for fully implicit non-linear systems of equations in reservoir simulation is an ongoing research area.

Example 6-9

Compare different solvers for large sparse matrices.

Solve the matrix equation: $\mathbf{Tx} = \mathbf{b}$ using different direct and iterative matrix solvers. $\mathbf{T}$ and $\mathbf{b}$ are generated from a 2D single-phase flow simulator. $\mathbf{T}$ is sparse and banded, while $\mathbf{x}$ is the pressure value (in Pa) at each grid location.

Solution

Run the MATLAB® code "matrix_solver.m," which can be found on this book's website (www.mhprofessional.com/PRMS), to load $\mathbf{T}$ and $\mathbf{b}$, which are stored in data files "T.dat" and "b.dat," as well as to solve for $\mathbf{x}$ using the following techniques:

- The built-in function "\" or "mldivide," for which MATLAB applies an optimization (or decision) workflow to choose the most optimal direct solution method;

- Conjugate gradients squared method (CGS);
- Bi-conjugate gradient stabilized method (BiCGSTAB); and
- Generalized minimal residual method (GMRES).

For the CGS method, both with or without preconditioning are compared. For all methods, Jacobi preconditioning, where $\mathbf{M}$ contains the diagonal elements of $\mathbf{A}$ (a_{ii}, $i = 1, 2, \ldots, n$), is applied.

First, run the code for the dataset generated from a 2D mesh of 61×61; the total number of grid blocks is 3721. Using the settings provided in the code, the CGS method without preconditioning does not converge properly (a non-zero "flag" value is returned). The solution of $\mathbf{x}$ is plotted.

Second, perform the following tasks:

- Adjust the maximum number of iterations and convergence tolerance (variables "maxit" and "tol").
- Remove the preconditioning.
- Examine the time and relative residual for each of the solution methods.

Some general conclusions can be obtained:

- The matrix size used in this example is relatively small. Large matrices would take significant CPU run time and memory on the desktop version of MATLAB. Therefore, for illustrative purposes, a smaller matrix generated from a simple 2D mesh is used. In such cases, the benefits of iterative methods may not be obvious. However, for much larger matrices, especially those generated from 3D meshes, direct solution techniques may be intractable.

- Iterative methods may not converge if "maxit" is too few or "tol" is too small.

- Preconditioning increases computational time, but the residual could be lower (with fewer iterations).

A larger dataset (generated from a 2D mesh of 100×100) is also provided on this book's website. The reader should repeat the exercise with this larger matrix. The code can be executed for both datasets without any modifications.

6.6   References

T. Ertekin, J.H. Abou-Kassem and G.R. King, Basic Applied Reservoir Simulation. Houston: Society of Petroleum Engineers, 2001.

M. R. Islam, S. Hossien Mousavizadegan, Shabbir Mustafiz, J. H. Abou-Kassem. Advanced Petroleum Reservoir Simulation. Wiley-Scrivener, 2010. ISBN: 978-0470625811

L.W. Lake, Enhanced Oil Recovery. Houston: Society of Petroleum Engineers, 2010. ISBN: 978-1555633059

M.P. Walsh, L.W. Lake, A Generalized Approach to Primary Hydrocarbon Recovery: Elsevier, 2003. ISBN: 978-0444506832

D. N. Dietz, "Determination of average reservoir pressure from build-up surveys," Journal of Petroleum Technology, vol. 17, no. 8, pp. 955–959, 1965.

J. H. Abou-Kassem, S.M. Farouq Ali, M. R. Islam. Petroleum Reservoir Simulation: A Basic Approach. Houston: Gulf, 2006. ISBN: 978-0127999746

D.W. Peaceman. "Interpretation of Well-Block Pressures in Numerical Reservoir Simulation," Society of Petroleum Engineers Journal, vol. 18, no. 3, pp. 183-194, 1978. doi:10.2118/6893-PA

J. Bear, Dynamics of Fluids in Porous Media Dynamics of Fluids in Porous Media. New York: Elsevier, 1972. ISBN: 978-0486656755

W. D. McCain Jr., The Properties of Petroleum Fluids (2nD ed.): PennWell, 1990. ISBN: 978-0878143351

M. J. Economides, D. A. Hill, C. Ehlig-Economides, Petroleum Production Systems. New Jersey: Prentice-Hall, 1994. ISBN: 978-0136586838

R. B. Bird, W. E. Stewart and E. N. Lightfoot, Transport Phenomena (Revised 2nd Ed.). New York: John Wiley & Sons, 2007. ISBN: 978-0470115398

Y. Wang, Implementation of a Two Pseudo-Component Approach for Variable Bubble Point Problems in GPRS, MSc Thesis, Stanford University, 2007.

L. K. Thomas, W. B. Lumpkin and G. M. Reheis, "Reservoir simulation of variable bubble-point problems." Society of Petroleum Engineers Journal, vol. 16, no. 01, pp. 10-16, 1976.

G. D. Shank and C. R. Vestal, "Practical techniques in two-pseudocomponent black-oil simulation," SPE reservoir engineering, vol. 4, no. 02, pp. 244-252, 1989.

R.E. Ewing. The Mathematics of Reservoir Simulation. Siam; 1983.

Z. Chen, G. Huan and Y. Ma, Computational methods for multiphase flows in porous media. Vol. 2. Siam, 2006.

Z. Chen, Finite Element Methods and Their Applications. Heidelberg, New York: Springer-Verlag, 2005. ISBN: 978-3-540-28078-1.

W.H. Reed and T.R. Hill. Triangular mesh methods for the neutron transport equation. Los Alamos Report LA-UR-73-479. 1973.

O.C. Zienkiewicz, R.L. Taylor and J.Z. Zhu, The Finite Element Method: Its Basis and Fundamentals (7th Edition). Oxford: Butterworth-Heinemann, 2013. ISBN: 978-1856176330

E. Kreyszig, Advanced Engineering Mathematics (10th Edition). Wiley, 2011. ISBN: 978-0-470-45836-5

W. Hackbusch, Multi-Grid Methods and Applications. Vol. 4. Springer Science & Business Media, 2013. ISBN: 9 78-3-540-12761-1

M.R. Hestenes and E. Stiefel, "Methods of conjugate gradients for solving linear systems," vol. 49. NBS, 1952.

P.K.W. Vinsome, "Orthomin, an iterative method for solving sparse banded sets of simultaneous linear equations." SPE 5729, 1976.

Y. Saad and M.H. Schultz, "GMRES: A generalized minimal residual algorithm for solving nonsymmetric linear systems," SIAM Journal on Scientific and Statistical Computing, vol. 7, no. 3, 856-869, 1986.

S.C. Eisenstat, H.C. Elman and M.H. Schultz, "Variational iterative methods for nonsymmetric systems of linear equations," SIAM Journal on Numerical Analysis, vol. 20, no. 2, 345-357, 1983.

D.M. Young and K.C. Jea, "Generalized conjugate-gradient acceleration of non-symmetrizable iterative methods," Linear Algebra and its applications, vol. 34, 159-194, 1980.

Y. Saad, Iterative methods for sparse linear systems (2nd Edition). SIAM, Philadelphia, PA, 2003. ISBN: 978-0-89871-534-7

H. A. van der Vorst, Bi-CGSTAB: A fast and smoothly converging variant of Bi-CG for the solution of nonsymmetric linear systems, SIAM J. Sci. Statist. Comput., 12, 631-644, 1992.

K. Stüben, "A review of algebraic multigrid," Journal of Computational and Applied Mathematics, vol. 128, no. 1, 281-309, 2001.

S. Sebastian, K. Stüben, G.L. Brown, D. Chen and D.A. Collins, "Preconditioning for efficiently applying algebraic multigrid in fully implicit reservoir simulations," SPE Journal, vol. 19, no. 4, 726-736, 2014.

J.R. Wallis, R.P. Kendall and T. E. Little, "Constrained residual acceleration of conjugate residual methods," In SPE Reservoir Simulation Symposium. Society of Petroleum Engineers, 1985. http://dx.doi.org/10.2118/13536-MS

A. Behie and P. K.W. Vinsome, "Block iterative methods for fully implicit reservoir simulation," Society of Petroleum Engineers Journal, vol. 22, no. 5, 658-668, 1982.

S.E. Buckley and M.C. Leverett (1942). "Mechanism of fluid displacements in sands". Transactions of the AIME (146): 107-116.

CHAPTER 7

Gridding Schemes for Flow Simulation

Theh chapter will start with a general discussion on gridding schemes. It is followed by an analysis of the accuracy and stability of numerical schemes corresponding to various structured and unstructured gridding schemes. More advanced numerical schemes for unstructured mesh, which is an extension of the formulation presented in Chap. 6, are presented next. Finally, numerical schemes for modeling media with dual- and triple-porosity systems, as in the case of fractured media, will be discussed. The chapter will end with some gridding examples in MATLAB® examples. The setting of numerical parameters will be presented and the corresponding results will be compared.

7.1 Gridding Schemes

7.1.1 Overview

As discussed in the previous chapters, geological models are constructed following fault and horizon modeling. Sections of the geological model may be subjected to flow simulation using the numerical schemes presented in Chap. 6. All numerical schemes, however, require a division of the computational domain into a set of sub-domains (i.e., grid cells). This procedure is referred to as grid (mesh) generation. Two types of grid systems are commonly employed in reservoir simulation: *block-centered* and *point-distributed* grids (Chen et al., 2006). When the grid is uniform, the difference between the two is insignificant (Aziz, 1993); however, when generating unstructured meshes, the use of point-distributed grids becomes more prominent.

A block-centered and point-distributed grid cell for a rectangular grid cell is shown in Fig. 7-1. A block-centered grid is specified by the centers of each grid cell. A point-distributed grid is specified by the points.

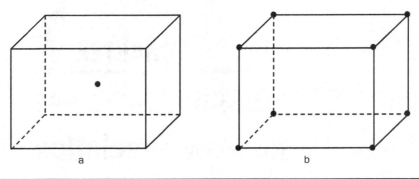

Figure 7-1 Illustration of **a)** a block-centered and **b)** point-distributed grid cell.

7.1.2 Cartesian Grid

The simplest grid structure is the regular Cartesian grid. It is structured, in the sense, that the cells are simply identified using their (i,j,k) index values. The regular grid defined in the normal Cartesian coordinates and the block-centered scheme is commonly adopted. A simple variation is to rotate it such that the layers are aligned along the bedding planes. This simple treatment may be sufficient if the dip angle is relatively constant.

In parts of the domain where the solution changes dramatically (e.g., pressure or saturation gradient is large), additional grid resolution is needed to improve the accuracy. An example of local grid refinement (LGR) within a Cartesian cell is shown in Fig. 7-2. Common examples of LGR may include near-well regions and around fractures (small features with high contrast in permeability with the background). It will be shown in Sec. 7.2 that accuracy is a function of grid size.

7.1.3 Corner-Point Grid

An improvement to the Cartesian grid is the corner-point grid (Wadsley, 1980; Ponting, 1989; Ding and Lemonnier, 1995). It is essentially a distorted grid that conforms better to an irregular computational domain. It involves the use of more arbitrarily defined quadrilaterals (2D) and hexahedra (3D). Each cell is defined by four coordinate lines and eight corner-point locations. It is great for modeling irregular geometries, such as sloping surfaces, fault planes, pinchouts, and erosion surfaces. A comparison between a regular Cartesian grid and a corner-point grid is shown in Fig. 7-3. Although it is possible to define any irregularly shaped object with corner points, a highly skewed grid may lead to distortion of fluid fronts and numerical stability issues (Ertekin et al., 2001).

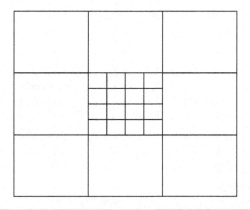

FIGURE 7-2   Illustration of a locally refined Cartesian mesh.

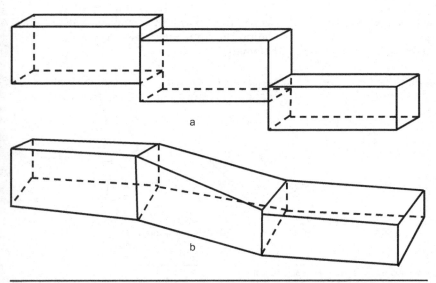

FIGURE 7-3   Illustration of **a)** a regular Cartesian grid and **b)** a corner-point grid.

7.1.4   Perpendicular Bisector Grid

Gridding schemes in Secs. 7.1.2 and 7.1.3 are considered to be *structured*, since an index system can be used to represent all grid cells. This convenience may limit its flexibility and accuracy. Unstructured grids are those constructed around a set of points with no particular indexing scheme.

The most popular unstructured grid in reservoir simulation is the perpendicular bisector (PEBI) (Heinemann et al., 1989). It is also called the Voronoï grid, in honor of the mathematician who developed this technique (Voronoï, 1908). The PEBI grid construction typically follows several steps:

1. Generate a distribution of points.

2. Use Delaunay triangles to find neighbors. A set of triangles is Delaunay if all the edges are Delaunay. An edge is Delaunay if there exists an empty circumcircle of that edge, such that no other vertices are enclosed (Delaunay, 1932; Cheng et al., 2012).

3. Construct PEBIs corresponding to each edge of the Delaunay triangle. A PEBI is a line that is perpendicular to the edge and is bisecting the edge into two equal parts.

4. Connect these bisectors around each point to create the block faces. Every point inside a Voronoï cell is closer to its center than to any other cell in the domain (Aziz, 1993).

The construction of a single PEBI block is shown in Fig. 7-4, while a more general example is shown in Fig. 7-5. In the vicinity of internal features and external boundaries, the circumcenter may lie outside the control volume; this situation can be rectified by applying grid refinement around these problematic regions.

The PEBI grid can be viewed as a generalization of a point-distributed grid (Palagi et al., 1994), with Cartesian, curvilinear, cylindrical, and hexagonal grids being some of its special cases

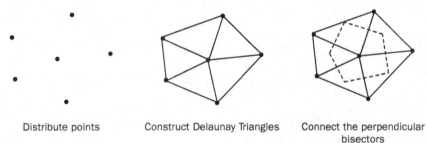

Distribute points   Construct Delaunay Triangles   Connect the perpendicular bisectors

Figure 7-4   Construction of a PEBI grid cell.

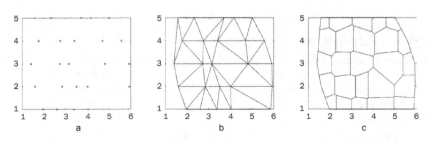

Figure 7-5   a) Illustration of a PEBI grid: grid centers. **b)** Delaunay triangular mesh. **c)** Corresponding PEBI grid. (The color version of this figure is available at www .mhprofessional.com/PRMS.)

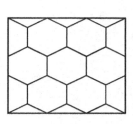 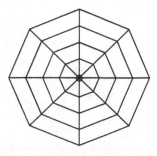

FIGURE 7-6   Illustration of a hexagonal and cylindrical mesh.

(Fig. 7-6). The following advantages with PEBI grid was discussed by Aziz (1993) and Palagi et al. (1994): (1) it is convenient to construct hybrid grids with Voronoï cells to align with wells and other complex geological structures; examples can be found in Palagi et al.'s work (1994); (2) since PEBI grid is locally orthogonal, grid-orientation effect (Sec. 7.2) is minimized; and (3) given that flux across an edge in control volume finite difference (CVFD) formulation is assumed to be proportional to the pressure or potential difference between the cell centers on either side of the boundary, the flux calculation is most accurate when the cell edge is equidistant from both cell centers. However, one obvious drawback is that the resultant Jacobian matrices are more difficult to solve than the simpler structured grids; convenient properties of banded matrices may no longer exist.

7.1.5   General Unstructured Grid

More general gridding schemes for unstructured meshes involve utilizing various geometric primitives, such as triangles in 2D and tetrahedra in 3D. An example of a tetrahedral mesh is shown in Fig. 7-7. The Delaunay technique can be applied to generate triangular and tetrahedral meshes (Shewchuk, 2002; Cheng et al., 2012; Hang, 2015). However, the Delaunay triangulation/tetrahedralization of a set of vertices does not guarantee a high-quality mesh, as it may not conform to the domain boundaries and contain poor-quality triangles/tetrahedra. To address the first issue, constrained Delaunay techniques can be applied to force certain required segments into the mesh. The second issue can be addressed with a Delaunay refinement technique, which is often employed to improve the mesh quality by inserting a vertex at the circumcenter of a triangle or a tetrahedron of poor quality (Shewchuk, 2002; Cheng et al., 2012).

Another common technique for generating unstructured mesh is the advancing front technique (George and Seveno, 1994; Möller and Hansbo, 1995; Schöberl, 1997). Starting with the boundary mesh, each element is cut off one-by-one, and the domain is reduced iteratively; hence, the algorithm is represented by an advancing boundary front.

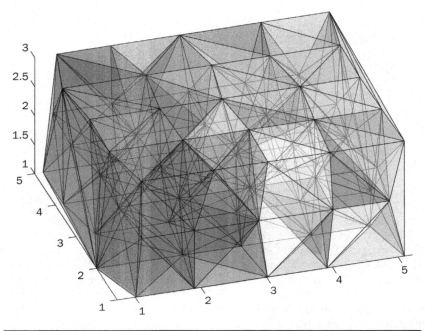

FIGURE 7-7 Illustration of a tetrahedral mesh.

Due to the computational requirement, the use of triangular or tetrahedral meshes is not very common in reservoir flow simulations, especially in CVFD schemes. One exception is the modeling of fractured reservoirs, where fracture planes can be readily incorporated as internal features in the mesh.

7.1.6 Other Specialized Gridding Options

More advanced gridding options include the use of multiblock grids, which facilitates the combination of different types of grids (structured or unstructured) in each sub-domain or block, with either matching or non-matching grid lines at block boundaries. In cell-cut/truncated grids, highly orthogonal, structured, and hexahedral cells are assigned in most parts of the computational domain, while geometric cutting operations are employed to create general polyhedra adjacent to the faults (or other discontinuities) and explicit contact polygons across the faults (Lasseter and Jackson, 2004; Mallison et al., 2014). This gridding option offers a number of benefits: it allows more precise modeling of the complex geological features, as compared to the regular corner-point grid, while maintaining an almost structured grid for the rest of the domain; it is also shown to be accurate. The main disadvantage is that the construction of a multiblock grid for complex reservoir geometries generally requires significant user expertise and is difficult to automate, in comparison to the more general unstructured meshing algorithms discussed in Sec. 7.1.5. Besides, the computational requirement is also increased for handling the more complex shapes around the block boundaries and faults (Mallison et al., 2014). Another issue that is often encountered when constructing multiblock grids is that the grid lines at the block boundaries are not matching. In other words, the grid lines are not continuous from one block to the next. To handle non-matching and non-conforming grid lines, special interpolation and refinement techniques are available to ensure flux continuity (Wheeler et al., 1999; Aavatsmark et al., 2001; Lee et al., 2003; Wolfsteiner et al., 2004).

Another area for advancement is that of dynamic gridding. It refers to a dynamic modification of the grid resolution during run time. It is usually employed to refine the grid locally in the vicinity of sharp displacement fronts and to coarsen the grid in regions with little change. It was first introduced by a number of researchers such as von Rosenberg (1982) and Heinemann et al. (1983). A summary of developments and challenges can be found in Hoteit and Chawathé's research (2016), which describes two general types of dynamic gridding algorithms: the first type is adaptively moving the mesh without significantly increasing the total number of grid cells, a technique often adopted in finite element (FE) formulation (Miller, 1981; Miller and Miller, 1981); the second type is to formulate a multi-level nested gridding framework and apply dynamic refinement and coarsening at run time. Two steps are involved in this type of formulation (Dong et al., 2011): (1) checking the flow/transport regime to identify if refinement or coarsening is needed; and (2) upscaling (merging) and downscaling (refining) all grid block properties and solution values. The dynamic grid updating would incur additional computational costs and memory requirements for handling the addition/elimination of grid blocks and to re-computation of the associated grid block property values. Many of the ongoing research efforts would focus on reducing the computational costs associated with dynamic gridding.

7.2   Consistency, Stability, and Convergence

Certain aspects of the numerical scheme should be examined in order to assess the quality of the solution obtained with these schemes. In this section, three particular aspects involving consistency, stability, and convergence are discussed.

7.2.1   Consistency

As shown in Chap. 6, the spatial and time derivatives are estimated using a truncated Taylor series expansion. The net result of this truncation error reflects the nature of this approximation. It means that even in the absence of round-off errors, the final solution would be different from the original governing equation.

As an example, let us analyze the IMPES simulation equations from Chap. 6 in 1D. A general form of the pressure equation can be written as:

$$L(P) = P_t - bP_{xx} = 0 \tag{7-1}$$

where b is a constant that depends on system properties. Equation 7-1 can be estimated with a first-order approximation to the Taylor series expansion:

$$L_D(P) = \frac{P_i^{n+1} - P_i^n}{\Delta t} - b\frac{\dfrac{P_{i+1/2}^m - P_i^m}{\Delta x / 2} - \dfrac{P_i^m - P_{i-1/2}^m}{\Delta x / 2}}{\Delta x} = 0 \tag{7-2}$$

$$L_D(P) = \frac{P_i^{n+1} - P_i^n}{\Delta t} - 2b\frac{P_{i+1/2}^m - 2P_i^m + P_{i-1/2}^m}{\Delta x^2} = 0 \tag{7-3}$$

A spectrum of schemes, depending on the terms weighted in space and time, is possible:

$$P_{i+1/2}^m = \varpi P_i^m + (1 - \varpi)P_{i+1}^m \tag{7-4a}$$

$$P_i^m = \theta P_i^{n+1} + (1 - \theta)P_i^n \tag{7-4b}$$

w and θ are the weighting factors in space and time, respectively. Forward (downstream weighting), backward (upstream weighting), and central (midpoint weighting) differencing in space would correspond to a value of w to be 0, 1, and 0.5, respectively. Similarly, forward (implicit), backward (explicit), and central differencing (Crank and Nicholson, 1947) in time would correspond to a value of θ to be 1, 0, and 0.5, respectively. Therefore, $P_{i+1/2}^m$ and $P_{i-1/2}^m$ can be expressed in terms of P_i^n, P_{i+1}^n, P_{i-1}^n, and the weighting factors:

$$P_{i+1/2}^m = \varpi\left[\theta P_i^{n+1} + (1 - \theta)P_i^n\right] + (1 - \varpi)\left[\theta P_{i+1}^{n+1} + (1 - \theta)P_{i+1}^n\right] \tag{7-5}$$

$$P_{i-1/2}^m = \varpi\left[\theta P_{i-1}^{n+1} + (1 - \theta)P_{i-1}^n\right] + (1 - \varpi)\left[\theta P_i^{n+1} + (1 - \theta)P_i^n\right] \tag{7-6}$$

Substituting Eqs. (7-5) and (7-6) into Eq. (7-3) would yield the following:

$$L_D(P) = \frac{P_i^{n+1} - P_i^n}{\Delta t}$$
$$- 2b\left\{\frac{\begin{array}{c}\theta\left[(1 - \varpi)P_{i+1}^{n+1} + \varpi P_{i-1}^{n+1} - P_i^{n+1}\right] \\ + (1 - \theta)\left[(1 - \varpi)P_{i+1}^n + \varpi P_{i-1}^n - P_i^n\right]\end{array}}{\Delta x^2}\right\} = 0 \tag{7-7}$$

The above equation can be rearranged as:

$$L_D(P) = \frac{P_i^{n+1} - P_i^n}{\Delta t} - \frac{2b\theta}{\Delta x^2}\left[\begin{array}{c}\left(\frac{1}{2} - \varpi\right)\left(P_{i+1}^{n+1} - P_{i-1}^{n+1}\right) \\ + \frac{1}{2}\left(P_{i+1}^{n+1} + P_{i-1}^{n+1} - 2P_i^{n+1}\right)\end{array}\right]$$

$$- \frac{2b(1 - \theta)}{\Delta x^2}\left[\begin{array}{c}\left(\frac{1}{2} - \varpi\right)\left(P_{i+1}^n - P_{i-1}^n\right) \\ + \frac{1}{2}\left(P_{i+1}^n + P_{i-1}^n - 2P_i^n\right)\end{array}\right] = 0 \tag{7-8}$$

$$L_D(P) = \frac{P_i^{n+1} - P_i^n}{\Delta t} - \frac{2b\left(\frac{1}{2} - \varpi\right)}{\Delta x^2}\left[\begin{array}{c}\theta\left(P_{i+1}^{n+1} - P_{i-1}^{n+1}\right) \\ + (1 - \theta)\left(P_{i+1}^n - P_{i-1}^n\right)\end{array}\right]$$

$$- \frac{b}{\Delta x^2}\left[\theta\left(P_{i+1}^{n+1} + P_{i-1}^{n+1} - 2P_i^{n+1}\right) + (1 - \theta)\left(P_{i+1}^n + P_{i-1}^n - 2P_i^n\right)\right] = 0 \tag{7-9}$$

Consider the Taylor series expansions for these variables:

$$P_{i+1}^{n+1} = P_i^{n+1} + (P_x)_i^{n+1}\Delta x + (P_{xx})_i^{n+1}\frac{\Delta x^2}{2!}$$
$$+ (P_{xxx})_i^{n+1}\frac{\Delta x^3}{3!} + (P_{xxxx})_i^{n+1}\frac{\Delta x^4}{4!} + O\left(\Delta x^5\right) \tag{7-10a}$$

$$P_{i-1}^{n+1} = P_i^{n+1} - (P_x)_i^{n+1}\Delta x + (P_{xx})_i^{n+1}\frac{\Delta x^2}{2!} - (P_{xxx})_i^{n+1}\frac{\Delta x^3}{3!}$$
$$+ (P_{xxxx})_i^{n+1}\frac{\Delta x^4}{4!} - O\left(\Delta x^5\right) \tag{7-10b}$$

$$P_i^{n+1} = P_i^m + (P_t)_i^m(1 - \theta)\Delta t + (P_{tt})_i^m(1 - \theta)^2\frac{\Delta t^2}{2!}$$
$$+ (P_{ttt})_i^m(1 - \theta)^3\frac{\Delta t^3}{3!} + (P_{tttt})_i^m(1 - \theta)^4\frac{\Delta t^4}{4!} + O\left(\Delta t^5\right) \tag{7-10c}$$

$$P_i^n = P_i^m - (P_t)_i^m\theta\Delta t + (P_{tt})_i^m\theta^2\frac{\Delta t^2}{2!}$$
$$- (P_{ttt})_i^m\theta^3\frac{\Delta t^3}{3!} + (P_{tttt})_i^m\theta^4\frac{\Delta t^4}{4!} - O\left(\Delta t^5\right) \tag{7-10d}$$

Similar expressions can be derived for the spatial derivatives:

$$(P_x)_i^{n+1} = (P_x)_i^m + (P_{xt})_i^m (1-\theta)\Delta t + (P_{xtt})_i^m (1-\theta)^2 \frac{\Delta t^2}{2!}$$
$$+ (P_{xttt})_i^m (1-\theta)^3 \frac{\Delta t^3}{3!} + (P_{xtttt})_i^m (1-\theta)^4 \frac{\Delta t^4}{4!} + O(\Delta t^5)$$

(7-10e)

$$(P_x)_i^n = (P_x)_i^m - (P_{xt})_i^m \theta \Delta t + (P_{xtt})_i^m \theta^2 \frac{\Delta t^2}{2!}$$
$$- (P_{xttt})_i^m \theta^3 \frac{\Delta t^3}{3!} + (P_{xtttt})_i^m \theta^4 \frac{\Delta t^4}{4!} - O(\Delta t^5)$$

(7-10f)

Substituting Eq. (7-10) into Eq. (7-9) would yield the following:

$$L_D(P) = (P_t)_i^m + \left(\frac{1}{2} - \theta\right)(P_{tt})_i^m \Delta t + O(\Delta t^2)$$
$$- \frac{2b\left(\frac{1}{2} - \varpi\right)}{\Delta x^2}\left[2(P_x)_i^m \Delta x + \frac{1}{3}(P_{xxx})_i^m \Delta x^3 + O(\Delta x^5)\right]$$
$$- \frac{b}{\Delta x^2}\left[(P_{xx})_i^m \Delta x^2 + \frac{1}{12}(P_{xxxx})_i^m \Delta x^4 + O(\Delta x^6)\right] = 0$$

(7-11)

Truncation error (ε_L) can be defined as the difference between the discretized solution and the original equation:

$$\varepsilon_L = L_D(P) - L(P) = L_D(P) - (P_t)_i^m + b(P_{xx})_i^m$$

(7-12)

Therefore,

$$\varepsilon_L = \left(\frac{1}{2} - \theta\right)(P_{tt})_i^m \Delta t + O(\Delta t^2)$$
$$- \frac{2b\left(\frac{1}{2} - \varpi\right)}{\Delta x^2}\left[2(P_x)_i^m \Delta x + \frac{1}{3}(P_{xxx})_i^m \Delta x^3 + O(\Delta x^5)\right]$$
$$- \frac{b}{\Delta x^2}\left[\frac{1}{12}(P_{xxxx})_i^m \Delta x^4 + O(\Delta x^6)\right]$$

(7-13)

A numerical scheme is considered to be consistent if the truncation error goes to 0 when the discretization size (i.e., grid size and time step) becomes 0 (i.e., $\varepsilon_L \to 0$, as $\Delta x \to 0$ and $\Delta t \to 0$).

Example 7-1

Derive the truncation error for an explicit scheme that employs central differencing in space.

Solution

An explicit scheme refers to $\theta = 0$, while central differencing in space refers to $\omega = 0.5$. Substituting these values in Eq. (7-13) gives:

$$\varepsilon_L = \left(\frac{1}{2}\right)(P_{tt})_i^m \Delta t + O(\Delta t^2) - \frac{b}{12}(P_{xxxx})_i^m \Delta x^2 + O(\Delta x^4)$$

Therefore, the error is first order in time and second order in space.

The saturation equation will be examined next. A general form of the saturation equation can be written in dimensionless form as

$$L(S) = S_t + S_x = 0 \tag{7-14}$$

Equation (7-14) can be estimated with a first-order approximation to the Taylor series expansion:

$$L_D(S) = \frac{S_i^{n+1} - S_i^n}{\Delta t} + \frac{S_{i+1/2}^m - S_{i-1/2}^m}{\Delta x} = 0 \tag{7-15}$$

Introducing the weighing factors:

$$S_{i+1/2}^m = \varpi S_i^m + (1 - \varpi) S_{i+1}^m \tag{7-16a}$$

$$S_i^m = \theta S_i^{n+1} + (1 - \theta) S_i^n \tag{7-16b}$$

Therefore, $S_{i+1/2}^m$ and $S_{i-1/2}^m$ can be expressed in terms of S_i^n, S_{i+1}^n, S_{i-1}^n, and the weighting factors:

$$S_{i+1/2}^m = \varpi\left[\theta S_i^{n+1} + (1 - \theta) S_i^n\right] + (1 - \varpi)\left[\theta S_{i+1}^{n+1} + (1 - \theta) S_{i+1}^n\right] \tag{7-17}$$

$$S_{i-1/2}^m = \varpi\left[\theta S_{i-1}^{n+1} + (1 - \theta) S_{i-1}^n\right] + (1 - \varpi)\left[\theta S_i^{n+1} + (1 - \theta) S_i^n\right] \tag{7-18}$$

Substituting Eqs. (7-5) and (7-6) into Eq. (7-3) would yield the following:

$$L_D(S) = \frac{S_i^{n+1} - S_i^n}{\Delta t}$$
$$+ \frac{1}{\Delta x}\left\{ \begin{array}{l} \theta\left[(1 - \varpi)S_{i+1}^{n+1} + (2\varpi - 1)S_i^{n+1} - \varpi S_{i-1}^{n+1}\right] \\ + (1 - \theta)\left[(1 - \varpi)S_{i+1}^n + (2\varpi - 1)S_i^n - \varpi S_{i-1}^n\right] \end{array} \right\} = 0$$
$$\tag{7-19}$$

The above equation can be rearranged as:

$$L_D(S) = \frac{S_i^{n+1} - S_i^n}{\Delta t} + \theta \left[\frac{\frac{S_{i+1}^{n+1} - S_{i-1}^{n+1}}{2\Delta x} + \Delta x \left(\frac{1}{2} - \varpi \right)}{\left(\frac{S_{i+1}^{n+1} - 2S_i^{n+1} + S_{i-1}^{n+1}}{\Delta x^2} \right)} \right] \qquad (7\text{-}20)$$

$$+ (1 - \theta) \left[\frac{S_{i+1}^n - S_{i-1}^n}{2\Delta x} + \Delta x \left(\frac{1}{2} - \varpi \right) \left(\frac{S_{i+1}^n - 2S_i^n + S_{i-1}^n}{\Delta x^2} \right) \right] = 0$$

Consider the Taylor series expansions for these variables:

$$S_{i+1}^{n+1} = S_i^{n+1} + (S_x)_i^{n+1} \Delta x + (S_{xx})_i^{n+1} \frac{\Delta x^2}{2!} + O(\Delta x^3) \qquad (7\text{-}21a)$$

$$S_{i-1}^{n+1} = S_i^{n+1} - (S_x)_i^{n+1} \Delta x + (S_{xx})_i^{n+1} \frac{\Delta x^2}{2!} - O(\Delta x^3) \qquad (7\text{-}21b)$$

$$S_i^{n+1} = S_i^m + (S_t)_i^m (1 - \theta)\Delta t + (S_{tt})_i^m (1 - \theta)^2 \frac{\Delta t^2}{2!} + O(\Delta t^3) \qquad (7\text{-}21c)$$

$$S_i^n = S_i^m - (S_t)_i^m \theta \Delta t + (S_{tt})_i^m \theta^2 \frac{\Delta t^2}{2!} - O(\Delta t^3) \qquad (7\text{-}21d)$$

Similar expressions can be derived for the spatial derivatives:

$$(S_x)_i^{n+1} = (S_x)_i^m + (S_{xt})_i^m (1 - \theta)\Delta t + (S_{xtt})_i^m (1 - \theta)^2 \frac{\Delta t^2}{2!} + O(\Delta t^3)$$
$$(7\text{-}21e)$$

$$(S_x)_i^n = (S_x)_i^m - (S_{xt})_i^m \theta \Delta t + (S_{xtt})_i^m \theta^2 \frac{\Delta t^2}{2!} - O(\Delta t^3) \qquad (7\text{-}21f)$$

Substituting Eq. (7-21) into Eq. (7-20) would yield the following:

$$L_D(S) = (S_t)_i^m + \left(\frac{1}{2} - \theta \right)(S_{tt})_i^m \Delta t + O(\Delta t^2)$$
$$+ \left[(S_x)_i^m + \frac{\Delta x^2}{6} (S_{xxx})_i^m + O(\Delta x^4) \right]$$
$$+ \left(\frac{1}{2} - \varpi \right) \left[(S_{xx})_i^m \Delta x + \frac{\Delta x^3}{12} (S_{xxxx})_i^m + O(\Delta x^5) \right] \qquad (7\text{-}22)$$

Truncation error (ε_L) can be defined as the difference between the discretized solution and the original equation:

$$\varepsilon_L = L_D(S) - L(S) = L_D(S) - (S_t)_i^m - (S_x)_i^m \qquad (7\text{-}23)$$

Therefore,

$$
\varepsilon_L = \left(\frac{1}{2} - \theta\right)(S_{tt})_i^m \Delta t + O(\Delta t^2)
$$
$$
+ \left[\frac{\Delta x^2}{6}(S_{xxx})_i^m + O(\Delta x^4)\right]
$$
$$
+ \left(\frac{1}{2} - \varpi\right)\left[(S_{xx})_i^m \Delta x + \frac{\Delta x^3}{12}(S_{xxxx})_i^m + O(\Delta x^5)\right] \quad (7\text{-}24)
$$

Collecting the higher-order terms in Eq. (7-24) can be simplified as:

$$
\varepsilon_L = \left(\frac{1}{2} - \theta\right)(S_{tt})_i^m \Delta t + \left(\frac{1}{2} - \varpi\right)(S_{xx})_i^m \Delta x + O(\Delta t^2) + O(\Delta x^2) \quad (7\text{-}25)
$$

It can be shown that $S_{tt} - S_{xx}$ by differentiating Eq. (7-14) with respect to t and x:

$$
(S_t + S_x)_t - (S_t + S_x)_x = S_{tt} + S_{xt} - S_{tx} - S_{xx} = 0 \quad (7\text{-}26)
$$

As a result, the truncation error can be expressed as:

$$
\varepsilon_L = \left[\left(\frac{1}{2} - \theta\right)\Delta t + \left(\frac{1}{2} - \varpi\right)\Delta x\right](S_{xx})_i^m + O(\Delta t^2) + O(\Delta x^2)
$$

Since $L_D(S) = L(S) + \varepsilon_L$, the above expression implies that truncation error has the same role as diffusion in the convective-diffusion equation. Given that the truncation error manifests itself as a diffusion-like term, it serves to smear the actual front, rendering breakthrough hard to predict accurately. Hence, $D_{num} = \left|\left(\theta - \frac{1}{2}\right)\Delta t + \left(\varpi - \frac{1}{2}\right)\Delta x\right|$ is often referred to as numerical (artificial) diffusion. Therefore, even if capillarity is ignored, the numerical solution does not produce a shock front, as in the analytical solution, due to D_{num} that is always greater than 0. Additional detail can be found in Lantz's work (1971).

Example 7-2

Compare the value of D_{num} for the implicit and explicit schemes. Assume upstream weighting (i.e., $\omega = 1$).

Solution

For the explicit scheme, $\theta = 0$, $D_{num} = \frac{1}{2}\Delta x - \frac{1}{2}\Delta t$

For the implicit scheme, $\theta = 1$, $D_{num} = \frac{1}{2}\Delta x + \frac{1}{2}\Delta t$

Though the implicit scheme is more stable (as shown in the next section), it involves more numerical diffusion and higher truncation error.

Treatment for multidimensional flow can be found in Fanchi's research work (1983 and 2005). Truncation error increases with heterogeneous properties and near the boundaries or discontinuities. Accuracy also degrades for irregular grid spacing. However, the analysis can be performed by setting $\Delta x = \Delta x_{max}$ to get a bound on ε_L.

In the case of unstructured grids, accuracy with two-point flux approximation (TPFA) diminishes; this is because the first-order Taylor series expansion for flux approximation is most accurate if the line connecting the two neighboring cell centers is bisecting the cell face perpendicularly (e.g., PEBI grid). TPFA is valid for K-orthogonal grids, where the permeability tensor is orthogonal to the normal vector of the cell surface. For other general unstructured meshes, multiple-point flux approximation (MPFA) that utilizes numerous grid cells near the cell face is required when formulating the flux expression. A discussion of the truncation error in terms of the FE method can be found in the references cited in Chap. 6.

7.2.2 Stability

Consistency compares the difference between the original equations with the discretized equation. However, a consistent scheme can still be considered unstable. In such cases, small errors in the initial conditions or those introduced in certain stages of the computations might grow uncontrollably to dominate the final solution. Therefore, a stable scheme, where small errors would result in small errors in the solution, is required. Instability tends to manifest as non-physical oscillations, overshoots, and overflows.

A number of procedures can be adopted to analyze stability conditions and to derive criteria for stable numerical solutions (e.g., unconditionally unstable, unconditionally stable, or conditionally stable). One technique is called perturbation analysis. The idea is to introduce a perturbation (error) to the solution and examine how the error will grow with time. It involves obtaining the Fourier series representation of the perturbed solution and deriving an expression for the amplification factor. A scheme is stable if the modulus of this factor does not exceed one, which is the von Neumann criterion for stability (Thomas, 2013). Further discussion and some examples can be found in the research works of Chen et al. (2006) and Ertekin et al. (2001). It should be noted that this method, however, neglects the effects of boundary conditions on stability (Smith, 1985; Ertekin et al., 2001).

Another technique that takes into account the boundary effects on stability is the matrix method (Siemieniuch and Gladwell, 1978; Eriksson and Rizzi, 1985). It examines the eigenvalues of the matrix equation (as a result of linearization). This method can be illustrated with an example involving the saturation equation. For an explicit scheme, ($\theta = 0$), Eq. (7-19) becomes:

$$S_i^{n+1} = -\frac{\Delta t}{\Delta x}(1 - \varpi)S_{i+1}^n + \left[1 + \frac{\Delta t}{\Delta x}(1 - 2\varpi)\right]S_i^n + \varpi\frac{\Delta t}{\Delta x}S_{i-1}^n \quad (7\text{-}27)$$

The corresponding matrix equation becomes $\vec{S}^{n+1} = \vec{\vec{A}}\vec{S}^n$.

$$
\vec{\vec{A}} = \begin{vmatrix}
1+\dfrac{\Delta t}{\Delta x}(1-2\varpi) & -\dfrac{\Delta t}{\Delta x}(1-\varpi) & 0 & \ldots & & \ldots \\[2ex]
\varpi\dfrac{\Delta t}{\Delta x} & 1+\dfrac{\Delta t}{\Delta x}(1-2\varpi) & -\dfrac{\Delta t}{\Delta x}(1-\varpi) & 0 & & \ldots \\[2ex]
0 & \varpi\dfrac{\Delta t}{\Delta x} & 1+\dfrac{\Delta t}{\Delta x}(1-2\varpi) & -\dfrac{\Delta t}{\Delta x}(1-\varpi) & 0 & \\[2ex]
\ldots & 0 & \varpi\dfrac{\Delta t}{\Delta x} & 1+\dfrac{\Delta t}{\Delta x}(1-2\varpi) & -\dfrac{\Delta t}{\Delta x}(1-\varpi) \\[2ex]
\ldots & \ldots & 0 & \varpi\dfrac{\Delta t}{\Delta x} & 1+\dfrac{\Delta t}{\Delta x}(1-2\varpi)
\end{vmatrix}
$$

$$(7\text{-}28)$$

For an implicit scheme ($\theta = 1$), Eq. (7-19) becomes:

$$
\frac{\Delta t}{\Delta x}(1-\varpi)S_{i+1}^{n+1} + \left[1+\frac{\Delta t}{\Delta x}(2\varpi-1)\right]S_i^{n+1} - \frac{\Delta t}{\Delta x}\varpi S_{i-1}^{n+1} = S_i^n \quad (7\text{-}29)
$$

The corresponding matrix equation becomes $\vec{S}^{n+1} = \vec{\vec{A}}^{-1}\vec{S}^n$.

$$
\vec{\vec{A}} = \begin{vmatrix}
1+\dfrac{\Delta t}{\Delta x}(2\varpi-1) & \dfrac{\Delta t}{\Delta x}(1-\varpi) & 0 & \ldots & & \ldots \\[2ex]
-\varpi\dfrac{\Delta t}{\Delta x} & 1+\dfrac{\Delta t}{\Delta x}(2\varpi-1) & \dfrac{\Delta t}{\Delta x}(1-\varpi) & 0 & & \ldots \\[2ex]
0 & -\varpi\dfrac{\Delta t}{\Delta x} & 1+\dfrac{\Delta t}{\Delta x}(2\varpi-1) & \dfrac{\Delta t}{\Delta x}(1-\varpi) & 0 & \\[2ex]
\ldots & 0 & -\varpi\dfrac{\Delta t}{\Delta x} & 1+\dfrac{\Delta t}{\Delta x}(2\varpi-1) & \dfrac{\Delta t}{\Delta x}(1-\varpi) \\[2ex]
\ldots & \ldots & 0 & -\varpi\dfrac{\Delta t}{\Delta x} & 1+\dfrac{\Delta t}{\Delta x}(2\varpi-1)
\end{vmatrix}
$$

$$(7\text{-}30)$$

If $\vec{S}^n$ is the unperturbed solution, a perturbed solution is obtained by adding a small perturbation: $\vec{\tilde{S}}^n = \vec{S}^n + \vec{\varepsilon}^n$. It is assumed that $\vec{\varepsilon}^{n+1} = \lambda\vec{\varepsilon}^n = \lambda^2\vec{\varepsilon}^{n-1} = \ldots = \lambda^{n+1}\vec{\varepsilon}^0$ and λ is a scalar constant. Taking the absolute values and applying Schwartz's inequality (Kreyszig, 2011) can establish the following:

$$
\left|\vec{\varepsilon}^{n+1}\right| = \left|\lambda\vec{\varepsilon}^n\right| = \left|\lambda^2\vec{\varepsilon}^{n-1}\right| = \ldots = \left|\lambda^{n+1}\vec{\varepsilon}^0\right|
$$

$$
\left|\vec{\varepsilon}^{n+1}\right| \leq \left|\lambda^{n+1}\right|\left|\vec{\varepsilon}^0\right|
$$

$$
\left|\vec{\varepsilon}^{n+1}\right| \leq \left|\lambda^n\right|\left|\lambda\right|\left|\vec{\varepsilon}^0\right|
$$

$$\vdots$$

$$
\left|\vec{\varepsilon}^{n+1}\right| \leq \left|\lambda\right|^{n+1}\left|\vec{\varepsilon}^0\right| \quad (7\text{-}31)
$$

Therefore, the error will approach 0, as $n \to \infty$, if the absolute value of λ is less than 1. The condition for stability against a perturbation can be stated mathematically as:

$$\lim_{n \to \infty} \left| \bar{\varepsilon}^{n+1} \right| \to 0, \quad |\lambda| < 1 \tag{7-32}$$

In fact, several statements can be made regarding the use of inequalities:

- Sufficient for stability: $|\lambda| < 1$
- Necessary for instability: $|\lambda| > 1$
- Neutral stability: $|\lambda| = 1$

To understand the nature of λ, the perturbed solution is substituted in the matrix equation. For the explicit case:

$$
\begin{aligned}
\vec{S}^{n+1} &= \vec{\bar{A}} \vec{S}^n \\
\vec{S}^{n+1} + \vec{\varepsilon}^{n+1} &= \vec{\bar{A}} \left(\vec{S}^n + \vec{\varepsilon}^n \right) \\
\vec{S}^{n+1} + \lambda^{n+1} \vec{\varepsilon}^0 &= \vec{\bar{A}} \left(\vec{S}^n + \lambda^n \vec{\varepsilon}^0 \right) \\
\lambda^{n+1} \vec{\varepsilon}^0 &= \vec{\bar{A}} \left(\lambda^n \vec{\varepsilon}^0 \right) \\
\lambda \vec{\varepsilon}^0 &= \vec{\bar{A}} \vec{\varepsilon}^0
\end{aligned}
\tag{7-33}
$$

Therefore, λ are the eigenvalues of $\vec{\bar{A}}$. The scheme is stable if the absolute value of the maximum eigenvalue (spectral radius of $\vec{\bar{A}}$) is less than 1. For the implicit case:

$$
\begin{aligned}
\vec{S}^{n+1} &= \vec{\bar{A}}^{-1} \vec{S}^n \\
\vec{S}^{n+1} + \vec{\varepsilon}^{n+1} &= \vec{\bar{A}}^{-1} \left(\vec{S}^n + \vec{\varepsilon}^n \right) \\
\vec{S}^{n+1} + \lambda^{n+1} \vec{\varepsilon}^0 &= \vec{\bar{A}}^{-1} \left(\vec{S}^n + \lambda^n \vec{\varepsilon}^0 \right) \\
\lambda^{n+1} \vec{\varepsilon}^0 &= \vec{\bar{A}}^{-1} \left(\lambda^n \vec{\varepsilon}^0 \right) \\
\lambda \vec{\varepsilon}^0 &= \vec{\bar{A}}^{-1} \vec{\varepsilon}^0 \\
\vec{\bar{A}} \vec{\varepsilon}^0 &= \lambda^{-1} \vec{\varepsilon}^0
\end{aligned}
\tag{7-34}
$$

Therefore, λ^{-1} is the eigenvalue of $\vec{\bar{A}}$, or λ is the inverse of the eigenvalues of $\vec{\bar{A}}$. The scheme is stable if the absolute value of the minimum eigenvalue is greater than 1.

Instead of computing all the eigenvalues directly (quite a tedious task), the Gershgorin circle theorem (Gershgorin, 1931; Varga, 2009) is often used to estimate the eigenvalues. The theorem states that eigenvalues of a $N \times N$ matrix are within the union of N circles in the complex plane defined by:

$$|z - a_{ii}| \leq \sum_{j=1}^{N} |a_{ij}| \qquad (7\text{-}35)$$

where a_{ii} and a_{ij} are the diagonal elements and off-diagonal elements, respectively. We can now apply this theorem to estimate the eigenvalues of $\vec{\vec{A}}$. Given the rows corresponding to the boundary cells are different than those corresponding to the interior cells, they have to be examined separately.

For example, consider an explicit scheme with upstream weighting ($\omega = 1$):

$$\vec{\vec{A}} = \begin{bmatrix} 1 - \dfrac{\Delta t}{\Delta x} & 0 & 0 & \cdots & & \cdots \\[2mm] \dfrac{\Delta t}{\Delta x} & 1 - \dfrac{\Delta t}{\Delta x} & 0 & 0 & & \cdots \\[2mm] 0 & \dfrac{\Delta t}{\Delta x} & 1 - \dfrac{\Delta t}{\Delta x} & 0 & & 0 \\[2mm] \cdots & 0 & \dfrac{\Delta t}{\Delta x} & 1 - \dfrac{\Delta t}{\Delta x} & & 0 \\[2mm] \cdots & \cdots & 0 & \dfrac{\Delta t}{\Delta x} & & 1 - \dfrac{\Delta t}{\Delta x} \end{bmatrix} \qquad (7\text{-}36)$$

$i = 1$ (boundary cell): $\left| z - \left(1 - \dfrac{\Delta t}{\Delta x}\right) \right| \leq 0$; it is a circle centered at $\left(1 - \dfrac{\Delta t}{\Delta x}\right)$ with a radius of 0.

$1 < i < N$ (interior cells): $\left| z - \left(1 - \dfrac{\Delta t}{\Delta x}\right) \right| \leq \left| \dfrac{\Delta t}{\Delta x} \right|$; it is a circle centered at $\left(1 - \dfrac{\Delta t}{\Delta x}\right)$ with a radius of $\dfrac{\Delta t}{\Delta x}$.

$i = N$ (boundary cell): $\left| z - \left(1 - \dfrac{\Delta t}{\Delta x}\right) \right| \leq \left| \dfrac{\Delta t}{\Delta x} \right|$; it is a circle centered at $\left(1 - \dfrac{\Delta t}{\Delta x}\right)$ with a radius of $\dfrac{\Delta t}{\Delta x}$.

The Gershgorin circle theorem indicates that the eigenvalues would lie within the union of these three circles, as shown in Fig. 7-8. As long as $\Delta t / \Delta x$ is less than 1, the dotted circle (and the eigenvalues) would reside within a *shaded* region of stability ($|\lambda| < 1$). The ratio of $\Delta t / \Delta x$ is sometimes referred to as the Courant number. The analysis reveals that this scheme is conditionally stable.

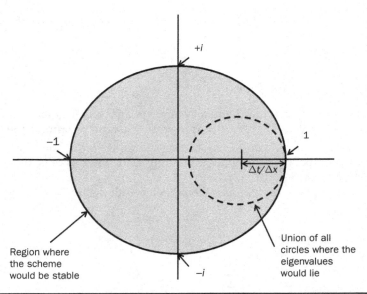

$+i$

-1

1

$\Delta t/\Delta x$

$-i$

Region where
the scheme
would be stable

Union of all
circles where the
eigenvalues
would lie

FIGURE 7-8   Illustration of the Gerschgorin circles and the region of stability. (The
color version of this figure is available at www.mhprofessional.com/PRMS.)

Next, consider an implicit scheme with upstream weighting ($\omega = 1$):

$$\vec{\vec{A}} = \begin{bmatrix} 1 + \dfrac{\Delta t}{\Delta x} & 0 & 0 & \dots & & \dots \\[2ex] -\dfrac{\Delta t}{\Delta x} & 1 + \dfrac{\Delta t}{\Delta x} & 0 & 0 & & \dots \\[2ex] 0 & -\dfrac{\Delta t}{\Delta x} & 1 + \dfrac{\Delta t}{\Delta x} & 0 & & 0 \\[2ex] \dots & 0 & -\dfrac{\Delta t}{\Delta x} & 1 + \dfrac{\Delta t}{\Delta x} & & 0 \\[2ex] \dots & \dots & & 0 & -\dfrac{\Delta t}{\Delta x} & 1 + \dfrac{\Delta t}{\Delta x} \end{bmatrix}$$

(7-37)

$i = 1$ (boundary cell): $\left| z - \left(1 + \dfrac{\Delta t}{\Delta x} \right) \right| \leq 0$; it is a circle centered at $\left(1 + \dfrac{\Delta t}{\Delta x} \right)$ with a radius of 0.

$1 < i < N$ (interior cells): $\left| z - \left(1 + \dfrac{\Delta t}{\Delta x} \right) \right| \leq \left| -\dfrac{\Delta t}{\Delta x} \right|$; it is a circle centered at $\left(1 + \dfrac{\Delta t}{\Delta x} \right)$ with a radius of $\dfrac{\Delta t}{\Delta x}$.

$i = N$ (boundary cell): $\left| z - \left(1 + \dfrac{\Delta t}{\Delta x} \right) \right| \leq \left| -\dfrac{\Delta t}{\Delta x} \right|$; it is a circle centered at $\left(1 + \dfrac{\Delta t}{\Delta x} \right)$ with a radius of $\dfrac{\Delta t}{\Delta x}$.

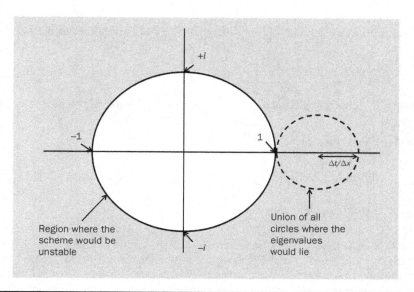

Figure 7-9 Illustration of the Gerschgorin circles and the region of instability. (The color version of this figure is available at www.mhprofessional.com/PRMS.)

The Gershgorin circle theorem indicates that the eigenvalues would lie within the union of these three circles, as shown in Fig. 7-9. For any non-zero $\Delta t/\Delta x$, the dotted circle (and the eigenvalues) would reside within a *shaded* region of stability ($|\lambda| < 1$ or $|\lambda^{-1}| < 1$). Therefore, this scheme is unconditionally stable.

The technique can be readily extended to 3D (though the equations are more complex).

Example 7-3
Show that downstream weighting (i.e., $\omega = 0$) is always unstsable.

Solution
For an explicit scheme, ($\theta = 0$), the corresponding matrix equation becomes:

$$\vec{\vec{A}} = \begin{bmatrix} 1 + \dfrac{\Delta t}{\Delta x} & -\dfrac{\Delta t}{\Delta x} & 0 & \cdots & & \cdots \\ 0 & 1 + \dfrac{\Delta t}{\Delta x} & -\dfrac{\Delta t}{\Delta x} & 0 & & \cdots \\ 0 & 0 & 1 + \dfrac{\Delta t}{\Delta x} & -\dfrac{\Delta t}{\Delta x} & & 0 \\ \cdots & 0 & 0 & 1 + \dfrac{\Delta t}{\Delta x} & & -\dfrac{\Delta t}{\Delta x} \\ \cdots & \cdots & 0 & 0 & & 1 + \dfrac{\Delta t}{\Delta x} \end{bmatrix} \quad (7\text{-}38)$$

$i = 1$ (boundary cell): $\left| z - \left(1 + \dfrac{\Delta t}{\Delta x} \right) \right| \leq \left| -\dfrac{\Delta t}{\Delta x} \right|$; it is a circle centered at

$\left(1 + \dfrac{\Delta t}{\Delta x} \right)$ with a radius of $\dfrac{\Delta t}{\Delta x}$.

$1 < i < N$ (interior cells): $\left| z - \left(1 + \dfrac{\Delta t}{\Delta x} \right) \right| \leq \left| -\dfrac{\Delta t}{\Delta x} \right|$; it is a circle centered

at $\left(1 + \dfrac{\Delta t}{\Delta x} \right)$ with a radius of $\dfrac{\Delta t}{\Delta x}$.

$i = N$ (boundary cell): $\left| z - \left(1 + \dfrac{\Delta t}{\Delta x} \right) \right| \leq 0$; it is a circle centered at

$\left(1 + \dfrac{\Delta t}{\Delta x} \right)$ with a radius of 0.

It is clear that, for any non-zero $\Delta t / \Delta x$, the union of these three circles would reside outside the *shaded* region of stability (Fig. 7-8). Therefore, this scheme is unconditionally unstable.

For an implicit scheme ($\theta = 1$), the corresponding matrix equation becomes:

$$\vec{\vec{A}} = \begin{bmatrix} 1 - \dfrac{\Delta t}{\Delta x} & \dfrac{\Delta t}{\Delta x} & 0 & \cdots & & \cdots \\ 0 & 1 - \dfrac{\Delta t}{\Delta x} & \dfrac{\Delta t}{\Delta x} & 0 & & \cdots \\ 0 & 0 & 1 - \dfrac{\Delta t}{\Delta x} & \dfrac{\Delta t}{\Delta x} & & 0 \\ \cdots & 0 & 0 & 1 - \dfrac{\Delta t}{\Delta x} & & \dfrac{\Delta t}{\Delta x} \\ \cdots & \cdots & 0 & 0 & & 1 - \dfrac{\Delta t}{\Delta x} \end{bmatrix} \tag{7-39}$$

$i = 1$ (boundary cell): $\left| z - \left(1 - \dfrac{\Delta t}{\Delta x} \right) \right| \leq \left| \dfrac{\Delta t}{\Delta x} \right|$; it is a circle centered at

$\left(1 - \dfrac{\Delta t}{\Delta x} \right)$ with a radius of $\dfrac{\Delta t}{\Delta x}$.

$1 < i < N$ (interior cells): $\left| z - \left(1 - \dfrac{\Delta t}{\Delta x} \right) \right| \leq \left| \dfrac{\Delta t}{\Delta x} \right|$; it is a circle centered

at $\left(1 - \dfrac{\Delta t}{\Delta x} \right)$ with a radius of $\dfrac{\Delta t}{\Delta x}$.

$i = N$ (boundary cell): $\left| z - \left(1 - \dfrac{\Delta t}{\Delta x} \right) \right| \leq 0$; it is a circle centered at

$\left(1 - \dfrac{\Delta t}{\Delta x} \right)$ with a radius of 0.

It is clear that, for any non-zero $\Delta t / \Delta x$, the union of these three circles would reside outside the *shaded* region of stability (Fig. 7-9). Therefore, this scheme is unconditionally unstable.

Remarks   A few remarks regarding stability should be made.

- First, all explicit schemes are at best conditionally stable.
- Second, greater stability can be achieved by increasing the degree of implicitness in the numerical schemes; however, as shown in the previous section, implicit scheme often involves larger numerical dispersion (more truncation error). Therefore, improved stability is often associated with larger numerical effort and increased truncation error.
- Finally, numerical dispersion essentially introduces a second-order term to the original equation, which enhances the stability of the resultant scheme. Therefore, in much of the FE literature, the use of artificial dispersion for stability control is widely described (Chen et al., 2006).

7.2.3   Convergence

As suggested at the end of the previous section, there is a relationship between consistency (e.g., numerical dispersion) and stability. An important theorem describing this relationship is the Lax–Richtmeyer equivalence theorem developed by Lax and Richtmeyer (1956), which states that for a consistent scheme, stability is a necessary and sufficient condition for convergence.

7.3   Advanced Numerical Schemes for Unstructured Grids

7.3.1   Generalized CVFD-TPFA Formulation

The transmissibility formulation presented in Chap. 6 is that of a TPFA on a structured grid; it takes into account the interaction between two neighboring cells. For a more general formulation on an unstructured grid, the geometric part of the transmissibility can be obtained as follows (Karimi-Fard et al., 2004):

$$T_{ij} = \frac{\alpha_i \alpha_j}{\alpha_i + \alpha_j}$$

$$\alpha_i = \frac{A_i k_i}{D_i} \mathbf{n}_i \cdot \mathbf{f}_i \tag{7-40}$$

A_i is the area of the interface between the two cells ($A_1 = A_2$); k_i is the permeability of cell i; D_i is the distance from the cell center to the

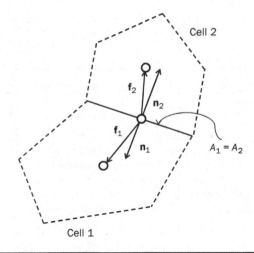

Figure 7-10 Illustration of various terms in the TPFA transmissibility formulation involving two neighboring cells.

centroid of the interface. These quantities are illustrated in Fig. 7-10 for a 2D configuration. The transmissibility formulation can be used directly for 3D configurations, where the interface is a polygon, instead of a line segment in 2D.

Karimi-Fard et al. (2004) extended this formulation to describe the domain with discrete fractures. The idea is to facilitate mesh generation by representing fractures as lower $(n-1)$-dimensional objects (i.e., a line in a 2D domain or a plane in a 3D domain); a volume correction is later employed in the computational domain to account for the actual fracture volume. In other words, fracture elements would appear as lower-dimensional objects in the mesh, but their actual volumes are used in the simulation equations. The transmissibility corresponding to the matrix-fracture (M-F) and matrix-matrix (M-M) fluxes can be computed using Eq. (7-40), while the transmissibility for the fracture-fracture (F-F) flux is computed as:

$$T_{ij} = \frac{\alpha_i \alpha_j}{\displaystyle\sum_{k=1}^{n} \alpha_k}$$

$$\alpha_i = \frac{A_i k_i}{D_i} \tag{7-41}$$

where T_{ij} is the transmissibility between two connecting fracture cells i and j and n is the total number of fractures at the intersection.

7.3.2 CVFD-MPFA Formulation

For general non-orthogonal grids, MPFA methods should be used to compute the transmissibility by taking into account the interaction of

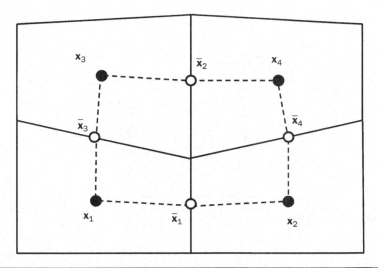

Figure 7-11   Illustration of four neighboring cells in the primal grid (solid line) and the interaction volume corresponding to the dual grid (dashed line).

multiple neighboring cells (Edwards and Rogers, 1998; Aavatsmark, 2002). An example involving the quadrilateral cells is discussed here.

For a grid shown with solid lines in Fig. 7-11, x_1, x_2, x_3, and x_4 are the centers of each of the four cells. A dual grid is constructed by connecting the cell centers to the centroids (midpoints) of the cell faces. The cells of the dual grid are referred to as the interaction volumes. Each interaction volume divides a particular cell face into two parts, which are called the sub-faces. The idea of MPFA is to construct the transmissibility and flux for all sub-faces within an interaction volume (which incorporates information from all four neighboring cells). The flux for a given cell face is then obtained by assembling the contributions from the corresponding sub-faces, incorporating contributions from two neighboring interaction volumes (i.e., six neighboring cells).

It is assumed that the potential (u) is described by a linear function in each sub-cell in the interaction volume:

$$u(\mathbf{x}) = \sum_{i=1}^{3} u_i \phi_i(\mathbf{x}) \tag{7-42}$$

where $\phi_i(\mathbf{x}_j) = \delta_{ij}$ is a linear basis function, with grad $\phi_i = -1/(2F)v_i$. F is the area of the triangle bounded by the three vertices, and v_i is the outer normal vector of the edge opposite from vertex i, as shown in Fig. 7-12. $u_i = u(\mathbf{x}_i)$ at vertex i and its gradient can be computed as:

$$\text{grad } u = \frac{1}{2F_k} \left[v_1^{(k)} (\bar{u}_1 - u_k) + v_2^{(k)} (\bar{u}_2 - u_k) \right] \tag{7-43}$$

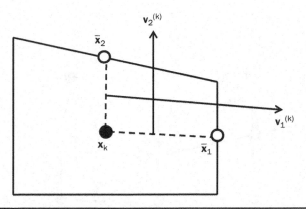

Figure 7-12   Illustration of the normal vectors in cell k.

where $u_k = u(\mathbf{x}_k)$ and $\bar{u}_i = u(\bar{\mathbf{x}}_i)$ for $i = 1, 2$. The flux through sub-face i as seen from cell k is denoted as $f_i^{(k)}$:

$$\begin{bmatrix} f_1^{(k)} \\ f_2^{(k)} \end{bmatrix} = -\begin{bmatrix} \Gamma_1 \mathbf{n}_1^{\mathrm{T}} \\ \Gamma_2 \mathbf{n}_2^{\mathrm{T}} \end{bmatrix} \mathbf{K}_k \mathrm{grad}\, u = -\mathbf{G}_k \begin{bmatrix} \bar{u}_1 - u_k \\ \bar{u}_2 - u_k \end{bmatrix}$$

$$\mathbf{G}_k = \frac{1}{2F_k} \begin{bmatrix} \Gamma_1 \mathbf{n}_1^{\mathrm{T}} \\ \Gamma_2 \mathbf{n}_2^{\mathrm{T}} \end{bmatrix} \begin{bmatrix} v_1^{(k)} & v_2^{(k)} \end{bmatrix} \tag{7-44}$$

Γ_i is the length of the half-edge i. Defining $\mathbf{G}_k$ for all four cells and using the normal vectors defined in Fig. 7-13, the following flux expressions can be obtained:

$$\begin{bmatrix} f_1^{(1)} \\ f_3^{(1)} \end{bmatrix} = -\mathbf{G}_1 \begin{bmatrix} \bar{u}_1 - u_1 \\ \bar{u}_3 - u_1 \end{bmatrix}; \begin{bmatrix} f_1^{(2)} \\ f_4^{(2)} \end{bmatrix} = -\mathbf{G}_2 \begin{bmatrix} u_2 - \bar{u}_1 \\ \bar{u}_4 - u_2 \end{bmatrix}; \begin{bmatrix} f_2^{(3)} \\ f_3^{(3)} \end{bmatrix}$$

$$= -\mathbf{G}_3 \begin{bmatrix} \bar{u}_2 - u_3 \\ u_3 - \bar{u}_3 \end{bmatrix}; \begin{bmatrix} f_2^{(4)} \\ f_4^{(4)} \end{bmatrix} = -\mathbf{G}_4 \begin{bmatrix} u_4 - \bar{u}_2 \\ u_4 - \bar{u}_4 \end{bmatrix} \tag{7-45}$$

Compared to Cell 1, the directions of $v_1^{(2)}$, $v_1^{(4)}$, $v_2^{(4)}$, and $v_2^{(3)}$ have been reversed; therefore, the differences in u appear in the opposite sign. Imposing the continuity condition in flux would yield the following:

$$\begin{aligned} f_1^{(1)} &= f_1^{(2)} = f_1 \\ f_2^{(3)} &= f_2^{(4)} = f_2 \\ f_3^{(1)} &= f_3^{(3)} = f_3 \\ f_4^{(2)} &= f_4^{(4)} = f_4 \end{aligned} \tag{7-46}$$

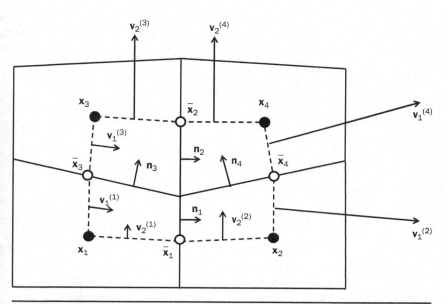

Figure 7-13   Illustration of the normal vectors for the primal and dual grids.

Therefore,

$$f_1 = -g_{11}^{(1)}(\bar{u}_1 - u_1) - g_{12}^{(1)}(\bar{u}_3 - u_1) = g_{11}^{(2)}(\bar{u}_1 - u_2) - g_{12}^{(2)}(\bar{u}_4 - u_2)$$
$$f_2 = -g_{11}^{(3)}(\bar{u}_2 - u_3) + g_{12}^{(3)}(\bar{u}_3 - u_3) = g_{11}^{(4)}(\bar{u}_2 - u_4) + g_{12}^{(4)}(\bar{u}_4 - u_4)$$
$$f_3 = -g_{21}^{(1)}(\bar{u}_1 - u_1) - g_{22}^{(1)}(\bar{u}_3 - u_1) = -g_{21}^{(3)}(\bar{u}_2 - u_3) + g_{22}^{(3)}(\bar{u}_3 - u_3)$$
$$f_4 = g_{21}^{(2)}(\bar{u}_1 - u_2) - g_{22}^{(2)}(\bar{u}_4 - u_2) = g_{21}^{(4)}(\bar{u}_2 - u_4) + g_{22}^{(4)}(\bar{u}_4 - u_4)$$

$$(7\text{-}47)$$

This formulation also ensures continuity in potential at $\bar{x}_1$, $\bar{x}_2$, $\bar{x}_3$, and $\bar{x}_4$. It is referred to as the O-method. Other methods involving different continuity conditions are possible (Aavatsmark, 2002). The expressions on each side of the first equality sign in the above equation can be written in the form of $\mathbf{f} = \mathbf{C}\bar{\mathbf{u}} + \mathbf{F}\mathbf{u}$, where $\mathbf{f} = [f_1\, f_2\, f_3\, f_4]^T$, $\mathbf{u} = [u_1\, u_2\, u_3\, u_4]^T$, and $\bar{\mathbf{u}} = [\bar{u}_1\, \bar{u}_2\, \bar{u}_3\, \bar{u}_4]^T$; similarly, the expressions on each side of the second equality sign in the above equation can be written in the form of $\mathbf{A}\bar{\mathbf{u}} = \mathbf{B}\mathbf{u}$. These two expressions can be combined to eliminate $\bar{\mathbf{u}}$.

$$\mathbf{f} = \mathbf{T}\mathbf{u}$$
$$\mathbf{T} = \mathbf{C}\mathbf{A}^{-1}\mathbf{B} + \mathbf{F} \qquad (7\text{-}48)$$

The matrix $\mathbf{T}$ contains the transmissibility. Equation (7-48) gives the flux through the sub-faces, and it involves only potential values at the cell centers.

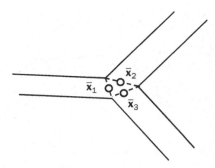

Figure 7-14 Computational domain for a fracture-fracture connection. An intermediate cell is bounded by the dashed line.

Convergence analysis for quadrilateral grids can be found in Klausen and Winther's research (2006). In the above example, the position of continuity is chosen at the midpoint of the cell faces. In practice, a family of MPFA schemes is available depending on the choice of the position of continuity (i.e., quadrature parametrization). Various families of schemes, as well as their convergence behavior, for different 2D and 3D grids are presented by Pal (2007). Formulations for triangular (2D), tetrahedral (3D), and general polygonal cells can also be found in Aavatsmark et al.'s research (1998a and 1998b). Others have also implemented hexahedral grids for reservoir flow simulation (Lee et al., 2002). Treatment of boundary conditions can be found in the references.

The MPFA formulation has gained much popularity in recent years for its application in fractured media, due to its flexibility in handling general unstructured grids with discrete fracture planes. Sandve et al. (2012) adopted an approach similar to that in the previous section, where fractures are represented as lower $(n - 1)$-dimensional objects. The transmissibility and fluxes corresponding to the matrix-fracture (M-F) and matrix-matrix (M-M) connections are computed using Eq. (7-48). The fracture-fracture (F-F) connections are computed by constructing an interaction region around a set of intersecting fractures and introducing hybrid faces (Fig. 7-14).

The pressure is assumed to be constant in the intermediate cell and continuous over the hybrid faces (i.e., $\bar{u}_1 = \bar{u}_2 = \bar{u}_3 = \bar{u}_4$); therefore, $C_h \bar{u}_h = C_0 u_0$, where C_h is the columns in C related to the unknowns at the continuity point on the hybrid faces, and C_0 is a vector containing the row sum of C_h. Flux is also assumed to be conservative at the intersection: $A_0 u_0 = B_0 u$. Imposing these conditions result in the following expressions for the fluxes in an interaction region with:

$$\mathbf{f} = \hat{\mathbf{T}} \mathbf{u};$$
$$\hat{\mathbf{T}} = \mathbf{T} + \frac{\mathbf{C}_0 \mathbf{B}_0}{A_0} \tag{7-49}$$

More sophisticated treatment and extensions to other grid types can be found in Ahmed et al.'s research (2015).

7.3.3   Control Volume Finite Element Formulation

The control volume finite element (CVFE) schemes are locally conservative, and their treatment of complex reservoirs with unstructured gridding requires MPFA formulation. Alternative FE-based approaches are available. The standard FE method presented in Chap. 6 is flexible for unstructured meshes. It is globally conservative, in the sense that the material balance is enforced globally over a summation of individual elements. However, it is not conservative locally over an individual element. A variation of the standard FE method, called CVFE, can be adopted (Lemonnier, 1979). It involves forming a control volume around each grid node by joining the midpoints of the edges of a triangle with the barycenter of the triangle. The method is locally conservative over each control volume. It should be noted that the CVFE grids are different from the PEBI grids discussed earlier in this chapter.

To illustrate the key elements of this technique, consider the steady-state problem in 2D. Equation (6-81) can be expressed as:

$$-\nabla \cdot \left(\vec{\vec{a}}(P)\nabla P \right) = f(P) \tag{7-50}$$

Once again, we seek the solution of $P_h \in V_h$ (the space of continuous piecewise linear functions) and integrate over a control volume V_i:

$$-\int_{V_i} \nabla \cdot \left(\vec{\vec{a}}(P_h)\nabla P_h \right) dx = \int_{V_i} f(P_h) dx \tag{7-51}$$

Applying the divergence theorem and denoting **n** as the normal vector:

$$-\int_{\partial V_i} \left(\vec{\vec{a}}(P_h)\nabla P_h \right) \cdot \mathbf{n} \, dl = \int_{V_i} f(P_h) dx \tag{7-52}$$

For a continuous function $\vec{\vec{a}}$, the integrand of the surface integral would also be continuous. Therefore, this scheme is flux continuous over the edges of the control volume. It is also clear that, by construction, the CVFE scheme is locally conservative (Chen et al., 2006). Next, the piecewise continuous linear basis function is introduced, and the approximation in Eq. (6-89) becomes:

$$P_h(\mathbf{x}) = P_1\varphi_1(\mathbf{x}) + P_2\varphi_2(\mathbf{x}) + P_3\varphi_3(\mathbf{x}) \tag{7-53}$$

where φ_i satisfies Eq. (6-87) with $\varphi_1 + \varphi_2 + \varphi_3 = 1$. It can be shown that:

$$\phi_i = \frac{1}{2|K|}\left(c_i + a_i x_1 + b_i x_2\right)$$

$$a_i = m_{2,x_2} - m_{3,x_2}; b_i = -\left(m_{2,x_1} - m_{3,x_1}\right); c_i = m_{2,x_1}m_{3,x_2} - m_{2,x_2}m_{3,x_1}$$

$$\tag{7-54}$$

x_1 and x_2 are the coordinates in 2D; $i = 1, 2, 3$; $\mathbf{m}_i = \left(m_{i,x_1}, m_{i,x_2} \right)$; $\mathbf{m}_2$ and $\mathbf{m}_3$ refer to the other two vertices; $|K|$ = area of the triangular element. As a result:

$$\frac{\partial \phi_i}{\partial x_1} = \frac{a_i}{2|K|}; \ \frac{\partial \phi_i}{\partial x_2} = \frac{b_i}{2|K|} \tag{7-55}$$

If we are formulating a control volume around $\mathbf{m}_1$ in Fig. 7-15, the flux terms across edge $\mathbf{m}_a - \mathbf{m}_c$ and $\mathbf{m}_c - \mathbf{m}_d$ in Eq. (7-52) become (Chen et al., 2006):

$$- \int_{\mathbf{m}_a \mathbf{m}_c, \mathbf{m}_c \mathbf{m}_d} \left(\vec{a}(P_h) \nabla P_h \right) \cdot \mathbf{n} dl = |K| \sum_{j=1}^{3} \vec{a}(P_h) \nabla \varphi_i \cdot \nabla \varphi_j P_j = \tilde{f} \tag{7-56}$$

It is interesting to note that CVFE and standard FE share the same stiffness matrix for piecewise continuous linear basis function (though the right-hand side is different). Equation (7-56) can be expanded and expressed as (Chen et al. 2006):

$$T_{12}(P_1 - P_2) + T_{13}(P_1 - P_3) = \tilde{f} \tag{7-57}$$

This formulation resembles that of the CVFD scheme with the transmissibility defined as:

$$T_{12} = -|K|\vec{a}(P_h) \nabla \varphi_1 \cdot \nabla \varphi_2$$
$$T_{13} = -|K|\vec{a}(P_h) \nabla \varphi_1 \cdot \nabla \varphi_3 \tag{7-58}$$

If the fluxes across all edges of the control volume are included, Eq. (7-52) would take the form of:

$$\sum_{V_i} T_{ij}\left(P_i - P_j\right) = \sum_{V_i} \tilde{f} \tag{7-59}$$

It should be noted that given every edge is shared between two neighboring elements (assuming that it is not located at the external boundary), the transmissibility should be the sum of contribution from both elements. Hence, Eq. (7-58) becomes:

$$T_{ij} = -|K|\vec{a}(P_h) \nabla \varphi_i \cdot \nabla \varphi_j \big|_{K_1} - |K|\vec{a}(P_h) \nabla \varphi_i \cdot \nabla \varphi_j \big|_{K_2} \tag{7-60}$$

This method can be readily extended to multiple dimensions. Treatment of transient problems is similar to the standard FE method presented in Chap. 6. Additional detail regarding the treatment of boundary conditions and upstream weighting can be found in Chen et al.'s work (2006). The CVFE scheme can be generalized to elements

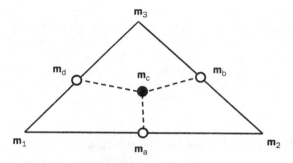

FIGURE 7-15   Example of a triangular element (adapted from Chen et al., 2006).

of higher orders (other than triangular/tetrahedral elements) by utilizing higher-order basis functions, and those techniques are usually referred to as control volume function approximation schemes (Li et al., 2003 and 2004). It should be noted that, as shown by Pal (2007), in comparison to CVFD-MPFA formulation that is both locally conservative and flux continuous, the CVFE formulation ensures only flux continuity across the control volume faces and does not guarantee flux continuity across the interior interfaces (i.e., between elements) over which the permeability can be discontinuous.

The CVFE approach is also popular for modeling fractured systems (Huber and Helmig, 2000; Monteagudo and Firoozabadi, 2004 and 2007a). The fracture element can be modeled as a lower $(n-1)$-dimensional object and coincided with a common edge between two neighboring triangles. Matrix-matrix flux is modeled in the same way as the non-fractured case. To account for fracture-fracture flow, an additional flux term along the fracture element is included in Eq. (7-59). Finally, to model matrix-fracture interaction, the cross-flow equilibrium is assumed:

$$\Phi_j^m = \Phi_j^f$$
$$\Phi_c^m \left(S_{wm} \right) = \Phi_c^f \left(S_{wf} \right) \tag{7-61}$$

where j is the phase index. In this scheme, the matrix-fracture flux is not explicitly modeled. Instead, the saturation in the fracture is related to the saturation in the matrix:

$$S_{wf} = \left(P_c^f \right)^{-1} P_c^m \left(S_{wm} \right) \tag{7-62}$$

Therefore, the time derivative can be formulated in terms of S_{wm}, and the CV formulation involves only matrix variables. A comparison between fully implicit and IMPES formulation of this CVFE scheme can be found in Monteagudo and Firoozabadi's research (2007b). Zhang et al. (2016) extended the technique by introducing a dual-medium matrix system that consists of "matrix" and "micro-scale fractures." The computational domain is composed of two sub-domains: each triangular element (in 2D) contains two sets of properties (one for the matrix and one for the micro-scale fractures). The two sets of governing equations are coupled in a fully implicit scheme with upstream weighting.

7.3.4 Mixed Finite Element Formulation

Another extension to the standard FE formulation is the mixed finite element (MFE) method. Early formulations were applied in solid mechanics (Malkus and Hughes, 1977), while later development has extended the approach for problems in fluid mechanics. The main idea is to approximate both a vector variable (e.g., flux across the interface) and a scalar variable (e.g., pressure in the cell center) simultaneously by employing two different FE spaces (i.e., basis functions) for the two variables; hence, this formulation is referred to as the *mixed* finite element method. The MFE approach is particularly useful for solving second-order partial differential equations (like the pressure equation), as it offers a high-order approximation of both variables (Brezzi and Fortin, 1991; Durlofsky, 1993; Arbogast et al., 1997; Masud and Hughes 2002).

The purpose of this section is to provide an overview of the theory and terminology involved in the MFE method. The pressure equation corresponding to steady-state flow in 1D was presented in Chap. 6, Eq. (6-82):

$$-\frac{\partial^2 P}{\partial x^2} = f(P), \, 0 < x < 1, \, P(0) = P(1) = 0$$

The above equation can be formulated in terms of u (flux):

$$u = -\frac{\partial P}{\partial x} \tag{7-63}$$

For simplicity, $k/\mu = 1$. The equation can be multiplied by a test function, $v \in V$, and integrated over $(0, 1)$:

$$(u,v) = -\left(\frac{\partial P}{\partial x}, v\right), \text{ where } (v,w)$$

$$= \int_0^1 v(x)w(x)dx, \text{ as shown in Chap. 6.} \tag{7-64}$$

Similar to the FE method, integrating by parts and incorporating the boundary conditions would yield:

$$(u,v) = \left(P, \frac{\partial v}{\partial x}\right) \tag{7-65}$$

Substituting Eq. (7-63) into Eq. (6-82) would give:

$$\frac{\partial u}{\partial x} = f(P) \tag{7-66}$$

It can be multiplied by another test function, $w \in W$, and integrated over $(0, 1)$:

$$\left(\frac{\partial u}{\partial x}, w\right) = (f, w) \tag{7-67}$$

Therefore, the *mixed variational* or weak form of Eq. (6-82) is:

$$(u, v) = \left(P, \frac{\partial v}{\partial x}\right), v \in V$$

$$\left(\frac{\partial u}{\partial x}, w\right) = (f, w), w \in W \tag{7-68}$$

It is clear that MFE spaces are involved: two different basis functions are employed for the two variables (u and P). For 1D problems, the domain is sub-divided into a series of elements, I_i, of size $h_i = x_i - x_{i-1}$, where $i = 2, \ldots, M$. The goal is to find $u_h \in V_h$ and $P_h \in W_h$, such that $(u_h, v) = \left(P_h, \frac{\partial v}{\partial x}\right)$ and $\left(\frac{\partial u_h}{\partial x}, w\right) = (f, w)$, $\forall v \in V_h$, $\forall w \in W_h$ (Chen et al., 2006). The next step is to introduce the first basis function: $\varphi_i \in V_h, i = 1, \ldots, M$:

$$\varphi_i(x_j) = \begin{cases} 1, i = j \\ 0, i \neq j \end{cases} \tag{7-69}$$

and the second basis function: $\psi_i \in W_h, i = 1, \ldots, M-1$:

$$\psi_i(x_j) = \begin{cases} 1, x \in I_i \\ 0, \text{else} \end{cases} \tag{7-70}$$

φ_i is a continuous function that is linear over each element I_i (with a value of 1 at the node and 0 everywhere else), while ψ_i is constant over each element I_i.

The functions of v and W can now be expressed as:

$$v(x) = \sum_{i=1}^{M} v_i \varphi_i(x) \quad \text{and} \quad W(x) = \sum_{k=1}^{M-1} w_i \psi_i(x) \tag{7-71}$$

The functions of u and P_h can now be expressed as:

$$u_h(x) = \sum_{i=1}^{M} u_i \varphi_i(x) \quad \text{and} \quad P_h(x) = \sum_{k=1}^{M-1} P_k \psi_k(x) \tag{7-72}$$

where $u_i = u_h(x_i)$ and $P_k = P_h|_{I_k}$. It makes sense that u is defined at the nodes (endpoint or edge of an element), while P is defined at the center of each element. Substituting these two functions in Eq. (7-68) would yield:

$$\sum_{i=1}^{M}\left(\varphi_i,\varphi_j\right)u_i - \sum_{k=1}^{M-1}\left(\frac{\partial\varphi_j}{\partial x},\psi_k\right)P_k = 0; \quad j = 1,\ldots,M$$

$$\sum_{i=1}^{M}\left(\frac{\partial\varphi_i}{\partial x},\psi_j\right)u_i = \left(f,\psi_j\right). \quad j = 1,\ldots,M-1 \qquad (7\text{-}73)$$

This is a system of equations with M unknowns of u_i and $(M-1)$ unknowns of P_k:

$$\begin{bmatrix} \mathbf{A} & \mathbf{B} \\ \mathbf{B}^{\mathsf{T}} & 0 \end{bmatrix}\begin{bmatrix} \mathbf{U} \\ \mathbf{P} \end{bmatrix} = \begin{bmatrix} 0 \\ -\mathbf{f} \end{bmatrix} \qquad (7\text{-}74)$$

where

$\mathbf{A}$ is a $M \times M$ matrix with entries $a_{ij} = \left(\varphi_i,\varphi_j\right)$;

$\mathbf{B}$ is a $M \times (M-1)$ matrix with entries $b_{jk} = -\left(\frac{\partial\phi_j}{\partial x},\psi_k\right)$;

$\mathbf{f}$ is a $(M-1) \times 1$ vector with entries $f_j = \left(f,\psi_j\right)$;

$\mathbf{U}$ is a $M \times 1$ vector with entries u_i;

$\mathbf{P}$ is a $(M-1) \times 1$ vector with entries P_k.

For non-zero Dirichlet boundary conditions (i.e., cell edge pressure is specified), there will be an additional term in Eq. (7-65) for the boundary edge (similar to the standard FE method). Neumann boundary can be imposed by explicitly specifying the flux values at the boundary nodes and including only the interior nodes in the matrix, such that $\vec{v} \cdot \vec{n} = 0$ at the boundary and $\int_{\Omega} w\,dx = 0$ (Chen et al., 2006).

Transient effects can be incorporated following the FE formulation discussed in Chap. 6; this would involve modifying the second equation in Eq. (7-68) [similar to Eq. (6-98) to Eq. (6-100)], and the corresponding matrix equation becomes:

$$\begin{bmatrix} \mathbf{A} & \mathbf{B} \\ \mathbf{B}^{\mathsf{T}} & \tilde{\mathbf{C}} \end{bmatrix}\begin{bmatrix} \mathbf{U} \\ \mathbf{P} \end{bmatrix} = \begin{bmatrix} 0 \\ -\tilde{\mathbf{f}} \end{bmatrix} \qquad (7\text{-}75)$$

where $\tilde{\mathbf{C}}$ and $\tilde{\mathbf{f}}$ have included the terms involving $\mathbf{C(P)}$ and Δt. The formulation can also be extended for a multi-dimensional problem. The mixed variational form of the governing equation becomes:

$$(\vec{u},\vec{v}) = (P,\nabla \cdot \vec{v}), \vec{v} \in \vec{V}$$
$$(\nabla \cdot \vec{v},w) = (f,w), w \in W \qquad (7\text{-}76)$$

The only difference with the higher-dimensional case is the choice of the MFE spaces. For instance, $w \in W_h$ is constant within the element, and the corresponding basis function is:

$$\psi_i(\vec{x}) = \begin{cases} 1, \vec{x} \in K_i \\ 0, \text{ else} \end{cases}$$

As for $\vec{v} \in \vec{V}_h$, Raviart and Thomas (1977) introduced the first family of MFE spaces in 2D, which was extended to 3D by Nédélec (1980). The lowest-order Raviart–Thomas triangular element (RT0) is:

$$\vec{\phi}_{k,i} = (\vec{n} \cdot \vec{n}_E) \frac{|e_i|}{2|K|} (\vec{x} - \vec{x}_i)$$

where $|e_i|$ is the length of the edge opposite to vertex i; $|K|$ is the area of the element; x_i is the coordinate vector of vertex i; $\vec{n}$ is the normal vector (pointing outward from the element), and $\vec{n}_E$ is the predefined normal vector of that edge (i.e., it is +1 for one element and –1 for the other element sharing the same edge). RT_0 space is formulated using the normal component at a single point along each edge (namely, the mid-point). Higher-order RT spaces are available, where the normal components at multiple points along each edge are incorporated.

The MFE approach is also popular for compositional flow modeling (Moortgat and Firoozabadi, 2016) and flow simulation in fractured systems (Hoteit and Firoozabadi, 2008; Erhel et al., 2009; Chen et al., 2016). Similar to the CVFE case, the fracture element is modeled as a lower $(n - 1)$-dimensional object and coincided with a common edge between two neighboring triangles. Two different mixed spaces are needed to model the matrix and fracture systems (e.g., RT_0 for the 2D matrix element and a linear space for the 1D fracture). Readers should refer to the references for detailed derivation and implementation. The general scheme is described here. The idea is to write Eq. (7-76) for the matrix system and the fracture system, meaning that there should be four matrix equations with four unknowns: $\vec{u}_m, \vec{u}_f, P_m,$ and P_f. To couple these two systems, the first equation in Eq. (7-76) for the matrix system is partitioned into two equations: one for the matrix elements that are not connected to any fracture and one for those that are connected to a fracture. The latter, as well as the first equation in Eq. (7-76) for the fracture system, are modified to incorporate Q_f, which is the transfer function between the matrix element and fracture element that share the same edge:

$$\nabla \cdot \vec{u}_f = f_f + Q_f \tag{7-77}$$

At the matrix-fracture edge, pressure continuity is imposed such that the edge pressure of the two matrix elements sharing a fracture in between must be equal to the pressure at the center of the fracture element. However, it appears that there are now four equations for five unknowns. Therefore, an additional constraint is introduced:

$$q_{K,E} + q_{K',E} = Q_f$$

where $q_{K,E}$ and $q_{K',E}$ are the total fluxes from elements K and K', across an edge E, which coincides with a fracture. If a fracture is not present, the sum of these two fluxes should be 0 (no accumulation at the edge). With the presence of a fracture, there can be accumulation ($Q_f > 0$) or depletion ($Q_f < 0$) into the fracture system.

7.4   Dual Media Models

A fractured medium consists of two-pore systems for flow: fracture (highly conductive conduits) and rock matrix. In theory, this can be modeled using a single-porosity system by assigning high porosity and permeability values in the fracture cells and low porosity and permeability values in the matrix cells. However, the vast contrast in these quantities, coupled with the small grid cells needed to represent the fracture aperture, often leads to significant convergence difficulties, despite the use of LGR around the fracture plane and small time steps.

An alternative approach is to consider the matrix and fracture as two separate, but overlapping, continua. Variants of a dual-porosity/dual-permeability (DP/DK) approach offer continuous representations of the dual media. Each continuum occupies the entire domain and may interact with each other. Two alternative formulations are generally adopted: dual-porosity and dual-permeability. The dual-porosity formulation allows for fluid flow from the rock matrix to the fractures, but the flow between matrix cells is not permitted. The dual-permeability formulation, on the other hand, would include flow between matrix cells.

7.4.1   Dual-Permeability Formulation

The simulation equations obtained from Chap. 6 are applied to both the matrix and fracture cells. The interaction between the two systems is represented with an additional flux term. For example, Eq. (6-60) is modified as:

Fracture cells:

$$\text{Water:}\quad \int_V \frac{\partial\left(\frac{\phi}{B_1}S_1\right)}{\partial t}dV = -\int_A \left(\frac{1}{B_1}\vec{u}_1 \cdot \vec{n}\right)dA + \frac{1}{\rho_1^0}\int_V R_1 dV + q_{1,mf}$$

$$(7\text{-}78a)$$

Oil:
$$\int_V \frac{\partial \left[\phi \left(\dfrac{S_2}{B_2} + \dfrac{R_v}{B_3} S_3 \right) \right]}{\partial t} dV = -\int_A \left(\frac{1}{B_2} \vec{u}_2 \cdot \vec{n} + \frac{R_v}{B_3} \vec{u}_3 \cdot \vec{n} \right) dA$$

$$+ \frac{1}{\rho_2^0} \int_V R_2 dV + q_{2,mf} \qquad (7\text{-}78\text{b})$$

Gas:
$$\int_V \frac{\partial \left[\phi \left(\dfrac{R_s}{B_2} S_2 + \dfrac{S_3}{B_3} \right) \right]}{\partial t} dV = -\int_A \left(\frac{R_s}{B_2} \vec{u}_2 \cdot \vec{n} + \frac{1}{B_3} \vec{u}_3 \cdot \vec{n} \right) dA$$

$$+ \frac{1}{\rho_3^0} \int_V R_3 dV + q_{3,mf} \qquad (7\text{-}78\text{c})$$

Matrix cells:

Water:
$$\int_V \frac{\partial \left(\dfrac{\varphi}{B_1} S_1 \right)}{\partial t} dV = -\int_A \left(\frac{1}{B_1} \vec{u}_1 \cdot \vec{n} \right) dA - q_{1,mf} \qquad (7\text{-}79\text{a})$$

Oil:
$$\int_V \frac{\partial \left[\phi \left(\dfrac{S_2}{B_2} + \dfrac{R_v}{B_3} S_3 \right) \right]}{\partial t} dV = -\int_A \left(\frac{1}{B_2} \vec{u}_2 \cdot \vec{n} + \frac{R_v}{B_3} \vec{u}_3 \cdot \vec{n} \right) dA - q_{2,mf}$$

$$(7\text{-}79\text{b})$$

Gas:
$$\int_V \frac{\partial \left[\phi \left(\dfrac{R_s}{B_2} S_2 + \dfrac{S_3}{B_3} \right) \right]}{\partial t} dV = -\int_A \left(\frac{R_s}{B_2} \vec{u}_2 \cdot \vec{n} + \frac{1}{B_3} \vec{u}_3 \cdot \vec{n} \right) dA - q_{3,mf}$$

$$(7\text{-}79\text{c})$$

It should be noted that the external source/sink terms are omitted in the matrix equations. It is often assumed that the external sources and sinks interact only with the fracture system; this assumption is reasonable given that the flow is much faster in the fractures than in the matrix.

The variable $q_{j,mf}$ ($j = 1$, 2, and 3) is a matrix-fracture transfer function. It can be interpreted as a source term that represents the flow of fluid from the matrix to the fractures. It serves to couple the two sets of equations together. With twice the unknowns, the size of the solution matrix (after linearization) is also doubled in size. There are a number of $q_{j,mf}$ formulations. One of these formulations is based on the shape factor (Warren and Root, 1963; Kazemi, 1969):

$$q_{mfj} = T_{mf} \frac{k_{rjm}}{B_{jm}\mu_{jm}}\left[\left(P_{jm} - \gamma D_m\right) - \left(P_{jf} - \gamma D_f\right)\right] \tag{7-80}$$

where T_{mf} is the matrix-fracture transmissibility. For example, for a Cartesian matrix block that is surrounded by fracture planes on all six sides, where L_x, L_y, and L_z are fracture spacings along the x, y, and z directions, respectively, T_{mf} can be expressed as:

$$T_{mf} = \sigma\left[\frac{k_x}{L_x^2} + \frac{k_y}{L_y^2} + \frac{k_z}{L_z^2}\right] \tag{7-81}$$

k_x, k_y, and k_z refer to the permeability of the matrix cell, and σ is the shape factor. Different values of T_{mf} have been derived in the literature according to different assumptions. For instance, assuming quasi steady-state conditions (at the late times) such that the pressure transient has penetrated to the middle of the matrix cell, Eq. (7-81) can be combined with Darcy's equation to yield:

$$
\begin{aligned}
T_{mf} &= 2\left[k_x\frac{L_zL_y}{L_x/2} + k_y\frac{L_xL_z}{L_y/2} + k_z\frac{L_xL_y}{L_z/2}\right] \\
&= 4\cdot\left[k_x\frac{L_xL_zL_y}{L_x^2} + k_y\frac{L_yL_xL_z}{L_y^2} + k_z\frac{L_zL_xL_y}{L_z^2}\right] \\
&= 4L_xL_yL_z\left[\frac{k_x}{L_x^2} + \frac{k_y}{L_y^2} + \frac{k_z}{L_z^2}\right]
\end{aligned} \tag{7-82}
$$

Exact values of σ for a variety of matrix block shapes have been derived analytically in a number of works (Zimmerman et al., 1993; Lim and Aziz, 1995; Mathias and Zimmerman, 2003).

If the pseudo steady-state approximation is inappropriate, time-dependent pseudo-capillary-pressure functions can be employed to facilitate the incorporation of transient effects. Rossen and Shen (1989) examined the mechanisms involved in gas/oil drainage and water/oil imbibition and noted that merely adding a simplified gravity term to the dual-porosity exchange and computing the fluid contact heights for the matrix and for the fracture was insufficient, because of the time-dependent nature of the gravity drainage process. They derived a set of pseudo-capillary-pressure curves to represent that time-dependent behavior. For instance, for water/oil imbibition process:

$$P_{cowf}\left(S_{wf}\right) = S_{wf}\frac{k_z\left(\rho_o - \rho_w\right)}{L_z F_{sK}} \tag{7-83a}$$

$$P_{cowma}\left(S_{wma}\right) = P_{cowf} + \frac{\phi_{ma}\dfrac{dS_{wma}(t)}{dt}S_{wma}(t)}{F_{sK}\left(\dfrac{k_{ro}}{\mu_o}\right)}\left(1 + \frac{k_{ro}\mu_w}{k_{rw}\mu_o}\right) \tag{7-83b}$$

$F_{sK} = 4[k_x/L_x^2 + k_y/L_y^2 + k_z/L_z^2]$. Equation (7-83) suggests that the P_{cowf} function can be determined directly from rock properties and matrix-block dimensions, while the P_{cowma} function can be computed after obtaining $dS_{wma}(t)/dt$ and $S_{wma}(t)$ from the results of one simulation of a fine-grid single-porosity model of a matrix block surrounded by fractures. Certain commercial simulation packages would take a table input of $dS_{wma}(t)/dt$ and $S_{wma}(t)$ data and generate the appropriate P_{cowf} and P_{cowm} functions. Since the effects of gravity and capillarity are captured in these pseudo-capillary-pressure functions, the gravity terms should be removed from Eq. (7-80) when following this approach. Similar sets of P_{cgof} and P_{cgom} functions can be found in Rossen and Shen's research (1989).

7.4.2   Dual-Porosity Formulation

The dual-porosity simulation equations for the fracture cells are the same as Eq. (7-84) for the dual-permeability case. For the matrix cells, the flux terms corresponding to the inter-matrix flow are now 0:

$$\text{Water:}\quad \int_V \frac{\partial\left(\dfrac{\phi}{B_1}S_1\right)}{\partial t}dV = -q_{1,mf} \tag{7-84a}$$

$$\text{Oil:}\quad \int_V \frac{\partial\left[\phi\left(\dfrac{S_2}{B_2} + \dfrac{R_v}{B_3}S_3\right)\right]}{\partial t}dV = -q_{2,mf} \tag{7-84b}$$

$$\text{Gas:}\quad \int_V \frac{\partial\left[\phi\left(\dfrac{R_s}{B_2}S_2 + \dfrac{S_3}{B_3}\right)\right]}{\partial t}dV = -q_{3,mf} \tag{7-84c}$$

In addition to the shape factor formulation for q_{mf}, another approach is to impose the boundary conditions explicitly on the matrix blocks (Pirson, 1953; Barenblatt et al., 1960). For example,

$$q_{1,mf} = -\sum_i \chi_i(\mathbf{x}) \frac{1}{|\Omega_i|} \int_{\Omega_i} \frac{\partial\left(\frac{\phi}{B_1}S_1\right)}{\partial t} d\mathbf{x} \qquad (7\text{-}85)$$

The idea is to discretize a matrix block into a number of sub-domains (Ω_i), where $|\Omega_i|$ denotes the volume of Ω_i and $\chi_i(\mathbf{x})$ is its characteristic function, where it is equal to 1, if $\mathbf{x} \in \Omega_i$ and is equal to 0 otherwise. Next, the boundary condition on the surface of each grid block is established (Arbogast, 1993):

$$\Phi_1' = \Phi_{1f}' - \Phi_1^o \text{ on } \partial\Omega_i, \qquad (7\text{-}86)$$

where $\Phi_1' = \int_{P_1^o}^{P_1} \frac{1}{\gamma_1(\xi)} d\xi - D$ is the phase pseudopotential; Φ_1^o refers to the reference value for each block determined by:

$$\frac{1}{|\Omega_i|} \int_{\Omega_i} \left(\frac{\phi}{B_1}S_1\right) \psi^{-1}\left(\Phi_{1f}' - \Phi_1^o + x_3\right) d\mathbf{x} = \left(\frac{\phi}{B_1}S_1\right) P_{1f} \qquad (7\text{-}87)$$

ψ^{-1} is the inverse of the integral of Φ_1' and a function of P_1; x_3 is vertical downward. The pseudopotential reference value and the corresponding distribution of P_1 in Ω_i can be solved using the above equation. Finally, $q_{1,mf}$ can be estimated based on Eq. (7-85).

7.4.3 Embedded Discrete Fracture Model

A key drawback of most continuum (effective) models, such as DP-DK, is that an idealized regular-fracture topology is required, rendering it inaccurate for most fracture networks with sparse and irregular topologies (Jiang and Younis, 2016). Hence, methods discussed in Sec. 7.3 are widely adopted for discrete fracture network models, where unstructured meshes are generally used and fracture segments are explicitly represented. The higher computational efforts associated with those techniques can be challenging. In recent years, a new technique, the embedded discrete fracture model (EDFM), has gained significant popularity. It offers an important improvement to the

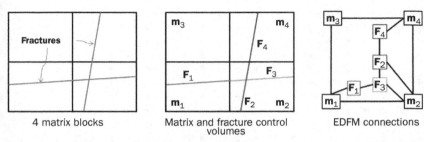

4 matrix blocks   Matrix and fracture control   EDFM connections
                              volumes

FIGURE 7-16   An example of EDFM connections. (The color version of this figure is available at www.mhprofessional.com/PRMS.)

DP-DK model by incorporating explicit representations of fracture segments in the formulation. It is different from typical DFN models, in that the effect of fracture is taken into account without constructing a mesh that is conformed to the fracture geometry (Lee et al., 2001; Li and Lee, 2008; Hajibeygi et al., 2011; Moinfar, 2013; Jiang and Younis, 2016). For instance, a Cartesian mesh can be superimposed on a set of arbitrarily oriented discrete fractures. The method utilizes the concept of non-neighboring connections to model flow between different fracture and matrix control volumes. Specific equations are used to compute the transmissibilities corresponding to these connections, and the readers should refer to the references for more details. These transmissibilities are generally proportional to the fracture length and fracture-matrix interface inside a certain grid block. The EDFM formulation can be readily implemented in most general-purpose reservoir simulators via a user-defined list of connections and transmissibilities. An example is shown in Fig. 7-16.

Example 7-4
Generate a 2D triangular mesh for the point set shown in Fig. 7-17.

Solution
Using the MATLAB code "GM.m" provided on this book's website, www.mhprofessional.com/PRMS, the mesh in Fig. 7-18 is generated using a built-in function of *Delaunay Triangulation*. The distance between each point is 0.25 in both x and y directions. If a larger cell size is selected (e.g., 0.5), it can be achieved by adjusting the variables "dx" and "dy" to 0.5, as shown in Fig. 7-18(b).

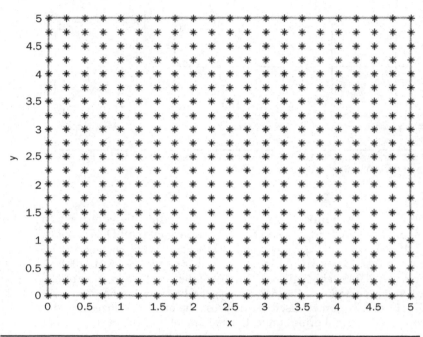

FIGURE 7-17 Point set for Example 7-4.

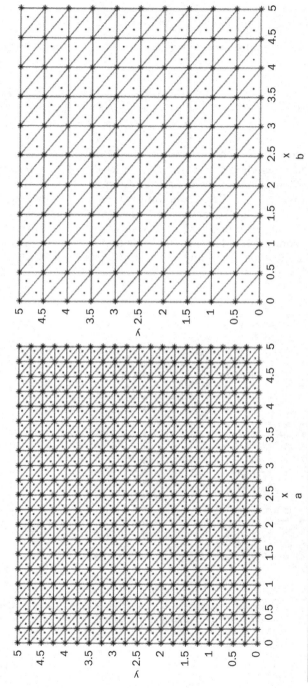

Figure 7-18 Generated mesh for Example 7-4: **a)** Cell size is 0.25. **b)** Cell size is 0.5. (The color version of this figure is available at www .mhprofessional.com/PRMS.)

Example 7-5

Modify the 2D triangular mesh in Example 7-4 by adding a few discrete fractures.

Solution

Using the MATLAB code "GM_frac.m" provided on this book's website, the mesh in Fig. 7-19 is generated. Fracture #3 (in the bottom left) is parallel to the x axis. Each fracture is treated as a lower-dimensional object (i.e., 1D line in a 2D domain).

The properties of each fracture are listed in the code. Let us try changing the orientation of fracture #3 by modifying the parameter "ang" from 0° to 20° (measured upward from the x axis). The new mesh is shown in Fig. 7-20. Examining the discretization around fracture #3 more closely, several small cells are noted. This situation arises when a mesh is imposed on a predefined (arbitrary) set of points. One commonly adopted solution is to post-process the mesh by removing cells that are smaller than a particular tolerance.

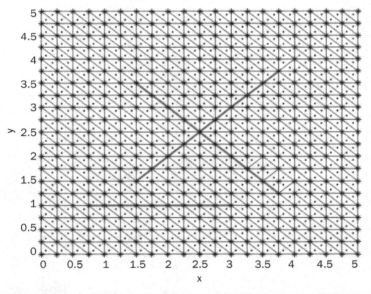

FIGURE 7-19 Generated mesh for Example 7-5 with fracture #3 (bottom left) being parallel to the x axis. (The color version of this figure is available at www.mhprofessional.com/PRMS.)

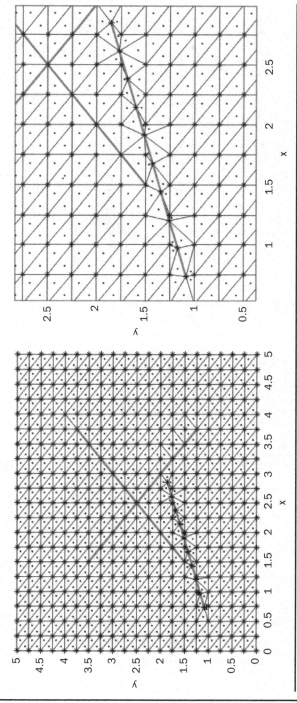

Figure 7-20 Generated mesh for Example 7-5 with fracture #3 at an azimuth angle of 20°. Discretization around fracture #3 is magnified. (The color version of this figure is available at www.mhprofessional.com/PRMS.)

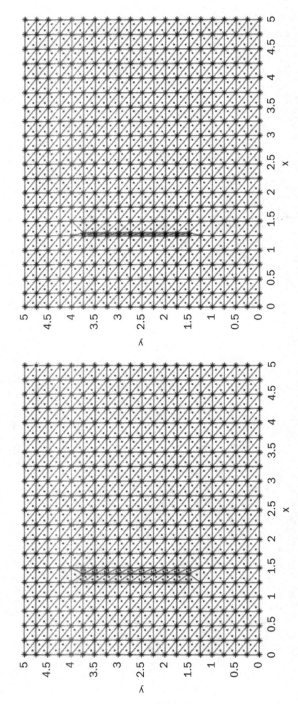

Figure 7-21 Generated mesh for Example 7-6 with different aperture sizes. (The color version of this figure is available at www .mhprofessional.com/PRMS.)

Example 7-6

Modify the 2D triangular mesh in Example 7-4 by adding a discrete fracture with a certain aperture. The fracture must be treated as an equi-dimensional object.

Solution

Using the MATLAB code "GM_frac_aperture.m" at www .mhprofessional.com/PRMS, the mesh is generated, as shown in Fig. 7-21. Two different aperture values (0.15 and 0.05) are tested.

7.5 References

Z. Chen, G. Huan and Y. Ma, Computational methods for multiphase flows in porous media. Vol. 2. Siam, 2006. SPE 25233

W.A. Wadsley. "Modelling reservoir geometry with non-rectangular coordinate grids." In SPE Annual Technical Conference and Exhibition. Society of Petroleum Engineers, 1980.

Ponting, David K. "Corner point geometry in reservoir simulation." In *ECMOR I-1st European Conference on the Mathematics of Oil Recovery*. 1989.

Y. Ding and and P. Lemonnier. "Use of corner point geometry in reservoir simulation." In International Meeting on Petroleum Engineering. Society of Petroleum Engineers, 1995.

Z.E. Heinemann, C. Brand, M. Munka and Y. M. Chen. "Modeling reservoir geometry with irregular grids." In SPE Symposium on Reservoir Simulation. Society of Petroleum Engineers, 1989.

G. Voronoï. "Nouvelles applications des paramètres continus à la théorie des formes quadratiques. Deuxième mémoire. Recherches sur les parallélloèdres primitifs." Journal für die reine und angewandte Mathematik 134 (1908): 198–287.

Delaunay, Boris. "Sur la sphere vide." Izv. Akad. Nauk SSSR, Otdelenie Matematicheskii i Estestvennyka Nauk 7, no. 793-800 (1934): 1–2.

Aziz, Khalid. "Reservoir simulation grids: opportunities and problems." *Journal of Petroleum Technology* 45, no. 07 (1993): 658–663.

Palagi, Cesar Luiz, and Khalid Aziz. "Use of Voronoi grid in reservoir simulation." SPE Advanced Technology Series 2, no. 02 (1994): 69–77.

Shewchuk, Jonathan Richard. "Delaunay refinement algorithms for triangular mesh generation." *Computational geometry* 22, no. 1 (2002): 21–74.

Si, Hang. "TetGen, a Delaunay-based quality tetrahedral mesh generator." ACM Transactions on Mathematical Software (TOMS) 41, no. 2 (2015): 11.

Cheng, Siu-Wing, Tamal K. Dey, and Jonathan Shewchuk. *Delaunay mesh generation.* CRC Press, 2012.

George, Paul Louis, and Eric Seveno. "The advancing-front mesh generation method revisited." *International Journal for Numerical Methods in Engineering* 37, no. 21 (1994): 3605–3619.

Gerschgorin, S. (1931), "Über die Abgrenzung der Eigenwerte einer Matrix", Izv. Akad. Nauk. USSR Otd. Fiz.-Mat. Nauk (in German), 6: 749–754.

Möller, Peter, and Peter Hansbo. "On advancing front mesh generation in three dimensions." International Journal for Numerical Methods in Engineering 38, no. 21 (1995): 3551–3569.

Schöberl, Joachim. "NETGEN An advancing front 2D/3D-mesh generator based on abstract rules." *Computing and visualization in science* 1, no. 1 (1997): 41–52.

Lasseter, Thomas J., and Stuart A. Jackson. "Improving integrated interpretation accuracy and efficiency using a single consistent reservoir model from seismic to simulation." *The Leading Edge* 23, no. 11 (2004): 1118–1121.

Mallison, Bradley, Charles Sword, Thomas Viard, William Milliken, and Amy Cheng. "Unstructured cut-cell grids for modeling complex reservoirs." *SPE Journal* 19, no. 02 (2014): 340–352.

Wheeler, Mary F., Todd Arbogast, Steven Bryant, Joe Eaton, Q. Lu, Malgorzata Peszynska, and Ivan Yotov. "A parallel multiblock/multidomain approach for reservoir simulation." In *Proceedings of the 15th Reservoir Simulation Symposium*, no. 51884, pp. 51–61. 1999.

Aavatsmark, I., E. Reiso, H. Reme, and R. Teigland. "MPFA for faults and local refinements in 3D quadrilateral grids with application to field simulations." In *SPE Reservoir Simulation Symposium*. Society of Petroleum Engineers, 2001.

Lee, S. H., C. Wolfsteiner, L. J. Durlofsky, P. Jenny, and H. A. Tchelepi. "New developments in multiblock reservoir simulation: Black oil modeling, nonmatching subdomains and near-well upscaling." In *SPE Reservoir Simulation Symposium*. Society of Petroleum Engineers, 2003.

Wolfsteiner, C., S. H. Lee, H. A. Tchelepi, P. Jenny, and W. H. Chen. "Unmatched multiblock grids for simulation of geometrically complex reservoirs." In *ECMOR IX-9th European Conference on the Mathematics of Oil Recovery*. 2004.

Hoteit, Hussein, and Adwait Chawathé. "Making field-scale chemical enhanced-oil-recovery simulations a practical reality with dynamic gridding." *SPE Journal* (2016).

von Rosenberg, Dale U. "Local mesh refinement for finite difference methods." In *SPE Annual Technical Conference and Exhibition*. Society of Petroleum Engineers, 1982.

Heinemann, Z. E., Gunter Gerken, and George von Hantelmann. "Using local grid refinement in a multiple-application reservoir simulator." In *SPE Reservoir Simulation Symposium*. Society of Petroleum Engineers, 1983.

Miller, Keith, and Robert N. Miller. "Moving finite elements. I." *SIAM Journal on Numerical Analysis* 18, no. 6 (1981): 1019–1032.

Miller, Keith. "Moving finite elements. II." *SIAM Journal on Numerical Analysis* 18, no. 6 (1981): 1033–1057.

Dong, Chao C., Mehdi Bahonar, Zhangxing John Chen, and Jalel Azaiez. "Development of an algorithm of dynamic gridding for multiphase flow calculation in wells." *Journal of Canadian Petroleum Technology* 50, no. 02 (2011): 35–44.

Lantz, R. B. "Quantitative evaluation of numerical diffusion (truncation error)." *Society of Petroleum Engineers Journal* 11, no. 03 (1971): 315–320.

Fanchi, John R. "Multidimensional numerical dispersion." *Society of Petroleum Engineers Journal* 23, no. 01 (1983): 143–151.

Fanchi, John R. *Principles of applied reservoir simulation*. Gulf Professional Publishing, 2005.

Thomas, James William. *Numerical partial differential equations: finite difference methods*. Vol. 22. Springer Science & Business Media, 2013.

Smith, Gordon D. *Numerical solution of partial differential equations: finite difference methods*. Oxford university press, 1985.

E. Kreyszig, Advanced Engineering Mathematics (10th Edition). Wiley, 2011. ISBN: 978-0-470-45836-5

Varga, Richard S. *Matrix iterative analysis*. Vol. 27. Springer Science & Business Media, 2009.

Lax, Peter D., and Robert D. Richtmeyer. "Survey of the stability of linear finite difference equations." *Communications on pure and applied mathematics* 9, no. 2 (1956): 267–293.

S. Gersgorin. "Uber die Abgrenzung der Eigenwerte einer Matrix", *Izv. Akad. Nauk*, SSSR 7 (1931), 749–754

Karimi-Fard, M., Durlofsky, L. J., & Aziz, K. "An efficient discrete-fracture model applicable for general-purpose reservoir simulators." *SPE Journal* 9, no. 2 (2004): 227–236.

Aavatsmark, Ivar. "An introduction to multipoint flux approximations for quadrilateral grids." *Computational Geosciences* 6, no. 3-4 (2002): 405–432.

Aavatsmark, Ivar, Tor Barkve, O. Bøe, and Trond Mannseth. "Discretization on unstructured grids for inhomogeneous, anisotropic media. Part I: Derivation of the methods." *SIAM Journal on Scientific Computing* 19, no. 5 (1998a): 1700–1716.

Aavatsmark, Ivar, Tor Barkve, O. Bøe, and Trond Mannseth. "Discretization on unstructured grids for inhomogeneous, anisotropic media. Part II: Discussion and numerical results." *SIAM Journal on Scientific Computing* 19, no. 5 (1998b): 1717–1736.

Pal, Mayur. "Families of control-volume distributed CVD (MPFA) finite volume schemes for the porous medium pressure equation on structured and unstructured grids." PhD diss., Ph. D. Thesis, University of Wales, Swansea, UK, 2007.

Klausen, Runhild A., and Ragnar Winther. "Convergence of multipoint flux approximations on quadrilateral grids." *Numerical methods for partial differential equations* 22, no. 6 (2006): 1438–1454.

Lee, Seong H., Hamdi A. Tchelepi, Patrick Jenny, and Larry J. DeChant. "Implementation of a flux-continuous finite-difference method for stratigraphic, hexahedron grids." *SPE Journal* 7, no. 03 (2002): 267–277.

Edwards, Michael G., and Clive F. Rogers. "Finite volume discretization with imposed flux continuity for the general tensor pressure equation." *Computational Geosciences* 2, no. 4 (1998): 259–290.

Sandve, T. H., I. Berre, and Jan M. Nordbotten. "An efficient multi-point flux approximation method for discrete fracture–matrix simulations." *Journal of Computational Physics* 231, no. 9 (2012): 3784–3800.

Ahmed, Raheel, Michael G. Edwards, Sadok Lamine, Bastiaan AH Huisman, and Mayur Pal. "Control-volume distributed multi-point flux approximation coupled with a lower-dimensional fracture model." *Journal of Computational Physics* 284 (2015): 462–489.

Lemonnier, Patrick A. "Improvement of reservoir simulation by a triangular discontinuous finite element method." In *SPE Annual Technical Conference and Exhibition*. Society of Petroleum Engineers, 1979.

Li, Baoyan, Zhangxin Chen, and Guanren Huan. "Control volume function approximation methods and their applications to modeling porous media flow." *Advances in Water Resources* 26, no. 4 (2003): 435–444.

Li, Baoyan, Zhangxin Chen, and Guanren Huan. "Control volume function approximation methods and their applications to modeling porous media flow II: the black oil model." *Advances in water resources* 27, no. 2 (2004): 99–120.

Zhang, Rui-han, Lie-hui Zhang, Jian-xin Luo, Zhi-dong Yang, and Ming-yang Xu. "Numerical simulation of water flooding in natural fractured reservoirs based on control volume finite element method." *Journal of Petroleum Science and Engineering* 146 (2016): 1211–1225.

Huber, Ralf, and Rainer Helmig. "Node-centered finite volume discretizations for the numerical simulation of multiphase flow in heterogeneous porous media." *Computational Geosciences* 4, no. 2 (2000): 141–164.

Monteagudo, J. E. P., and Abbas Firoozabadi. "Control-volume method for numerical simulation of two-phase immiscible flow in two-and three-dimensional discrete-fractured media." *Water resources research* 40, no. 7 (2004).

Monteagudo, Jorge EP, and Abbas Firoozabadi. "Control-volume model for simulation of water injection in fractured media: incorporating matrix heterogeneity and reservoir wettability effects." *SPE journal* 12, no. 03 (2007a): 355–366.

Monteagudo, J. E. P., and A. Firoozabadi. "Comparison of fully implicit and IMPES formulations for simulation of water injection in fractured and unfractured media." *International journal for numerical methods in engineering* 69, no. 4 (2007b): 698–728.

Malkus, David S., and Thomas JR Hughes. "Mixed finite element methods—reduced and selective integration techniques: a unification of concepts." *Computer Methods in Applied Mechanics and Engineering* 15, no. 1 (1978): 63–81.

Masud, Arif, and Thomas JR Hughes. "A stabilized mixed finite element method for Darcy flow." *Computer methods in applied mechanics and engineering* 191, no. 39 (2002): 4341–4370.

Raviart, Pierre-Arnaud, and Jean-Marie Thomas. "A mixed finite element method for 2-nd order elliptic problems." In *Mathematical aspects of finite element methods*, pp. 292–315. Springer Berlin Heidelberg, 1977.

Nédélec, Jean-Claude. "Mixed finite elements in R3." *Numerische Mathematik* 35, no. 3 (1980): 315–341.

Hoteit, Hussein, and Abbas Firoozabadi. "An efficient numerical model for incompressible two-phase flow in fractured media." *Advances in Water Resources* 31, no. 6 (2008): 891–905.

Erhel, Jocelyne, Jean-Raynald De Dreuzy, and Baptiste Poirriez. "Flow simulation in three-dimensional discrete fracture networks." *SIAM Journal on Scientific Computing* 31, no. 4 (2009): 2688–2705.

Moortgat, Joachim, and Abbas Firoozabadi. "Mixed-hybrid and vertex-discontinuous-Galerkin finite element modeling of multiphase compositional flow on 3D unstructured grids." Journal of Computational Physics 315 (2016): 476–500.

Arbogast, Todd, Mary F. Wheeler, and Ivan Yotov. "Mixed finite elements for elliptic problems with tensor coefficients as cell-centered finite differences." *SIAM Journal on Numerical Analysis* 34, no. 2 (1997): 828–852.

Brezzi, Franco, and Michel Fortin. "Mixed and Hybrid Finite Element Methods, no. 15 in Springer Series in Computational Mathematics." (1991).

Chen, Huangxin, Amgad Salama, and Shuyu Sun. "Adaptive mixed finite element methods for Darcy flow in fractured porous media." *Water Resources Research* 52, no. 10 (2016): 7851–7868.

Zimmerman, Robert W., Gang Chen, Teklu Hadgu, and Gudmundur S. Bodvarsson. "A numerical dual-porosity model with semianalytical treatment of fracture/matrix flow." *Water resources research* 29, no. 7 (1993): 2127–2137.

Mathias, Simon A., and Robert W. Zimmerman. "Laplace transform inversion for late-time behavior of groundwater flow problems." *Water resources research* 39, no. 10 (2003).

MATLAB. (2018). version 9.5.0 (R2018b). Natick, Massachusetts: The MathWorks Inc.

Lim, K. T., and K. Aziz. "Matrix-fracture transfer shape factors for dual-porosity simulators." *Journal of Petroleum Science and Engineering* 13, no. 3–4 (1995): 169–178.

Warren, J. E., and P. J. Root. "The behavior of naturally fractured reservoirs." *Society of Petroleum Engineers Journal* 3, no. 03 (1963): 245–255.

Kazemi, Hossein. "Pressure transient analysis of naturally fractured reservoirs with uniform fracture distribution." *SPE Journal* 9, no. 04 (1969): 451–462.

Rossen, R. H., and E. I. C. Shen. "Simulation of gas/oil drainage and water/oil imbibition in naturally fractured reservoirs." *SPE Reservoir Engineering* 4, no. 04 (1989): 464–470.

Arbogast, T. "Gravitational forces in dual-porosity systems: I. Model derivation by homogenization." *Transport in Porous Media* 13, no. 2 (1993): 179–203.

Pirson, S.J. "Performance of fractured oil reservoirs." *AAPG Bulletin* 37, no. 2 (1953): 232–244.

Barenblatt, G. I., Iu P. Zheltov, and I. N. Kochina. "Basic concepts in the theory of seepage of homogeneous liquids in fissured rocks [strata]." *Journal of applied mathematics and mechanics* 24, no. 5 (1960): 1286–1303.

Durlofsky, L.J. "A triangle based mixed finite element—finite volume technique for modeling two phase flow through porous media." *Journal of Computational Physics* 105, no. 2 (1993): 252–266.

Lee, S. H., Lough, M. F. and Jensen, C. L. "Hierarchical modeling of flow in naturally fractured formations
with multiple length scales." *Water Resour. Res.* 37, no. 3 (2001): 443–455.

Li, L. and Lee, S. H. "Efficient field-scale simulation of black oil in a naturally fractured reservoir through discrete
fracture networks and homogenized media." *SPE Res Eval & Eng* 11, no. 4 (2008): 750–758.

Moinfar, A. "Development of an efficient embedded discrete fracture model for 3D compositional reservoir
simulation in fractured reservoirs." Ph.D. Thesis, The University of Texas at Austin, USA (2013).

Hajibeygi, Hadi, Dimitris Karvounis, and Patrick Jenny. "A hierarchical fracture model for the iterative multiscale finite volume method." *Journal of Computational Physics* 230, no. 24 (2011): 8729–8743.

Jiang, Jiamin, and Rami M. Younis. "Hybrid coupled discrete-fracture/matrix and multicontinuum models for unconventional-reservoir simulation." SPE Journal 21, no. 03 (2016): 1-009.

Eriksson, L. E., and Rizzi, A. 'Computer-aided analysis of the convergence to steady state of discrete pproximations to the Euler Equations." *Journal of Computational Physics*, 57 (1985): 90–128.

Siemieniuch, J., and Gladwell, I. "Analysis of explicit difference methods for the diffusion-convection equation." Int. Journal for Numerical Methods in Engineering,12 (1978): 899–916.

Ertekin, T., Abou-Kassem, J.H., and King, G.R. *Basic Applied Reservoir Simulation.* Society of Petroleum Engineers, 2001.

Crank, John, and Phyllis Nicolson. "A practical method for numerical evaluation of solutions of partial differential equations of the heat-conduction type." *Mathematical Proceedings of the Cambridge Philosophical Society*, 43, no. 1 (1947): 50–67.

CHAPTER 8

Upscaling of Reservoir Models

Upscaling is an essential bridge between reservoir geologic modeling and reservoir flow modeling. This chapter will begin with an overview of basic statistical upscaling methods, such as power averaging and statistical re-normalization. Subsequently, the concept of flow-based upscaling will be introduced in order to derive the effective medium properties for the reservoir. The effect of boundary conditions on the effective medium approximations will be expounded. The chapter will end with a brief look at techniques for computing the effective flow and transport properties of a medium corresponding to complex processes involving dispersive and reactive transport.

8.1 Statistical Upscaling

Upscaling is a process that aims to replace a fine-scale detailed description of reservoir properties with a coarse-scale description that has equivalent properties (Christie, 1996; Renard and De Marsily, 1997). Upscaling techniques can generally be classified in terms of the types of parameters being upscaled: single- or multi-phase flow parameters (Durlofsky, 2003). Single-phase upscaling produces effective reservoir properties, such as absolute permeability. Two-phase upscaling formulates effective multi-phase flow parameters, such as relative permeability and capillary pressure functions.

8.1.1 Power Average

For a random variable, $Z(\mathbf{x})$, its spatial mean $\bar{Z}$ over a volume support V can be computed via a generalized or power average.

$$\bar{Z} = \left[\frac{1}{V} \int_V [Z(\mathbf{x})]^\omega d\mathbf{x} \right]^{1/\omega} \tag{8-1}$$

The power ω is a non-zero real number. It encompasses several special cases: harmonic average ($\omega = -1$), arithmetic average ($\omega = 1$), and geometric average ($\omega \to 0$). Generally speaking, arithmetic averaging is appropriate for attributes including primary porosity and fluid saturation. Journel et al. (1986) applied the power average to compute effective permeability. The value of ω depends on many factors including the structure of heterogeneity within V, as well as the size and shape of V. Besides, given that permeability is a flow-related variable, effective permeability would be a function of flow conditions (e.g., local flow velocities and global boundary conditions). Therefore, ω should be direction-dependent, with effective permeability being a full tensor. To apply the power average approach, the principal directions of the permeability are supposed to be parallel to the block sides (off-diagonal elements are 0), and a different value of ω is used to determine each principal component. As pointed out by Wen and Gómez-Hernández (1996), it is not practical to assume that all blocks would have the same orientation and principal directions. Therefore, flow-based techniques are often needed to upscale permeability and flow-related variables.

8.1.2 Statistical Re-Normalization

The idea of re-normalization is to compute an effective permeability for a set of 2×2 cells (in 2D), which is then placed back into the grid to replace the original set of cells. The procedure is repeated successively for many levels until the final cell size is achieved. To compute the effective permeability for the 2×2 cells quickly, King (1989) proposed an analytical formulation based on a resistor-network analogy. With the advances in computing technology, others, such as Peaceman (1997), proposed computing the effective permeability by solving the flow equations directly.

Following the derivation of King (1989), a 2D example is shown here. Formulae for the 3D case can be found in the references. Let us consider a group of four cells with different permeability (k_1, k_2, k_3, and k_4). The objective is to find an analytical expression for the effective permeability k_e. If permeability is viewed as the "conductance" to flow, the equivalent "resistance" between the edge midpoint and the cell center is $1/(2k)$, since the conductance and the resistance are the inverses of each other. After imposing a uniform pressure gradient across the opposite edges, an equivalent resistor network is shown in Fig. 8-1. Please note that the effective permeability calculated here is that along the x direction. The calculation can be easily repeated for the transverse direction along the y direction.

The network is further simplified following the star-triangle transformation (King, 1989), and the final network, as well as the resistance corresponding to each segment, are displayed in Fig. 8-2.

The network essentially consists of three segments:

- Left segment: conductance $= 4(k_1 + k_3)$

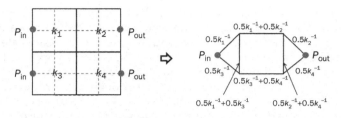

FIGURE 8-1 Construction of the equivalent resistor network. (The color version of this figure is available at www.mhprofessional.com/PRMS.)

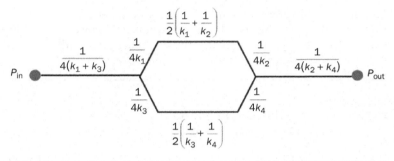

FIGURE 8-2 Equivalent resistor network and the corresponding resistances after star-triangle transformation. (The color version of this figure is available at www.mhprofessional.com/PRMS.)

- Middle segment (the hexagon): conductance $= \dfrac{4k_1 k_2}{3(k_1 + k_2)} +$

$\dfrac{4k_3 k_4}{3(k_3 + k_4)}$

- Right segment: conductance $= 4(k_2 + k_4)$

The equivalent conductance or effective permeability is:

$$\frac{1}{k_e} = \frac{1}{4(k_1 + k_3)} + \frac{1}{\dfrac{4k_1 k_2}{3(k_1 + k_2)} + \dfrac{4k_3 k_4}{3(k_3 + k_4)}} + \frac{1}{4(k_2 + k_4)} \quad (8\text{-}2)$$

King (1989) also shows that with each level of re-normalization, the mean of the permeability distribution is unchanged, while the variance is reduced. In particular, for a Gaussian distribution of k, the variance is reduced by a factor of 4 in 2D and a factor of 8 in 3D. The big advantage of this implementation is its computational efficiency. A major drawback, however, is that the boundary conditions applied to the 2 × 2 networks may be too restrictive and unrealistic. The use of local boundary conditions for upscaling can be an issue, even if flow-based upscaling is adopted, as shown in the next section. Furthermore, re-applying the same boundary conditions repeatedly at multiple levels in the re-normalization scheme may exaggerate the error. The method provides excellent results (when comparing flow simulations at the measurement and block scales) for statistically isotropic, log-normal conductivity fields; for strongly anisotropic media, such as sand–shale formations, the method did not perform well due to the poor resolution of the successive upscaling around the edges of the shales (Wen and Gómez-Hernández, 1996). Another issue is that effective transmissibility, instead of effective permeability, is generally computed in flow simulations (see Chaps. 6 and 7). Therefore, Peaceman (1997) proposed a modification to yield effective half-block transmissibility for a general grid with non-uniform cells (varying size and geometry). In his work, the equivalent network is solved directly via numerical solutions of the flow equations.

Example 8-1

Calculate k_e for the following two cases.
Case 1: Dispersed permeability field: $k_1 = 100$ mD, $k_2 = 600$ mD, $k_3 = 500$ mD, $k_4 = 200$ mD
Case 2: Layered permeability field: $k_1 = k_2 = 600$ mD, $k_3 = k_4 = 200$ mD

Solution
Case 1: $k_e = 249$ mD
Case 2: $k_e = 400$ mD

8.1.3   Facies Upscaling

In many reservoirs, the flow characteristics are dominated by the architecture of the reservoir in terms of the distribution of the lithofacies in the reservoirs. Properties, such as porosity and permeability, are assumed to be constant within lithofacies. Consequently, it is meaningful to examine the upscaling of lithofacies. In some cases, multivariate simulation can be performed to simulate both facie indicators and other variables simultaneously.

The most common approach for facies upscaling is the "majority rule." Within a coarsened interval, the facies with the highest proportion is considered as the most representative of the corresponding upscaled interval. However, this process renders a loss of information regarding the other facies in the interval. In the case of categorical variables, upscaling entails mixing characteristics of different individual intervals; for instance, the quality of sand often diminishes in proximity to lower-quality facies (e.g., shale) (Lajevardi and Deutsch, 2016).

Given that the lithofacies distribution is often used as supporting information for primary variable modeling and it is not utilized directly in flow simulation, other variables that are more representative of this degree of mixing and information loss have been proposed to replace the upscaled lithofacies; for instance, entropy is a widely adopted measure of the information loss introduced by Shannon (1948):

$$E(\mathbf{u}) = -\sum_{k=1}^{K} p_k(\mathbf{u})\log_2\left(p_k(\mathbf{u})\right) \qquad (8\text{-}3)$$

$p_k(\mathbf{u})$ is the proportions of facie k at location $\mathbf{u}$. The base value of the logarithmic term 2 is commonly used. It originates from the binary variables in information theory. Entropy is a bounded value between 0 (no mixing, no information loss) and a maximum value that varies depending on the total number of facies. The maximum entropy occurs when there is a uniform distribution (i.e., equal facies proportions of $1/K$ for all facies in the interval). Though E captures the amount of information loss, it fails to identify the specific facies in the upscaling interval and the corresponding proportions. Imagine two cases, each consisting of two facies, with sand being the major facie (60%). The minor facie (40%) is shaly sand in one case and pure shale in the other. The entropy in both cases would be $[0.4 \times \log_2(0.4) + 0.6 \times \log_2(0.6)] = 0.971$. However, in terms of rock properties, the first case is more similar to sand than the second case. These differences in properties of the two upscaled intervals are not reflected in the entropy value. Therefore, Babak et al. (2013) proposed the use of conditional entropy:

$$E_c(\mathbf{u}) = -\sum_{k=1}^{K} p_k(\mathbf{u} \mid k_m)\log_2\left(p_k(\mathbf{u} \mid k_m)\right)$$

where $p_k(\mathbf{u} \mid k_m) = \left(1 - S_{k_m k}\right)p_k(\mathbf{u}); p_{k_m}(\mathbf{u} \mid k_m) = \sum_{k=1}^{K} S_{k_m k}p_k(\mathbf{u})$ (8-4)

k_m = the major facie. They have defined a symmetric similarity matrix with entries S_{ij} $(i,j = 1, \ldots, K)$. When $i \neq j$, S_{ij} represents the similarity between two facies and ranges between 0 and 1. It is used to "redistribute" the proportions within the interval based on how similar are the individual facies (i.e., the fraction of facie i that is identical to facie j). For instance, in the previous example, sand is the major facie (k_m); if shaly sand consists of 80% sand and 20% shale, a value of 0.8 will be assigned to S_{ij}, with $i = $ sand and $j = $ shaly sand.

Example 8-2
Calculate the conditional entropy for the aforementioned sand–shale example.

Solution
Case 1: Sand and shaly sand ($S_{sand\text{-}shaly\ sand} = 0.8$; $S_{sand\text{-}sand} = 1$; $p_{sand} = 0.6$; $p_{shaly\ sand} = 0.4$)

$$p_{shaly\ sand}(\mathbf{u} \mid sand) = (1 - 0.8)0.4 = 0.08$$

$$p_{sand}(\mathbf{u} \mid sand) = (0.8)(0.4) + (1)(0.6) = 0.92$$

$$E_c(\mathbf{u}) = -0.08\log_2(0.08) - 0.92\log_2(0.92) = 0.402$$

Case 2: Sand and pure shale ($S_{sand\text{-}shale} = 0$; $S_{sand\text{-}sand} = 1$; $p_{sand} = 0.6$; $p_{shale} = 0.4$)

$$p_{shale}(\mathbf{u} \mid sand) = (1 - 0)0.4 = 0.4$$

$$p_{sand}(\mathbf{u} \mid sand) = (0)(0.4) + (1)(0.6) = 0.6$$

$$E_c(\mathbf{u}) = -0.4\log_2(0.4) - 0.6\log_2(0.6) = 0.971$$

Though condition entropy is more sensitive to the types of facies in the mixing, there are still some deficiencies with this approach. Most importantly, the same value for Case 2 in the previous example would be computed if the mixture contains 60% shale and 40% sand. In the end, it does not offer specific information about what types of facies are included in the mix. Similar to that of conventional entropy, it only highlights the variability between the facies that are mixed in the upscaled interval. Lajevardi and Deutsch (2016) proposed a new variable called the facies mixing measure (FMM). It incorporates specific properties of the facies and the distance to the nearest dissimilar facies, as well as its types.

$$FMM(\mathbf{u}) = p_k(\mathbf{u})\tilde{\phi}_k(d_{k,k'}) + \sum_{k=1}^{K-1} p_k(\mathbf{u})\tilde{\phi}_k \qquad (8\text{-}5)$$

where the first term on the right-hand side refers to the major facie, while the second term refers to a summation over all the non-major facies. The average porosity, $\tilde{\phi}_k$, is a function of distance, d, to the nearest dissimilar facie. In the limit of 100% of a given facie, FMM is equivalent to the average porosity of the corresponding facie. Several parameterizations of $\tilde{\phi}_k(d_{k,k'})$ based on the exponential distributions were proposed.

It should be noted that upscaling facie indicators may not be very meaningful or useful in many cases. Ultimately, it is the information carried by a facie indicator that is relevant in reservoir modeling. For example, a facie indicator is not a direct input in flow simulation, rather the reservoir property (e.g., porosity) associated with that facie indicator is used for flow calculations. Therefore, FMM or E_c can be considered as a secondary variable: instead of computing an upscaled facie indicator, the upscaled secondary variable is included directly in a multivariate modeling workflow that simulates the primary reservoir variables (e.g., facie, porosity, and saturation).

8.2   Flow-Based Upscaling

8.2.1   Effective Medium Approximations

In the previous section, effective properties are computed explicitly in terms of their fine-scale values within the averaging volume or block. However, in the case of flow-related variables, such as permeability and dispersivity, it is expected that the resultant coarse-scale values would depend on the flow conditions within the block and the associated boundary conditions. It means that effective properties are non-local and should not be considered as an intrinsic property of the block.

8.2.2   Single-Phase Flow

For example, to scale-up single-phase flow and transport, it is important to compute effective permeability. Though the discussion here focuses on absolute permeability, the same techniques can be applied to the transport equation (Chap. 6) and used to obtain effective dispersivity. At the fine scale, combining Darcy's law and the conservation of mass for a steady-state, incompressible system would yield:

$$\nabla \cdot \vec{\vec{k}} \nabla P = 0 \tag{8-6}$$

where $\vec{\vec{k}}$ is a tensor and is equal to $\begin{bmatrix} k_{xx} & k_{xy} \\ k_{yx} & k_{yy} \end{bmatrix}$ in 2D. It should be symmetric and positive definite. P is the local pressure, and it can be modified in the usual way to include the gravitational potential (see Chap. 6). Upscaling means finding an effective $\vec{\vec{k}}^*$ such that:

$$\nabla \cdot \vec{\vec{k}}^* \nabla P = 0 \tag{8-7}$$

The idea is to solve Eq. (8-6) directly on a fine-scale grid to determine P (and velocity), which are then used to define $\vec{\vec{k}}^*$ in Eq. (8-7). Though many techniques are available, they differ primarily on two aspects: (1) the boundary conditions used for the solution and (2) the extent of the area within which flow is numerically solved at the fine scale.

For instance, the simplest setup in 2D is illustrated in Fig. 8-3. The block being upscaled is isolated from the rest of the grid. The principal components of $\vec{\vec{k}}^*$ are assumed to be parallel to the block sides, and each principal component is computed by numerically solving a flow problem with a prescribed pressure gradient across the faces of the block that are orthogonal to the principal direction and impermeable otherwise. Figure 8-3 shows the setup for computing k_{xx}^*:

$$k_{xx}^* = -\frac{Q}{y_1 - y_0} \frac{\mu(x_1 - x_0)}{P\big|_{x_1} - P\big|_{x_0}} \tag{8-8}$$

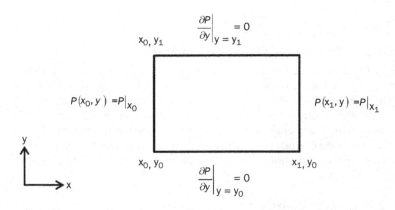

Figure 8-3   Simple boundary conditions for upscaling permeability in 2D.

where Q is the total flow through a cross-section parallel to the y axis. k_{yy}^* can be solved by solving the same flow problem after rotating the boundary conditions by 90°. This method is useful for determining a diagonal upscaled permeability tensor. The main criticism of this technique is the assumption that the principal components are parallel to the block sides. This assumption is acceptable if the fine-scale permeability values are randomly distributed (Warren and Price, 1961).

It would be better to use the boundary conditions that are as close as possible to those existing around the block when the flow equation is solved originally at the fine scale over the entire domain. However, knowledge of the precise boundary conditions is impossible without solving the fine-scale flow equation, which is what upscaling tries to avoid in the first place. There could be one potential justification for performing the fine-scale single-phase flow simulation over the entire domain: when effective permeability is used as input parameters for multi-phase flow modeling. With the advances of computing technology, the solution of single-phase flow is quite efficient. Improving the accuracy of effective absolute permeability would significantly improve the quality of the coarse-scale multi-phase flow simulation.

White and Horne (1987) proposed a way to determine the full permeability tensors from flow solution at the fine scale over the entire domain. Instead of using Eq. (8-8) to obtain the diagonal elements of the permeability tensor, all elements must satisfy the following:

$$
\begin{bmatrix} \dfrac{Q_x}{(y_1 - y_0)} \\[2ex] \dfrac{Q_y}{(x_1 - x_0)} \end{bmatrix} = -\dfrac{1}{\mu} \begin{bmatrix} k_{xx}^* & k_{xy}^* \\[1ex] k_{yx}^* & k_{yy}^* \end{bmatrix} \begin{bmatrix} \dfrac{\left(P|_{x_1} - P|_{x_0} \right)}{(x_1 - x_0)} \\[2ex] \dfrac{\left(P|_{y_1} - P|_{y_0} \right)}{(y_1 - y_0)} \end{bmatrix}
\tag{8-9}
$$

The difficulty with the system of equations in Eq. (8-9) is that it is underdetermined, involving four unknowns with two equations. One solution is to repeat the fine-scale solution numerous (n) times with different boundary conditions each time. This would yield a total of $2n$ linear equations, constituting an over-determined inverse linear problem. The major drawback is that it is computationally intensive.

Example 8-3

Calculate the diagonal upscaled permeability tensor for the following 2×2 fine-scale, isotropic, permeability field:

$$\begin{bmatrix} k_{1,1} & k_{2,1} \\ k_{2,1} & k_{2,2} \end{bmatrix} = \begin{bmatrix} 1 & 5 \\ 10 & 2 \end{bmatrix}$$

Solution

To calculate an upscaled diagonal permeability tensor $\begin{bmatrix} k^*_{xx} & 0 \\ 0 & k^*_{yy} \end{bmatrix}$, we can first utilize the setup in Fig. 8-3 to compute k^*_{xx} via using Eq. (8-8) and then rotate the boundary conditions to compute k^*_{yy}. The local pressure, P, is the solution to a steady-state, single-phase, flow simulation. The constant pressure boundary can be implemented following the same idea presented in Example 6-6. Steady-state condition is achieved if $c_t = 0$. The following system of equation is obtained:

$$\begin{bmatrix} \left(T_{\frac{1}{2},1} + T_{\frac{3}{2},1} + T_{1,\frac{3}{2}}\right) & -T_{\frac{3}{2},1} & -T_{1,\frac{3}{2}} & 0 \\ -T_{\frac{3}{2},1} & \left(T_{\frac{3}{2},1} + T_{2,\frac{3}{2}} + T_{\frac{5}{2},1}\right) & 0 & -T_{2,\frac{3}{2}} \\ -T_{1,\frac{3}{2}} & 0 & \left(T_{\frac{1}{2},2} + T_{\frac{3}{2},2} + T_{1,\frac{3}{2}}\right) & -T_{\frac{3}{2},2} \\ 0 & -T_{2,\frac{3}{2}} & -T_{\frac{3}{2},2} & \left(T_{\frac{3}{2},2} + T_{2,\frac{3}{2}} + T_{\frac{5}{2},2}\right) \end{bmatrix}$$

$$\begin{bmatrix} P_{11} \\ P_{12} \\ P_{21} \\ P_{22} \end{bmatrix} = \begin{bmatrix} T_{\frac{1}{2},1}P_{boundary\_1} \\ T_{\frac{5}{2},1}P_{boundary\_2} \\ T_{\frac{1}{2},2}P_{boundary\_1} \\ T_{\frac{5}{2},2}P_{boundary\_2} \end{bmatrix}$$

Assuming $\Delta x = \Delta y = $ constant, constant Δz and μ, the system of equations can be simplified as follows:

$$
\begin{bmatrix}
\left(k_{\frac{1}{2},1} + k_{\frac{3}{2},1} + k_{1,\frac{3}{2}}\right) & -k_{\frac{3}{2},1} & -k_{1,\frac{3}{2}} & 0 \\
-k_{\frac{3}{2},1} & \left(k_{\frac{3}{2},1} + k_{2,\frac{3}{2}} + k_{\frac{5}{2},1}\right) & 0 & -k_{2,\frac{3}{2}} \\
-k_{1,\frac{3}{2}} & 0 & \left(k_{\frac{1}{2},2} + k_{\frac{3}{2},2} + k_{1,\frac{3}{2}}\right) & -k_{\frac{3}{2},2} \\
0 & -k_{2,\frac{3}{2}} & -k_{\frac{3}{2},2} & \left(k_{\frac{3}{2},2} + k_{2,\frac{3}{2}} + k_{\frac{5}{2},2}\right)
\end{bmatrix}
$$

$$
\begin{bmatrix} P_{11} \\ P_{12} \\ P_{21} \\ P_{22} \end{bmatrix} =
\begin{bmatrix}
k_{\frac{1}{2},1} P_{boundary\_1} \\
k_{\frac{5}{2},1} P_{boundary\_2} \\
k_{\frac{1}{2},2} P_{boundary\_1} \\
k_{\frac{5}{2},2} P_{boundary\_2}
\end{bmatrix}.
$$

We should recall from Example 6-6 that $k_{1/2} = 2k_1$, $k_{7/2} = 2k_3$, $P_{boundary\_1} = P|_{x_0}$, and $P_{boundary\_2} = P|_{x_1}$. It should be emphasized that these boundary pressures are arbitrary; given that it is a steady-state problem, the resultant upscaled permeability tensor does not depend on the choice of the imposed pressure gradient. The solution is provided in the MATLAB® code "upscaling_1.m," which can be found on this book's website, www.mhprofessional .com/PRMS. The code utilizes a function called "k_upscale" to perform the above upscaling calculations.

Example 8-4

Upscale a 40 × 40 fine-scale isotropic, permeability field. The permeability values are stored in "k.dat." Compare the different levels of upscaling.

Solution

The solution is provided in the MATLAB code "upscaling_2.m," which can be found on this book's website. The default level of upscaling is 2 (i.e., the upscaled permeability field is 20 × 20). This can be adjusted by modifying the variables "nx" and "ny"; for example, changing both values to 4 would yield an upscaled field of 10 × 10. Once again, the code utilizes the same function "k_upscale" to perform the upscaling calculations. The results are shown in Figs. 8-4 and 8-5.

Solving the fine-scale flow equations over the entire domain is certainly computationally expensive; therefore, a more efficient alternative was proposed by Gomez-Hernandez (1991), where the fine-scale flow equations are solved over an area that contains the block and a small border region (e.g., half the block size in each direction). The idea is to impose more realistic boundary conditions on the sides of the block while reducing the size of the computational domain. Formulations in 3D can be found in Holden and Lia's research work (1992).

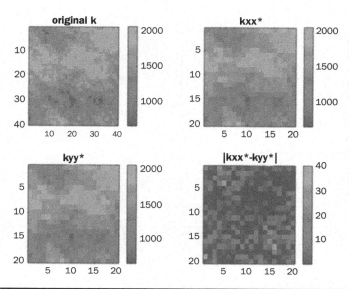

FIGURE 8-4 Comparison of original permeability (40 × 40) and upscaled permeability (20 × 20). The unit of permeability is mD. (The color version of this figure is available at www.mhprofessional.com/PRMS.)

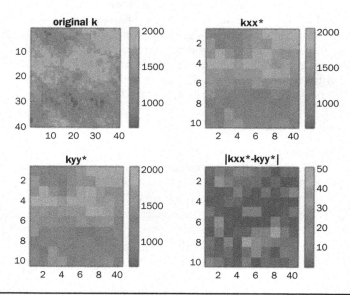

FIGURE 8-5 Comparison of original permeability (40 × 40) and upscaled permeability (10 × 10). The unit of permeability is mD. (The color version of this figure is available at www.mhprofessional.com/PRMS.)

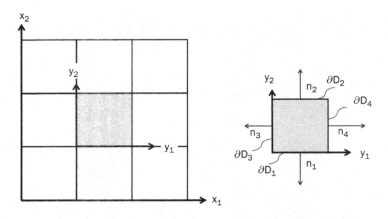

FIGURE 8-6 Representation of a periodic medium in 2D and the corresponding unit cell.

However, these formulations thus far do not guarantee a positive-definite permeability tensor at the coarse scale, which is essential for ensuring that fluid flows do not occur in a direction of decreasing pressure (or potential). Durlofsky (1991) proposed a formulation to compute positive-definite block permeability tensors via the use of periodic boundary conditions. The key assumption is that the heterogeneities of fine-scale permeability can be decomposed into two scales: a large-scale variability or trend plus a small-scale variability that is periodic. This concept is illustrated in Fig. 8-6, where x represents the large scale, whereas y represents the local scale.

The pressure at the x scale is represented as:

$$P = P_0 + \vec{G} \cdot (\vec{x} - \vec{x}_0) \tag{8-10}$$

where $\vec{x}_0$ represents the location vector to the center of the domain, and P_0 is the corresponding pressure. $\vec{G} = G_1\mathbf{i} + G_2\mathbf{j}$ is the local pressure gradient (**i** and **j** are the unit coordinate directions). To solve for $\vec{\vec{k}}^*$, Eq. (8-7) is solved over the y scale. The boundary conditions are based on the periodicity of the system and an imposed pressure gradient. As a result, both the pressure and local velocity fields are periodic over the unit cell. To completely define all components of $\vec{\vec{k}}^*$, two sets of problems should be solved. Assume the dimensions are normalized to a length of one (i.e., we are dealing with a unit cell). First, set $G_2 = 0$ and specify these boundary conditions:

$$P(y_1, y_2 = 0) = P(y_1, y_2 = 1)$$
$$\vec{u}(y_1, y_2 = 0) \cdot \vec{n}_1 = -\vec{u}(y_1, y_2 = 1) \cdot \vec{n}_2$$
$$G_1 = P(y_1 = 1, y_2) - P(y_1 = 0, y_2)$$
$$\vec{u}(y_1 = 0, y_2) \cdot \vec{n}_3 = -\vec{u}(y_1 = 1, y_2) \cdot \vec{n}_4 \tag{8-11}$$

Along with the specification of pressure at any single point inside the domain, a unique solution of Eq. (8-7) can be obtained for this set of boundary conditions. This solution is used to compute an average flux across ∂D_3 and ∂D_4:

$$\langle u \rangle_{\partial D_3} = -\int_{\partial D_3} \vec{u} \cdot \vec{n}_3 dy_2 = -\left(k_{11}^* G_1 + k_{12}^* G_2\right)$$

$$\langle u \rangle_{\partial D_4} = -\int_{\partial D_4} \vec{u} \cdot \vec{n}_4 dy_2 = -\left(k_{21}^* G_1 + k_{22}^* G_2\right) \qquad (8\text{-}12)$$

Since $G_2 = 0$, k_{11}^* and k_{21}^* can be readily determined. Next, set $G_1 = 0$ and specify these conditions:

$$P(y_1 = 0, y_2) = P(y_1 = 1, y_2)$$
$$\vec{u}(y_1 = 0, y_2) \cdot \vec{n}_3 = -\vec{u}(y_1 = 1, y_2) \cdot \vec{n}_4$$
$$G_2 = P(y_1, y_2 = 1) - P(y_1, y_2 = 0)$$
$$\vec{u}(y_1, y_2 = 0) \cdot \vec{n}_1 = -\vec{u}(y_1, y_2 = 1) \cdot \vec{n}_2 \qquad (8\text{-}13)$$

Similarly, since $G_1 = 0$, k_{22}^* and k_{12}^* can be readily determined. The symmetric nature of the unit cell would ensure $k_{12}^* = k_{21}^*$.

Durlofsky (1991) demonstrates that the solution of these sets of equations would always yield a symmetric and positive-definite tensor. This approach provides exact analytical results for truly periodic media, in which two distinct length scales of heterogeneity exist. Pickup et al. (1994) showed that this approach is quite robust even if periodicity does not apply. In particular, this method is still valid if the block size is much larger than the correlation scale of heterogeneity [i.e., the block size is greater than the representative elementary volume (REV)].

In the case of a highly heterogeneous domain, solving the flow equation over the entire domain is the only way to impose local boundary conditions close to those existing around the block. To avoid solving the fine-scale flow equation over the entire domain, Chen et al. (2003) proposed a more efficient coupled global-local upscaling scheme. The idea is to (1) compute upscaled permeabilities using local boundary conditions derived from the coarse-scale global flow solution and (2) compute the global flow solution using the upscaled permeabilities from the previous step. The two steps are repeated until the global solution and the upscaled permeabilities are consistent. The steps are summarized as follow:

1. Obtain an initial estimate of $\vec{\vec{k}}^*$ using any standard upscaling procedure.

2. Compute coarse-scale global flow solution.

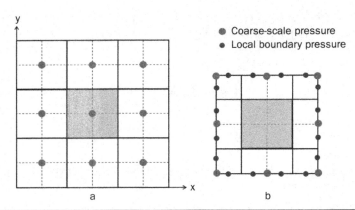

Figure 8-7 Coupling of the two domains: **a)** Global flow domain. **b)** Local flow domain. (The color version of this figure is available at www .mhprofessional.com/PRMS.)

3. Extract the local boundary conditions via interpolation (Fig. 8-7) and compute the corresponding local flow solution

4. Upscale $\vec{\vec{k}}^*$ from the local flow solution.

5. Steps 2 to 4 are repeated until the value of $\vec{\vec{k}}^*$ has been stabilized.

As discussed earlier, to fully resolve all four unknown components of $\vec{\vec{k}}^*$ using Eq. (8-6), at least two sets of flow problems with different boundary conditions must be solved at the local scale. Since the local boundary conditions are determined from the global flow, two global flow problems must also be solved. These global flow problems could be specified in several different ways. One option is to specify some generic global flows that are nominally in the x and y directions, respectively (as in Fig. 8-3). The method was later extended to more general global flows that are driven by wells (Chen and Durlofsky, 2006).

To assign the local boundary pressures from the global flow solution, either linear or harmonic interpolations were suggested. For example, considering two coarse-scale pressures P_1^c and P_2^c at x_1^c and x_2^c, respectively, along the top edge, the local boundary pressure P^l at x^l can be obtained via linear interpolation:

$$P^l = P_1^c + \frac{\left(P_2^c - P_1^c\right)\left(x^l - x_1^c\right)}{x_2^c - x_1^c} \tag{8-14}$$

It should be noted that global flow-based upscaling might cause negative elements, and positive definiteness is not guaranteed. It was suggested that the negative values can be replaced individually with a local upscaling scheme with periodicity.

8.2.3 Two-Phase Flow (Relative Permeability)

The non-linearity of flow functions also leads to some interesting scale-up characteristics of relative permeability. Scale dependency of these functions is well documented. It is widely observed that the curvature of relative permeability curves decreases, capillary pressure decreases, and residual saturation increases with scale (Ringrose et al., 1993; Kumar and Jerauld, 1996; Crotti and Cobeñas, 2001). The cross-over saturation and endpoints of the oil relative permeability are also impacted by averaging.

Prior to upscaling multi-phase flow functions, absolute permeability should be upscaled according to the methods described in the previous section. Methods for two-phase upscaling can be classified into two main categories: steady-state and dynamic. For example, Pickup and Stephen (2000) provided a summary of steady-state approaches for upscaling multi-phase flow functions. The general idea is to perform a two-phase flow simulation at a steady state. Numerous cases, such as the following, are possible depending on the specific flow regimes:

- Capillary equilibrium—The flow rate is low, and capillary equilibrium is valid approximately (Pickup and Sorbie, 1996):

  1. Select a capillary pressure (P_c) value.

  2. Using the P_c–S functions for each region, compute the distribution of water saturation. Use the fine-scale relative permeability functions to determine the relative permeabilities and phase permeabilities ($k \times k_r$) for each fine grid cell.

  3. Perform single-phase steady-state simulations separately on each phase using the corresponding phase permeabilities. Next, compute the effective phase permeabilities and the effective relative permeabilities (effective phase permeabilities divided by the effective absolute permeabilities).

  4. Calculate the average water saturation for the coarse block based on porosity weighting.

  5. Steps 1 to 4 would yield one point in the coarse-scale relative permeability tables. Repeat these steps with another capillary pressure value.

- Viscous-dominated steady state—The fraction flow is constant with time (e.g., a constant fractional flow of water is injected into a medium with isotropic relative permeabilities); a method proposed by Kumar and Jerauld (1996) can be adopted:

  1. Select a fractional flow (f_w) value.

2. Using the f_w–S functions for each region, compute the distribution of water saturation.

$$f_w = \frac{\dfrac{k_{rw}}{\mu_w}}{\dfrac{k_{rw}}{\mu_w} + \dfrac{k_{ro}}{\mu_o}} \tag{8-15}$$

Use the fine-scale relative permeability functions to determine the relative permeabilities and total mobility for each fine grid cell.

$$\lambda_t = k\left(\frac{k_{rw}}{\mu_w} + \frac{k_{ro}}{\mu_o}\right) \tag{8-16}$$

3. Perform the single-phase steady-state simulations using the total mobilities. Next, compute the effective total mobilities. Finally, compute effective relative permeabilities using the following equation:

$$\bar{f}_w = \frac{\dfrac{\bar{k}_{rw}}{\bar{\mu}_w}}{\dfrac{\bar{\lambda}_t}{\bar{k}}}; \left(1 - \bar{f}_w\right) = \frac{\dfrac{\bar{k}_{ro}}{\bar{\mu}_o}}{\dfrac{\bar{\lambda}_t}{\bar{k}}} \tag{8-17}$$

4. Calculate the average water saturation for the coarse block based on porosity weighting.

5. Steps 1 to 4 would yield one point in the coarse-scale relative permeability tables. Repeat these steps with another fractional flow value.

Kumar and Jerauld (1996) demonstrated with a field example that the effective upscaled relative permeability curves for the viscous and capillary limits can differ substantially. The message was that it is important to select a technique based on the knowledge of the relevant flow regimes.

For anisotropic relative permeabilities, it is suggested to assume constant fractional flow along a particular direction and repeat the calculations over all principal axes (Pickup et al., 2000). For other intermediate (arbitrary) viscous/capillary ratios, there would be no easy way to compute the fine-scale saturation and relative permeabilities (Step 2). The following procedure was proposed by Saad et al. (1995):

1. Normalize individual fine-scale relative permeability functions in terms of $S_n = (S_w - S_{wi})/(1 - S_{or} - S_{wi})$.

2. At a particular value of S_n, determine the phase permeabilities for each direction for each fine grid cell.

3. Inject a certain fractional flow of water and the simulation is continued until a steady-state is reached. Calculate effective phase permeabilities and relative permeabilities for the coarse grid block in each direction.

4. Compute the average water saturation: $S_w = \bar{S}_w + S_n(1 - \bar{S}_{or} - \bar{S}_w)$. Average endpoint saturations can be estimated via a simple volume average of the fine-scale values.

5. Repeat for all S_n levels.

An important drawback of the steady-state assumption is that, in most cases, reservoirs are not in a steady state. Besides, the steady-state methods do not account for numerical dispersion (Kyte & Berry, 1975). A review of various techniques can be found in Artus and Noetinger's research work (2004). The idea is to perform fine-scale simulation over the domain of a coarse block. At certain time intervals, fluxes, pressures, and saturations are computed at every fine grid cell. The total flow of each phase at the outlet is summed up, while the averages of these quantities are calculated over the coarse grid block. There are different variations in how the averaged quantities are computed. For example, the average pressure of each phase can be computed from values in the middle of the coarse block (e.g., a column in 2D or a plane in 3D). Fine-scale parameters, including cell height, pore volume, permeability, and relative permeability, can be used as weighting factors in this averaging. Effective P_c can be computed from the difference in average pressures between phases (Kumar and Jerauld, 1996). In the end, two main approaches have been adopted to compute the pseudo functions associated with the mean saturation. One option is to define the average pressure gradients over neighboring coarse blocks (e.g., Kyte and Berry, 1975) and apply Darcy's law to compute the pseudo relative permeabilities. The second option is to compute the pseudo-fractional flows on the coarse blocks and the corresponding pseudo relative permeabilities (e.g., Stone, 1991):

1. Compute an average fractional flow:

$$\bar{f}_w = \frac{\sum q f_w}{\sum q}; \quad \bar{\lambda}_t = \frac{\sum T \lambda_t}{\sum T} \tag{8-18}$$

where q, T, and l_t are evaluated at the outlet edge.

2. Estimate effective relative permeabilities using Eq. (8-17).

As pointed out by Christie (1996), a few issues with two-phase upscaling are:

- Rate dependence of effective multi-phase flow functions— These pseudo functions are not general, in the sense that different pseudos are required for different velocity fields. It is recommended to pre-compute a local velocity field using known well data, which is used to approximate more realistic rates for the upscaling calculations.

- Strategies for grouping a large number of effective functions into a manageable number—It is important to recognize that using separate effective functions over the entire domain is impractical. For example, in order to model two-phase, a coarse-scale model consisting of 100,000 cells would require six sets of relative permeability functions (2 phase × 3 directions) at each cell. This represents a total of 600,000 functions. Therefore, some strategies must be implemented to lump these functions into a more manageable subset of functions. Christie (1996) suggested grouping them based on several physical characteristics, such as the corresponding Buckley-Leverett shock height, the slope of the fractional flow curve, and the minimum of the total mobility curve. Techniques of data analytics (e.g., cluster analysis) can also be used to group these functions (Christie and Clifford, 1998).

- Computational costs for flow simulations—This is particularly challenging in dynamic simulations. Techniques that integrate analytical flow calculations have been developed. For example, Li et al. (1996) presented a procedure to derive pseudo-shock-fronts from a superposition of the shock velocities for the residual field and the fine-grid field. In their approach, a look-up table of values of residual shock velocities is determined from numerical simulations of 2D and 3D uncorrelated permeability fields with different heterogeneity structures. The fine-scale shock velocities are computed analytically. Finally, pseudo relative permeabilities are computed from the pseudo-shock-fronts.

- Similar to the single-phase case, negative values may result. Some post-processing is required.

Many researchers have proposed multi-stage workflows to obtain geologically based coarse-scale models (Alpak, 2015). For instance, Pickup et al. (2005) discussed upscaling models with millimeter-sized grid cells to full-field models with grid cells of several tens of meters in length be carried out in stages, which are defined by geologically based length scales. This type of procedure is also referred to as the Geopseudo method developed by Corbett et al. (1992) and Corbett and Jensen (1993). Effective flow properties (pseudos) are calculated progressively at larger models based on field-specific geological length scales; for instance, Ringrose et al. (1993) described these four main scales for laminated and cross-bedded sandstone formations:

1. Microscale—This is the sub-lamina scale, over which properties are locally homogeneous and isotropic.

2. Lamina scale—Lamina thickness may range between 5 mm and 500 mm, and pseudo functions should capture the lamination characteristics at this scale. Results at this scale may correspond to core-plug measurements in some cases.

3. Bedform scale—Pseudos at the lamina scale are placed in an appropriate geological framework to derive pseudos at the bedform scale.

4. Formation scale—Pseudos at the bedform scale are placed in an appropriate geological framework to derive pseudos at the formation scale.

8.2.4　Compositional Flow Simulation

Though details of compositional flow simulation are outside of the scope of this book, some general concepts of compositional upscaling are included here for completeness. The idea is that in addition to permeability and multi-phase flow functions upscaling, molar flux (velocity of a certain component i in phase j) should also be upscaled. Barker and Fayers (1994) introduced the formulation of α-factors, which are transport modifiers in the flow terms for a coarse block, to capture the effects of sub-grid heterogeneities. Fine-scale simulations are performed over the coarse block and the corresponding α-factors are calibrated (Christie and Clifford, 1998):

$$\alpha_i^j = \frac{\sum \rho_j x_{ij} u_j}{\bar{\rho} \bar{x}_{ij} \bar{u}_j} \tag{8-19}$$

where x_{ij} is the mole fraction of component i in phase j. The overbar signifies an average property. The average velocity $\bar{u}_j$ is calculated over the outlet face, while $\bar{\rho}$ and $\bar{x}_{ij}$ are averages over the entire coarse block. The α-factors are substituted into the following expression for calculating the flux at the coarse scale:

$$F_i = \sum \alpha_i^j \rho_j x_{ij} u_j \tag{8-20}$$

It should be clear that $\alpha_i^j = 1$ if there is no sub-grid heterogeneity (i.e., no upscaling is needed). Typically, a table of values is needed for $\alpha_i^j \ (i = 1, \ldots, N_c; j = 1, \ldots, N_p)$.

8.3 Scale-Up

We have, thus far, focused on the topic of upscaling. There is an important distinction between upscaling and scale-up. Recall that upscaling aims to replace a fine-scale detailed description of reservoir properties with a coarser-scale description that has equivalent properties. Scale-up, on the other hand, is a process of relating phenomena observable at one scale to another scale. Upscaling forces a correspondence between fine- and coarse-scale flow responses (e.g., pressure drop or flow rate); however, no such correspondence is assumed in scale-up. Instead, scale-up attempts to translate the flow response at a given volume support to a larger (different) volume support.

These two concepts are certainly related to each other, but not identical, as illustrated in Fig. 8-8. If one is to construct a suite of coarse-scale reservoir models; there are two possible options: (1) build a suite of fine-scale realizations and perform upscaling and (2) scale up the fine-scale conditioning data to the coarse-scale volume support and infer the coarse-scale multivariate statistics, which are then used to generate the coarse-scale realizations; as will be shown shortly, the previously discussed upscaling techniques are useful for the scale-up step. Therefore, one may conclude that upscaling entails "smoothing a high-resolution model to a lower-resolution one," while scale-up involves "computing the multivariate statistics for a different volume support."

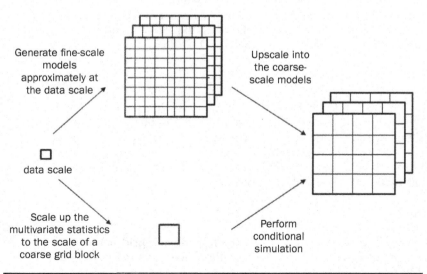

Figure 8-8   Relationship between scale-up and upscaling in the context of reservoir modeling.

8.3.1 General Concepts

Scale-up of reservoir properties is often considered in various contexts. First, measurement scales vary among different data types and sources; for instance, seismic data resolution is generally much larger than the log or core resolution. Second, volume support for measured data is typically much smaller than the geo-modeling scale. Finally, the flow simulation scale is generally larger than the geo-modeling scale. These changes in volume support and the resultant averaging or smoothing of heterogeneity below the flow modeling scale lead to a loss of information; this increase in uncertainty is referred to as the sub-scale variability.

The following thought-experiment serves to illustrate this crucial point. Imagine taking a small piece of rock and measuring its average porosity. This process of porosity measurement is repeated with larger and larger pieces of rock. A schematic of average porosity as a function of the size of rock or scale is shown in Fig. 8-9. At a very small scale, there are significant fluctuations due to microscopic effects. The averaged quantity starts to stabilize at a length scale referred to as *representative elementary volume* or REV. At the infinitesimal scale, there is either a grain ($\phi = 0$) or a pore ($\phi = 1$); hence, the extreme fluctuations can be observed. At the reservoir scale, however, heterogeneity occurs at many scales. The REV in some cases (where reservoir features exhibit long-range correlation) could be larger than the inter-well distance.

When reservoir attributes are modeled as random variables, the associated uncertainty is understood as the ensemble variance, which should not be confused with heterogeneity that is related to spatial

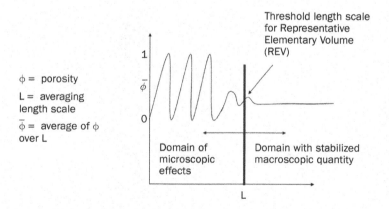

FIGURE 8-9 Variability of spatial average as a function of scale and its stabilization at the representative elementary volume (adapted from Bear, 1972). (The color version of this figure is available at www.mhprofessional .com/PRMS.)

variance. Moreover, the average of a set of outcomes of a random variable is also a random variable. The corresponding variance of mean is indicative of the variability of the spatial average (or sub-scale variability) for that particular averaging length scale L. A plot of the variance of the mean as a function of L would, therefore, be useful for estimating the REV scale. This variance of mean is illustrated in Fig. 8-10, and it is related to the REV representation in Fig. 8-9. Significant fluctuations are observed at small scales, and the corresponding variance of mean is large. This variance of mean decreases slowly with L when L is small, and it decreases drastically after the REV scale is reached. In fact, the plot reaches a constant negative unit slope on a log-log plot, provided that the attribute is stationary. This negative unit slope originates from the law of large numbers, which states that the sample variance is equal to its fine-scale variance divided by the number of samples. At the REV threshold, the samples act independent of each other, and, consequently, the mean of sample values and the variance of that mean follow the law of large numbers.

This variance of mean is directly related to the sub-scale variability—when the averaging scale is large, the sub-scale variability within the averaging scale might be high, but the process of averaging dampens the influence of that variability, causing the variability of the mean (across different samples) to be small; therefore, both the variance of mean and the sub-scale variability are small at large scales. For $V \geq$ REV, the sub-scale variability is considered negligible (Leung and Srinivasan, 2011). On the other hand, averages computed over tiny blocks would still show the influence of sub-scale variability. The impact of sub-scale variability on the effective/average property decreases as the averaging scale increases. As the correlation length

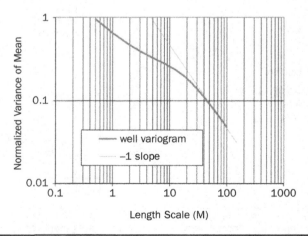

FigURE 8-10 Variance of mean as a function of scale. (The color version of this figure is available at www.mhprofessional.com/PRMS.)

of heterogeneity increases, REV becomes larger. For a given length scale, the variance of the mean increases with correlation length; in other words, it takes a much larger averaging volume for the mean or the variance of mean to stabilize.

In order to compute the variance of mean for a given autocorrelation model, one can refer to the formulation given by Lake and Srinivasan (2004). If Z is a Gaussian random variable with a variance σ^2, with a multivariate distribution described by a second-order stationary spatial correlation function ρ_{corr}, $Var(\overline{Z})$ can be computed by integrating ρ_{corr} over all possible lag distance η over the averaging length L in 1D or a volume V in 3D:

$$Var\left(\overline{Z}\right) = \frac{2\sigma^2}{L^2}\left(\int_{\xi=0}^{\xi=L}\int_{\eta=0}^{\eta=\xi} \rho_{corr}(\eta)\,d\eta\,d\xi\right) \qquad (8\text{-}21)$$

$$Var\left(\overline{Z}\right) = \frac{2\sigma^2}{V^2}\left(\int_{V}\int_{\eta} \rho_{corr}(\eta)\,d\eta\,d\xi\right) \qquad (8\text{-}22)$$

It is assumed that linear average applies. As L or V increases, the variance of mean $(\overline{Z})$ decreases, while as L approaches 0, the variance of mean becomes the population mean at the point-scale. Another interpretation of this variance of mean is in terms of Krige's relation (Journel and Huijbregts, 1978):

$$Var\left(\overline{Z}\right) = \sigma^2_{L/D} = \sigma^2_{O/D} - \sigma^2_{O/L} \qquad (8\text{-}23)$$

In the above equation, $\sigma^2_{O/D} = \sigma^2$ is the variance of the attribute computed over points (O) embedded in a large domain D; $\sigma^2_{O/L}$ is the variance of points within a length or volume L smaller than D, and $\sigma^2_{L/D}$ is the same as the variance at the support of $\overline{Z}$. The variance of the mean referred to in the earlier discussion is exactly $\sigma^2_{L/D}$ and Krige's relation prescribes its limits. The general procedure for calculating the variance of the mean and the REV scale is to infer and model the variogram using data at the point support and then compute the variance of mean numerically by summing the model $\rho_{corr}(\eta)$ over all possible lag distances within a given length scale L.

Another alternative technique for calculating $Var(\overline{Z})$, without any explicit assumption of the multivariate distribution, is to compute the spatial mean over a particular averaging window for a large (or multiple) spatial realization(s) of the chosen attribute. Different averaging schemes corresponding to different values of ω [Eq. (8-1)] are possible. Repeating this calculation by systematically changing the averaging window would yield a plot similar to that shown Fig. 8-10. Note that as ω decreases, the variance of mean increases.

Example 8-5

Data from two different wells ("porosity1.dat" and "porosity2.dat") are provided. Compute the variance mean and estimate the REV scale.

Solution

The solution is provided in the MATLAB code "Var_Mean.m," which can be found on this book's website, www.mhprofessional .com/PRMS. First, we use the code to analyze "porosity1.dat," and the results are shown in Fig. 8-11. Variogram analysis of this dataset reveals a range (or correlation length) of 100 blocks. The REV scale is approximately 250 blocks, where the slope of the variance of mean reaches a constant negative unit slope on a log-log plot. This example illustrates how the REV scale is generally much larger than the correlation length. If the distance between each data point is 0.1 m, the REV would be approximately 250 × 0.1 m = 25 m.

Next, we repeat the analysis for "porosity2.dat." For this dataset, we change the variable "lag" from 5 to 50, in order to increase the maximum number of averaging blocks to 15,000. The results are shown in Fig. 8-12. The decline of variance of mean with averaging scale is much more gradual for this dataset. Despite a decreasing variance of mean with averaging length, the determination of an appropriate REV scale is challenging. If the slope keeps decreasing, the assumption of stationarity may not be applicable. If that is the case, a possible solution is to remove the trend or non-stationary component from the attribute prior to the variance of mean analysis. More detail is provided in the next section.

To compute the volume-averaged (or block-to-block) semivariogram or gamma bar, the following linear averaging can be adopted:

$$\bar{\gamma}(V,V') = \frac{1}{|V||V'|}\sum_{v=1}^{V}\sum_{v'=1}^{V'}\gamma(v,v') \tag{8-24}$$

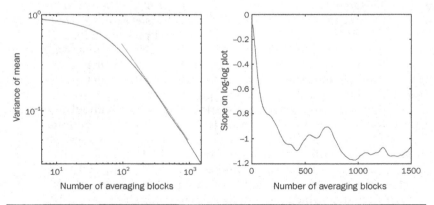

Figure 8-11   Variance of mean and REV scale analysis for Dataset 1 (the red line has a slope of −1). (The color version of this figure is available at www .mhprofessional.com/PRMS.)

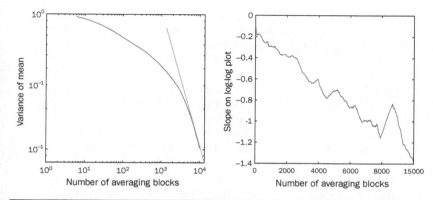

FIGURE 8-12   Variance of mean and REV scale analysis for Dataset 2 (the red line has a slope of −1). (The color version of this figure is available at www .mhprofessional.com/PRMS.)

Finally, the concept of REV is particularly important in scale-up analysis:

- Case 1: Grid size ≤ data resolution → no scale-up is needed
- Case 2: Grid size > data resolution
    - Grid size > REV → negligible sub-scale variability
    - Grid size < REV → sub-scale variability exists and should be accounted for. This case is most common in practical applications.

Unfortunately, most reservoir attributes, except for porosity, do not average linearly; however, many attributes, such as permeability or fracture porosity, are non-linearly related to porosity. Computing the corresponding variance of mean and REV directly from its auto-correlation may not be feasible. One option is to transform these attributes into a space where linear averaging is applicable. For example, if permeability is log-normally distributed, a logarithmic transform can be performed to convert the variable to one where linear averaging is permissible. If such transformation does not exist, the cloud transform technique can be adopted, as will be discussed in the next section.

8.3.2   Scale-Up of Linearly Averaged Attributes

Multivariate statistics of multi-scale heterogeneous petrophysical properties would often exhibit non-stationarity, characterized by ever-increasing variability with scale. As a result, its variance of mean would also increase with scale, and the determination of REV is impossible. The modeling of such variables is facilitated by decomposing its variability into the sum of a non-stationary trend and fine-scale stationary residual: $Z(x) - \bar{Z}_T(x) + Z_R(x)$, where $\bar{Z}_T(x)$ and $Z_R(x)$

refer to the trend and residual, respectively. The overbar denotes a quantity defined at the coarse scale. It is assumed that the trend component is already defined at the transport modeling scale.

Given that the volume support for the conditioning data is typically much less than that of the modeling scale, sub-scale variability in the conditioning value should be captured by sampling multiple sets of conditioning data via bootstrapping (Leung and Srinivasan, 2011). A plausible implementation is based on parametric bootstrapping of a likelihood function (whose variance is the variance of mean and the mean is the block average of the actual measured values). For a Gaussian random variable, a Gaussian likelihood function is proposed. Finally, conditional simulation is performed on all sets of the conditioning data (Leung and Srinivasan, 2011). The procedure is summarized as follows:

1. Estimate $Z_R(x)$ and $\bar{Z}_T(x)$ at conditioning locations.

2. Estimate γ_R. Compute $Var(\bar{Z}_R)$ and $\bar{\gamma}_R$.

3. Draw multiple realizations of conditioning data of $\bar{Z}_R(x)$ via bootstrapping. For example, a Gaussian likelihood function, whose mean is the block average of the actual measured values and the variance is the variance of mean calculated in Step 2, can be adopted.

4. Construct realizations of $\bar{Z}_R(x)$ via sequential simulation for each conditioning dataset from Step 3: $\bar{\gamma}_R$ is obtained from Step 2; coarse-scale histogram is formulated with variance = $Var(\bar{Z}_R)$ and mean = fine-scale global mean of Z_R.

5. Construct realizations of $\bar{Z}_T(x)$.

6. Reconstruct realizations of $\bar{Z}(x) = \bar{Z}_T(x) + \bar{Z}_R(x)$ by combining the results from Steps 4 and 5.

7. Repeat Steps 4 to 6 for other conditioning datasets obtained in Step 3.

If the random variable does not follow Gaussian statistics, other sequential techniques for simulating continuous variables can be applied to generate the sub-grid and coarse-scale models. Instead of formulating γ_R and $\bar{\gamma}_R$, alternative multivariate statistics description can be adopted. Therefore, this workflow is general, in the sense, that no explicit assumption of the multivariate distribution is required.

8.3.3 Scale-Up of Non-Linearly Averaged Attributes

As discussed in previous sections, averaging of flow and transport variables (e.g., permeability) typically can be highly non-linear. However, many of these variables are correlated to another linearly averaged variable, at least at the fine scale. As an example, let us assume that permeability is correlated to porosity at the fine scale:

$k = a \times \phi^b$. The objective is to calibrate the bivariate relationship between $\bar{k}$ and $\bar{\phi}$, which is used in a cloud transform to construct models of $\bar{k}$ that are correlated to the $\bar{\phi}$ models (Vishal and Leung, 2015, 2017a-b, 2018).

To calibrate the bivariate relationship between $\bar{k}$ and $\bar{\phi}$, the following procedure is followed:

1. Divide the histograms of $\bar{\phi}_R$ and $\bar{\phi}_T$ into many bins n_{bR} and n_{bT}, respectively. $\bar{\phi}_T$ and $\bar{\phi}_R$ refer to the trend and residual component at the coarse scale, respectively (e.g., see Steps 4 and 5 in the previous section). A total of $n_{bR} \times n_{bT}$ bin combinations are generated.

2. For a particular bin combination, perform an unconditional sequential simulation to construct n_s detailed fine-grid realizations of ϕ_R. If ϕ_R follows Gaussian statistics, realizations of $\phi_R \sim N(\bar{\phi}_R, \sigma_R)$ can be constructed using the fine-scale variogram of γ_R, and ϕ is constructed as $\phi = \phi_R + \bar{\phi}_T$. Once ϕ is computed, $k = a \times \phi^b$. These realizations are of the same physical size as of the transport modeling grid block, and they describe the spatial heterogeneity below the transport modeling scale.

3. For each n_s sub-grid realization of ϕ obtained in Step 2, construct an equivalent homogeneous model (e.g., by linearly averaging ϕ).

4. Simulate flow and transport by imposing certain initial and boundary conditions for all n_s heterogeneous models and n_s homogeneous models. Effective permeability $\bar{k}$ is estimated by minimizing the difference in effluent profile (Q) for each realization and that of an equivalent average medium. For example, the root mean square error, or *RMSE* (Nash and Sutcliffe, 1970), can be minimized using a non-linear regression scheme:

$$RMSE = \sqrt{\frac{1}{N-1} \sum_{i=1}^{N} \left[Q_{i,(k,\phi)i} - Q_{i,(\bar{k},\bar{\phi}_R,\bar{\phi}_T)} \right]^2} \qquad (8\text{-}25)$$

where $i = 1, \dots, N$ refers to the number of uniform time steps along with the Q profile.

5. Steps 2 to 4 are repeated for all bin combinations. Results are aggregated to construct $P\{\bar{k} \mid (\bar{\phi}_R, \bar{\phi}_T)\}$.

6. For each of the coarse-scale $\bar{\phi}$ models (Sec. 8.3.1), assign effective permeability at each location by sampling from $P\{\bar{k} \mid (\bar{\phi}_R, \bar{\phi}_T)\}$.

Remarks

- The last step in the workflow is essentially adopting a cloud transform to reproduce the bivariate relationship between the linearly averaged variable and the variable of interest. Typical issues may include binning artifacts and anomalous values near the tails (Pyrcz and Deutsch, 2014). Binning effects can be dampened by densifying the cloud model with more conditional bins. Similarly, more values near tails should be added if necessary.

- A set of correlated random numbers can be used in the sampling to preserve the spatial correlation structure of the permeability field.

- The workflow can be used to scale up other flow or transport properties. If numerical dispersion is an important concern (e.g., trying to isolate the effects of numerical dispersion when estimating physical dispersivities at the coarse scale), a particle-based technique is recommended for the flow or transport modeling (Vishal and Leung, 2015; 2017a-b; 2018).

8.4 Scale-Up of Flow and Transport Equations

Popular methods for scale-up of flow are presented here: application of dimensionless scaling groups and deriving flow/mass transfer equations at the macroscopic scale as averages of fine-scale quantities. A discussion on each approach and some key limitations with current implementations of that method is also highlighted.

8.4.1 Dimensionless Scaling Groups

Dimensionless groups are often used to relate processes at different scales. The underlying assumption is that physical processes or relationships describing physical processes are independent of length scales. Two processes may have different absolute scales, but their dimensionless response would be the same if the corresponding dimensionless groups remain the same (i.e., the relative importance of various parts of the full governing equations does not vary).

Flow is assumed to scale according to several dimensionless groups (Shook et al., 1992). Some dimensionless groups presented by Li and Lake (1995) are:

- Effective aspect ratio: $R_L = \dfrac{L}{H}\sqrt{\dfrac{\overline{\overline{k_z}}}{\overline{\overline{k_x}}}}$

- Dip angle group: $N_\alpha = \dfrac{L}{H}\tan(\alpha)$

- Endpoint mobility ratio: $M^o = \dfrac{k_{r1}^o \mu_2}{k_{r2}^o \mu_1}$

- Buoyancy number (gravity vs. viscous forces):

$$N_g = \frac{\overline{k_x}\lambda_{r2}^o \Delta \rho g \cos\alpha H}{u_t L}$$

- Capillary number (capillary vs. viscous forces):

$$N_{Pc} = \frac{\lambda_{r2}^o \sigma \cos\theta \sqrt{\overline{k_x}\phi}}{u_t L}$$

L and H = global window length and width, respectively; α = dip angle; and u_t = total Darcy's velocity. The first two relate to the scaling of reservoir geometry, while the last three relate to the scaling of physical mechanisms. In some cases, the dimensionless Leverett J function and phase saturations are also included.

Li and Lake (1995) generated "type curves" of flow response (e.g., recovery) to be functions of various flow scaling groups. The major limitation with dimensionless groups is that detailed information regarding the spatial correlation of reservoir parameters is not considered. All fine-scale geologic models are essentially lumped into a single dimensionless number that is used within the scaling groups. As a result, they can only offer some aggregated scaling behavior.

8.4.2 Stochastic Perturbation Methods

Another approach is to derive the relevant flow/mass transfer equations at the macroscopic scale as averages of fine-scale quantities. In stochastic formalism, uncertainty is represented by probability or other related quantities like statistical moments. Boundary conditions, initial conditions, and reservoir attributes (model parameters) are treated as random fields depicted by probability distributions. Dependent variables, like pressure (or hydraulic head) and flux, are also modeled as random fields. Finally, the resultant governing equations become stochastic differential equations (SDEs), whose solutions are probability distributions of pressure and flux. Generally speaking, we cannot solve an SDE exactly, as we can only estimate the first few moments of the corresponding probability distribution (e.g., its mean, variance, and covariances). These ensemble or moment statistics are used subsequently to describe average transport at the coarse scale and to define the corresponding effective transport coefficients. In the framework of volume averaging, instead of ensemble statistics, the concept of spatial averaging is employed. Applying the spatial averaging theorem (Cushman, 1982; Howes and Whitaker, 1985), the fine-scale governing equations are spatially averaged over a particular volume to produce the scaled-up equations.

Stochastic Perturbation or Ensemble Averaging

It is based on the statistical moment equations governing transport phenomena. The majority of this work has been done in the area of hydrology: Gelhar and Axness (1983), Dagan (1984, 1987, and 1989), Neuman et al. (1987), Cvetkovic et al. (1992), Rubin and Gomez-Hernandez (1990), Newman and Orr (1993).

Though earlier works tended to focus on single-phase steady-state flow, these assumptions have since been relaxed. For illustrative purposes, let us derive the general mathematical statements for isothermal unsaturated flow by following the notation and assumptions of Zhang (2001) and Vereecken et al. (2007). We begin with the general governing equations presented in Sec. 6.1. If we ignore the interaction of solids and assume complete immiscibility between water ($j = 1$) and air ($j = 2$), Eq. (6-1) becomes:

$$\int_V \frac{\partial\left(\phi\rho_j S_j\right)}{\partial t} dV = -\int_A \left(\rho_j \vec{u}_j \cdot \vec{n}\right) dA + \int_V R_i dV \qquad (8\text{-}26)$$

Upon application of the divergence theorem:

$$\frac{\partial\left(\phi\rho_j S_j\right)}{\partial t} = -\nabla \cdot \left(\rho_j \vec{u}_j\right) + R_j \qquad (8\text{-}27)$$

For the water phase, the flux is given by:

$$\vec{u}_1 = -\vec{\vec{\lambda}}_1 \cdot \left(\nabla P_1 + \rho_1 \vec{g}\right) = -\vec{\vec{\lambda}}_1 \cdot \nabla(P_1 + \rho_1 g z) \qquad (8\text{-}28)$$

If the pressure gradient in the gas phase is negligible:

$$\vec{u}_1 = -\vec{\vec{\lambda}}_1 \cdot \nabla(P_1 + \rho_1 g z) \qquad (8\text{-}29)$$

Therefore,

$$\frac{\partial\left(\phi\rho_j S_j\right)}{\partial t} - \nabla \cdot \left(\rho_1 \vec{\vec{\lambda}}_1 \cdot \nabla(P_1 + \rho_1 g z)\right) = R_1. \qquad (8\text{-}30)$$

Assuming negligible air density and incompressible water phase, we define the following:

$\theta = \phi S_j$ = Moisture content

$\vec{\vec{K}} = \rho_1 \vec{\vec{\lambda}}_1 g$ = Hydraulic conductivity

$\psi = \dfrac{\nabla P_1}{\rho_1 g}$ = Pressure head

$g = \dfrac{R_1}{\rho_1}$ = Sink/source function

The mass balance equation can be expressed as follows:

$$\frac{\partial \theta}{\partial t} - \nabla \cdot \left(\vec{\vec{K}}_1 \cdot \nabla(\psi + z) \right) = g \tag{8-31}$$

If we assume locally isotropic $\vec{\vec{K}}$ and drop the subscript for water phase, we can obtain the Richards equation that is commonly used in groundwater literature:

$$C(\mathbf{x},\psi)\frac{\partial \psi}{\partial t} - \nabla \cdot \left(K(\psi,\mathbf{x}) \cdot \nabla(\psi + z) \right) = g(\psi,\mathbf{x},t) \tag{8-32}$$

$C(\mathbf{x},\psi)$ = soil water capacity = θ/ψ. The functional dependencies of various variables are also noted. Certain models are indeed needed to describe the constitutive relationships between K vs. ψ. One of the simplest empirical models is that of Gardner−Russo (Gardner, 1958; Russo, 1988):

$$K(\mathbf{x}) = K_s(\mathbf{x})\exp[\alpha(\mathbf{x})\psi(\mathbf{x})] \tag{8-33}$$

α is a property of the medium depending on the pore size distribution. If we further assume that α and $ln(K_s)$ are Gaussian random functions, each parameter can be expressed as the sum of their mean (uppercase symbols) and a perturbation (lowercase symbols):

$$\ln K_s(\mathbf{x}) = F + f(\mathbf{x})$$
$$\alpha(\mathbf{x}) = A + a(\mathbf{x})$$
$$\psi(\mathbf{x}) = H(\mathbf{x}) + h(\mathbf{x}) \tag{8-34}$$

It is assumed that F and A are constant (i.e., stationarity applies). We can manipulate the expression for K in a few different ways:

$$\ln K(\mathbf{x}) = \ln K_s(\mathbf{x}) + \alpha(\mathbf{x})\psi(\mathbf{x})$$
$$\ln K(\mathbf{x}) = F + f(\mathbf{x}) + [A + a(\mathbf{x})][H(\mathbf{x}) + h(\mathbf{x})]$$
$$K(\mathbf{x}) = e^{F+AH(\mathbf{x})}e^{f(\mathbf{x})+a(\mathbf{x})H(\mathbf{x})+Ah(\mathbf{x})+a(\mathbf{x})h(\mathbf{x})} \tag{8-35}$$

Ignoring the higher-order term of $a(\mathbf{x})h(\mathbf{x})$ and setting $K_m(\mathbf{x}) = e^{F+AH(\mathbf{x})}$:

$$K(\mathbf{x}) = K_m(\mathbf{x})e^{f(\mathbf{x})+a(\mathbf{x})H(\mathbf{x})+Ah(\mathbf{x})}$$

We can introduce a new variable, $Y(\mathbf{x})$:

$$Y(\mathbf{x}) = \ln[K(\mathbf{x})] = Y(\mathbf{x}) + y(\mathbf{x})$$
$$Y = F + AH(\mathbf{x})$$
$$y = f(\mathbf{x}) + a(\mathbf{x})H(\mathbf{x}) + Ah(\mathbf{x}) \tag{8-36}$$

First, considering the steady state, we will derive the linear perturbation equation of Eq. (8-32):

$$\nabla \cdot \left(K(\mathbf{x}) \cdot \nabla(\psi + z) \right) = -g(\mathbf{x}, \psi, t) \tag{8-37}$$

$$\nabla \cdot K(\mathbf{x}) \cdot \nabla(\psi + z) + K(\mathbf{x}) \cdot \nabla^2(\psi + z) = -g(\mathbf{x}, \psi, t) \tag{8-38}$$

Divide both sides by $K(\mathbf{x})$:

$$\nabla \cdot Y(\mathbf{x}) \cdot \nabla(\psi + z) + \nabla^2(\psi + z) = -\frac{g(\mathbf{x}, \psi, t)}{K(\mathbf{x})} \tag{8-39}$$

For the moment, we will drop the notation of functional dependencies:

$$\begin{aligned}
\nabla \cdot (Y + y) &\cdot \nabla(H + h + z) + \nabla^2(H + h + z) \\
&= -\frac{g}{K_m} e^{-(f + aH + Ah)} \approx -\frac{g}{K_m}(1 - f - aH - Ah)
\end{aligned} \tag{8-40}$$

Collecting all the *first-order terms* gives:

$$\nabla \cdot Y \cdot \nabla(H + z) + \nabla^2(H + z) = -\frac{g}{K_m} \tag{8-41}$$

Collecting all the *remaining terms* gives:

$$\nabla^2 h + \nabla \cdot Y \cdot \nabla h + \nabla \cdot y \cdot \nabla(H + h + z) = -\frac{g}{K_m}(-f - aH - Ah)$$

$$\tag{8-42}$$

Ignoring the higher-order term of $\Delta \cdot y \cdot \Delta h$ and setting $J = \Delta(H + z)$, Eq. (8-42) becomes:

$$\nabla^2 h + \nabla \cdot Y \cdot \nabla h = -\nabla \cdot y \cdot J - \frac{g}{K_m}(-f - aH - Ah) \tag{8-43}$$

Substitute Eq. (8-36) and simplify:

$$\begin{aligned}
\nabla^2 h + \nabla \cdot (F + AH) \cdot \nabla h &= -\nabla \cdot (f + aH + Ah) \cdot J \\
&\quad + \frac{g}{K_m}(f + aH + Ah) \\
\nabla^2 h + A\nabla H \cdot \nabla h &= -(\nabla f + H\nabla a) \cdot J - a\nabla H \cdot J - AJ \cdot \nabla h \\
&\quad + \frac{g}{K_m}(f + aH + Ah) \\
\nabla^2 h + A(2J - \nabla z) \cdot \nabla h &= -(\nabla f + H\nabla a) \cdot J - a\nabla H \cdot J \\
&\quad + \frac{g}{K_m}(f + aH + Ah)
\end{aligned} \tag{8-44}$$

We have worked with the equation containing the "remaining terms" since the idea was to subtract the expected value of the equation from the original fine-scale equation. For completeness, the final form of the linear perturbation equation [same as Eq. (5-20) in Zhang's research (2001)] is:

$$
\nabla^2 h(\mathbf{x}) + A\big(2J(\mathbf{x}) - \nabla z\big)^T \nabla h(\mathbf{x}) - A\frac{g(\mathbf{x})}{K_m(\mathbf{x})}h(\mathbf{x})
$$
$$
= -J^T(\mathbf{x})\big(\nabla f(\mathbf{x}) + H(\mathbf{x})\nabla a(\mathbf{x})\big) + \frac{g(\mathbf{x})}{K_m(\mathbf{x})}f(\mathbf{x})
$$
$$
+ \left[\frac{g(\mathbf{x})}{K_m(\mathbf{x})}H(\mathbf{x}) - J^T(\mathbf{x})\nabla H(\mathbf{x})\right]a(\mathbf{x})
$$

$$(8\text{-}45)$$

To obtain the moments equations, we multiply the linear perturbation equation with a perturbation $h(\chi)$ at a location χ, and take the ensemble mean. Cov_{ij} represents the covariance between variable i and j.

$$
\nabla^2 Cov_{hh}(\mathbf{x},\chi) + A\big(2J(\mathbf{x}) - \nabla z\big)^T \nabla Cov_{hh}(\mathbf{x},\chi) - A\frac{g(\mathbf{x})}{K_m(\mathbf{x})}Cov_{hh}(\mathbf{x},\chi)
$$
$$
= -J^T(\mathbf{x})\big(\nabla Cov_{fh}(\mathbf{x},\chi) + H(\mathbf{x})\nabla Cov_{ah}(\mathbf{x},\chi)\big)
$$
$$
+ \frac{g(\mathbf{x})}{K_m(\mathbf{x})}Cov_{fh}(\mathbf{x},\chi)
$$
$$
+ \left[\frac{g(\mathbf{x})}{K_m(\mathbf{x})}H(\mathbf{x}) - J^T(\mathbf{x})\nabla H(\mathbf{x})\right]Cov_{ah}(\mathbf{x},\chi)
$$

$$(8\text{-}46)$$

Next, we multiply the linear perturbation equation with $f(\chi)$ and take the ensemble mean:

$$
\nabla^2 Cov_{fh}(\mathbf{x},\chi) + A\big(2J(\mathbf{x}) - \nabla z\big)^T \nabla Cov_{fh}(\mathbf{x},\chi) - A\frac{g(\mathbf{x})}{K_m(\mathbf{x})}Cov_{fh}(\mathbf{x},\chi)
$$
$$
= -J^T(\mathbf{x})\big(\nabla Cov_{ff}(\mathbf{x},\chi) + H(\mathbf{x})\nabla Cov_{af}(\mathbf{x},\chi)\big)
$$
$$
+ \frac{g(\mathbf{x})}{K_m(\mathbf{x})}Cov_{ff}(\mathbf{x},\chi)
$$
$$
+ \left[\frac{g(\mathbf{x})}{K_m(\mathbf{x})}H(\mathbf{x}) - J^T(\mathbf{x})\nabla H(\mathbf{x})\right]Cov_{af}(\mathbf{x},\chi)
$$

$$(8\text{-}47)$$

Finally, we multiply the linear perturbation equation with $a(\chi)$ and take the ensemble mean:

$$\nabla^2 Cov_{ah}(\mathbf{x},\chi) + A(2J(\mathbf{x}) - \nabla z)^T \nabla Cov_{ah}(\mathbf{x},\chi) - A\frac{g(\mathbf{x})}{K_m(\mathbf{x})}Cov_{ah}(\mathbf{x},\chi)$$
$$= -J^T(\mathbf{x})\big(\nabla Cov_{af}(\mathbf{x},\chi) + H(\mathbf{x})\nabla Cov_{aa}(\mathbf{x},\chi)\big)$$
$$+ \frac{g(\mathbf{x})}{K_m(\mathbf{x})}Cov_{af}(\mathbf{x},\chi)$$
$$+ \left[\frac{g(\mathbf{x})}{K_m(\mathbf{x})}H(\mathbf{x}) - J^T(\mathbf{x})\nabla H(\mathbf{x})\right]Cov_{aa}(\mathbf{x},\chi)$$

$$(8\text{-}48)$$

Cov_{hh} can be obtained from Eq. (8-46), provided that Cov_{ah} and Cov_{fh} can be solved from Eqs. (8-47) and (8-48). In order to attain a complete solution, the following are needed:

- Cov_{aa}, Cov_{af}, Cov_{ff};
- Boundary conditions;
- $H(x)$, which can be obtained by solving the steady-state solution using the expected values of all the parameters; however, such an approach neglects the effects of spatial variability of hydraulic parameters that may have on the mean flow variables.

These moments in the partial differential equations can be solved numerically via finite differences or finite elements. Finally, to obtain the ensemble average of flux, the following equation is considered:

$$\bar{q}_1 = -K(\psi,\mathbf{x}) \cdot \nabla(\psi + z) \qquad (8\text{-}49)$$

Substituting $K(\mathbf{x}) = e^{Y(\mathbf{x})+y(\mathbf{x})} = K_m(\mathbf{x})e^{y(\mathbf{x})}$

$$= K_m(\mathbf{x})\left[1 + y(\mathbf{x}) + \frac{1}{2}(y(\mathbf{x}))^2 + ...\right] \text{ would give:}$$

$$\bar{q}_1 = -K_m(\mathbf{x})\left[1 + y(\mathbf{x}) + \frac{1}{2}(y(\mathbf{x}))^2\right] \cdot \nabla(H(\mathbf{x}) + h(\mathbf{x}) + z) \quad (8\text{-}50)$$

Collecting the terms of different orders:

$$\bar{q}_1^{(0)} = -K_m(\mathbf{x}) \cdot \nabla(H(\mathbf{x}) + z)$$
$$\bar{q}_1^{(1)} = -K_m(\mathbf{x}) \cdot \left[\nabla h(\mathbf{x}) + y(\mathbf{x}) \cdot J(\mathbf{x})\right]$$
$$\bar{q}_1^{(2)} = -K_m(\mathbf{x})\left[\frac{1}{2}J(\mathbf{x}) \cdot (y(\mathbf{x}))^2 + y(\mathbf{x}) \cdot \nabla h(\mathbf{x})\right] \qquad (8\text{-}51)$$

The mean flux is, to the second order, equal to $\langle \vec{q}_1 \rangle = \vec{q}_1^{(0)} + \langle \vec{q}_1^{(2)} \rangle$ (Zhang, 2001). Therefore, the average flux along the direction i (Vereecken et al., 2007):

$$\langle q_1 \rangle_i \approx -K_m(\mathbf{x}) \left(\left| J_i \left\{ 1 + C_{ah}(\mathbf{x},\mathbf{x}) + \frac{1}{2} \left[\begin{matrix} C_{ff}(\mathbf{x},\mathbf{x}) + A^2 C_{hh}(\mathbf{x},\mathbf{x}) + H^2 C_{aa}(\mathbf{x},\mathbf{x}) \\ + 2AC_{fh}(\mathbf{x},\mathbf{x}) + 2HC_{fh}(\mathbf{x},\mathbf{x}) + \\ + 2AHC_{ah}(\mathbf{x},\mathbf{x}) \end{matrix} \right] \right\} \right. \right.$$
$$\left. \left. + E \left[(f + Ah + Ha) \frac{dh}{d\mathbf{x}_i} \right] \right| \right)$$

$$(8\text{-}52)$$

The effective hydraulic conductivity $K_{eff,ij}$ can be obtained as:

$$K_{eff,ij} = \frac{\langle q_1 \rangle_i}{J_j} \qquad (8\text{-}53)$$

Remarks

- The above formulation involves only the first two orders of moments; hence multivariate Gaussian distributions are assumed. A more complex formulation involving higher-order moments is possible.

- Ergodicity, which implies the equivalence of spatial mean and ensemble mean, is assumed: this is valid if the length scale of interest is much larger than the correlation length. Non-stationarity or trend model can be incorporated if needed. For such formulations, a well-defined separation of scales is required.

- The above formulation does not consider data conditioning. Readers should refer to the cited references for additional information. For example, Rubin and Gomez-Hernandez (1990) presented a formulation for upscaling block conductivities conditional to log-normally distributed point conductivities.

8.4.3 Volume Averaging Methods

As opposed to the ensemble averaging methods, the volume averaging methods deal with the spatial mean. Governing equations are spatially averaged to produce equations that are valid everywhere in the system. General treatments of various types of transport problems using the volume averaging techniques can be found in Whitaker's book (1999). A coupled set of large-scale average and small-scale deviation equations are formulated based on the concept of spatial stationarity (Wood, 2013). This method is particularly useful for deriving statistics (mean or variance) of effective transport parameters (provided that reservoir heterogeneity is accounted for) at the coarse scale.

Common types of transport problems studied with this method are solute transport (Zanotti and Carbonell, 1984), fluid flow (Quintard and Whitaker, 1998), where scaled-up versions of various transport parameters such as dispersivity, mass transfer coefficients, and permeability tensor can be derived. For volume averaging, spatial averaging (spatial moments), instead of ensemble averaging (moments of the statistical distribution), is employed. An important requirement is that the length scale of heterogeneities must be much smaller than the averaging volume (Cushman, 1983). In theory, if the restriction on the heterogeneity length scale is satisfied, both volume averaging and stochastic perturbation would yield the same results (Wang and Kitanidis, 1999). Similar to the stochastic perturbation method, if an explicit separation of scales is defined, a hierarchical framework can be applied to upscale properties sequentially through multiple scales.

Let us derive the general mathematical statements for upscaling hydraulic conductivity (or permeability if we ignore variation in viscosity) by following the notation and assumptions of Whitaker (1999). The governing equation for a single-phase, incompressible flow [Eq. (8-37)] at steady state can be described as follows:

$$\nabla \cdot \left(\vec{\vec{K}}(\mathbf{x}) \cdot \nabla(\psi + z)\right) = -g(\mathbf{x}, \psi, t) \tag{8-54}$$

It is further assumed that the source/sink function is 0 and $\Psi = \psi + z$ (we will drop the functional dependency for now):

$$\nabla \cdot \left(\vec{\vec{K}} \cdot \nabla\Psi\right) = 0 \tag{8-55}$$

Consider an averaging volume V with a length scale of R_o. Within the volume, there is a mixture of two distinct regions, whose length scale is l_ϖ and l_η, respectively. It is expected that $l_\varpi, l_\eta \ll R_o$. It is assumed that hydraulic conductivity is homogeneous within each region; in other words, the variability within the region is negligible in comparison to the difference between regions. These regions may represent rock types or facies. The physical process of single-phase flow can be described as:

In the ω-region: $\nabla \cdot \left(\vec{\vec{K}}_\varpi \cdot \nabla\Psi_\varpi\right) = 0 \tag{8-56}$

In the η-region: $\nabla \cdot \left(\vec{\vec{K}}_\eta \cdot \nabla\Psi_\eta\right) = 0 \tag{8-57}$

B.C. 1 at $A_{\omega\eta}$: $\vec{n}_{\varpi\eta} \cdot \vec{\vec{K}}_\eta \cdot \nabla\Psi_\eta = \vec{n}_{\varpi\eta} \cdot \vec{\vec{K}}_\varpi \cdot \nabla\Psi_\varpi$ (8-58)

B.C. 2 at $A_{\omega\eta}$: $\Psi_\eta = \Psi_\varpi$. (8-59)

The two boundary conditions are needed to ensure pressure and flux continuity at the region interfaces. Besides, $\vec{n}_{\varpi\eta} = \vec{n}_{\eta\varpi}$. Applying the spatial averaging theorem to the ω-region would yield:

$$\left\langle \nabla \cdot \left(\vec{\vec{K}}_\varpi \cdot \nabla\Psi_\varpi \right) \right\rangle = 0$$

$$\nabla \cdot \left\langle \vec{\vec{K}}_\varpi \cdot \nabla\Psi_\varpi \right\rangle + \frac{1}{V} \int_{A_{\varpi\eta}} \vec{n}_{\varpi\eta} \cdot \vec{\vec{K}}_\varpi \cdot \nabla\Psi_\varpi dA = 0 \quad (8\text{-}60)$$

The angle bracket represents spatial averaging. Assuming constant $\vec{\vec{K}}_\varpi$ and applying the spatial averaging theorem for the second time gives:

$$\nabla \cdot \left[\vec{\vec{K}}_\varpi \cdot \langle \nabla\Psi_\varpi \rangle \right] + \frac{1}{V} \int_{A_{\varpi\eta}} \vec{n}_{\varpi\eta} \cdot \vec{\vec{K}}_\varpi \cdot \nabla\Psi_\varpi dA = 0 \quad (8\text{-}61)$$

The spatial average of $\nabla\Psi_\varpi$ is:

$$\langle \nabla\Psi_\varpi \rangle = \nabla \langle \Psi_\varpi \rangle + \frac{1}{V} \int_{A_{\varpi\eta}} \vec{n}_{\varpi\eta} \Psi_\varpi dA \quad (8\text{-}62)$$

$\langle \Psi_\varpi \rangle$ is the superficial average of Ψ_ϖ:

$$\langle \Psi_\varpi \rangle = \frac{\varphi_\varpi}{V} \int_{A_{\varpi\eta}} \Psi_\varpi dA \quad (8\text{-}63)$$

It is related to the intrinsic average $\langle \Psi_\varpi \rangle^\varpi$:

$$\langle \Psi_\varpi \rangle^\varpi = \frac{\langle \Psi_\varpi \rangle}{\varphi_\varpi} = \frac{1}{V} \int_{A_{\varpi\eta}} \Psi_\varpi dA \quad (8\text{-}64)$$

Therefore, Eq. (8-61) becomes:

$$\nabla \cdot \left[\vec{\vec{K}}_\varpi \cdot \left(\nabla \langle \Psi_\varpi \rangle + \frac{1}{V} \int_{A_{\varpi\eta}} \vec{n}_{\varpi\eta} \Psi_\varpi dA \right) \right] + \frac{1}{V} \int_{A_{\varpi\eta}} \vec{n}_{\varpi\eta} \cdot \vec{\vec{K}}_\varpi \cdot \nabla\Psi_\varpi dA = 0$$

$$\nabla \cdot \left[\vec{\vec{K}}_{\varpi} \cdot \left(\varphi_{\varpi} \nabla \langle \Psi_{\varpi} \rangle^{\omega} + \langle \Psi_{\varpi} \rangle^{\omega} \nabla \varphi_{\varpi} + \frac{1}{V} \int_{A_{\varpi\eta}} \vec{n}_{\varpi\eta} \Psi_{\varpi} dA \right) \right] \tag{8-65}$$

$$+ \frac{1}{V} \int_{A_{\varpi\eta}} \vec{n}_{\varpi\eta} \cdot \vec{\vec{K}}_{\varpi} \cdot \nabla \Psi_{\varpi} dA = 0$$

Next, the following decomposition is applied to describe each quantity as a sum of an average and a deviation (tilde):

$$\Psi_{\varpi} = \langle \Psi_{\varpi} \rangle^{\varpi} + \tilde{\Psi}_{\varpi} \tag{8-66}$$

Substituting this decomposition into (8-65) yields:

$$\nabla \cdot \left[\vec{\vec{K}}_{\varpi} \cdot \left(\phi_{\varpi} \nabla \langle \Psi_{\varpi} \rangle^{\omega} + \langle \Psi_{\varpi} \rangle^{\omega} \nabla \phi_{\varpi} + \frac{1}{V} \int_{A_{\varpi\eta}} \vec{n}_{\varpi\eta} \Psi_{\varpi} dA \right) \right]$$

$$+ \frac{1}{V} \int_{A_{\varpi\eta}} \vec{n}_{\varpi\eta} \cdot \vec{\vec{K}}_{\varpi} \cdot \nabla \Psi_{\varpi} dA = 0$$

$$\nabla \cdot \left[\vec{\vec{K}}_{\varpi} \cdot \left(\begin{array}{c} \varphi_{\varpi} \nabla \langle \Psi_{\varpi} \rangle^{\omega} + \langle \Psi_{\varpi} \rangle^{\omega} \nabla \varphi_{\varpi} + \frac{1}{V} \int_{A_{\varpi\eta}} \vec{n}_{\varpi\eta} \langle \Psi_{\varpi} \rangle^{\varpi} dA \\ + \frac{1}{V} \int_{A_{\varpi\eta}} \vec{n}_{\varpi\eta} \tilde{\Psi}_{\varpi} dA \end{array} \right) \right]$$

$$+ \frac{1}{V} \int_{A_{\varpi\eta}} \vec{n}_{\varpi\eta} \cdot \vec{\vec{K}}_{\varpi} \cdot \nabla \Psi_{\varpi} dA = 0 \tag{8-67}$$

If the length scale restriction applies:

$$\frac{1}{V} \int_{A_{\varpi\eta}} \vec{n}_{\varpi\eta} \langle \Psi_{\varpi} \rangle^{\varpi} dA = -\nabla \varphi_{\varpi} \langle \Psi_{\varpi} \rangle^{\varpi} \tag{8-68}$$

Therefore, Eq. (8-67) becomes:

$$\nabla \cdot \left[\vec{\vec{K}}_{\varpi} \cdot \left(\varphi_{\varpi} \nabla \langle \Psi_{\varpi} \rangle^{\omega} + \frac{1}{V} \int_{A_{\varpi\eta}} \vec{n}_{\varpi\eta} \tilde{\Psi}_{\varpi} dA \right) \right] + \frac{1}{V} \int_{A_{\varpi\eta}} \vec{n}_{\varpi\eta} \cdot \vec{\vec{K}}_{\varpi} \cdot \nabla \Psi_{\varpi} dA = 0$$

$$\tag{8-69}$$

The same procedure can be applied to derive the equation for η-region:

$$\nabla \cdot \left[\vec{\vec{K}}_{\eta} \cdot \left(\varphi_{\eta} \nabla \langle \Psi_{\eta} \rangle^{\eta} + \frac{1}{V} \int_{A_{\eta\varpi}} \vec{n}_{\eta\varpi} \tilde{\Psi}_{\eta} dA \right) \right] + \frac{1}{V} \int_{A_{\eta\varpi}} \vec{n}_{\eta\varpi} \cdot \vec{\vec{K}}_{\eta} \cdot \nabla \Psi_{\eta} dA = 0$$

$$\tag{8-70}$$

In order to obtain a single-equation model, the *large-scale mechanical equilibrium* assumption is invoked:

$$\langle \Psi \rangle = \varphi_\varpi \langle \Psi_\varpi \rangle^\varpi + \varphi_\eta \langle \Psi_\eta \rangle^\eta$$
$$\langle \Psi_\eta \rangle^\eta = \langle \Psi \rangle + \hat{\Psi}_\eta$$
$$\langle \Psi_\varpi \rangle^\varpi = \langle \Psi \rangle + \hat{\Psi}_\varpi \tag{8-71}$$

$\hat{\Psi}_\eta$ and $\hat{\Psi}_\varpi$ refer to large-scale deviations and $\langle \Psi \rangle$ refers to large-scale intrinsic average. Therefore, the single-equation model can be derived as:

$$\nabla \cdot \left[\vec{\vec{K}}_\eta \cdot \left(\varphi_\eta \nabla \big(\langle \Psi \rangle + \hat{\Psi}_\eta \big) + \frac{1}{V} \int_{A_{\eta\varpi}} \vec{n}_{\eta\varpi} \tilde{\Psi}_\eta dA \right) \right]$$
$$+ \frac{1}{V} \int_{A_{\eta\varpi}} \vec{n}_{\eta\varpi} \cdot \vec{\vec{K}}_\eta \cdot \nabla \Psi_\eta dA$$
$$+ \nabla \cdot \left[\vec{\vec{K}}_\varpi \cdot \left(\varphi_\varpi \nabla \big(\langle \Psi \rangle + \hat{\Psi}_\varpi \big) + \frac{1}{V} \int_{A_{\varpi\eta}} \vec{n}_{\varpi\eta} \tilde{\Psi}_\varpi dA \right) \right]$$
$$+ \frac{1}{V} \int_{A_{\varpi\eta}} \vec{n}_{\varpi\eta} \cdot \vec{\vec{K}}_\varpi \cdot \nabla \Psi_\varpi dA = 0 \tag{8-72}$$

Applying B.C. 1 and re-arranging would yield:

$$\nabla \cdot \left[\begin{array}{c} \vec{\vec{K}}_\eta \cdot \left(\varphi_\eta \nabla \big(\langle \Psi \rangle + \hat{\Psi}_\eta \big) \right) + \vec{\vec{K}}_\varpi \cdot \left(\varphi_\varpi \nabla \big(\langle \Psi \rangle + \hat{\Psi}_\varpi \big) \right) \\ + \dfrac{\vec{\vec{K}}_\eta}{V} \int_{A_{\eta\varpi}} \vec{n}_{\eta\varpi} \tilde{\Psi}_\eta dA + \dfrac{\vec{\vec{K}}_\varpi}{V} \int_{A_{\varpi\eta}} \vec{n}_{\varpi\eta} \tilde{\Psi}_\varpi dA \end{array} \right] = 0 \tag{8-73}$$

$$\nabla \cdot \left[\begin{array}{c} \left(\varphi_\eta \vec{\vec{K}}_\eta + \varphi_\varpi \vec{\vec{K}}_\varpi \right) \cdot \nabla \langle \Psi \rangle + \dfrac{\vec{\vec{K}}_\eta}{V} \int_{A_{\eta\varpi}} \vec{n}_{\eta\varpi} \tilde{\Psi}_\eta dA \\ + \dfrac{\vec{\vec{K}}_\varpi}{V} \int_{A_{\varpi\eta}} \vec{n}_{\varpi\eta} \tilde{\Psi}_\varpi dA \end{array} \right]$$
$$+ \nabla \cdot \left(\varphi_\eta \vec{\vec{K}}_\eta \cdot \nabla \hat{\Psi}_\eta + \varphi_\varpi \vec{\vec{K}}_\varpi \cdot \nabla \hat{\Psi}_\varpi \right) = 0 \tag{8-74}$$

Introducing the large-scale hydraulic conductivity tensor, $\vec{\vec{K}}^*$, Eq. (8-74) becomes:

$$\nabla \cdot \left[\left(\varphi_\eta \vec{\vec{K}}_\eta + \varphi_\varpi \vec{\vec{K}}_\varpi \right) \cdot \nabla \langle \Psi \rangle + \frac{\vec{\vec{K}}_\eta}{V} \int\limits_{A_{\eta\varpi}} \vec{n}_{\eta\varpi} \tilde{\Psi}_\eta dA + \frac{\vec{\vec{K}}_\varpi}{V} \int\limits_{A_{\varpi\eta}} \vec{n}_{\varpi\eta} \tilde{\Psi}_\varpi dA \right]$$

$$+ \nabla \cdot \left(\varphi_\eta \vec{\vec{K}}_\eta \cdot \nabla \hat{\Psi}_\eta + \varphi_\varpi \vec{\vec{K}}_\varpi \cdot \nabla \hat{\Psi}_\varpi \right) = \nabla \cdot \vec{\vec{K}}^* \cdot \nabla \langle \Psi \rangle$$

$$(8\text{-}75)$$

If the following constraint applies:

$$\varphi_\eta \vec{\vec{K}}_\eta \cdot \nabla \hat{\Psi}_\eta + \varphi_\varpi \vec{\vec{K}}_\varpi \cdot \nabla \hat{\Psi}_\varpi << \vec{\vec{K}}^* \cdot \nabla \langle \Psi \rangle \qquad (8\text{-}76)$$

we obtain the following:

$$\left(\varphi_\eta \vec{\vec{K}}_\eta + \varphi_\varpi \vec{\vec{K}}_\varpi \right) \cdot \nabla \langle \Psi \rangle + \frac{\vec{\vec{K}}_\eta}{V} \int\limits_{A_{\eta\varpi}} \vec{n}_{\eta\varpi} \tilde{\Psi}_\eta dA$$

$$+ \frac{\vec{\vec{K}}_\varpi}{V} \int\limits_{A_{\varpi\eta}} \vec{n}_{\varpi\eta} \tilde{\Psi}_\varpi dA = \vec{\vec{K}}^* \cdot \nabla \langle \Psi \rangle \qquad (8\text{-}77)$$

Besides, the average governing equation becomes:

$$\nabla \cdot \left[\left(\varphi_\eta \vec{\vec{K}}_\eta + \varphi_\varpi \vec{\vec{K}}_\varpi \right) \cdot \nabla \langle \Psi \rangle + \frac{\vec{\vec{K}}_\eta}{V} \int\limits_{A_{\eta\varpi}} \vec{n}_{\eta\varpi} \tilde{\Psi}_\eta dA + \frac{\vec{\vec{K}}_\varpi}{V} \int\limits_{A_{\varpi\eta}} \vec{n}_{\varpi\eta} \tilde{\Psi}_\varpi dA \right] = 0$$

$$(8\text{-}78)$$

To obtain a closed-form of the above equation, we need to develop boundary value problems for $\tilde{\Psi}_\eta$ and $\tilde{\Psi}_\varpi$. This is achieved by first deriving a deviation equation (subtracting the average equation from the fine-scale equation) in each region:

$$\nabla \cdot \left(\vec{\vec{K}}_\varpi \cdot \nabla \Psi_\varpi \right) - \nabla \cdot \left[\begin{array}{l} \left(\varphi_\eta \vec{\vec{K}}_\eta + \varphi_\varpi \vec{\vec{K}}_\varpi \right) \cdot \nabla \langle \Psi \rangle + \dfrac{\vec{\vec{K}}_\eta}{V} \int\limits_{A_{\eta\varpi}} \vec{n}_{\eta\varpi} \tilde{\Psi}_\eta dA \\[3mm] + \dfrac{\vec{\vec{K}}_\varpi}{V} \int\limits_{A_{\varpi\eta}} \vec{n}_{\varpi\eta} \tilde{\Psi}_\varpi dA \end{array} \right] = 0$$

$$\nabla \cdot \left(\vec{\vec{K}}_\eta \cdot \nabla \Psi_\eta \right) - \nabla \cdot \left[\begin{array}{l} \left(\varphi_\eta \vec{\vec{K}}_\eta + \varphi_\varpi \vec{\vec{K}}_\varpi \right) \cdot \nabla \langle \Psi \rangle + \dfrac{\vec{\vec{K}}_\eta}{V} \int\limits_{A_{\eta\varpi}} \vec{n}_{\eta\varpi} \tilde{\Psi}_\eta dA \\[3mm] + \dfrac{\vec{\vec{K}}_\varpi}{V} \int\limits_{A_{\varpi\eta}} \vec{n}_{\varpi\eta} \tilde{\Psi}_\varpi dA \end{array} \right] = 0$$

$$(8\text{-}79)$$

Assuming that average quantity varies smoothly over V:

$$\vec{\tilde{K}}_{\varpi} \cdot \nabla \tilde{\Psi}_{\varpi} = \frac{1}{V} \int_{A_{\varpi\eta}} \vec{n}_{\varpi\eta} \cdot \vec{\tilde{K}}_{\varpi} \tilde{\Psi}_{\varpi} dA$$

$$\vec{\tilde{K}}_{\eta} \cdot \nabla \tilde{\Psi}_{\eta} = \frac{1}{V} \int_{A_{\eta\varpi}} \vec{n}_{\eta\varpi} \cdot \vec{\tilde{K}}_{\eta} \cdot \tilde{\Psi}_{\eta} dA$$

$$(8\text{-}80)$$

Next, we introduce a set of closure variables ($\vec{b}_{\varpi}$ and $\vec{b}_{\eta}$) that aim to relate $\tilde{\Psi}_{\eta}$ and $\tilde{\Psi}_{\varpi}$ to $\langle \Psi \rangle$:

$$\tilde{\Psi}_{\varpi} = \vec{b}_{\varpi} \cdot \nabla \langle \Psi \rangle$$
$$\tilde{\Psi}_{\eta} = \vec{b}_{\eta} \cdot \nabla \langle \Psi \rangle$$

$$(8\text{-}81)$$

The closure variables are determined by the following boundary value problems, which are defined to satisfy the deviation equation. As a result, the original fine-scale problem is *closed*. In addition, the integrals on the right-hand side of Eq. (8-80) are generally negligible in comparison to the left-hand side. Therefore, the closure variables should satisfy the following set of boundary value problems:

$$\text{In the } \omega\text{-region: } \vec{\tilde{K}}_{\varpi} : \nabla\nabla \vec{b}_{\varpi} = 0 \qquad (8\text{-}82)$$

$$\text{In the } \eta\text{-region: } \vec{\tilde{K}}_{\eta} : \nabla\nabla \vec{b}_{\eta} = 0 \qquad (8\text{-}83)$$

$$\text{B.C. 1 at } A_{\varpi\eta}\text{: } \vec{n}_{\varpi\eta} \cdot \vec{\tilde{K}}_{\eta} \cdot \nabla \vec{b}_{\eta} + \vec{n}_{\varpi\eta} \cdot \left(\vec{\tilde{K}}_{\eta} - \vec{\tilde{K}}_{\varpi} \right) = \vec{n}_{\varpi\eta} \cdot \vec{\tilde{K}}_{\varpi} \cdot \nabla \vec{b}_{\varpi}$$

$$(8\text{-}84)$$

$$\text{B.C. 2 at } A_{\varpi\eta}\text{: } \vec{b}_{\eta} = \vec{b}_{\varpi} \qquad (8\text{-}85)$$

$$\text{Periodicity: } \vec{b}_{\eta}(\mathbf{r} + l_i) = \vec{b}_{\eta}(\mathbf{r}); \vec{b}_{\varpi}(\mathbf{r} + l_i) = \vec{b}_{\varpi}(\mathbf{r}). \ i = 1, 2, 3 \quad (8\text{-}86)$$

The periodic outer boundary is needed to complete the problem definition. Periodic boundary conditions are commonly adopted, and they are consistent with the assumption that a representative unit cell exists which accurately describes the geometry of the porous medium. It is also consistent with the length scale restriction.

The large-scale Darcy's law:

$$\nabla \cdot \left[\vec{\tilde{K}}^{*} \cdot \nabla \langle \Psi \rangle \right] = 0 \qquad (8\text{-}87)$$

Substituting the closure formulation into Eq. (8-78) would give:

$$\nabla \cdot \left[\left(\varphi_{\eta} \vec{\tilde{K}}_{\eta} + \varphi_{\varpi} \vec{\tilde{K}}_{\varpi} + \frac{\vec{\tilde{K}}_{\eta}}{V} \int_{A_{\eta\varpi}} \vec{n}_{\eta\varpi} \vec{b}_{\eta} dA + \frac{\vec{\tilde{K}}_{\varpi}}{V} \int_{A_{\varpi\eta}} \vec{n}_{\varpi\eta} \vec{b}_{\varpi} dA \right) \cdot \nabla \langle \Psi \rangle \right] = 0$$

$$(8\text{-}88)$$

Invoking B.C. 2, the large-scale hydraulic conductivity is:

$$\vec{\vec{K}}^{*} = \left| \varphi_{\eta}\vec{\vec{K}}_{\eta} + \varphi_{\varpi}\vec{\vec{K}}_{\varpi} + \frac{\vec{\vec{K}}_{\varpi} - \vec{\vec{K}}_{\eta}}{V} \int_{A_{\varpi\eta}} \vec{n}_{\varpi\eta}\vec{b}_{\varpi}dA \right| \qquad (8\text{-}89)$$

It should be noted that the first two terms give rise to a volume-weighted average of $\vec{\vec{K}}_{\eta}$ and $\vec{\vec{K}}_{\varpi}$. The area integral essentially represents a filter that controls the characteristics of ϖ- and η-regions that would contribute to $\vec{\vec{K}}^{*}$. Another way to understand the closure procedure is to recall that the average and deviation equations are coupled, but the true fine-scale solution is not known a priori. Therefore, a "closure condition" is imposed such that the two equations are decoupled, while ensuring that the summation of the average and deviation quantities would satisfy the fine-scale mass balance condition (Kechagia et al., 2002).

An alternative approach for solving the two coupled equations was developed by Leung and Srinivasan (2012). It is based on the availability of fine-scale quantities, which is commonly the case (Sec. 8.2). The method was applied to study a variety of transport problems: mass transfer (Leung, 2014; Leung and Srinivasan, 2012 and 2016) and fluid flow (Leung et al., 2010). The main idea is to incorporate the fine-scale quantities (e.g., Y_{ϖ} and Y_{η}), which can be obtained from detailed numerical simulation, into an optimization procedure to tune the values of $\langle\Psi\rangle$, $\tilde{\Psi}_{\varpi}$, and $\tilde{\Psi}_{\eta}$ until a converged solution to Eqs. (8-78) and (8-79) are reached (i.e., the difference between the left- and right-hand sides of both equations are minimized). A major advantage of incorporating flow simulation results is that it facilitates the application of the volume averaging techniques to realistic geologic media, as opposed to those periodic media employed in conventional formulations. The region definition can be stochastic, as long as the proportion of each region within the averaging scale is known. Knowledge of the fine-scale solution relaxes many of the restrictive assumptions associated with the traditional volume averaging formulations. The procedure allows the deviation and averaged equations to remain coupled. The key steps are outlined below:

1. Construct a fine-scale heterogeneous reservoir model (porosity and permeability) for a section of the reservoir and subject to flow simulation. Obtain $\mathbf{u}$, Ψ_{ϖ}, and Ψ_{η}. Choose a sub-domain for volume averaging calculations to avoid any near-well boundary effects.

2. Propose initial estimates of $\langle\mathbf{u}\rangle$ and $\langle\Psi\rangle$ at a particular coarse scale and a given snapshot of time t.

3. Using the initial estimate of $\langle\Psi\rangle$, compute $\tilde{\Psi}_{\varpi}$ and $\tilde{\Psi}_{\eta}$. Perturb $\langle\Psi\rangle$ to minimize the mismatch between the left- and

right-hand sides of Eqs. (8-78) and (8-79) using a non-linear minimization scheme. Repeat this step until the convergence criterion is reached.

4. Compute $\vec{\vec{K}}^*$ for each coarse-scale grid block, and a probability distribution of $\vec{\vec{K}}^*$ is constructed by aggregating its corresponding value from all coarse-scale grid blocks.

5. Repeat Steps 2 to 5 for other averaging length scales and values of t.

6. The procedure yields a set of probability distributions of $\vec{\vec{K}}^*$ at different length scales corresponding to different times during the flow and transport process.

The method was applied to a variety of heterogeneity distribution and fractal media (Leung and Srinivasan, 2016). The results highlighted the impacts of higher-order multivariate distributions on the scaling characteristics. The continuity of a spatially varying property, as characterized by the higher-order multivariate statistics, could considerably influence the scale-up results. An example is shown in Fig. 8-13; the scaling characteristics of the effective mass transfer coefficient for a channelized model are compared against those for a covariance model. Despite being constrained to the same univariate (histogram) and bivariate statistics, the two models are exhibiting different scaling characteristics and transport behavior.

In the context of scale-up, wherever the length scale of heterogeneity is larger than the averaging scale, the assumption of stationary

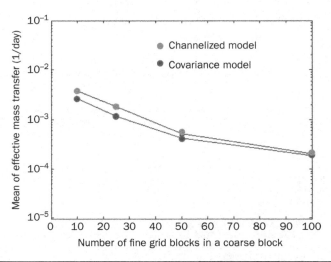

FIGURE 8-13 Scaling characteristics of effective mass transfer coefficient. (The color version of this figure is available at www.mhprofessional.com/PRMS.)

proportions within an averaging volume would not be exact. Applying any averaging scheme at scales smaller than REV introduces variability in the corresponding averaged attribute. In such cases, the averaged quantity would be characterized by a probability distribution (Sec. 8.3). Pooling the results from multiple coarse-scale grid blocks can establish the corresponding probability distributions. The assignment of coarse-scale properties is facilitated by (1) sampling from the calibrated probability distributions and assigning the corresponding value to a particular coarse-grid location and (2) repeating Step 1 until the entire grid is populated. If large-scale non-stationarity is present, probability distributions should be estimated over numerous sub-regions.

8.5 Final Remarks

Generally speaking, the most appropriate methods for scaling permeability, relative permeability, or other transport-related parameters, such as mass transfer coefficients, are those that take into the dynamics of the underlying flow and transport processes. Therefore, flow-based upscaling or scale-up procedures tend to better embody the interplay between heterogeneity and the physical processes.

Reservoir heterogeneity exists over multiple scales, and special care should be taken to capture the uncertainty associated with that heterogeneity into the final flow simulation scale. When handling an arbitrary medium that exhibits a spatial structure described by a semi-variogram, a statistical scale-up procedure is recommended, where the uncertainty due to sub-scale variability is captured in the final coarse-scale models.

8.6 References

Journel, A. G., C. Deutsch, and A. J. Desbarats. "Power averaging for block effective permeability." In SPE California Regional Meeting. Society of Petroleum Engineers, 1986.

Wen, Xian-Huan, and J. Jaime Gómez-Hernández. "Upscaling hydraulic conductivities in heterogeneous media: An overview." Journal of Hydrology 183, no. 1-2 (1996): ix-xxxii.

King, P. R. "The use of renormalization for calculating effective permeability." *Transport in porous media* 4, no. 1 (1989): 37-58.

Christie, Michael A. "Upscaling for reservoir simulation." *Journal of Petroleum Technology* 48, no. 11 (1996): 1-004.

Peaceman, D. W. "Effective transmissibilities of a gridblock by upscaling-comparison of direct methods with renormalization." *SPE Journal* 2, no. 03 (1997): 338-349.

Bear, J. *Dynamics of Fluids in Porous Media*. Courier Corporation, 1972.

Leung, J.Y., and S. Srinivasan. "Analysis of uncertainty introduced by scaleup of reservoir attributes and flow response in heterogeneous reservoirs." *SPE Journal* 16, no. 03 (2011): 713-724.

Leung, J.Y., and S. Srinivasan. "Scale-up of mass transfer and recovery performance in heterogeneous reservoirs." *Journal of Petroleum Science and Engineering* 86 (2012): 71-86.

Leung, J.Y., and S. Srinivasan. "Effects of reservoir heterogeneity on scaling of effective mass transfer coefficient for solute transport." *Journal of Contaminant Hydrology* 192 (2016): 181-193.

Leung, J.Y., S. Srinivasan, and C. Huh. "Accounting for Heterogeneity in Scale-Up of Apparent Polymer Viscosity for Field Scale Application." In *SPE Improved Oil Recovery Symposium.* Society of Petroleum Engineers, 2010.

Leung, J.Y. "Scale-up of effective mass transfer in vapor extraction process accounting for field-scale reservoir heterogeneities." *Journal of Canadian Petroleum Technology* 53, no. 5 (2014): 275-289.

Vishal, V. and J.Y. Leung. "Modeling impacts of subscale heterogeneities on dispersive solute transport in subsurface systems." Journal of contaminant hydrology 182 (2015): 63-77.

Vishal, V., Leung, J.Y.. "A multi-scale particle-tracking framework for dispersive solute transport modeling." *Computational Geosciences* 22, no. 2 (2017a): 485–503.

Vishal, V., Leung, J.Y. "Statistical framework for scale-up of dispersivity in multi-scale heterogeneous media." *Environmental Earth Sciences* 76 (2017b): 624.

Vishal, V. and Leung, J.Y. " Statistical scale-up of 3D particle-tracking simulation for non-Fickian dispersive solute transport modeling." *Stochastic Environmental Research and Risk Assessment* 32, no. 7 (2018): 2075-2091.

Lake, Larry W., and Sanjay Srinivasan. "Statistical scale-up of reservoir properties: concepts and applications." *Journal of Petroleum Science and Engineering* 44, no. 1 (2004): 27-39.

Journel, A.G. and C. Hujbregts, *Mining Geostatistics,* Academic Press, New York City, 1978.

Pyrcz, Michael J., and Clayton V. Deutsch. *Geostatistical reservoir modeling.* Oxford University Press, 2014.

Nash, J. Eamonn, and Jonh V. Sutcliffe. "River flow forecasting through conceptual models part I—A discussion of principles." *Journal of hydrology* 10, no. 3 (1970): 282-290.

Warren, J. E., and H. S. Price. "Flow in heterogeneous porous media." *Society of Petroleum Engineers Journal* 1, no. 03 (1961): 153-169.

White, C. D., and R. N. Horne. "Computing absolute transmissibility in the presence of fine-scale heterogeneity." In *SPE symposium* on reservoir simulation. Society of Petroleum Engineers, 1987.

Gomez-Hernandez, J.J., 1991. A stochastic approach to the simulation of block conductivity fields conditioned upon data measured at a smaller scale. Ph.D. thesis, Stanford University, CA

Holden, Lars, and Oddvar Lia. "A tensor estimator for the homogenization of absolute permeability." *Transport in Porous Media* 8, no. 1 (1992): 37-46.

Durlofsky, L. J. "Numerical calculation of equivalent grid block permeability tensors of heterogeneous porous media: Water Resour Res 27, no.5, (1991): 299–708.

Durlofsky, L.J. "Upscaling of geocellular models for reservoir flow simulation: a review of recent progress." In *7th International Forum on Reservoir Simulation Bühl/Baden-Baden, Germany,* pp. 23-27. 2003.

Chen, Y., Louis J. Durlofsky, M. Gerritsen, and Xian-Huan Wen. "A coupled local–global upscaling approach for simulating flow in highly heterogeneous formations." *Advances in Water Resources* 26, no. 10 (2003): 1041-1060.

Chen, Y. and L.J. Durlofsky. "Adaptive local–global upscaling for general flow scenarios in heterogeneous formations." *Transport in porous Media* 62, no. 2 (2006): 157-185.

Renard, Ph, and G. De Marsily. "Calculating equivalent permeability: a review." Advances in water resources 20, no. 5 (1997): 253-278.

Lajevardi, Saina, and Clayton V. Deutsch. "A measure of facies mixing in data upscaling to account for information loss in the estimation of petrophysical variables." *Petroleum Geoscience* 22, no. 2 (2016): 191-202.

Babak, Olena, John G. Manchuk, and Clayton V. Deutsch. "Accounting for non-exclusivity in sequential indicator simulation of categorical variables." *Computers & Geosciences* 51 (2013): 118-128.

Pickup, G. E., P. S. Ringrose, J. L. Jensen, and K. S. Sorbie. "Permeability tensors for sedimentary structures." *Mathematical Geology* 26, no. 2 (1994): 227-250.

Pickup, Gillian E., and Karl D. Stephen. "An assessment of steady-state scale-up for small-scale geological models." *Petroleum Geoscience* 6, no. 3 (2000): 203-210.

Pickup, G.E., and K.S. Sorbie. "The scaleup of two-phase flow in porous media using phase permeability tensors." *SPE Journal* 1, no. 04 (1996): 369-382.

Pickup, G., P. S. Ringrose, and A. Sharif. "Steady-state upscaling: from lamina-scale to full-field model." *SPE Journal* 5, no. 02 (2000): 208-217.

Pickup, G. E., K. D. Stephen, J. Ma, P. Zhang, and J. D. Clark. "Multi-stage upscaling: selection of suitable methods." In *Upscaling Multiphase Flow in Porous Media*, pp. 191-216. Springer Netherlands, 2005.

Kumar, Arun TA, and G. R. Jerauld. "Impacts of scale-up on fluid flow from plug to gridblock scale in reservoir rock." In *SPE/DOE Improved Oil Recovery Symposium*. Society of Petroleum Engineers, 1996.

Saad, Naji, A. S. Cullick, and M. M. Honarpour. "Effective relative permeability in scale-up and simulation." In *Low Permeability Reservoirs Symposium*. Society of Petroleum Engineers, 1995.

Kyte, J.R., and D. W. Berry. "New pseudo functions to control numerical dispersion." *Society of Petroleum Engineers Journal* 15, no. 04 (1975): 269-276.

Artus, V., and B. Noetinger. "Up-scaling two-phase flow in heterogeneous reservoirs: current trends." *Oil & gas science and technology* 59, no. 2 (2004): 185-195.

Stone, H. L. "Rigorous black oil pseudo functions." In *SPE Symposium on reservoir simulation*. Society of Petroleum Engineers, 1991.

Corbett, P.W.M., and J. L. Jensen. "Application of probe permeametry to the prediction of two-phase flow performance in laminated sandstones (lower Brent Group, North Sea)." *Marine and Petroleum Geology* 10, no. 4 (1993): 335-346.

Corbett, P. W. M., P. S. Ringrose, J. L. Jensen, and K. S. Sorbie. "Laminated clastic reservoirs: The interplay of capillary pressure and sedimentary architecture." In *SPE Annual Technical Conference and Exhibition*. Society of Petroleum Engineers, 1992.

Ringrose, P. S., K. S. Sorbie, P. W. M. Corbett, and J. L. Jensen. "Immiscible flow behaviour in laminated and cross-bedded sandstones." *Journal of Petroleum Science and Engineering* 9, no. 2 (1993): 103-124.

Alpak, F.O. "Quasiglobal Multiphase Upscaling of Reservoir Models With Nonlocal Stratigraphic Heterogeneities." *SPE Journal* 20, no. 02 (2015): 277-293.

Christie, M. A., and P. J. Clifford. "Fast procedure for upscaling compositional simulation." *SPE Journal* 3, no. 03 (1998): 272-278.

Barker, J. W., and F. J. Fayers. "Transport coefficients for compositional simulation with coarse grids in heterogeneous media." *SPE Advanced Technology Series* 2, no. 02 (1994): 103-112.

Li, D., and L. W. Lake. "Scaling fluid flow through heterogeneous permeable media." *SPE Advanced Technology Series* 3, no. 01 (1995): 188-197.

Shook, M., D. Li, and L.W. Lake. "Scaling immiscible flow through permeable media by inspectional analysis." *IN SITU-NEW YORK-* 16 (1992): 311-311.

Zhang, D. *Stochastic methods for flow in porous media: coping with uncertainties.* Academic press, 2001.

Vereecken, H., R. Kasteel, J. Vanderborght, and T. Harter. "Upscaling hydraulic properties and soil water flow processes in heterogeneous soils." Vadose Zone Journal 6, no. 1 (2007): 1-28. Howes, F.A., and S. Whitaker. "The spatial averaging theorem revisited." *Chemical engineering science* 40, no. 8 (1985): 1387-1392.

Cushman, J.H. "Proofs of the volume averaging theorems for multiphase flow." *Advances in Water Resources* 5, no. 4 (1982): 248-253.

Gelhar, L.W., and C.L. Axness. "Three-dimensional stochastic analysis of macrodispersion in aquifers." *Water Resources Research* 19, no. 1 (1983): 161-180.

Dagan, G. "Solute transport in heterogeneous porous formations." *Journal of Fluid Mechanics*, 145 (1984): 151-177.

Dagan, G. "Theory of solute transport by groundwater." *Annual Review of Fluid Mechanics*, 19 (1987): 183-215.

Dagan, G. *Flow and Transport in Porous Formations.* New York: Springer-Verlag, 1989.

Neuman, S.P., Winter, C.L., and Newman, C.M. (1987). "Stochastic theory of field-scale Fickian dispersion in anisotropic porous media." *Water Resources Research*, 23, no.3 (1987): 453-466.

Neuman, S.P. and Orr, S. (1993). "Prediction of steady state flow in nonuniform Geologic Media by Conditional Moments: Exact Nonlocal Formalism, Effective Conductivities, and Weak Approximation." *Water Resources Research*, 29, no.2 (1993): 341-364.

Cvetkovic, V., Shapiro, A.M., and Dagan, G. "A solute flux approach to transport in heterogeneous formations 2. uncertainty analysis." *Water Resources Research*, 28, no. 5 (1992): 1377-1388.

Rubin, Y. and Gomez-Hernandez, J.J. "A stochastic approach to the problem of upscaling of conductivity in disordered media: Theory and unconditional numerical simulations." *Water Resources Research*, 26, no. 4 (1990): 691-701.

Gardner, W. R. "Some steady-state solutions of the unsaturated moisture flow equation with application to evaporation from a water table." *Soil science* 85, no. 4 (1958): 228-232.

Russo, D. "Determining soil hydraulic properties by parameter estimation: On the selection of a model for the hydraulic properties." *Water Resour. Res* 24, no. 3 (1988): 453-459.

Cushman, J. H. "Volume averaging, probabilistic averaging, and ergodicity." *Advances in Water Resources* 6, no. 3 (1983): 182-184.

Whitaker, S. "The method of volume averaging, theory and applications of transport in porous media." *Dordrecht: Kluwer Academic* (1999).

Kechagia, P.E., I.N. Tsimpanogiannis, Y.C. Yortsos, and P.C. Lichtner. "On the upscaling of reaction-transport processes in porous media with fast or finite kinetics." *Chemical engineering science* 57, no. 13 (2002): 2565-2577.

Wood, B.D. "Technical note: Revisiting the geometric theorems for volume averaging." *Advances in Water Resources* 62 (2013): 340-352.

Zanotti, F., and R. G. Carbonell. "Development of transport equations for multiphase system—I: General Development for two phase system." *Chemical Engineering Science* 39, no. 2 (1984): 263-278.

Quintard, M., and S. Whitaker. "Transport in chemically and mechanically heterogeneous porous media—III. Large-scale mechanical equilibrium and the regional form of Darcy's law." *Advances in Water Resources* 21, no. 7 (1998): 617-629.

Wang, J., and P. K. Kitanidis. "Analysis of macrodispersion through volume averaging: comparison with stochastic theory." *Stochastic Environmental Research and Risk Assessment* 13, no. 1 (1999): 66-84.

Shannon, C.E. "A mathematical theory of communication." *The Bell System Technical Journal* 27, no. 3 (1948): 379-423.

Neuman, S.P., and S. Orr. "Prediction of steady state flow in nonuniform geologic media by conditional moments: Exact nonlocal formalism, effective conductivities, and weak approximation." *Water Resources Research* 29, no. 2 (1993): 341-364.

Crotti, M.A., and R.H. Cobeñas. "Relative Permeability Curves. An advantageous approach based on realistic average water saturations." In *SPE Latin American and Caribbean Petroleum Engineering Conference*. Buenos Aires, Argentina, Mar. 25-28, 2001.

CHAPTER 9

History Matching–Dynamic Data Integration

Istory matching is the process of calibrating reservoir models to reflect the observed production characteristics. Uncertain model parameters are perturbed until the simulation prediction is consistent with the actual production observations. Traditionally, history matching has been studied as a data-based inversion process; various optimization strategies, including both local (gradient-based) and global schemes, have been widely adopted. The chapter will begin with a review of these traditional schemes. History matching can also be viewed as a data-assimilation process. In this context, several probabilistic history matching methods have also been proposed. The chapter will review these methods and present workflows summarizing the application of these methods. The optimization methods would yield a single assimilated reservoir model in most cases. It is impossible to quantify the residual uncertainty after the optimization process. Probabilistic methods, such as the probability perturbation method (PPM), do quantify the uncertainty associated with the prediction at a location given the production data, but in the end, a realization of the property field is sampled from the underlying uncertainty distribution. In order to quantify the global uncertainty associated with production forecasts, several realizations have to be sampled and processed through a complex transfer function, such as a flow simulator, and that is cumbersome. In this context, the chapter will conclude with a discussion on ensemble-based techniques for history matching that realize the uncertainty through the ensemble of updated models. The ensemble Kalman filter and ensemble pattern search algorithms, as well as the alternative paradigm of model selection instead of history matching, will be presented.

9.1 History Matching as an Inverse Problem

A mathematical model (g) that describes the relationship between a set of model parameters (**m**) and the corresponding observable outcomes of the system, i.e., data (**d**), can be denoted as follows:

$$\mathbf{d} = g(\mathbf{m}) \qquad (9\text{-}1)$$

A *forward* problem is one where **m** is known, and Eq. (9-1) is used to predict **d**. On the other hand, an *inverse* problem is one where **d** is known, and Eq. (9-1) is used to predict **m**. A case in point is history matching, where observable production performance is used to infer the unknown reservoir model parameters. Unfortunately, real observations are affected by errors from various sources and several other factors/parameters (such as fluid properties, boundary conditions, and well-operating conditions) affect the production response of a reservoir. In the end, the ultimate goal of history matching is not merely to match the production history; it is desired to infer the unknown model parameters such that they are consistent with all available data (both static and dynamic), as well as offering the most accurate predictions of future behavior.

For the sake of simplicity, the boldfaced will be dropped from now on. If $g(\mathrm{m})$ can be represented as Gm, where G is an $N \times M$ matrix (N is the number of observations and M is the number of model parameters), the problem can be considered as a discrete linear inverse problem. In the case of history matching, the function g would be a reservoir simulator, which entails a discretized set of nonlinear partial differential equations and constraints, as discussed in Chaps. 6 and 7. This aspect renders the inverse problem to be nonlinear. However, it is still instructional to present a brief overview of the linear inverse theory (Tarantola, 2005; Menke, 2012), which helps to illustrate several salient features of an inverse problem.

A *purely overdetermined* problem has several characteristics:

- There exists an observation (d_{obs}) for which there is no solution.
- The rank of G is equal to M, and the dimension of the null space of G is 0.
- $G^{T}G$ is a non-singular positive-definite matrix, or $(G^{T}G)^{-1}$ exists.
- The number of unknowns is less than the number of equations (i.e., $M < N$).

The functional (O) to be minimized and the corresponding solution (m^{est}) are:

$$O(m) = (Gm - d_{obs})^{T}(Gm - d_{obs}) \qquad (9\text{-}2)$$

$$m^{est} = (G^{T}G)^{-1}G^{T}d_{obs} \qquad (9\text{-}3)$$

The objective is simply to minimize the sum of the squared difference between the predicted response and the observed data.

A *purely underdetermined* problem has several characteristics:

- For every observation (d_{obs}), there exists more than one solution vector m.

- The rank of G is equal to N, and the dimension of the null space of G is non-zero.

- GGT is a non-singular positive-definite matrix, or $(GGT)^{-1}$ exists.

- The number of unknowns is greater than the number of equations (i.e., $M > N$)

In order to reach a unique solution, an additional condition is needed to constrain the problem. One option is to find the simplest solution (e.g., closest to the a priori model m_{pr}) such that the error is minimized. As a constrained optimization problem with λ being the Lagrange multiplier, the functional to be minimized and the corresponding solution are:

$$O(m, \lambda) = (m - m_{pr})^T(m - m_{pr}) + \lambda(Gm - d_{obs}) \tag{9-4}$$

$$m^{est} = m_{pr} + G^T(GG^T)^{-1}(d_{obs} - Gm_{pr}) \tag{9-5}$$

The estimate for the model parameter is updated by an amount regulated by the extent of the mismatch between the predicted response and the observation.

A *mixed-determined* problem is one that is neither overdetermined nor underdetermined. It represents a situation where some of the model space is overdetermined, while the remaining is underdetermined. For example, suppose a case where $N = M = 5$; however, 3 of the measurements are redundant. Therefore, the number of linearly independent equations, as well as the rank of G, is 3. This problem is not overdetermined because the rank of G is less than M and the dimension of its null space is greater than 0. On the other hand, the problem is also not underdetermined because M is not greater than N. Another way to visualize a mixed-determined problem in a history-matching application is to consider the following: history matching is performed to infer permeability distribution across the entire reservoir domain; wells have been drilled preferentially in areas with high productivity (high kh). Regions with low kh and far away from the wells are essentially left undrained. It would be easy to speculate that the well production data would offer little information about these far-away regions, which experience little fluid movement. Therefore, the inference of permeability distribution is likely to be an overdetermined problem for the near-well regions, but an underdetermined problem for the far-away regions.

The functional to be minimized is:

$$O(m) = (Gm - d_{obs})^T(Gm - d_{obs}) + \varepsilon^2(m - m_{pr})^T(m - m_{pr}) \quad (9\text{-}6)$$

The solution is often referred to as the regularized least-squares solution, where the second term $(m - m_{pr})^T(m - m_{pr})$ offers the regularization. The corresponding solution is:

$$m^{est} = m_{pr} + (G^TG + \varepsilon^2I)^{-1}G^T(d_{obs} - Gm_{pr}) \quad (9\text{-}7)$$

The parameter ε should be selected depending on the problem. If a more precise data match is desired or if there is considerable uncertainty about the prior model, a small value should be chosen; conversely, if the model parameters are expected to be similar to the a priori model, a larger value should be selected.

In some cases, a more sophisticated weighting scheme is desired. The functional to be minimized becomes:

$$O(m) = (Gm - d_{obs})^TW_e(Gm - d_{obs}) + \varepsilon^2(m - m_{pr})^TW_m(m - m_{pr}) \quad (9\text{-}8)$$

The corresponding solution is:

$$m = m_{pr} + (G^TW_eG + \varepsilon^2W_m)^{-1}G^TW_e(d_{obs} - Gm_{pr}) \quad (9\text{-}9)$$

Two relevant concepts are those of model resolution and data resolution. Model resolution refers to how well the model vector can be resolved from the data. Data resolution refers to how well $d^{predict} = Gm^{est}$ would reproduce the data. For a purely overdetermined problem, the least-squares solution gives perfect model resolution, while the simplest solution for a purely underdetermined problem offers perfect data resolution. This is because, for a purely overdetermined problem, there is sufficient data to resolve the entire model vector completely; contrarily, for a purely underdetermined problem, there is not enough data to resolve all parts of the model vector, and the simplest solution is merely one of the many solutions that would match the data perfectly.

9.2   Optimization Schemes

9.2.1   Gradient-Based Methods

To find the minimum of a given objective function $O(m)$, we seek to find the model vector such that its gradient with respect to m is equal to an $M \times 1$ zero vector:

$$\nabla_m O(m) = 0 \quad (9\text{-}10)$$

or

$$\frac{\partial O(m)}{\partial m_i} = 0, i = 1, 2, ..., M \quad (9\text{-}11)$$

The solution can be obtained with the Newton–Raphson algorithm (Press et al., 1989), with m^0 as the initial guess and l as the iteration counter:

$$m^{l+1} = m^l + \delta m^{l+1} \quad (9\text{-}12)$$

The update (as in Newton–Raphson methods) is calculated using the gradient or the Hessian:

$$\nabla\left[\left(\nabla_m O\left(m^l\right)\right)^T\right]\delta m^{l+1} = H\left(m^l\right)\delta m^{l+1} = -\nabla_m O\left(m^l\right) \qquad (9\text{-}13)$$

$H(m)$ is the Hessian matrix with an element at row i and column j:

$$h_{ij} = \frac{\partial^2 O(m)}{\partial m_i \partial m_j} \qquad (9\text{-}14)$$

Next, we will discuss a few aspects of the gradient-based methods that should be considered for practical application.

Improving Downhill Search

Equation (9-12) can be modified with a different step size, μ_i:

$$m^{l+1} = m^l + \mu_l \delta m^{l+1} \qquad (9\text{-}15)$$

Techniques such as line search (Nocedal and Wright, 2006) can be implemented to estimate the step size. Another technique to obtain an improved downhill search direction is the Levenberg–Marquardt algorithm, which modifies the left-hand side of Eq. (9-13) and, in effect, controls both the search direction and step size:

$$(H(m^l) + \lambda_l I)\delta m^{l+1} = -\nabla_m O(m^l) \qquad (9\text{-}16)$$

During each iteration, a temporary value m_c^{l+1} is defined:

$$m_c^{l+1} = m^l + \delta m^{l+1} \qquad (9\text{-}17)$$

If $O(m_c^{l+1}) < O(m^l)$, then set $m_c^{l+1} = m^{l+1}$ and $\lambda_{l+1} = \lambda_l / a_1$, where $a_1 > 1$; the iteration is complete. Otherwise, multiply λ_l by a_2, where $a_2 > 1$, and repeat the iteration.

Computing H^{-1}

Another obvious issue with these Newton methods is the need for H^{-1}, which can be computationally expensive. There are alternative methods where the inverse is estimated approximately. For instance, in the Broyden–Fletcher–Goldfarb–Shanno scheme, an initial estimate of H^{-1} is proposed, and it is continuously updated at each iteration (Nocedal and Wright, 2006).

$$H_{l+1}^{-1} = V_l^T H_l^{-1} V_l + \rho_l s_l s_l^T \qquad (9\text{-}18)$$

where

$$V_l = I - \rho_l y_l s_l^T$$
$$\rho_l = \frac{1}{y_l^T s_l} \qquad (9\text{-}19)$$

$y_l = \nabla_m O_{l+1} - \nabla_m O_l$ and $s_l = m^{l+1} - m^l$; $\nabla_m O_l$ = gradient of the objective function evaluated at m^l. It is an $M \times 1$ column vector with the ith element $\partial O(m)/\partial m_i$. Additional detail and theoretical background can be found in the references.

9.2.2 Global Methods

Gradient-based techniques generally involve updating the model vector based on the local gradient. The extent to which the parameter is updated depends on the magnitude of the gradient of the objective function. For a highly non-linear problem, the objective function value may go through several local minima and in the vicinity of each of these local minima, the gradient may have a low value. This can cause the estimate to get stuck and a truly optimal solution may not be found. Many global methods, on the other hand, utilize probabilistic schemes to approximate the global minimum by performing stochastic searches in a large discrete model space (Linde et al., 2015) and because the update is randomized, there is a chance that the estimates escape local minima and converge to the "true" minima. Several techniques, including simulated annealing (SA), genetic algorithm (GA), and particle swarm, are discussed next.

Simulated Annealing

The idea of simulated annealing was first proposed by Kirkpatrick et al. (1983). It borrows from statistical physics; it draws an analogy between model parameters of an optimization problem and particles in a physical system. A physical annealing process occurs when a solid in a heat bath is initially heated by increasing the temperature; this is followed by slow cooling such that all the particles arrange themselves in a low energy state where crystallization occurs. A slow cooling schedule would yield the lowest energy state (global minimum), while a fast cooling schedule would yield a local minimum. The steps involved in a general SA scheme can be summarized as follow:

1. Start with a random guess of the solution m_0 with energy $E(m_0)$. This energy can also represent an objective function that is to be minimized.

2. Loop over a series of decreasing temperature (T):

Loop over a series of random moves:

   i. Propose a *trial model*—loop over model parameters $i = 1,\ldots,M$; for metropolis sampling:

$$r_i = u_i; u_i \in U[0,1]$$

$$m_i^{new} = m_i^{old} + r_i\left(m_i^{max} - m_i^{nin}\right)$$

End loop

ii. Calculate $\Delta E = E\left(m_i^{new}\right) - E\left(m_i^{old}\right)$ and *acceptance probability*
$P = \exp(-\Delta E/T)$

iii. If $\Delta E \leq 0$,

$$m_0 = m^{new}; E(m_0) = E(m^{new})$$

Else

Draw a random number $s \in U[0, 1]$

If $P > s$, $m_0 = m^{new}$; $E(m_0) = E(m^{new})$

End if
End loop
End loop

Remarks

- The number of T and number of random moves per T describe the *cooling schedule*. A slow cooling schedule would involve gradually reducing the temperature and incorporating more random moves per state. The idea is to explore as many trial models as possible, such that a global minimum can be found, instead of quickly reducing the temperature, which may lead to being stuck at a local minimum.

- Step 2(iii) shows that there is a finite probability that the algorithm will accept a proposed trial model, even if $\Delta E > 0$, despite as ΔE increases, the probability of acceptance would decrease. The rationale is that to reach the global minimum, one must not be "greedy" and should accept a worse proposal occasionally to avoid descending to a local minimum too quickly. It is an adaption of the Metropolis–Hastings algorithm for Monte Carlo sampling (Van Laarhoven et al., 1987). The probability of accepting an adverse update is high initially when the temperature is high and as the temperature reduces, that probability decreases.

- Formulations with different cooling schedule, acceptance criteria, and choice of trial models are possible.

Example 9-1

The flowing bottom-hole pressure collected during a single-well drawdown test is provided in the file "history.dat," which can be found on this book's website, www.mhprofessional.com/PRMS, and which contains the daily bottom-hole pressure over 365 days. Use the SA technique to estimate both permeability (k) and skin (s) from the data.

flow rate, $q = 0.025$ m³/s
viscosity, $\mu = 0.001$ Pa·s
formation volume factor, $B = 1$ m³/std. m³
initial pressure, $P_i = 1 \times 10^7$ Pa

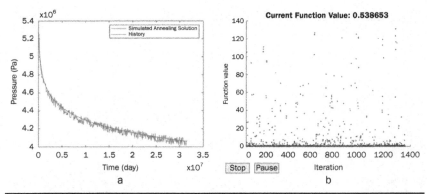

Figure 9-1 History matching using the simulated annealing method: **a)** Final solution. **b)** Objective function evolution. (The color version of this figure is available at www.mhprofessional.com/PRMS.)

net pay thickness, $h = 10$ m
well radius, $r_w = 0.1$ m
porosity, $\phi = 0.1$
total compressibility, $c = 3 \times 10^{-9}$ Pa^{-1}

Solution

The solution is provided on this book's website in the MATLAB® code "SA.m." Two additional functions are called to compute the objective function:

1. "f_model"—computes the forward model $\mathbf{d} = g(\mathbf{m})$. In this case, it computes $P_{wf}(t)$ according to the following solution to the diffusivity equation for an infinite-acting reservoir with a line-source wellbore: $P_{wf} = P_i - \dfrac{q\mu}{4\pi kh}\left[-Ei\left(-\dfrac{\phi\mu c r_w^2}{4kt}\right) + 2s\right].$

2. "mismatch"—computes the objective function: $\|\mathbf{d}-\mathbf{d}_{obs}\|_2/\|\mathbf{d}_{obs}\|_2$, where $\|\cdot\|_2$ refers to the l_2-norm.

The built-in SA function "simulannealbnd" is used. The following bounds are imposed on the two unknowns: $0.01 \leq k$ (Darcy) ≤ 2 and $-5 \leq s \leq 5$. An initial guess of $k = 0.5$ D and $s = 0$ is used. The initial temperature ($T_{initial}$) is set at 1000, and a cooling schedule is defined using the function "cooling," where $T = T_{initial} \times 0.95k$ (where $k =$ iteration count). Finally, an acceptance function is included in "accept_fun," which is used to define the acceptance probability $P = \exp(-\Delta E/T)$. The results are presented in Fig. 9-1. The convergence behavior can be erratic depending on the acceptance criteria and the cooling schedule.

Genetic Algorithm

Genetic algorithm is another stochastic search technique based on principles of natural selection and evolutionary genetics (Whitley,

1994; Man et al., 1996; Mitchell, 1998; Haupt and Haupt, 2004). Examples of history-matching applications can be found in numerous references (Schulze-Riegert et al., 2002; Maschio, 2015).

The algorithm starts with a set of initial solutions, which is called a *population*. Each individual in the population is called a *chromosome*. In the history-matching context, these individuals can be realizations of model parameters sampled from the prior distribution. A procedure called *encoding* is used to represent each chromosome as a binary bit string: a binary bit consists of a value of 0 or 1, and a given model parameter can be represented by a series of binary bits or a *gene*. The procedure aims to evolve the population through a series of iterations or *generations*. During each generation, every individual is evaluated for its *fitness*, which can be understood as the inverse of an objective function. New individuals (*offspring* or *children*) are created from certain members of the population (*parents*). Genetic operators such as crossover and mutation are employed. A new population (next generation) is created by selecting from the previous generation and the newly constructed offspring according to their fitness (while keeping the population size constant). Fitter individuals would have a higher chance of being selected. This process repeats until certain stopping criteria are reached. Here are a few useful additional terminologies:

- Locus: position of a gene on the chromosome
- Allele: value of a gene
- Phenotype: decoded solution
- Genotype: encoded solution

The general GA workflow can be described as follows:

1. **Initialization**—Generate an initial ensemble of N model vectors (initial population, generation $g = 1$). For example, if the unknown model vector contains permeability values at different locations, the initial ensemble would then consist of N realizations of the permeability field.

2. **Binary Encoding and Fitness Evaluation**—Each individual is encoded into binary strings, and the corresponding fitness is evaluated.

3. **Construction of Offspring**

   a. **Reproduction**—Select pairs of models (parents) for reproduction. A common technique is the roulette wheel selection or fitness proportionate selection (Lipowski and Lipowska, 2012):

$$P_s^j = \frac{f_{gj}}{\sum_{j=1}^{N} f_{gj}}$$

P_s^j = probability of *jth* model to be selected. The solutions with better fitness values will have a bigger chance to be selected.

b. **Crossover**—For each pair, a user-defined crossover probability would control whether a crossover would occur. The crossover would select genes from both parents to create a new individual. A few possible mechanisms can be adopted:

    i. One-point crossover—Values of the two strings are exchanged up to a single randomly selected point to form two new individuals.

    ii. Two-point crossover—Values of the two strings are exchanged between two randomly selected points to form two new individuals.

    iii. Uniform crossover—Genes of the chromosomes are exchanged randomly based on a specific mixing ratio.

c. **Mutation**—A portion of the binary bits of the new individuals will be changed randomly ($0 \rightarrow 1$ and $1 \rightarrow 0$).

4. **Replacement**—A new population is created by selecting individuals from the current generation and the offspring. Each offspring is decoded, and the corresponding fitness is computed. Population size should be kept constant. The same roulette wheel concept can be applied here. Another alternative is to apply the Boltzmann selection if the number of offspring is the same as the original population size:

    i. For $j = 1, \ldots, N$, calculate $\Delta F = f_{gj}^{new} - f_{gj}^{old}$ and *acceptance probability* $P = \exp(\Delta F / T)$

    ii. If $\Delta F \geq 0$,

$$\text{Accept } m_{gj}^{new}$$

Else

Draw a random number $s \in U[0, 1]$

$$\text{If } P > s, \text{ accept } m_{gj}^{new}$$

End if

T can be lowered at every generation. This is similar to the SA approach, in which the probability of accepting the offspring increases with each generation.

5. **Assessing Stopping Criterion**—If the stop criterion (e.g. maximum iterations, best fitness below a threshold) is encountered, select the solution with the highest fitness from the latest population to be the optimized solution. Otherwise, repeat from Step 3.

A schematic of the procedure is shown in Fig. 9-2.

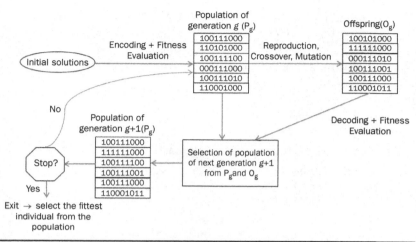

FIGURE 9-2 Genetic algorithm workflow. (The color version of this figure is available at www.mhprofessional.com/PRMS.)

Example 9-2

Consider a model vector consisting of a single parameter of permeability (k), with values ranging between 1490 mD and 1810 mD.

a. How many binary bits are needed to represent a set of values within the designated range, assuming a resolution of 10?

b. Encode the following three individuals into a set of binary strings with a resolution of 10 mD:

- Individual 1: 1760 mD
- Individual 2: 1510 mD
- Individual 3: 1600 mD

c. If the fitness function $= 5^{-10} \times k^3$, calculate the fitness function of the three individuals from Part b and select two individuals for reproduction. Use a random number of 0.38 and 0.66 in the roulette wheel approach.

d. Perform a one-point crossover of the parents from Part c at Position 3 and mutate the resultant offspring at Position 1.

Solution

a. For a resolution or interval of 10, the possible values are 1490, 1500, 1510, 1520, ... , 1800, 1810. There are a total of $(1810 - 1490)/10 = 32$ values. The number of bits required is 5, since $2^5 = 32$.

b. The three individuals can be encoded as follows:

Individual 1: $1760 = 1490 + (2^4 + 2^3 + 0 + 2^1 + 2^0) \times 10 \rightarrow$ 11011

Individual 2: $1550 = 1490 + (0 + 0 + 2^2 + 2^1 + 0) \times 10 \rightarrow 00110$
Individual 3: $1600 = 1490 + (0 + 2^3 + 0 + 2^1 + 2^0) \times 10 \rightarrow$ 01011

c. The fitness functions for the three individuals are,

$f_1 = 5^{-10} \times 1760^3 = 2.73$
$f_2 = 5^{-10} \times 1550^3 = 1.86$
$f_3 = 5^{-10} \times 1600^3 = 2.05$
$f_1 + f_2 + f_3 = 6.64$

Following the roulette wheel approach,

$p_1 = 2.73/6.64 = 0.41$
$p_2 = 1.86/6.64 = 0.28$
$p_3 = 2.05/6.64 = 0.31$

Keeping the same ordering, Individual 1 will be selected if the random draw $r \leq 0.41$, Individual 2 will be selected if $0.41 < r \leq 0.69$, Individual 3 will be selected if $r > 0.69$. Therefore, Individuals 1 and 2 are selected for reproduction.

d. The two selected parents from the previous step are:

Parent 1: 11011
Parent 2: 00110

Crossover (swapping the gene between parents) at
Position 3:   Offspring 1: 11111
                     Offspring 2: 00010

Mutation at Position 1:   Offspring 1: 01111
                                     Offspring 2: 10010

After decoding: Offspring 1: 01111 $\rightarrow$ 1490 + (0 + 2^3 + $2^2 + 2^1 + 20) \times 10 = 1640$
                     Offspring 2: 10010 $\rightarrow$ 1490 + ($2^4 + 0 + 0 + 2^1 + 0) \times 10 = 1670$

Example 9-3
Solve the problem in Example 9-1 using the GA.

Solution
The solution is provided at www.mhprofessional.com/PRMS in the MATLAB code "GA.m," which calls the same functions "f_model" and "mismatch," as shown in Example 9-1. The results are presented in Fig. 9-3. The built-in GA function "ga" is used. The same bounds are imposed on the two unknowns: $0.01 \leq k$ (Darcy) ≤ 2 and $-5 \leq s \leq 5$. Once again, an initial guess of $k = 0.5$ D and $s = 0$ is used. A population size of 1000 is used. The following specifications are employed:

Reproduction—Selection by the roulette wheel.
Crossover—Single-point crossover, crossover probability $= 0.8$.

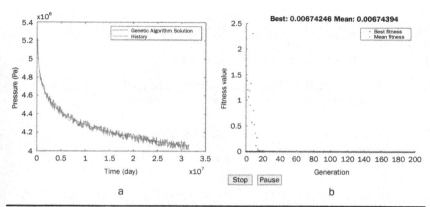

Figure 9-3   History matching using genetic algorithm: **a)** Final solution. **b)** Objective function evolution. (The color version of this figure is available at www.mhprofessional .com/PRMS.)

Mutation—Adaptive scheme is a default option that randomly generates directions in an adaptive fashion, according to the previous generations; it ensures the specified bounds for k and s are satisfied. **Replacement**—Elite count = 5% of population size (number of individuals from the current generation that will be selected and advanced to the next generation).

When compared to the results of SA, the objective function decreases more quickly with the number of iterations/generations.

Particle Swarm Optimization

Particle swarm optimization (PSO) is a stochastic search technique inspired by the social behavior of bird flocking or fish schooling (Eberhart and Kennedy, 1995). Since its introduction, many variants and improvements to the original formulation have been published (Kennedy et al., 2001; Engelbrecht, 2006; Poli et al., 2007). Its application in history matching can be found in many references (Mohamed, 2010).

Similar to other evolutionary computation methods, such as GA, it is also a population-based algorithm. An initial group of solutions, called particles, would fly through the model parameter space in search of the global minimum by following the current optimum ("best") particles. A particle can be considered as a realization of the unknown reservoir parameter.

Instead of genetic operations, the velocity of each particle v is updated. The velocity essentially controls the amount of perturbation to be made to each particle. During every iteration, a given particle i would update its position by following two "best" values: (1) $\mathbf{p}_i^{best}$—the best solution (highest fitness) it has achieved over all the past iterations up to now (i.e., its own personal best) and (2) $\mathbf{g}^{best}$—the best solution achieved by any particle over all the past iterations up to now (i.e., a global best).

The general PSO workflow can be described as follows:

1. For each particle, $i = 1, ..., S$, initialize the particle's position to be $\mathbf{x}_i$ and velocity to be $\mathbf{v}_i$, calculate its fitness $F(\mathbf{x}_i)$, and set $\mathbf{p}_i^{best} = \mathbf{x}_i$.

2. Choose the particle with the highest fitness value and assign its position to $\mathbf{g}^{best}$.

3. For each particle:

   • Calculate its velocity in each dimension (d), which is the number of model parameters:

$$v_{i,d}^{new} = v_{i,d}^{old} + c_1 r_{ip,d}\left(p_{i,d}^{best} - x_{i,d}\right) + c_2 r_{ig,d}\left(g_d^{best} - x_{i,d}\right)$$

$$r_{ip,d}, r_{ig,d} \in U[0, 1]$$

   c_1 and c_2 are user-defined learning factors. The idea is to determine its velocity (i.e., amount of updating) based on the distance between the particle's current position and the best solutions ($\mathbf{p}_i^{best}$ and $\mathbf{g}^{best}$).

   • Update its position:

$$\mathbf{x}_i^{new} = \mathbf{x}_i^{old} + \mathbf{v}_i$$

4. If $F(\mathbf{x}_i^{new}) < F(\mathbf{p}_i^{best})$, $\mathbf{p}_i^{best} \leftarrow \mathbf{x}_i^{new}$

   If $F(\mathbf{p}_i^{best}) < F(\mathbf{g}_i^{best})$, $\mathbf{g}^{best} \leftarrow \mathbf{p}_i^{best}$

5. Repeat from Step 3 until the stop criterion (e.g. maximum iterations, best fitness below a threshold) is encountered. Select the solution with the highest fitness from the latest population to be the optimized solution.

Example 9-4

Solve the problem in Example 9-1 using the PSO method.

Solution

The solution is provided on this book's website in the MATLAB code "PSO.m," which calls the same functions "f_model" and "mismatch," as in Example 9-1. The built-in particle swarm function "particleswarm" is used. The same bounds are imposed on the two unknowns: $0.01 \leq k$ (Darcy) ≤ 2 and $-5 \leq s \leq 5$. Once again, an initial guess of $k = 0.5$ D and $s = 0$ is used. A swarm size of $100 \times$ number of model parameters $= 200$ is used. The two user-defined learning factors (c_1 and c_2) are assumed to be 1.5. The results are presented in Fig. 9-4.

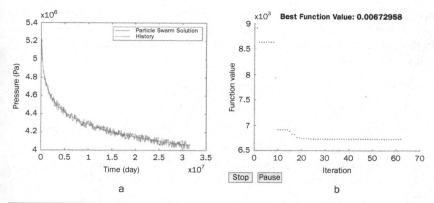

FIGURE 9-4   History matching using the particle swarm optimization method: **a)** Final solution. **b)** Objective function evolution. (The color version of this figure is available at www.mhprofessional.com/PRMS.)

Remark   All the optimization techniques covered in Sec. 9.2 aim to progressively minimize a given objective function over many iterations. During each updating, there is no mechanism to enforce that the reference or prior distribution, as well as conditioning data informed by static data, be honored. If consistency with static data is desired, several choices are available:

- Incorporate mismatch in static data as part of the objective function.

- Re-parameterize the model parameter vector into a set of secondary variables, which are updated using the aforementioned techniques. At the end of each iteration, a new realization of the original model parameter vector is generated based on the conditioning data and the updated secondary variables.

- Use master or pilot point scheme (will be discussed in the next section).

9.3   Probabilistic Schemes

9.3.1   Optimization-Based Bayesian Methods

Characterization of the full a posteriori probability density function (PDF) for the model parameters is often challenging. According to Bayes' rule:

$$p(m \mid d_{obs}) \propto p(d_{obs} \mid m)p(m) \tag{9-20}$$

$p(d_{obs} \mid m)$ is the likelihood function, and $p(m)$ is the prior probability. However, if both the prior and noise are Gaussian, the posterior PDF will also be Gaussian. In particular, for a linear inverse problem (Oliver et al., 2008):

$$p(m \mid d_{obs})$$
$$\propto \exp\left[-\frac{1}{2}(d_{obs} - Gm)^T C_D^{-1}(d_{obs} - Gm) - \frac{1}{2}(m - m_{pr})^T C_M^{-1}(m - m_{pr})\right]$$

$$(9\text{-}21)$$

C_D and C_M refer to the data and model covariance matrices, respectively. The model parameters with the maximum probability density can be obtained by minimizing the following objective function:

$$O(m) = \frac{1}{2}(Gm - d_{obs})^T C_D^{-1}(Gm - d_{obs}) + \frac{1}{2}(m - m_{pr})^T C_M^{-1}(m - m_{pr})$$

$$(9\text{-}22)$$

$\nabla O(m)$ and $H(m)$ are expressed as:

$$\nabla O(m) = C_M^{-1}(m - m_{pr}) + G^T C_D^{-1}(Gm - d_{obs}) \qquad (9\text{-}23)$$

$$H(m) = C_M^{-1} + G^T C_D^{-1}G \qquad (9\text{-}24)$$

Given that $H(m)$ is positive definite, a unique global minimum of $O(m)$ can be found. The model vector that would yield the maximum a posteriori probability density, or the MAP estimate, can be found by setting :

$$m_{map} = m_{pr} + \left(C_M^{-1} + G^T C_D^{-1}G\right)^{-1} G^T C_D^{-1}\left(d_{obs} - Gm_{pr}\right) \quad (9\text{-}25)$$

It is easy to see that Eq. (9-25) is identical to Eq. (9-9) if we set $W_e = CD^{-1}$ and $\varepsilon^2 W_m = CM^{-1}$. However, for non-linear inverse problems, an analytical expression for m_{map} is impractical. Recast Eq. (9-22) for a general non-linear forward model $d = g(m)$:

$$O(m) = \frac{1}{2}(g(m) - d_{obs})^T C_D^{-1}(g(m) - d_{obs})$$
$$+ \frac{1}{2}(m - m_{pr})^T C_M^{-1}(m - m_{pr}) \qquad (9\text{-}26)$$

The MAP estimate is one that minimizes the above objective function. $\nabla O(m)$ and $H(m)$ are expressed as:

$$\nabla O(m) = C_M^{-1}(m - m_{pr}) + G^T C_D^{-1}(g(m) - d_{obs}) \qquad (9\text{-}27)$$

$$H(m) = C_M^{-1} + G^T C_D^{-1}G + \left(\nabla G^T\right)C_D^{-1}(g(m) - d_{obs}) \quad (9\text{-}28)$$

G is the sensitivity matrix with a dimension of $N \times M$:

$$G(m) = \begin{bmatrix} (\nabla_m g_1)^T \\ (\nabla_m g_2)^T \\ \cdots \\ (\nabla_m g_N)^T \end{bmatrix} \tag{9-29}$$

Its element at row i and column j refers to the sensitivity of ith data (d_i) with respect to the jth model parameter (m_j):

$$g_{ij} = \frac{\partial g_i(m)}{\partial m_j} \tag{9-30}$$

If we ignore the last term in Eq. (9-28) and substitute that, together with Eq. (9-27), into Eq. (9-13), we obtain the Gauss–Newton approximation:

$$\left(C_M^{-1} + G^T C_D^{-1} G \right) \delta m^{l+1}$$
$$= -\left[C_M^{-1} \left(m^l - m_{pr} \right) + G^T C_D^{-1} \left(g(m^l) - d_{obs} \right) \right]$$

If the prior model, $p(m)$, is known, the Bayesian framework can be used to integrate this knowledge:

$$p(m \mid d_{obs}) = \frac{p(d_{obs} \mid m) p(m)}{p(d_{obs})} \tag{9-31}$$

The explicit knowledge of $p(d_{obs})$ is unnecessary, as it can be treated as a constant:

$$p(m \mid d_{obs}) = C \times p(d_{obs} \mid m) p(m) = C \times L(m \mid d_{obs}) p(m) \tag{9-32}$$

The likelihood function, $L(m \mid dobs)$, represents the likelihood of an actual observation being a realization of $dobs$, which is a function of the true model vector. The MAP estimate is one that maximizes the a posteriori PDF in Eq. (9-32). In many cases, instead of generating a single MAP estimate, it is desired to sample multiple realizations by sampling approximately from the a posteriori PDF. In the randomized maximum likelihood (RML) scheme, a different set of (mu, du) is sampled for each realization (r): du is obtained by adding a realization of the measurement error to the actual observation, while mu is an unconditional realization sampled from the prior distribution. The following objective function, which is a slight modification of Eq. (9-26), is minimized:

$$O_r(m) = \frac{1}{2}(g(m) - d_u)^T C_D^{-1}(g(m) - d_u) + \frac{1}{2}(m - m_u)^T C_M^{-1}(m - m_u) \tag{9-33}$$

This particular implementation can be referred to as rough plus smooth (Oliver et al., 2008): a "rough" or unconditional realization is first generated, and it is subsequently subjected to a "smooth" correction that forces the realization to pass through the data.

Any important consideration, which has not been addressed so far, is the computation of G. It consists of sensitivity coefficients describing the derivative of each data with respect to each model parameter. Even in those formulations that do not involve G explicitly, the calculation of the gradient of the objective function requires the knowledge of these sensitivity coefficients. For instance, if the objective function is defined as the mismatch in well pressure, its gradient would entail computing the derivative of data (well pressure in this case) with respect to the model parameters.

One option is to write the discretized simulation equations as $f(y, m) = 0$, where y is a $N_y \times 1$ vector of dependent variables (e.g., the pressure at each grid location) and m is an $M \times 1$ vector of model variables, and solve the following system of equations for the $Ny \times 1$ vector of $(i = 1, \dots, M)$:

$$\left(\nabla_y f^T\right)^T \frac{\partial y}{\partial m_i} = -\left(\nabla_m f^T\right)^T \frac{\partial m}{\partial m_i} \tag{9-34}$$

$\left(\nabla_m f^T\right)^T$ and $\left(\nabla_y f^T\right)^T$ are $N_y \times M$ and $N_y \times N_y$ matrices, respectively:

$$\left(\nabla_m f^T\right)^T = \begin{bmatrix} \left(\nabla_m f_1\right)^T \\ \left(\nabla_m f_2\right)^T \\ \dots \\ \left(\nabla_m f_{N_y}\right)^T \end{bmatrix} ; \left(\nabla_y f^T\right)^T = \begin{bmatrix} \left(\nabla_y f_1\right)^T \\ \left(\nabla_y f_2\right)^T \\ \dots \\ \left(\nabla_y f_{N_y}\right)^T \end{bmatrix} \tag{9-35}$$

Once $\partial y / \partial m_i$ are solved, they are substituted in the following equations to compute dF / dm_i:

$$\frac{dF}{dm_i} = \left(\nabla_y h\right)^T \frac{\partial y}{\partial m_i} \tag{9-36}$$

where $F = h(y)$ is a real-valued function dependent on y (e.g., objective function or predicted data). It is usually quite simple, e.g., saturation at well location is the same as well block saturation; therefore, $(\nabla_y h)^T$ should be straightforward to compute.

This method of calculating G is rather cumbersome and computationally intensive, as it requires solving a system of equations with the same dimension as the actual simulation system for each $\partial y / \partial m_i$. This is particularly undesirable if there are many model parameters, especially when using stochastic realizations of porosity and permeability distribution, where the properties are different everywhere.

Another alternative is to compute $\partial y/\partial m_i$ numerically via finite differences.

$$\frac{\partial y}{\partial m_i}$$
$$\approx \frac{y(m_1, m_2, ..., m_{i-1}, m_i, m_{i+1}, ..., m_M) - y(m_1, m_2, ..., m_{i-1}, m_i + \delta m_i, m_{i+1}, ..., m_M)}{\delta m_i} \quad (9\text{-}37)$$

Choosing the correct perturbation size for m_i is difficult. Besides, the computational demand is similar to the previous approach.

The final approach commonly adopted in history matching is the adjoint method. The idea is to find a way to eliminate the need to compute $\partial y/\partial m_i$ in the intermediate step when computing $\partial y/\partial m_j$. Before introducing the method, consider the following: if we rearrange Eq. (9-34) and then add to Eq. (9-36), we would obtain the total derivative of F as:

$$\frac{dF}{dm_i} = (\nabla_y h)^T \frac{\partial y}{\partial m_i} + (\nabla_y f^T)^T \frac{\partial y}{\partial m_i} + (\nabla_m f^T)^T \frac{\partial m}{\partial m_i} \quad (9\text{-}38)$$

Now, let us adjoin the function F as follows:

$$F = h(y) + \lambda^T f(y, m) \quad (9\text{-}39)$$

The adjoint vector λ consists of a set of arbitrary multipliers because, as stated earlier, $f(y, m) = 0$. The total derivative of F becomes:

$$\frac{dF}{dm_i} = (\nabla_y h)^T \frac{\partial y}{\partial m_i} + \lambda^T (\nabla_y f^T)^T \frac{\partial y}{\partial m_i} + \lambda^T (\nabla_m f^T)^T \frac{\partial m}{\partial m_i} \quad (9\text{-}40)$$

It can be readily shown that if λ satisfies the following equation, the terms involving $\partial y/\partial m_i$ would disappear from Eq. (9-40):

$$-(\nabla_y h) = (\nabla_y f^T)^T \lambda \quad (9\text{-}41)$$

λ has a dimension of $N_y \times 1$. For this method, Eq. (9-41) needs to be solved once for λ. Its solution does not depend on m at all. If F is used to compute specific gi, Eq. (9-41) needs to be repeated for as many adjoint systems as there are data observations.

$$\frac{dF}{dm_i} = \lambda^T (\nabla_m f^T)^T \frac{\partial m}{\partial m_i} \quad (9\text{-}42)$$

The method can be summarized as follows: first, solve the forward simulation equation for y; next, use the calculated y in Eq. (9-41) to solve for λ; and finally, apply Eq. (9-42) to solve for $\partial F/\partial m_i$. The advantage of the adjoint method is to "screen out" the $\partial y/\partial m_i$ terms (derivative of every dependent variable with respect to every model parameter). It saves us from solving Eq. (9-34) repeatedly for all vectors of $\partial y/\partial m_i$. Computation of $(\nabla_y f^T)$ and $(\nabla_m f^T)^T$ is not so difficult if we recall that f is a discretized linear system of equations in y, as implemented in reservoir simulations. Detail of this method for transient multi-phase flow can be found in Oliver et al.'s book (2008). For transient problems, the adjoint field is a function of time, and it can be solved backward in time (once the forward simulation is computed). Besides, given that $(\nabla_y f^T)$ in Eq. (9-41) are the same as those required for the fully implicit forward simulation. Therefore, the solution of the adjoint system can be achieved rather efficiently (Li et al., 2003). The adjoint formulation for compositional flow can be found in Kourounis et al.'s research (2014).

9.3.2   Sampling Algorithms

The techniques presented in this section aim to draw samples or realizations from the a posteriori PDF for the model parameters directly. In the case of history matching, the dimension of model parameter space is large, so a simple implementation of a rejection sampling scheme is generally not capable of sampling from the PDF efficiently.

Markov Chain Monte Carlo

The idea is to get an ensemble of possible states or realizations of the reservoir permeability distribution, which can be thought of as a Markov chain if the probability of generating a new realization depends on the previous realization in the sequence (Gilks, 1995). The last realization of the sequence is one that is sampled from the correct PDF.

Assume each realization is associated with a finite probability, π_i for ith realization (m_i). If we want to perform a Monte Carlo simulation, we must draw these realizations with the correct probability. However, in real applications of history matching, it is extremely difficult to characterize π_i precisely. For instance, the proportionality constant in Eq. (9-21) is not defined. Therefore, it is difficult to ensure that the ith realization is being sampled at the correct π_i. The idea of Markov chain Monte Carlo (McMC) is to make use of the fact that the relative ratio of probabilities between two realizations is easy to compute (e.g., the proportionality constant is canceled out), and that ratio can be used in the specification of $p_{ij} = p(m_j \mid m_i)$, which is the conditional probability of transitioning from state/realization m_i to m_j. It offers a way to transition a chain of possible states, such that the final state is sampled from the correct probability distribution:

$$\pi_j = \sum_i \pi_i p_{ij} \qquad (9\text{-}43)$$

The chain is stationary and ergodic if there exist a finite number of states between any initial and the final state (Feller, 1968). Metropolis et al. (1953) proposed the following equations to simplify the problem:

$$\pi_i p_{ij} = \pi_i p_{ij} \qquad (9\text{-}44)$$

In addition,

$$p_{ij} = \alpha_{ij} q_{ij} \qquad (9\text{-}45)$$

where q_{ij} = probability of proposing a transition from state i to state j, and α_{ij} = probability of acceptance. Hastings (1970) proposed the following formulation of α_{ij}:

$$\alpha_{ij} = \min\left(1, \frac{\pi_j q_{ji}}{\pi_i q_{ij}}\right) \tag{9-46}$$

This is referred to as the Metropolis–Hastings sampling, while Gibbs sampling is one where $\alpha_{ij} = 1$ (a proposal is always accepted) (Gilks, 1995). The choice of q_{ij} is somewhat arbitrary. Another option is to make q_{ij} a function of m_i, such that the probability of proposing the next state depends on the current state. If $q_{ij} = q_{ji}$ (symmetry), then Eq. (9-46) becomes:

$$\alpha_{ij} = \min\left(1, \frac{\pi_j}{\pi_i}\right) \tag{9-47}$$

Furthermore, if π_i follows Eq. (9-21),

$$\pi_i = p(m_i \mid d_{obs}) \tag{9-48}$$

$$\propto \exp\left[-\frac{1}{2}(d_{obs} - Gm_i)^T C_D^{-1}(d_{obs} - Gm_i) - \frac{1}{2}(m_i - m_{pr})^T C_M^{-1}(m_i - m_{pr})\right]$$

Equation (9-47) becomes:

$$\alpha_{ij} = \min \tag{9-49}$$

$$\left(1, \frac{\exp\left[-\frac{1}{2}(d_{obs} - Gm_j)^T C_D^{-1}(d_{obs} - Gm_j) - \frac{1}{2}(m_j - m_{pr})^T C_M^{-1}(m_j - m_{pr})\right]}{\exp\left[-\frac{1}{2}(d_{obs} - Gm_i)^T C_D^{-1}(d_{obs} - Gm_i) - \frac{1}{2}(m_i - m_{pr})^T C_M^{-1}(m_i - m_{pr})\right]}\right)$$

$$\alpha_{ij} \min\left(1, \frac{\exp[-E_j]}{\exp[-E_i]}\right) = \min\{1, \exp[-(E_j - E_i)]\}$$

$$= \min\{1, \exp[-\Delta E]\} \tag{9-50}$$

Equation (9-50) essentially describes the acceptance criterion adopted in the SA implementation shown in Sec. 9.2.2. If the error of the proposed state is less than that of the current state (i.e., $E_j < E_i$), $\alpha_{ij} = 1$ (the proposal is accepted); if the error of the proposed state is greater than that of the current state (i.e., $E_j > E_i$), the acceptance probability, a_{ij}, becomes $\exp(-\Delta E)$. This illustrates that SA is a special implementation of McMC with Metropolis–Hastings sampling.

Gradual Deformation

The idea of gradual deformation (GD) is to propose gradual perturbations of an initial realization, such that the perturbed realization matches the data better (Hu, 2000; Hu et al., 2001). Although the perturbations are continuous or gradual, the variable being perturbed can also be categorical (Caers, 2007). The method searches for an optimal linear combination of two Gaussian reservoir models, with identical covariances, to produce a new model that has the same covariance as the original models and matches the data better (Le Ravalec-Dupin and Nœtinger, 2002). Recent examples of GD in history matching can be found in Perozzi et al.'s research (2016).

Consider a standard Gaussian random vector **m** can be written as a linear combination of two other independent standard Gaussian vectors, whose components can be correlated according to a covariance matrix **C**:

$$\mathbf{m} = \mathbf{m}_1 \cos r = \mathbf{m}_2 \sin r \qquad (9\text{-}51)$$

As r increases, any outcome of m will change "gradually" from m1 ($r = 0$) to $\mathbf{m}_2$ ($r = \pi/2$). In the context of history matching, an objective function can be formulated as:

$$O(\mathbf{m}) = ||\mathbf{d} - g(\mathbf{m}(r))|| \qquad (9\text{-}52)$$

The objective function is minimized by modifying the value of r (e.g., using any technique described in Sec. 9.2) or even a simpler 1D optimization scheme such as the golden section search method (Kiefer, 1953) or its variant, the Dekker–Brent method (Brent, 1971). A single optimization of r is typically not sufficient to achieve a reasonable match; therefore, the process is performed iteratively. A histogram transformation can be used to transform the standard Gaussian realization to honor the marginal distribution of the prior model. The basic algorithm can be summarized as:

1. Generate two Gaussian realizations $\mathbf{m}_1$ and $\mathbf{m}_2$; sample a random realization of **m**.

2. Repeat the following until **d** is matched:

   a. Minimize Eq. (9-52) $\rightarrow$ get r_{opt}

   $$r_{opt} = \min_r \left\{ ||\mathbf{d} - g(\mathbf{m}_1 \cos r + \mathbf{m}_2 \sin r)|| \right\} \qquad (9\text{-}53)$$

   b. Set $\mathbf{m}_1 \leftarrow \mathbf{m}(r_{opt})$

   Sample a random realization of **m**

   Set $\mathbf{m}_2 \leftarrow \mathbf{m}$

For more complex problems, the values of r may vary at each location. However, the simplicity of a 1D optimization problem [Eq. (9-53)] is lost (Caers, 2007). Hu and Le Ravalec-Dupin (2004) presented a gradient-based technique for solving this multi-dimensional extension.

The formulation can also be adopted for conditional sequential simulation. Hu et al. (2001) described a GD technique for sequential simulations. Consider a vector **u**, which consists of random numbers for drawing from the local conditional cumulative density functions. It is related to the Gaussian noise as follows:

$$G^{-1}(\mathbf{u}) = \mathbf{y} \qquad (9\text{-}54)$$

Therefore, $\mathbf{y}$ can be perturbed following the GD algorithm to yield $\mathbf{y}(r)$, and back transform can be used to obtain $\mathbf{u}(r) = G(\mathbf{y}(r)) = G(\mathbf{y}_1 \cos r + \mathbf{y}_2 \sin r)$, which is used, together with a fixed random path, in the sequential simulation to generate a perturbed model. Alternative adaption of this global perturbation technique can be found in the research works of Mata-Lima (2008) and Caeiro et al. (2015), which implemented direct sequential image transformation to perturb permeability maps.

Probability Perturbation

The probability perturbation method is a perturbation approach based on the Bayesian updating framework. The method was originally developed for categorical variables (Caers and Hoffman, 2006), but it has since been extended to modeling continuous variables in Caers (2007). Examples of PPM in history-matching applications can be found in Kashib and Srinivasan's research (2006). Additional examples of fractured reservoirs can be found in Suzuki et al.'s research (2007).

If $\mathbf{m}$ is a model vector of categorical variables, its probability distribution conditional to hard data and previously simulated values $(\mathbf{b})$ can be expressed as $P(\mathbf{m} \mid \mathbf{b})$. This probability is relatively easy to obtain (e.g., via sequential simulation). Similarly, the probability distribution conditional to dynamic data can be expressed as $P(\mathbf{m} \mid \mathbf{d})$ The initial realization of indicators is represented by i^0. To generate a perturbation,

$$P(\mathbf{m} \mid \mathbf{d}) = (1 - r)i^0 = rP(\mathbf{m}) \qquad (9\text{-}55)$$

Both probabilities can be combined using the "tau model," which is based on the hypothesis of permanence of ratios (Journel, 2002) to obtain $P(\mathbf{m} \mid \mathbf{b}, \mathbf{d})$:

$$P(\mathbf{m} \mid \mathbf{b}, \mathbf{d}) = \frac{1}{1 + x}$$

$$\frac{x}{a} = \left(\frac{b}{a}\right)^{\tau_1} \left(\frac{d}{a}\right)^{\tau_2} \qquad (9\text{-}56)$$

where $b = \dfrac{1 - P(\mathbf{m} \mid \mathbf{b})}{P(\mathbf{m} \mid \mathbf{b})}; d = \dfrac{1 - P(\mathbf{m} \mid \mathbf{d})}{P(\mathbf{m} \mid \mathbf{d})}; a = \dfrac{1 - P(\mathbf{m})}{P(\mathbf{m})}$

The parameters τ_1 and τ_2 capture the redundancy between the two data information. The method is called probability perturbation because $P(\mathbf{m} \mid \mathbf{b})$ is perturbed using $P(\mathbf{m} \mid \mathbf{d})$ (Caers, 2007). The distributions $P(\mathbf{m} \mid \mathbf{b})$ and $P(\mathbf{m} \mid \mathbf{d})$ are termed pre-posterior, as they inform

$P(\mathbf{m})$ prior to considering the other data. At each node to be simulated, a new value is drawn from $P(\mathbf{m} \mid \mathbf{b}, \mathbf{d})$. The algorithm can be described as follows:

1. Choose a random seed s and generate an initial realization $i^{0(s)}$ based on $\mathbf{b}$; compute the corresponding objective function.

2. Repeat the following until $\mathbf{d}$ is matched:

    a. Choose a random seed s' and generate another realization.

    b. Repeat the following until r is optimized: $r_{opt} = \min\limits_{r}$ $\left\{ \left\| \mathbf{d} - g\left(i_r^{(s')} \right) \right\| \right\}$

        i. Set an initial value of r.

        ii. Sequentially visiting all estimation nodes $(j = 1, \dots, M)$, compute $P(m_j \mid \mathbf{b}_j)$ and $P(m_j \mid \mathbf{b}_j, \mathbf{d})$ according to Eqs. (9-55) and (9-56), respectively. A subscript j is added to $\mathbf{b}_j$ to emphasize that the hard data keep expanding when the previously simulated node is added. A new realization, $i_r^{(s')}$, which is based on a particular value of r, is generated.

        iii. Evaluate the objective function.

    c. If the objective function decreases, set $i^{0(s)} \leftarrow i_{r_{opt}}^{(s')}$; else, $i^{0(s)}$ remains unchanged.

A few remarks should be made. First, the entire Step 2 is often referred to as the outer loop, while the sub-step (b) is referred to an inner loop. Second, the new realization is a function of r [Eq. (9-55)], since r controls the perturbation of the initial realization: when $r = 0$, $P(\mathbf{m} \mid \mathbf{d}) = P(\mathbf{m} \mid \mathbf{b}, \mathbf{d}) = i^0 \rightarrow$ no perturbation; when $r = 1$, $P(\mathbf{m} \mid \mathbf{d}) = = P(\mathbf{m})$, $P(\mathbf{m} \mid \mathbf{b}, \mathbf{d}) = P(\mathbf{m} \mid \mathbf{b})$.

Kashib and Srinivasan (2006) presented a modified approach for handling indicator-based variables, where Eq. (9-55) is modified to describe the probability of transitioning from the indicator category k at step l to the category k' at step $l + 1$:

$$P\left(I^{l+1}(\vec{u}) = k' \mid I^l(\vec{u}) = k, \mathbf{d} \right) = rP\left(I^l(\vec{u}) = k' \right); \forall k' \neq k$$
$$P\left(I^{l+1}(\vec{u}) = k \mid I^l(\vec{u}) = k, \mathbf{d} \right) = 1 - \sum_{k' \neq k} rP\left(I^l(\vec{u}) = k' \right) \quad (9\text{-}57)$$

Caers (2007) offered an in-depth comparison between the GD and PPM approaches. In contrast to GD, PPM does not assume Gaussian distributions. A random path needs to be fixed in GD when perturbing realizations using sequential simulation; this is to ensure that the perturbations are small (gradual) and not completely random. However, this requirement is not necessary for PPM. Similar to GD, a multi-parameter PPM can be implemented by setting r to be spatially varying.

For a continuous variable, a co-kriging framework can be adopted to replace Eqs. (9-55) and (9-56) with Eq. (9-58) (Caers, 2007). Assuming the model vector is described by a Gaussian vector $\mathbf{Y}$ and its prior model is defined by the correlogram of $\mathbf{Y}$ (i.e., ρ_Y), a perturbation of an initial realization y^0 can be obtained via sequential simulation, where the local conditional cumulative density function at a location $\mathbf{u}$ is characterized by the mean and SK variance:

$$y_{SK}^*(\mathbf{u}) = \sum_{\alpha=1}^{N_\alpha} \lambda_\alpha y(\mathbf{u}_\alpha) + \lambda y^0(\mathbf{u}) \tag{9-58}$$

The correlogram between $y^0(\mathbf{u})$ and $y(\mathbf{u})$ is:

$$\rho_{YY^0} = (1 - r)\rho_Y \tag{9-59}$$

The idea is to utilize the initial realization as soft data. The redundancy between the two data is captured via the cross-variogram (not the tau model) here. Since the soft data follow the same correlogram as the hard data, the generated realization will reproduce the variogram. The algorithm can be described as follows:

1. Choose a random seed s and generate an initial realization $y^{0(s)}$ based on $\mathbf{b}$; compute the corresponding objective function.

2. Repeat the following until $\mathbf{d}$ is matched:

   a. Choose a random seed s' and generate another realization;

   b. Repeat the following until r is optimized: $r_{opt} = \min_r \left\{ \left\| \mathbf{d} - g\left(y_r^{(s')}\right) \right\| \right\}$

      i. Set an initial value of r.

      ii. Sequentially visit all estimation nodes and generate a new realization, $y_r^{(s')}$, according to Eq. (9-58).

      iii. Evaluate the objective function.

   c. If the objective function decreases, set $y^0 \leftarrow y_{r_{opt}}^{(s')}$; else, $y^{0(s)}$ remains unchanged.

Example 9-5

Solve the problem in Example 9-1 using PPM.

Solution

The solution is provided in the MATLAB code "PPM.m," which can be found on this book's website, www.mhprofessional.com/PRMS, and which calls the same functions "f_model" and "mismatch," as in Example 9-1. The results are shown in Figs. 9-5 and 9-6.

The same bounds are imposed on the two unknowns: $0.01 \leq k$ (Darcy) ≤ 2 and $-5 \leq s \leq 5$. Both model parameters are represented

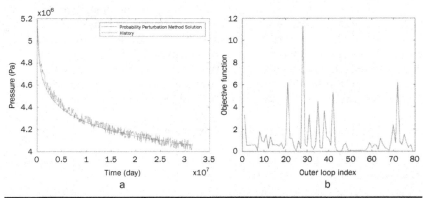

FIGURE 9-5   History matching using PPM: **a)** Final solution. **b)** Objective function evolution. (The color version of this figure is available at www.mhprofessional.com /PRMS.)

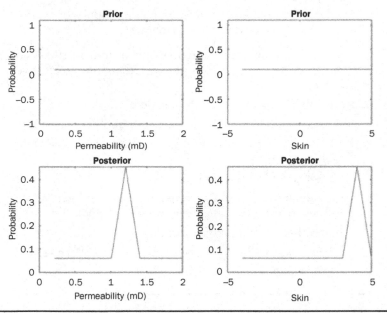

FIGURE 9-6   Prior and posterior probability density functions for the PPM example. (The color version of this figure is available at www.mhprofessional.com/PRMS.)

by indicators (10 bins are selected). $P(\mathbf{m})$ is assumed to be uniformly distributed. There is no conditioning data for this example, so $\mathbf{b}$ can be ignored. Therefore, Step 2 in the PPM workflow is simplified as:
    Outer loop—Repeat the following until $\mathbf{d}$ is matched:
        a. Choose a random seed s' and generate another realization.

b. **Inner loop**—Repeat the following until r is optimized:

$$r_{opt} = \min_r \left\{ \left\| \mathbf{d} - g\left(i_r^{(s')}\right) \right\| \right\}$$

i. Set an initial value of r.

ii. Compute $P(m_j \mid \mathbf{d})$ according to Eq. (9-55). A new realization, $i_r^{(s')}$, which is based on a particular value of r, is generated.

iii. Evaluate the objective function.

c. If the objective function decreases, set $i^{0(s)} \leftarrow i_{r_{opt}}^{(s')}$; else, $i^{0(s)}$ remains unchanged.

For the inner loop, a simple built-in constrained optimization function called "fminbnd" is used to find r_{opt}. It utilizes the function "r_search" to carry out Step b(ii) for a particular r value: (1) compute $P(m_j \mid \mathbf{d})$, (2) generate a new realization, and (3) evaluate the objective function. The convergence behavior is quite erratic, since, unlike other methods, the one deformation parameter bears the load of optimizing the objective function.

Various parameterization techniques are often adopted. For example, a reduced-order representation of the model parameter space can be achieved with orthogonal decomposition methods such as principal component analysis (PCA), Karhunen–Loève (K-L) transform, and singular value decomposition (SVD) (Oliver, 1996; Sarma et al., 2006, 2008; Thomas, 2010; Zeng and Zhang, 2010; Chang and Zhang, 2014; Khaninezhad and Jafarpour, 2014; Vo and Durlofsky, 2015; Afra and Gildin, 2016). Bhark et al. (2011) explained that K-L basis functions are calculated from the covariance of the prior (which must be assumed a priori), but the scheme is flexible on grid structure; on the other hand, in discrete cosine transform (DCT) and discrete wavelet transform (DWT), basis functions are independent of the prior covariance, but they are limited to structured meshes; therefore, the authors proposed a generalized grid-connectivity-based transform, whose basis functions are independent of prior covariance and flexible regarding grid structure. Yadav et al. (2005) presented a domain decomposition method for delineating different regions of the reservoir for which the deformation parameter r is calibrated. They implemented the probability perturbation algorithm in parallel CPUs with each node performing the parameter optimization independently and the final solution put together without any artifacts. In Barrera's dissertation (2007), geological realism is ensured by updating effective rock type and connectivity index (CI), while porosity, permeability, and multi-phase flow functions are assigned to each CI category. A reduced-order representation can also be achieved with the use of pilot points (RamaRao et al., 1995; LaVenue et al., 1995). A subset of the model parameters is updated during history matching, and conditional simulation is performed afterward to estimate the remaining model parameters, with the assimilated model parameters being the conditioning data.

9.4 Ensemble-Based Schemes

9.4.1 Ensemble Kalman Filters

Ensemble Kalman filter, or EnKF, is a sequential data assimilation scheme. It provides an ensemble of history-matched models, as well as calibration of both state variables (e.g., saturation and pressure) and model variables. Different types of assimilation data can be included. A review of recent developments can be found in a number of references (Aanonsen et al., 2009).

The main idea about sequential data assimilation is that, instead of assimilating all the data (e.g., the entire production history) simultaneously, the data are partitioned into various sets (e.g., production history at a certain instance of time), which are being assimilated in a sequential fashion. The advantage is that whenever new data become available, the model is updated to assimilate only the new data, without considering other data that have been previously assimilated. This feature can be particularly useful in applications where new data are being collected continuously. It can be shown that for linear inverse problems with independent Gaussian errors, the consequence of assimilating all data is the same as sequentially assimilating individual datasets (Kalman, 1960).

The original Kalman filter (Kalman, 1960) consists of two steps: (1) forecast and (2) analysis. The objective is to estimate the current state (x_k) of a given system. The first step entails forecasting the probability $p(x_k \mid D_{k-1})$, while the second step is to conditioning the prior $p(x_k \mid D_{k-1})$ to the observation at t_k and estimate $p(x_k \mid D_k)$.

Forecast

$$x_k^f = A_k x_{k-1}^a$$
$$C_k^f = A_k C_{k-1} A_k^T \tag{9-60}$$

Analysis   Consider the following:

$$p(x_k \mid D_k) \propto p(d_{obs,k} \mid x_k, D_{k-1}) p(x_k \mid D_{k-1}) \tag{9-61}$$

$$p(x_k \mid D_k) \propto \exp\left[-\frac{1}{2}(G_k x_k - d_{obs,k})^T C_{D,k}^{-1}(G_k x_k - d_{obs,k})\right]$$
$$\exp\left[-\frac{1}{2}(x_k - x_k^f)^T (C_k^f)^{-1}(x_k - x_k^f)\right],$$

where $D_k = \begin{bmatrix} d_{obs,1} & d_{obs,2} & \cdots & d_{obs,k} \end{bmatrix}^T$. Maximizing the above probability gives the optimal estimate of x_k:

$$x_k^a = x_k^f + K\left(d_{obs,k} - G_k x_k^f\right) \qquad (9\text{-}62)$$

K is the Kalman gain:

$$K = C_k^f G_k^T \left(G_k C_k^f G_k^T + C_{D,k}\right)^{-1} \qquad (9\text{-}63)$$

The posterior covariance becomes:

$$C_k^a = C_k^f - C_k^f G_k^T \left(G_k C_k^f G_k^T + C_{D,k}\right)^{-1} G_k C_k^f \qquad (9\text{-}64)$$

C_k can be very large and difficult to estimate directly. EnKF is a Monte Carlo implementation of the Kalman filter, as it utilizes an ensemble of realizations to estimate the covariance matrix. The original formulation was developed by Evensen (1994), while many of the subsequent developments were in the application of weather prediction. Since its inception, applications in the subsurface and environmental fields, such as petroleum engineering (Reinlie, 2006; Kim et al., 2016) and hydrology (Li et al., 2012a-b; Vogt et al., 2012) are common. The method involves an ensemble of N_e initial models (typically ≤ 100), which should be constructed to honor any prior information and static data:

$$\Psi_k = \{y_{k,1}, y_{k,2}, \ldots, y_{k',N_e}\} \qquad (9\text{-}65)$$

$y_{k,i}$ denotes the state vector of the ith model at t_k:

$$y_{k,i} = \left[\left(x_{k,i}\right)^T, \left[g\left(x_{k,i}\right)\right]^T\right]^T \qquad (9\text{-}66)$$

$x_{k,i}$ denotes all model variables and the associated rock and dynamic (e.g., pressure and saturation) variables at every location, while $g(x_{k,i})$ refers to the simulated data observations. The Kalman filter scheme is modified as:

Forecast

$$x_{k,i}^f = f(y_{k-1,i}); \text{ for } i = 1, \ldots, N_e \qquad (9\text{-}67)$$

Analysis

$$y_{k,i} = y_{k,i}^f + K_{e,k}(d_{k,i} - H_k y_{k,i}^f); \text{ for } i = 1, \ldots, N_e \qquad (9\text{-}68)$$

where $d_k = d_{obs,k} + \varepsilon_k$, representing a perturbed data vector with measurement error. Finally, the ensemble Kalman gain (K_e) and the covariance can be estimated as:

$$K_{e,k} = C_{\Psi,k}^f H_k^T \left(H_k C_{\Psi,k}^f H_k^T + C_{D,k}\right)^{-1} \qquad (9\text{-}69)$$

$$C_{\Psi,k}^f = \frac{1}{N_e - 1} \sum_{i,j=1}^{N_e} \left(y_{k,i}^f - \overline{y}_k^f\right)\left(y_{k,j}^f - \overline{y}_k^f\right)^T \qquad (9\text{-}70)$$

$\overline{y}_k^f$ is the ensemble mean and is estimated from the ensemble members at t_k. The matrix $H_k = [0\ I]$ is called the measurement operator that extracts the data observations from the state vector.

The original EnKF, as well as its ensemble-based variant, assumes that the data mismatch can be linearly distributed to update the state variables. This can cause the updates made at a particular step to violate the match that may have been obtained at previous steps. Several modifications have been proposed to address this limitation. One strategy is to implement EnKF in an iterative manner (Evensen, 2003; 2009), which essentially involves repeating the analysis step iteratively. Different algorithms are available, and they primarily differ in how iteration is formulated (Gu and Oliver, 2007; Li and Reynolds, 2009; Sakov et al., 2011; Ungarala, 2012; Zhang et al., 2014; Chang and Zhang, 2015). One example is that of the ensemble randomized maximum likelihood (EnRML) method (Gu and Oliver, 2007). The analysis equation is recast as the following updating equation:

$$m_k^{l+1} = m_k^f - C_{M,k}^f G_{k,l}^T \left(G_{k,l} C_{M,k}^f G_{k,l}^T + C_{D,k}\right)^{-1}$$
$$\left[g\left(m_k^l\right) - d_{obs,k} - G_{k,l}\left(m_k^l - m_k^f\right)\right] \qquad (9\text{-}71)$$

The idea is to compute an average sensitivity, G_l, by solving the changes in computed data $\Delta d^l = G_l \Delta m^l$ with the ensemble. This G_l is used to update the model parameters with the Gauss–Newton formulation:

$$m_k^{l+1} = \beta_l m_k^f + (1 - \beta_l) m_k^l - \beta_l C_{M,k}^f G_{k,l}^T \left(G_{k,l} C_{M,k}^f G_{k,l}^T + C_{D,k}\right)^{-1}$$
$$\left[g\left(m_k^l\right) - d_{obs,k} - G_{k,l}\left(m_k^l - m_k^f\right)\right] \qquad (9\text{-}72)$$

β_i is an adjustment to the step size. It should be noted that only the model parameters are updated, and the state variables are computed by rerunning the simulation using m_k^{l+1} after each iteration. In a practical implementation of the EnRML, a standard EnKF update is employed, and the method would switch to the iterative updating scheme if the changes in the state variables are large. The EnKF update is equivalent to the first iteration of the EnRML with a full step length (Gu and Oliver, 2007).

The linear estimator for the model parameter is an optimal estimator for the conditional expectation of the distribution $P(x_k \mid d_k)$ only if the distribution is Gaussian. In addition, restricting the Kalman gain to only the covariance function is sufficient only to describe the variability associated with a multi-Gaussian distribution. Several alternative parameterization schemes to transform the non-Gaussian distributions into Gaussian ones have been proposed. The choice of an appropriate parameterization scheme is problem-dependent. Several schemes have been implemented for modeling channelized reservoirs or non-Gaussian geological features or objects: normal score (Zhou et al., 2012), truncated pluri-Gaussian (Astrakova and Oliver, 2015); Gaussian mixture models (Dovera and Della Rossa, 2011); level set functions (Chang et al., 2010; Moreno and Aanonsen, 2011; Lorentzen et al., 2012); DCT (Bhark et al., 2011; Jafarpour and McLaughlin, 2009) and the DWT (Chen and Oliver, 2012; Jafarpour, 2011; Gentilhomme et al., 2015); and curvelet transform (Zhang et al., 2015). Various techniques for updating MPS simulations have been developed. A probability conditioning method is proposed in Jafarpour and Khodabakhshi (2011), where the log-permeability ensemble mean is used to generate a facies probability map at the end of the analysis step, which is subsequently used for conditional sampling in a common MPS algorithm. In Hu et al.'s research (2013), the uniform random field used to sample for the conditional distribution (derived from conditioning data and multi-point statistics inferred from a training image) is assimilated using EnKF.

For fractured reservoirs, some recent studies have employed the Hough transform for fracture networks in 2D on a Cartesian grid (Lu and Zhang, 2015). Others have proposed employing EnKF to assimilate grid-cell average properties, such as fracture intensity, which can be easily parameterized in a dual-media flow simulation; the updated cell averages are used to condition a new discrete fracture model at the end of each assimilation step (Nejadi et al., 2015a, 2017). In the end, there is still a debate that though these extensions/hybrid formulations may retain the idea of utilizing an ensemble, they do not rely on linear update or transformation of space; they could offer only a partial approximation of the true distribution (Linde et al., 2015).

Kumar and Srinivasan (2019) recognize that EnKF is a regression scheme in which the required statistics are established using the ensemble of realizations. They establish a bridge between the EnKF update equation and kriging and, on that basis, propose a nonparametric regression scheme based on the concept of indicators. Such a scheme has the advantages of being optimal despite the underlying random function model being non-Gaussian or the updating scheme being non-linear. They refer to the data assimilation technique as indicator-based data assimilation (InDA). In the present from of its formulation, InDA updates are performed on individual components of the state vector, guaranteeing marginal non-Gaussian distribution of the components to be preserved. In a subsequent paper, Kumar and Srinivasan (2020) extend the formulation to multi-point statistics. The update equation can be used to update specific multi-point statistics that characterize the spatial features exhibited in the reservoir.

The ensemble size is generally quite small, and that could lead to a few issues (Kepert, 2004; Petrie, 2008). First, the covariance estimated from the ensemble may contain spurious correlation, which causes unnecessary updates and ensemble collapse (reduction in ensemble variability). Second, if the dataset is large and independent, it would be difficult to assimilate them with a small ensemble size (fewer degrees of freedom) (Oliver and Chen, 2011). Therefore, covariance localization or regularization is often needed. A localization function (ρ) is applied to individual elements of the covariance matrix, as shown in Eq. (9-69), using the Schur product:

$$K_{e,k} = \left(\rho \circ C_{\Psi,k}^f\right)H_k^T\left(H_k\left(\rho \circ C_{\Psi,k}^f\right)H_k^T + C_{D,k}\right)^{-1} \qquad (9\text{-}73)$$

A popular choice is distance-based localization (Hamill et al., 2001; Emerick and Reynolds, 2011). It assumes that the correlation between a model parameter and the observation data is a function of the distance between the grid cell and the well location. This relationship can be described by a correlation function with a specific range (e.g., beyond a certain distance, $\rho = 0$). Others argued that distance alone is not sufficient to describe the sensitivity of the model parameter to a given data observation. Various forms of streamline-based schemes have been proposed to introduce flow dynamics into the localization function (Arroyo et al., 2008; Watanabe and Datta-Gupta, 2012; Jung and Choe, 2012). In these methods, streamline trajectories are calculated, and the localization region for a particular ensemble member consists of grid cells intersected by the streamlines at the given instance. Finally, the covariance calculation is performed only in the common localization region shared by all members. A weighting function can also be incorporated to further define parameter sensitivity obtained from the streamline computations (Devegowda et al., 2010). As discussed by Chen and Oliver (2010a), the choice of localization function can be quite complex in practical applications, as there should be a reasonable balance between excluding spurious covariance and retaining the real correlations.

When the ensemble size is small, there is also the issue of under-sampling, as the ensemble is not statistically representative of the system's state and leads to inbreeding (Hamill et al., 2001). After the forecast step, the deviation of each member from the mean is inflated by a factor r (≥ 1):

$$\tilde{y}_{k,i}^f = r\left(y_{k,i}^f - \overline{y}_k^f\right) + \overline{y}_k^f \cdot \text{for } i = 1, \dots, N_e \qquad (9\text{-}74)$$

$\tilde{y}_{k,i}^f$ is the inflated member. Examples in history matching can be found in Chen and Oliver's research (2010b). The inflation strategy can vary spatially and temporally, such that regions that experience the greatest reduction in variability would receive the largest inflation. Myrseth et al. (2013) explained that ensemble collapse is attributed to the positive coupling of ensemble members as a result of using K to update all members for all time steps. They proposed schemes for resampling K for general non-linear examples. The resampling strategy involves (semi)parametric bootstrapping of the ensemble and Monte Carlo simulation of the likelihood model. Other adaptive schemes have been proposed, such that the localization functions are determined by minimizing the difference between the modified K and the true K, which is also estimated from bootstrap resampling (Chen and Oliver, 2012).

In recent years, multi-scale EnKF, which has its origin in image processing, has been proposed to impose spatial correlation and reduce spurious updating (Zhou et al., 2008; Lawniczak et al., 2009). Without any additional conditioning in the original EnKF, prior geological knowledge in the initial ensemble would often disappear as the update progresses. A multi-scale tree is used to represent the spatial correlations, and the analysis step is replaced by a set of upward sweep and downward sweeps. The mathematics involved are related to the EnKF analysis equations, and the statistics of the state are derived from the ensemble. The key point is that data observation at a particular grid node is passed through to other nodes (through the sweep operations) based on the tree structure, which essentially controls the correlation between all the nodes. Not surprisingly, the most difficult (and subjective) part of this technique is the definition or construction of this multi-scale tree structure to reflect the correct spatial correlation.

Nejadi et al. (2014, 2015b) have proposed various resampling schemes to improve ensemble diversity and to maintain reference facies proportions inferred from static geologic information. The idea is that during the assimilation process when data mismatch is reduced substantially but the ensemble variance is still high, an improved ensemble can be resampled. In a probability-weighted resampling procedure (Nejadi et al., 2014), a set of pilot points is selected from the model parameters that have experienced the most reduction in variance thus far; values of the model parameters at these pilot points are then sampled from the probability distributions using the most recently updated ensemble. Along with the hard data, they are regarded as conditioning data for generating a new ensemble via conditional sequential simulation. An alternative is to calibrate the conditional PDFs of the model parameters from the most recently updated ensemble, which can be employed in a p-field simulation to resample a new ensemble (Nejadi et al., 2015). In both cases, the new ensemble incorporates the model-updating information of the initial steps and is consistent with the static geological data; it is then subjected to the forward model from the beginning to the last update step prior to re-sampling, and updating would resume for the remaining production history using EnKF. Performing EnKF to model non-Gaussian facie indicators with non-linear process dynamics is important. A resampling scheme is also proposed by Sebacher et al. (2015) to model channelized reservoirs with an adaptive Gaussian mixture (AGM) filter; AGM is similar to the EnKF update, except that a dampening factor is introduced to the measurement error covariance matrix; the importance weights (i.e., weights assigned to individual ensemble member) are based on the Gaussian mixture and updated at each assimilation step (Stordal et al., 2011).

Many studies have also examined factors influencing the estimation results, including initial ensemble members, ensemble size, observation error, model error, and assimilation interval (Huber et al., 2011; Li and Li, 2011; Kovalenko et al., 2012; Kang, and Choe, 2017). When dealing with a large number of observations, in order to avoid storing and inverting large matrices of the Kalman filter equations, observations can be organized into batches that are assimilated sequentially (Houtekamer and Herschel, 2001). The first batch of observations is assimilated, and the resultant analysis fields are then used as initial fields when the next batch of observations is assimilated. The process is repeated until all batches are assimilated. The idea is to gradually evolve the original fields as more and more batches of observations are assimilated. Strictly speaking, the sequential technique is valid when observational errors are correlated with each other and processed in the same batch. This type of sequential updating is particularly useful when handling multi-scale data: fine-scale data (e.g., production rates) can be first assimilated, and the estimated ensemble is used as a prior for the second assimilation of the coarse-scale data (e.g., 4D seismic interpretations). If the coarse-scale data are permeability at a larger scale, then the forward model in the second assimilation is merely an upscaling procedure (Akella et al., 2011).

Another variation of the ensemble-based technique is the ensemble smoother (ES) (Skjervheim and Evensen, 2011). It is the same as EnKF, except that ES includes all data (e.g., observations at all time steps) in single assimilation. Chen and Oliver (2012) proposed implementing a batch-EnRML method as an ensemble smoother, while others have formulated an ES-MDA scheme (where MDA stands for multiple data assimilation) for assimilating the same data N_a times (Emerick and Reynolds, 2013a; Emerick, 2017). Theory regarding MDA in ensemble-based methods can be found in Emerick and Reynolds's research work (2012a). Chen and Oliver (2017) discussed the use of truncated SVD for regularization and localization in ES schemes.

Here is a simple summary of these different variations of ensemble-based approaches:

- EnKF—Sequential data assimilation, non-iterative
- EnRML—Sequential data assimilation, iterative
- ES—Batch data assimilation, non-iterative

- Batch-EnRML—Batch data assimilation, iterative
- ES-MDA—Multiple batch data assimilation, non-iterative

Emerick and Reynolds (2013b) compare the sampling performance of different ensemble-based methods with a synthetic reservoir model. The methods tested in the study include McMC, RML, various forms of iterative and non-iterative EnKF, ES, and ES-MDA. For that particular test problem, superior performance is observed for ES-MDA and RML. Hybrid techniques that involve more than one algorithm are also possible; for instance, the ensemble mean is used to propose new states in an McMC implementation (Emerick and Reynolds, 2012b).

Example 9-6
Solve the problem in Example 9-1 using the basic EnKF method.

Solution
The solution is provided at www.mhprofessional.com/PRMS in the MATLAB code "EnKF.m," which calls the same functions "f_model," as shown in Example 9-1. The "mismatch" function is not needed because EnKF updates the ensemble member based on the Kalman gain.

The same bounds are imposed on the two unknowns: $0.01 \leq k$ (Darcy) ≤ 2 and $-5 \leq s \leq 5$ when generating the initial ensemble. However, during each update, there are no explicit constraints to ensure that the model parameters would stay within these bounds. However, assuming that the bounds are wide enough, the ensemble would converge toward the true solution over successive updates. A population size (N_e) of 200 is selected. In addition, instead of assimilating all 365 data points, only the initial 300 data points are used.

In this example, we specified y to be a 3×1 state vector $= [k\, s\, P_{wf}]^T$. The first two elements represent the two unknowns x, while the last element is d. Therefore, H $= [0\ 0\ 1]$. The assimilation results are shown in Fig. 9-7. The variability within the ensemble reduces quickly after just a few update steps.

9.4.2   Ensemble Pattern Search and Model Selection
Most of the inverse methods, including EnKF, aim to assimilate the observed data by modifying the model parameters through an optimization algorithm. However, during this optimization process, the spatial structure exhibited by the model parameters in the prior is often drastically altered. As discussed previously, techniques such as pilot point, resampling, parameterization, regularization, or constrained optimization are needed to ensure consistency with the static information incorporated in the initial models. To avoid parameterizing the complex geological spatial structure with a set of

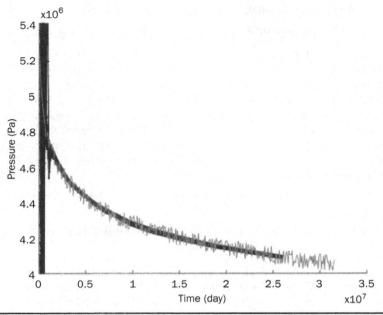

FIGURE 9-7 History matching using the EnKF method (dark: ensemble members; light: history). (The color version of this figure is available at www.mhprofessional .com/PRMS.)

model parameters, an alternative is to search from a large set of prior model realizations (which can be constructed with any method such as MPS) ones that are consistent with the observed production data. It is important to note that during this search or selection process, the spatial structure of the model parameters is not modified.

First, a metric space, which is a mathematical set equipped with only a distance function, is defined to represent the similarity between any two model realizations. The most common example is the Euclidean 3D space. Next, a stochastic search is performed to identify the inverse problem solutions in the metric space (Suzuki et al., 2008). A static distance, namely, the Hausdorff distance (Dubuisson and Jain, 1994), is used to measure the similarity between different channelized objects. Flow-based measures that are based on connectivity or simplified proxy (e.g., streamline) computations are also possible (Scheidt and Caers, 2009a; Bhowmik et al., 2013; Nwachukwu et al., 2017). Stochastic search algorithms, such as neighborhood algorithm or NA (Suzuki and Caers, 2008; Suzuki et al., 2008), can be used to search the metric space. Details can be found in the references, while the key steps are summarized here:

1. Select a subset of initial models that are "far away" from each other in the metric space (based on the distance function). Subject these models to flow simulation and evaluate the objective function $O(m_i)$.

2. Partition the metric space—cluster the remaining models to the "closest" model that has been flow-simulated in Step 1. This partitioning step is equivalent to partitioning a Cartesian space into Voronoi cells, as only distance is considered.

3. Randomly select one of the Voronoi cells according to this selection probability:

$$p(m_i) = \exp\left\{-\frac{1}{T}\frac{O(m_i)}{N_i}\right\} \tag{9-75}$$

N_i is the number of models included in the Voronoi cell i that have not yet been flow-simulated; T is a normalization parameter. Next, randomly select one of the models, perform flow simulation, and evaluate the objective function. Add this newly simulated model to the subset of models described in Step 1. Repeat from Step 2 until the desired numbers of model realizations with sufficiently low values of objective functions are achieved.

An ensemble pattern search technique, which shares some similarities with EnKF, was proposed by Zhou et al. (2012). An extended formulation incorporating pilot points was presented by Li et al. (2013). Instead of sampling from the posterior distribution, a pattern search algorithm is implemented to solve the inverse problem. It is assumed that the unknown model parameter (e.g., hydraulic conductivities at a given location) is related to the geologic structure and the flow behavior within its vicinity. Therefore, in order to estimate its value at a particular grid cell, an ensemble of realizations is searched to identify matches to the pattern consisting of previously simulated data (including local hard data) and observed dynamic data (e.g., hydraulic head measurements) within its neighborhood. It is similar to EnKF, in a sense, that data are assimilated sequentially; a different ensemble is created after each assimilation step. Following the terminology presented in Zhou et al. (2012), the key steps can be summarized:

1. For each assimilation step, forecast the dependent state variables for each ensemble member. Gather the resultant ensemble of training image couples (hydraulic conductivity and piezometric head).

2. For each realization, select a random path to loop over all the cells with unknown hydraulic conductivity. At each cell, identify a joint conditioning data pattern, which includes M hydraulic conductivity values (measured hard data and/or previously-simulated values) and N piezometric head values extracted from the neighborhood. Since the pattern's areal extent is controlled by M and N, and not by physical distance, it will gradually decrease as the random path is traveled: near the end when most of the conductivities are updated, the pattern will cover a limited area and local variability is controlled. This

flexibility helps to avoid spurious updating and offers similar advantages to the multi-grid approach (Zhou et al., 2012).

3. Search the ensemble of training image couples for a match to the conditioning data pattern. In each realization, only the region close to the grid cell is searched to ensure similar boundary effects are accounted for. The goodness of match is defined by a distance or measure (e.g., Euclidean for l_2-norm or other arbitrary distance function).

Instead of implementing a stochastic search in the metric space, other model-selection frameworks have been proposed. Two steps are generally involved: (1) performing multi-dimensional scaling (MDS) (Kruskal, 1964) to map all realizations into a Euclidean space **R** and (2) clustering realizations in **R** into groups and selecting the group with the best match to the observed data. Step 2 is repeated until a sufficiently close match is achieved. Additional realizations can also be sampled in the group that exhibits the best match to the observed data (Scheidt and Caers, 2009a, 2009b; Mantilla, 2010; Srinivasan and Mantilla, 2012; Park et al., 2013; Bhowmik, 2014).

To perform MDS, a $N_R \times N_R$ matrix containing the distance or dissimilarity (as discussed earlier) between any two realizations is assembled for a total of N_R realizations (Cox and Ferry, 1993). The goal of MDS is, given a dissimilarity matrix **D**, to find a set of vectors $\mathbf{x}_1, \ldots, \mathbf{x}_{N_R} \in \mathbb{R}^m$ such that $||\mathbf{x}_i - \mathbf{x}_j|| \approx d_{ij}$, where d_{ij} is the distance between object i and object j (i and $j = 1, \ldots, N_R$).

$$
\mathbf{D} = \begin{bmatrix} d_{11} & \cdots & d_{1N_R} \\ \vdots & & \\ d_{1N_R} & \cdots & d_{N_R N_R} \end{bmatrix}
\tag{9-76}
$$

$|| \cdot ||$ is the vector norm, in classical MDS, this is the Euclidean distance, but it may be any arbitrary distance function. Mathematically, metric MDS aims to minimize the following stress function:

$$
\text{stress} = \sqrt{\frac{\sum_{i,j}\left(d_{ij} - \left\|\mathbf{x}_i - \mathbf{x}_j\right\|\right)^2}{\sum_{i,j}\left(d_{ij}\right)^2}}
\tag{9-77}
$$

A more general non-metric formulation of the above equation can be written by replacing the vector norm with a monotonic transformation of **x**:

$$
\text{stress} = \sqrt{\frac{\sum_{i,j}\left(d_{ij} - f(\mathbf{x})\right)^2}{\sum_{i,j}\left(d_{ij}\right)^2}}
\tag{9-78}
$$

Returning to the classical metric formulation, MDS aims to map these distances into a Euclidean space, where each realization is denoted as a point. The points are arranged such that their distances in **R** correspond to their dissimilarity matrix: similar realizations are represented by points that are close to each other (Cox and Cox, 2000; Borg and Groenen, 2005). Imagine we are given a set of distances between five spatial locations; MDS can map these distances into five points in a 2D/3D Euclidean space, where the distance between the points is honored. The mapping is invariant to rotation and reflection, since the absolute coordinates are irrelevant, as long as the relative configuration of points is maintained. In the context of history matching, these distances no longer refer to the spatial distance but instead represent the production performance difference between two realizations. As noted by Park et al. (2013), the goal of MDS is to estimate a non-Euclidean distance between model responses with a Euclidean one; in fact, it looks for a low-dimensional space in which such approximation exists; therefore, it is not implemented for the purpose of dimension reduction. An alternative to MDS would be to apply PCA or SVD. However, those schemes would only be optimal if the distribution of dissimilarity measures is Gaussian and characterized by the covariance.

The classical metric MDS algorithm (using Euclidean norm) for N objects can be summarized as:

1. Set up the matrix of squared distance $\mathbf{D}^{(2)} = [\mathbf{d}^2]$.

2. Perform double centering:

$$\mathbf{J} = \mathbf{I} - N^{-1}\mathbf{1}\mathbf{1}'$$

$$\mathbf{B} = -\frac{1}{2}\mathbf{J}\mathbf{D}^{(2)}\mathbf{J} \tag{9-79}$$

$\mathbf{1}\mathbf{1}'$ is a $N \times N$ matrix of unity.

3. Perform eigenvalue decomposition of **B** and retain m largest eigenvalues (λ_m) and the corresponding eigenvectors ($\mathbf{e}_m$).

4. The coordinate matrix for these N objects in the m-dimensional Euclidean space (**X**) is:

$$\mathbf{X} = \mathbf{E}_m\mathbf{\Lambda}_m^{1/2} \tag{9-80}$$

where $\mathbf{E}_m$ is the diagonal matrix of λ_m and $\mathbf{\Lambda}_m$ is a matrix of $\mathbf{e}_m$.

Cluster analysis is performed next. In the simplest form, the realization closest to the cluster centroid can be regarded as a representative model for that cluster. The objective function corresponding to each representative model is evaluated and compared. The cluster with the lowest objective function is selected and additional models are sampled from that cluster. In a probabilistic framework

(Mantilla, 2010; Scheidt and Caers, 2010), the general procedure can be described as:

1. Model $p(d \mid m_k)$, where m_k refers to a representative model vector for cluster k. Kernel density estimation can be employed in this step.
2. Define $p(d_{obs} \mid m_k)$ from the region near the observed data.
3. Employ Bayes' rule:

$$p(m_k \mid d_{obs}) = \frac{p(d_{obs} \mid m_k)p(m_k)}{p(d_{obs})} \qquad (9\text{-}81)$$

$p(m_k)$ is the probability of cluster k, which is the number of realizations in cluster k divided by the total number of realizations. $p(d_{obs})$ is obtained by summation over all clusters. The cluster with the best match is sampled based on the posterior probability $p(m_k \mid d_{obs})$. This procedure can be repeated multiple times until a refined cluster with a sufficiently close match to the observed data is achieved. Additional realizations can be sampled from the history-matched cluster via different bootstrapping and perturbation techniques. In Park et al.'s research (2013), the clustering step is skipped, but a similar procedure is used to calibrate the posterior distribution, which is used in a rejection sampling scheme to screen out realizations that are inconsistent with the production data.

Example 9-7
The distance matrix for the four reservoir models is given below. Construct the 2D spatial configuration of these models using MDS.

$$\mathbf{d} = \begin{bmatrix} 0 & 5 & 100 & 105 \\ 5 & 0 & 90 & 90 \\ 100 & 90 & 0 & 5 \\ 105 & 90 & 5 & 0 \end{bmatrix}$$

Solution

Matrix of squared distance: $\mathbf{D}^{(2)} = [\mathbf{d}^2] = \begin{bmatrix} 0 & 5^2 & 100^2 & 105^2 \\ 5^2 & 0 & 90^2 & 90^2 \\ 100^2 & 90^2 & 0 & 5^2 \\ 105^2 & 90^2 & 5^2 & 0 \end{bmatrix}$

$$
= \begin{bmatrix} 0 & 25 & 10000 & 11025 \\ 25 & 0 & 8100 & 8100 \\ 10000 & 8100 & 0 & 25 \\ 11025 & 8100 & 25 & 0 \end{bmatrix}
$$

$$
\mathbf{J} = \begin{bmatrix} 1 & 0 & 0 & 0 \\ 0 & 1 & 0 & 0 \\ 0 & 0 & 1 & 0 \\ 0 & 0 & 0 & 1 \end{bmatrix} - \frac{1}{4}\begin{bmatrix} 1 & 1 & 1 & 1 \\ 1 & 1 & 1 & 1 \\ 1 & 1 & 1 & 1 \\ 1 & 1 & 1 & 1 \end{bmatrix} = \begin{bmatrix} 0.75 & -0.25 & -0.25 & -0.25 \\ -0.25 & 0.75 & -0.25 & -0.25 \\ -0.25 & -0.25 & 0.75 & -0.25 \\ -0.25 & -0.25 & -0.25 & 0.75 \end{bmatrix}
$$

$$
\mathbf{B} = -\frac{1}{2}\mathbf{J}\mathbf{D}^{(2)}\mathbf{J} = -\frac{1}{2}\begin{bmatrix} 0.75 & -0.25 & -0.25 & -0.25 \\ -0.25 & 0.75 & -0.25 & -0.25 \\ -0.25 & -0.25 & 0.75 & -0.25 \\ -0.25 & -0.25 & -0.25 & 0.75 \end{bmatrix}\begin{bmatrix} 0 & 25 & 10000 & 11025 \\ 25 & 0 & 8100 & 8100 \\ 10000 & 8100 & 0 & 25 \\ 11025 & 8100 & 25 & 0 \end{bmatrix}
$$

$$
\begin{bmatrix} 0.75 & -0.25 & -0.25 & -0.25 \\ -0.25 & 0.75 & -0.25 & -0.25 \\ -0.25 & -0.25 & 0.75 & -0.25 \\ -0.25 & -0.25 & -0.25 & 0.75 \end{bmatrix} = \begin{bmatrix} 2932.8 & 2317.2 & -2432.8 & -2817.2 \\ 2317.2 & 1726.6 & -2085.9 & -1957.8 \\ -2432.8 & -2085.9 & 2201.6 & -2317.2 \\ -2817.2 & -1957.8 & -2317.2 & 2457.8 \end{bmatrix}
$$

Extracting the two largest eigenvalues of **B** and their corresponding eigenvectors:

$$
\Lambda_2 = \begin{bmatrix} 9375.9 & 0 \\ 0 & 212.9 \end{bmatrix}; \quad \mathbf{E}_2 = \begin{bmatrix} -0.5623 & -0.4196 \\ -0.4332 & 0.4955 \\ 0.4824 & -0.5744 \\ 0.5132 & 0.4985 \end{bmatrix}
$$

Finally, the location matrix becomes:

$$
\mathbf{X} = \mathbf{E}_m \Lambda_m^{1/2} = \begin{bmatrix} -0.5623 & -0.4196 \\ -0.4332 & 0.4955 \\ 0.4824 & -0.5744 \\ 0.5132 & 0.4985 \end{bmatrix}\begin{bmatrix} 9375.9^{\frac{1}{2}} & 0 \\ 0 & 212.9^{\frac{1}{2}} \end{bmatrix} = \begin{bmatrix} -54.4513 & -6.1218 \\ -41.9507 & 7.2295 \\ 46.7108 & -8.3816 \\ 49.6912 & 7.2740 \end{bmatrix}
$$

The coordinates of these points are mapped and shown in Fig. 9-8. An inspection of the distance matrix would suggest that the distances between Models 1 and 2 and 3 and 4 are small. The MDS results corroborate this observation and reveal two distinct groupings.

If the structure of the points in mapping space **R** is highly non-linear, the performance of typical pattern detection methods such as PCA and cluster analysis would be sub-optimal. Kernel methods can

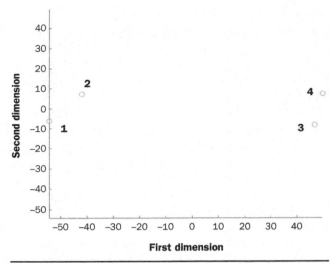

Figure 9-8   2D mapping of the models in Example 9-7. (The color version of this figure is available at www.mhprofessional.com /PRMS.)

be used to transform **R** into a feature space F, where linear techniques can be used (Scheidt and Caers, 2009a).

9.5 References

Caers, Jef. "Comparing the gradual deformation with the probability perturbation method for solving inverse problems." *Mathematical Geology* 39, no. 1 (2007): 27-52.

Caers, Jef, and Todd Hoffman. "The probability perturbation method: A new look at Bayesian inverse modeling." *Mathematical geology* 38, no. 1 (2006): 81-100.

Nocedal, Jorge, and Stephen Wright. *Numerical optimization.* Springer Science & Business Media, 2006.

Li, Ruijian, A. C. Reynolds, and D. S. Oliver. "History Matching of Three-Phase Flow Production Data." *SPE Journal* 8, no. 04 (2003): 328-340.

N. Linde, P. Renard, T. Mukerji, and J. Caers, "Geological realism in hydrogeological and geophysical inverse modeling: A review," Adv. Water Resour., vol. 86, 86-101, 2015.

Kirkpatrick, Scott, C. Daniel Gelatt, and Mario P. Vecchi. "Optimization by simulated annealing." *science* 220, no. 4598 (1983): 671-680.

Van Laarhoven, Peter JM, and Emile HL Aarts. "Simulated annealing." In *Simulated Annealing: Theory and Applications,* pp. 7-15. Springer Netherlands, 1987.

Whitley, Darrell. "A genetic algorithm tutorial." *Statistics and computing* 4, no. 2 (1994): 65-85.

Mitchell, Melanie. *An introduction to genetic algorithms.* MIT press, 1998.

Man, Kim-Fung, Kit-Sang Tang, and Sam Kwong. "Genetic algorithms: concepts and applications [in engineering design]." *IEEE transactions on Industrial Electronics* 43, no. 5 (1996): 519-534.

Haupt, Randy L., and Sue Ellen Haupt. *Practical genetic algorithms.* John Wiley & Sons, 2004.

Eberhart, Russell, and James Kennedy. "A new optimizer using particle swarm theory." In *Micro Machine and Human Science, 1995. MHS'95., Proceedings of the Sixth International Symposium on,* pp. 39-43. IEEE, 1995.

Kennedy, James F., Russell C. Eberhart, and Yuhui Shi. *Swarm intelligence*. Morgan Kaufmann, 2001.

Engelbrecht, Andries P. *Fundamentals of computational swarm intelligence*. John Wiley & Sons, 2006.

Hastings, W. Keith. "Monte Carlo sampling methods using Markov chains and their applications." *Biometrika* 57, no. 1 (1970): 97-109.

Hu, Lin Y. "Gradual deformation and iterative calibration of Gaussian-related stochastic models." *Mathematical Geology* 32, no. 1 (2000): 87-108.

Hu, Lin Y., Georges Blanc, and Benoît Noetinger. "Gradual deformation and iterative calibration of sequential stochastic simulations." *Mathematical Geology* 33, no. 4 (2001): 475-489.

Hu, Lin Y., and Mickaele Le Ravalec-Dupin. "An improved gradual deformation method for reconciling random and gradient searches in stochastic optimizations." *Mathematical Geology* 36, no. 6 (2004): 703-719.

Le Ravalec-Dupin, Mickaële, and Benoît Nœtinger. "Optimization with the gradual deformation method." *Mathematical Geology* 34, no. 2 (2002): 125-142.

Journel, A. G. "Combining knowledge from diverse sources: An alternative to traditional data independence hypotheses." *Mathematical geology* 34, no. 5 (2002): 573-596.

Feller, William. *An introduction to probability theory and its applications: volume I*. Vol. 3. New York: John Wiley & Sons, 1968.

Gilks, Walter R., Sylvia Richardson, and David Spiegelhalter, eds. *Markov chain Monte Carlo in practice*. CRC press, 1995.

Menke, William. Geophysical data analysis: Discrete inverse theory. Vol. 45. Academic press, 2012.

Metropolis, Nicholas, Arianna W. Rosenbluth, Marshall N. Rosenbluth, Augusta H. Teller, and Edward Teller. "Equation of state calculations by fast computing machines." *The journal of chemical physics* 21, no. 6 (1953): 1087-1092.

Mata-Lima, Herlander. "Reservoir characterization with iterative direct sequential co-simulation: integrating fluid dynamic data into stochastic model." *Journal of Petroleum Science and Engineering* 62, no. 3 (2008): 59-72.

Oliver, Dean S., Albert C. Reynolds, and Ning Liu. *Inverse theory for petroleum reservoir characterization and history matching*. Cambridge University Press, 2008.

Oliver, Dean S., and Yan Chen. "Recent progress on reservoir history matching: a review." *Computational Geosciences* 15, no. 1 (2011): 185-221.

Poli, Riccardo, James Kennedy, and Tim Blackwell. "Particle swarm optimization." *Swarm intelligence* 1, no. 1 (2007): 33-57.

Press, William H., Brian P. Flannery, Saul A. Teukolsky, and William T. Vetterling. *Numerical recipes*. Vol. 3. cambridge University Press, cambridge, 1989.

Kang, Byeongcheol, and Jonggeun Choe. "Initial model selection for efficient history matching of channel reservoirs using Ensemble Smoother." *Journal of Petroleum Science and Engineering* 152 (2017): 294-308.

Kashib, Tarun, and Sanjay Srinivasan. "A probabilistic approach to integrating dynamic data in reservoir models." *Journal of Petroleum Science and Engineering* 50, no. 3 (2006): 241-257.

Barrera, Alvaro Enrique. *History matching by simultaneous calibration of flow functions*. The University of Texas at Austin, 2007.

Reinlie, Shinta Tjahyaningtyas. "Analysis of continuous monitoring data and rapid, stochastic updating of reservoir models." PhD diss., 2006.

Houtekamer, Peter L., and Herschel L. Mitchell. "A sequential ensemble Kalman filter for atmospheric data assimilation." *Monthly Weather Review* 129, no. 1 (2001): 123-137.

Li, Chao, and Li Ren. "Estimation of unsaturated soil hydraulic parameters using the ensemble Kalman filter." *Vadose Zone Journal* 10, no. 4 (2011): 1205-1227.

Romary, Thomas. "History matching of approximated lithofacies models under uncertainty." *Computational Geosciences* 14, no. 2 (2010): 343-355.

Huber, E., H. J. Hendricks-Franssen, H. P. Kaiser, and F. Stauffer. "The role of prior model calibration on predictions with ensemble Kalman filter." *Groundwater* 49, no. 6 (2011): 845-858.

Dovera, Laura, and Ernesto Della Rossa. "Multimodal ensemble Kalman filtering using Gaussian mixture models." *Computational Geosciences* 15, no. 2 (2011): 307-323.

Chang, Haibin, Dongxiao Zhang, and Zhiming Lu. "History matching of facies distribution with the EnKF and level set parameterization." *Journal of Computational Physics* 229, no. 20 (2010): 8011-8030.

Li, Liangping, Haiyan Zhou, J. Jaime Gómez-Hernández, and Harrie-Jan Hendricks Franssen. "Jointly mapping hydraulic conductivity and porosity by assimilating concentration data via ensemble Kalman filter." *Journal of hydrology* 428 (2012a): 152-169.

Li, Liangping, Haiyan Zhou, Harrie-Jan Hendricks Franssen, and J. Jaime Gómez-Hernández. "Modeling transient groundwater flow by coupling ensemble Kalman filtering and upscaling." *Water Resources Research* 48, no. 1 (2012b).

Vogt, C., G. Marquart, Ch Kosack, A. Wolf, and Ch Clauser. "Estimating the permeability distribution and its uncertainty at the EGS demonstration reservoir Soultz-sous-Forêts using the ensemble Kalman filter." *Water Resources Research* 48, no. 8 (2012).

Aanonsen, Sigurd I., Geir Nævdal, Dean S. Oliver, Albert C. Reynolds, and Brice Vallès. "The ensemble Kalman filter in reservoir engineering – a review." *SPE Journal* 14, no. 03 (2009): 393-412.

Kalman, Rudolph Emil. "A new approach to linear filtering and prediction problems." *Journal of basic Engineering* 82, no. 1 (1960): 35-45.

Tarantola, Albert. *Inverse problem theory and methods for model parameter estimation.* Society for Industrial and Applied Mathematics, 2005.

Caeiro, M. Helena, Vasily Demyanov, and Amilcar Soares. "Optimized history matching with direct sequential image transforming for non-stationary reservoirs." *Mathematical Geosciences* 47, no. 8 (2015): 975

Kim, Sungil, Choongho Lee, Kyungbook Lee, and Jonggeun Choe. "Aquifer characterization of gas reservoirs using Ensemble Kalman filter and covariance localization." *Journal of Petroleum Science and Engineering* 146 (2016): 446-456.

Suzuki, Satomi, Colin Daly, Jef Karel Caers, and Dietmar Mueller. "History Matching of Naturally Fractured Reservoirs Using Elastic Stress Simulation and Probability Perturbation Method." *SPE Journal* 12, no. 01 (2007): 118-129.

Suzuki, Satomi, and Jef Caers. "A distance-based prior model parameterization for constraining solutions of spatial inverse problems." *Mathematical Geosciences* 40, no. 4 (2008): 445-469.

Suzuki, Satomi, Guillaume Caumon, and Jef Caers. "Dynamic data integration for structural modeling: model screening approach using a distance-based model parameterization." *Computational Geosciences* 12, no. 1 (2008): 105-119.

Evensen, Geir. "Sequential data assimilation with a non-linear quasi-geostrophic model using Monte Carlo methods to forecast error statistics." *Journal of Geophysical Research: Oceans* 99, no. C5 (1994): 10143-10162.

Evensen, Geir. "The ensemble Kalman filter: Theoretical formulation and practical implementation." *Ocean dynamics* 53, no. 4 (2003): 343-367.

Evensen, Geir. *Data assimilation: the ensemble Kalman filter.* Springer Science & Business Media, 2009.

Gu, Yaqing, and Dean S. Oliver. "An iterative ensemble Kalman filter for multiphase fluid flow data assimilation." *Spe Journal* 12, no. 04 (2007): 438-446.

G. Li and A.C. Reynolds, "Iterative ensemble Kalman filters for data assimilation," SPE J., vol. 14, no.3, pp. 496–505. 2009.

Sakov, Pavel, Dean S. Oliver, and Laurent Bertino. "An iterative EnKF for strongly nonlinear systems." *Monthly Weather Review* 140, no. 6 (2012): 1988-2004.

Mohamed, Linah, Michael A. Christie, and Vasily Demyanov. "Comparison of stochastic sampling algorithms for uncertainty quantification." *SPE Journal* 15, no. 01 (2010): 31-38.

Sarma, Pallav, Louis J. Durlofsky, and Khalid Aziz. "Kernel principal component analysis for efficient, differentiable parameterization of multipoint geostatistics." *Mathematical Geosciences* 40, no. 1 (2008): 3-32.

Sarma, Pallav, Louis J. Durlofsky, Khalid Aziz, and Wen H. Chen. "Efficient real-time reservoir management using adjoint-based optimal control and model updating." *Computational Geosciences* 10, no. 1 (2006): 3-36.

Oliver, Dean S. "Multiple realizations of the permeability field from well test data." *SPE Journal* 1, no. 02 (1996): 145-154.

Afra, Sardar, and Eduardo Gildin. "Tensor based geology preserving reservoir parameterization with Higher Order Singular Value Decomposition (HOSVD)." *Computers & Geosciences* 94 (2016): 110-120.

Ungarala, Sridhar. "On the iterated forms of Kalman filters using statistical linearization." *Journal of Process Control* 22, no. 5 (2012): 935-943.

Zhang, Yanhui, Dean S. Oliver, Hervé Chauris, and Daniela Donno. "Ensemble-based data assimilation with curvelets regularization." *Journal of Petroleum Science and Engineering* 136 (2015): 55-67.

Bhark, E., Jafarpour, B., Datta-Gupta, A., 2011. An adaptively scaled frequency-domain parameterization for history matching. J. Pet. Sci. Eng. 75 (3), 289–303.

Jafarpour, Behnam. "Wavelet reconstruction of geologic facies from nonlinear dynamic flow measurements." *IEEE Transactions on Geoscience and Remote Sensing* 49, no. 5 (2011): 1520-1535.

Jafarpour, Behnam, and Dennis B. McLaughlin. "Reservoir characterization with the discrete cosine transform." *SPE Journal* 14, no. 01 (2009): 182-201.

Kovalenko, Andrey, Trond Mannseth, and Geir Nævdal. "Sampling error distribution for the ensemble Kalman filter update step." *Computational Geosciences* 16, no. 2 (2012): 455-466.

Zeng, Lingzao, and Dongxiao Zhang. "A stochastic collocation based Kalman filter for data assimilation." *Computational Geosciences* 14, no. 4 (2010): 721-744.

Chang, Haibin, and Dongxiao Zhang. "History matching of statistically anisotropic fields using the Karhunen-Loeve expansion-based global parameterization technique." *Computational Geosciences* 18, no. 2 (2014): 265-282.

M Khaninezhad, Mohammadreza, and Behnam Jafarpour. "Hybrid parameterization for robust history matching." *SPE Journal* 19, no. 03 (2014): 487-499.

Lorentzen, Rolf J., Kristin M. Flornes, and Geir Nævdal. "History matching channelized reservoirs using the ensemble Kalman filter." *SPE Journal* 17, no. 01 (2012): 137-151.

Gentilhomme, Théophile, Dean S. Oliver, Trond Mannseth, Guillaume Caumon, Rémi Moyen, and Philippe Doyen. "Ensemble-based multi-scale history-matching using second-generation wavelet transform." *Computational Geosciences* 19, no. 5 (2015): 999-1025.

Moreno, David L., and Sigurd I. Aanonsen. "Continuous facies updating using the ensemble Kalman filter and the level set method." *Mathematical Geosciences* 43, no. 8 (2011): 951-970.

Maschio, Célio, Alessandra Davolio, Manuel Gomes Correia, and Denis José Schiozer. "A new framework for geostatistics-based history matching using genetic algorithm with adaptive bounds." *Journal of Petroleum Science and Engineering* 127 (2015): 387-397.

Schulze-Riegert, R. W., J. K. Axmann, O. Haase, D. T. Rian, and Y-L. You. "Evolutionary algorithms applied to history matching of complex reservoirs." *SPE Reservoir Evaluation & Engineering* 5, no. 02 (2002): 163-173.

Astrakova, Alina, and Dean S. Oliver. "Conditioning truncated pluri-Gaussian models to facies observations in ensemble-Kalman-based data assimilation." *Mathematical Geosciences* 47, no. 3 (2015): 345-367.

Kourounis, Drosos, Louis J. Durlofsky, Jan Dirk Jansen, and Khalid Aziz. "Adjoint formulation and constraint handling for gradient-based optimization of compositional reservoir flow." *Computational Geosciences* 18, no. 2 (2014): 117.

Chang, Haibin, and Dongxiao Zhang. "Jointly updating the mean size and spatial distribution of facies in reservoir history matching." *Computational Geosciences* 19, no. 4 (2015): 727-746.

Bhark, Eric W., Behnam Jafarpour, and Akhil Datta-Gupta. "A generalized grid connectivity–based parameterization for subsurface flow model calibration." *Water Resources Research* 47, no. 6 (2011).

Vo, Hai X., and Louis J. Durlofsky. "Data assimilation and uncertainty assessment for complex geological models using a new PCA-based parameterization." *Computational Geosciences* 19, no. 4 (2015): 747-767.

Zhou, Haiyan, Liangping Li, Harrie-Jan Hendricks Franssen, and J. Jaime Gómez-Hernández. "Pattern recognition in a bimodal aquifer using the normal-score ensemble Kalman filter." *Mathematical Geosciences* 44, no. 2 (2012): 169-185.

Perozzi, Lorenzo, Erwan Gloaguen, Bernard Giroux, and Klaus Holliger. "A stochastic inversion workflow for monitoring the distribution of CO2 injected into deep saline aquifers." *Computational Geosciences* 20, no. 6 (2016): 1287-1300.

Lu, Le, and Dongxiao Zhang. "Assisted history matching for fractured reservoirs by use of Hough-transform-based parameterization." *SPE Journal* 20, no. 05 (2015): 942-961.

Petrie, R. "Localization in the ensemble Kalman filter." *MSc Atmosphere, Ocean and Climate University of Reading* (2008).

Hamill, Thomas M., Jeffrey S. Whitaker, and Chris Snyder. "Distance-dependent filtering of background error covariance estimates in an ensemble Kalman filter." *Monthly Weather Review* 129, no. 11 (2001): 2776-2790.

Arroyo, Elkin, Deepak Devegowda, Akhil Datta-Gupta, and Jonggeun Choe. "Streamline-assisted ensemble Kalman filter for rapid and continuous reservoir model updating." *SPE Reservoir Evaluation & Engineering* 11, no. 06 (2008): 1-046.

Watanabe, Shingo, and Akhil Datta-Gupta. "Use of phase streamlines for covariance localization in ensemble Kalman filter for three-phase history matching." *SPE Reservoir Evaluation & Engineering* 15, no. 03 (2012): 273-289.

Jung, SeungPil, and Jonggeun Choe. "Reservoir characterization using a streamline-assisted ensemble Kalman filter with covariance localization." *Energy Exploration & Exploitation* 30, no. 4 (2012): 645-660.

Devegowda, Deepak, Elkin Arroyo-Negrete, and Akhil Datta-Gupta. "Flow relevant covariance localization during dynamic data assimilation using EnKF." *Advances in Water Resources* 33, no. 2 (2010): 129-145.

Chen, Yan, and Dean S. Oliver. "Cross-covariances and localization for EnKF in multiphase flow data assimilation." *Computational Geosciences* 14, no. 4 (2010a): 579-601.

Chen, Yan, and Dean S. Oliver. "Ensemble-based closed-loop optimization applied to Brugge field." *SPE Reservoir Evaluation & Engineering* 13, no. 01 (2010b): 56-71.

Myrseth, Inge, Jon Sætrom, and Henning Omre. "Resampling the ensemble Kalman filter." *Computers & geosciences* 55 (2013): 44-53.

Kepert, Jeffrey D. "On ensemble representation of the observation-error covariance in the Ensemble Kalman Filter." *Ocean Dynamics* 54, no. 6 (2004): 561-569.

Chen, Yan, and Dean S. Oliver. "Multiscale parameterization with adaptive regularization for improved assimilation of nonlocal observation." *Water Resources Research* 48, no. 4 (2012).

Akella, Santha, A. Datta-Gupta, and Yalchin Efendiev. "ASSIMILATION OF COARSE-SCALEDATAUSINGTHE ENSEMBLE KALMAN FILTER." *International Journal for Uncertainty Quantification* 1, no. 1 (2011).

Lawniczak, Wiktoria, Remus Hanea, Arnold Heemink, and Dennis McLaughlin. "Multiscale ensemble filtering for reservoir engineering applications." *Computational Geosciences* 13, no. 2 (2009): 245-254.

Zhou, Yuhua, Dennis McLaughlin, Dara Entekhabi, and Gene-Hua Crystal Ng. "An ensemble multiscale filter for large nonlinear data assimilation problems." *Monthly Weather Review* 136, no. 2 (2008): 678-698.

Skjervheim, Jan-Arild, and Geir Evensen. "An ensemble smoother for assisted history matching." In *SPE Reservoir Simulation Symposium*. Society of Petroleum Engineers, 2011.

Chen, Yan, and Dean S. Oliver. "Ensemble randomized maximum likelihood method as an iterative ensemble smoother." *Mathematical Geosciences* 44, no. 1 (2012): 1-26.

Emerick, Alexandre, and Albert Reynolds. "Combining sensitivities and prior information for covariance localization in the ensemble Kalman filter for petroleum reservoir applications." *Computational Geosciences* 15, no. 2 (2011): 251-269.

Emerick, Alexandre A., and Albert C. Reynolds. "History matching time-lapse seismic data using the ensemble Kalman filter with multiple data assimilations." *Computational Geosciences* 16, no. 3 (2012a): 639-659.

Emerick, Alexandre A., and Albert C. Reynolds. "Combining the Ensemble Kalman Filter With Markov-Chain Monte Carlo for Improved History Matching and Uncertainty Characterization." *SPE Journal* 17, no. 02 (2012b): 418-440.

Emerick, Alexandre A., and Albert C. Reynolds. "Ensemble smoother with multiple data assimilation." *Computers & Geosciences* 55 (2013a): 3-15.

Emerick, Alexandre A., and Albert C. Reynolds. "Investigation of the sampling performance of ensemble-based methods with a simple reservoir model." *Computational Geosciences* 17, no. 2 (2013b): 325.

Emerick, Alexandre A. "Investigation on Principal Component Analysis Parameterizations for History Matching Channelized Facies Models with Ensemble-Based Data Assimilation." *Mathematical Geosciences* 49, no. 1 (2017): 85-120.

Zhang, Yanhui, Dean S. Oliver, Yan Chen, and Hans J. Skaug. "Data assimilation by use of the iterative ensemble smoother for 2D facies models." *SPE Journal* 20, no. 01 (2014): 169-185.

Chen, Yan, and Dean S. Oliver. "Localization and regularization for iterative ensemble smoothers." *Computational Geosciences* 21, no. 1 (2017): 13-30.

Nejadi, Siavash, Juliana Leung, and Japan Trivedi. "Characterization of non-Gaussian geologic facies distribution using ensemble Kalman filter with probability weighted re-sampling." *Mathematical Geosciences* 47, no. 2 (2015): 193-225.

Nejadi, Siavash, Juliana Y. Leung, Japan J. Trivedi, and Claudio Virués. "Integrated Characterization of Hydraulically Fractured Shale-Gas Reservoirs—Production History Matching." *SPE Reservoir Evaluation & Engineering* 18, no. 04 (2015a): 481-494.

Nejadi, Siavash, and Japan Trivedi, and Juliana Y. Leung. "Estimation of facies boundaries using categorical indicators with P-Field simulation and ensemble Kalman filter (EnKF)." *Natural Resources Research* 24, no. 2 (2015b): 121-138.

Nejadi, Siavash, Japan J. Trivedi, and Juliana Leung. "History matching and uncertainty quantification of discrete fracture network models in fractured reservoirs." *Journal of Petroleum Science and Engineering* 152 (2017): 21-32.

Sebacher, Bogdan, Andreas S. Stordal, and Remus Hanea. "Bridging multipoint statistics and truncated Gaussian fields for improved estimation of channelized reservoirs with ensemble methods." *Computational Geosciences* 19, no. 2 (2015): 341.

Stordal, Andreas S., Hans A. Karlsen, Geir Nævdal, Hans J. Skaug, and Brice Vallès. "Bridging the ensemble Kalman filter and particle filters: the adaptive Gaussian mixture filter." *Computational Geosciences* 15, no. 2 (2011): 293-305.

Jafarpour, Behnam, and Morteza Khodabakhshi. "A probability conditioning method (PCM) for nonlinear flow data integration into multipoint statistical facies simulation." *Mathematical Geosciences* 43, no. 2 (2011): 133-164.

Hu, L. Y., Y. Zhao, Y. Liu, C. Scheepens, and A. Bouchard. "Updating multipoint simulations using the ensemble Kalman filter." *Computers & Geosciences* 51 (2013): 7-15.

Li, Liangping, Sanjay Srinivasan, Haiyan Zhou, and J. J. Gómez-Hernández. "A pilot point guided pattern matching approach to integrate dynamic data into geological modeling." *Advances in water resources* 62 (2013): 125-138.

Zhou, Haiyan, J. Jaime Gómez-Hernández, and Liangping Li. "A pattern-search-based inverse method." *Water Resources Research* 48, no. 3 (2012).

RamaRao, Banda S., A. Marsh LaVenue, Ghislain De Marsily, and Melvin G. Marietta. "Pilot point methodology for automated calibration of an ensemble of conditionally simulated transmissivity fields: 1. Theory and computational experiments." *Water Resources Research* 31, no. 3 (1995): 475-493.

LaVenue, A. Marsh, Banda S. RamaRao, Ghislain De Marsily, and Melvin G. Marietta. "Pilot point methodology for automated calibration of an ensemble of conditionally simulated transmissivity fields: 2. Application." *Water Resources Research* 31, no. 3 (1995): 495-516.

Dubuisson, M-P., and Anil K. Jain. "A modified Hausdorff distance for object matching." In *Pattern Recognition, 1994. Vol. 1-Conference A: Computer Vision & Image Processing., Proceedings of the 12th IAPR International Conference on*, vol. 1, pp. 566-568. IEEE, 1994.

Park, Hyucksoo, Céline Scheidt, Darryl Fenwick, Alexandre Boucher, and Jef Caers. "History matching and uncertainty quantification of facies models with multiple geological interpretations." *Computational Geosciences* 17, no. 4 (2013): 609-621.

Kruskal, Joseph B. "Multidimensional scaling by optimizing goodness of fit to a nonmetric hypothesis." *Psychometrika* 29, no. 1 (1964): 1-27.

Borg, Ingwer, and Patrick JF Groenen. *Modern multidimensional scaling: Theory and applications*. Springer Science & Business Media, 2005.

Cox, Trevor F., and Gillian Ferry. (1993) "Discriminant analysis using non-metric multidimensional scaling." *Pattern Recognition* 26, no.1: 145-153.

Cox, Trevor F., and Michael AA Cox. *Multidimensional scaling*. CRC press, 2000.

Nwachukwu, Azor, Baehyun Min, and Sanjay Srinivasan. "Model selection for CO 2 sequestration using surface deformation and injection data." *International Journal of Greenhouse Gas Control* 56 (2017): 67-92.

Bhowmik, Sayantan, Sanjay Srinivasan, and Steven Bryant. "Prediction of Plume Migration Using Injection Data and a Model Selection Approach." *Energy Procedia* 37 (2013): 3672-3679.

Bhowmik, Sayantan. "Particle tracking proxies for prediction of CO_2 plume migration within a model selection framework." PhD diss., 2014.

Srinivasan, Sanjay, and Cesar Mantilla. "Uncertainty Quantification and Feedback Control Using a Model Selection Approach Applied to a Polymer Flooding Process." In *Geostatistics Oslo 2012*, pp. 197-208. Springer Netherlands, 2012.

Mantilla, Cesar A. "Feedback control of polymer flooding process considering geologic uncertainty." PhD diss., 2010.

Scheidt, Céline, and Jef Caers. "Representing spatial uncertainty using distances and kernels." *Mathematical Geosciences* 41, no. 4 (2009a): 397-419.

Scheidt, Céline, and Jef Caers. "Uncertainty Quantification in Reservoir Performance Using Distances and Kernel Methods--Application to a West Africa Deepwater Turbidite Reservoir." *SPE Journal* 14, no. 04 (2009b): 680-692.

Scheidt, Céline, and Jef Caers. "Bootstrap confidence intervals for reservoir model selection techniques." *Computational Geosciences* 14, no. 2 (2010): 369-382.

Yadav, S., Srinivasan, S., Bryant, S.L., and A. Barrera. "History Matching Using Probabilistic Approach in a Distributed Computing Environment." Paper presented at the SPE Reservoir Simulation Symposium, The Woodlands, Texas, January 2005.

Lipowski, Adam, and Dorota Lipowska. "Roulette-wheel selection via stochastic acceptance." *Physica A: Statistical Mechanics and its Applications* 391 no. 6 (2012): 2193-2196.

Kiefer, J. (1953), "Sequential minimax search for a maximum", Proceedings of the American Mathematical Society, 4 no. 3: 502–506.

Brent, Richard P. (1971). "An algorithm with guaranteed convergence for finding a zero of a function." *The Computer Journal* 14 no. 4: 422-425.

Kumar, D. and Srinivasan, S. (2019). "Ensemble-Based Assimilation of Nonlinearly Related Dynamic Data in Reservoir Models Exhibiting Non-Gaussian Characteristics," *Mathematical Geosciences*, 51 no. 1: 75–107.

Kumar, D. and Srinivasan, S. (2020) "Indicator-based data assimilation with multiple-point statistics for updating an ensemble of models with non-Gaussian parameter distributions," *Advances in Water Resources Research*, 141: 163611.

APPENDIX A

Quantile Variograms

The classical semi-variogram estimator is written as:

$$\hat{\gamma}(\mathbf{x}_\alpha - \mathbf{x}_\beta) = \frac{1}{2}E\left\{\left[Z(\mathbf{x}_\alpha) - Z(\mathbf{x}_\beta)\right]^2\right\} \qquad \text{(A-1)}$$

That is, if a distribution of square differences $D(\mathbf{x}_\alpha - \mathbf{x}_\beta) = [Z(\mathbf{x}_\alpha) - Z(\mathbf{x}_\beta)]^2$ is constructed, D is a stochastic random variable (RV) with individual outcomes $d(\mathbf{x}_\alpha - \mathbf{x}_\beta)$. The mean of the distribution for D is related to the classical semi-variogram estimator written in Eq. (A-1).

If the RVs $Z(\mathbf{x}_\alpha)$ and $Z(\mathbf{x}_\beta)$ are Gaussian, then it can be proved that the difference $[Z(\mathbf{x}_\alpha) - Z(\mathbf{x}_\beta)]$ is also Gaussian. Furthermore, since the estimation proceeds under the decision of stationarity:

$$E\left\{Z(\mathbf{x}_\alpha)\right\} = E\left\{Z(\mathbf{x}_\beta)\right\} = E\left\{Z(\mathbf{x}_\alpha + \mathbf{h}_{\alpha\beta})\right\}$$

That is, the distribution of the RV $Y(\mathbf{x}, \mathbf{x} + \mathbf{h}) = [Z(\mathbf{x}) - Z(\mathbf{x} + \mathbf{h})]$ is Gaussian with mean equal to 0. The variance of the distribution of Y is given as:

$$Var\{Y\} = E\left\{Y^2\right\} - \left[E\{Y\}\right]^2 = E\left\{\left[Z(\mathbf{x}) - Z(\mathbf{x} + \mathbf{h})\right]^2\right\} - 0^2$$

This implies that $Var\{Y\} = 2 \cdot \hat{\gamma}(\mathbf{h})$ utilizing the definition of the estimator in Eq. (A-1). We have thus established that the RV Y is Gaussian with mean 0 and variance equal to $2 \cdot \hat{\gamma}(\mathbf{h})$. Let us define the standard normal RV $Y' = \dfrac{Y}{\sqrt{2 \cdot \hat{\gamma}(\mathbf{h})}}$ (i.e., we did the standardization $Y' = \dfrac{(Y - E\{Y\})}{\sqrt{Var\{Y\}}}$). The distribution of $Y'' = Y'^2$ is given as:

$$
\begin{aligned}
F_{Y''}(y'') = \text{Prob}\left\{Y'' \le y''\right\} &= \text{Prob}\left\{Y'^2 \le y''\right\} \\
&= \text{Prob}\left\{Y' \le \pm\sqrt{y''}\right\} \\
&= \text{Prob}\left\{-\sqrt{y''} \le Y' \le \sqrt{y''}\right\}
\end{aligned}
$$

As Y' is standard normal and denoting the standard normal *cdf* as $G_{Y'}$, we can write the probability that Y' will lie in the interval $\left[-\sqrt{y''}, \sqrt{y''}\right]$ as:

$$F_{Y''}(y'') = G_{Y'}\left(\sqrt{y''}\right) - G_{Y'}\left(-\sqrt{y''}\right)$$

The corresponding probability density is:

$$g_{Y''}(y'') = \frac{\partial G_{Y''}}{\partial y''} = \frac{\partial G_{Y'}\left(\sqrt{y''}\right)}{\partial y''} - \frac{\partial G_{Y'}\left(-\sqrt{y''}\right)}{\partial y''}$$

$$= \frac{1}{2\sqrt{y''}} g_{Y'}\left(\sqrt{y''}\right) + \frac{1}{2\sqrt{y''}} g_{Y'}\left(-\sqrt{y''}\right)$$

This last step is obtained by applying the chain rule of differentiation. Using the symmetry of the standard normal distribution, we get:

$$g_{Y'}\left(\sqrt{y''}\right) = \frac{1}{\sqrt{2\pi}} e^{-\frac{y''}{2}} = g_{Y'}\left(-\sqrt{y''}\right)$$

Therefore:

$$g_{Y''}(y'') = \frac{1}{\sqrt{y''}\sqrt{2\pi}} e^{-\frac{y''}{2}} \tag{A-2}$$

Equation (A-2) is referred to as the *pdf* corresponding to a χ^2 distribution with one degree of freedom. The squared difference distribution of standardized differences, i.e., $Y'' = Y'^2$, where $Y' = \dfrac{\left[Z(\mathbf{x}_\alpha) - Z(\mathbf{x}_\beta)\right]}{\sqrt{2 \cdot \hat{\gamma}(\mathbf{h})}}$, therefore follows a χ^2 distribution with one degree of freedom. The variance of Y'' is given as:

$$E\{Y''\} = E\{Y'^2\} = E\left\{\left[\frac{\left[Z(\mathbf{x}_\alpha) - Z(\mathbf{x}_\beta)\right]}{\sqrt{2 \cdot \hat{\gamma}(\mathbf{h})}}\right]^2\right\}$$

$$= \frac{1}{2 \cdot \hat{\gamma}(\mathbf{h})} E\left\{\left[Z(\mathbf{x}_\alpha) - Z(\mathbf{x}_\beta)\right]^2\right\}$$

$$= \frac{\hat{\gamma}(\mathbf{h})}{\hat{\gamma}(\mathbf{h})}$$

$$= 1$$

The traditional semi-variogram defined by Eq. (A-1) can therefore be interpreted to be $\hat{\gamma}(\mathbf{h}) \cdot Mean(\chi^2)$. Extending this notion, we can define a quantile variogram:

$$\hat{\gamma}_p(\mathbf{h}) = \hat{\gamma}(\mathbf{h}) \cdot Q_p(\chi^2) \qquad \text{(A-3)}$$

i.e., the traditional semi-variogram scaled by the pth quantile sampled from an χ^2 distribution with 1 degree of freedom.

Remark The quantile variogram can also be used to verify if the spatial variability exhibited by a phenomenon is bi-Gaussian. For such a phenomenon, the traditional semi-variogram $\hat{\gamma}(\mathbf{h})$ can be computed and then scaled by the corresponding pth quantile from an χ^2 distribution with 1 degree of freedom. The *cdf* corresponding to half the squared difference $[Z(\mathbf{u}) - Z(\mathbf{u} + \mathbf{h})]^2$ for various lags $\mathbf{h}$ can be computed and the same pth quantile values can be read from these *cdfs*. If the phenomenon is indeed bi-Gaussian, the scaled semi-variogram value at lag $\mathbf{h}$ corresponding to the pth quantile should be close to the quantile value read from the *cdf* of half the squared differences.

Consider the two descriptions of reservoir heterogeneity shown in Fig. A-1. The model on the left depicts diffused heterogeneity typical of a Gaussian model. The model on the right on the other hand exhibits continuity of the black and white regions, i.e., of the extremities suggesting that the model may be non-Gaussian.

The semi-variograms for these two models were computed in the azimuth 0° (North) and 90° directions for 10 lags. These semi-variograms were then scaled by the quantiles from the χ^2 distribution with one degree of freedom corresponding to $p = 0.20, 0.60$, and 0.80. The pairs making up the semi-variogram value corresponding to $|\mathbf{h}| = 1, 2$, and 5 were extracted and used to compute the *cdf* of half the squared difference $[Z(\mathbf{u}) - Z(\mathbf{u} + \mathbf{h})]^2$. The comparison of these values for the computations corresponding to unit lag in the 0° direction is presented in Table A-1.

The significant deviation between the quantile variogram value especially corresponding to $p = 0.8$ for the non-Gaussian model while the closeness of the same values for the Gaussian model points to the utility of this approach to checking for bi-Gaussianity.

The check for the variogram corresponding to $|\mathbf{h}| = 2$ in the 90° direction is presented in Table A-2. Once again, the results confirm that the quantiles drawn from the distribution of half the squared differences match with the scaled semi-variogram values for the Gaussian case, while they differ significantly for the non-Gaussian case.

The last check is corresponding to lag $|\mathbf{h}| = 5$ in the 0° direction. These results are presented in Table A-3. The results again point to the

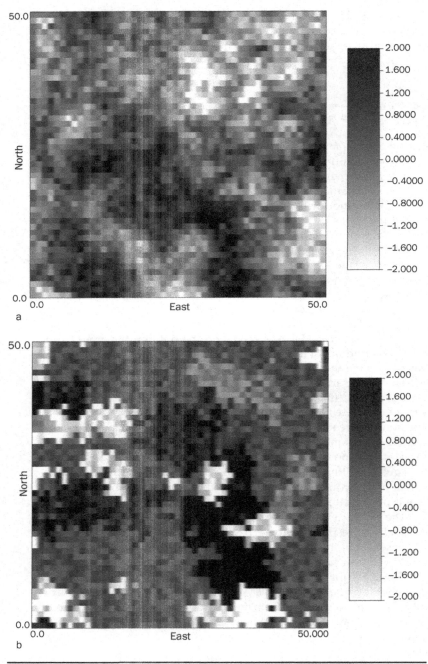

FIGURE A-1 **a)** A model exhibiting multi-Gaussian characteristics. **b)** A model exhibiting continuity of extremes typical of a non-Gaussian realization.

	(1) Gaussian [Fig. A-1(a)]	(2) Non-Gaussian [Fig. A-1(b)]
Semi-variogram value $\hat{\gamma}(\mathbf{h})$	0.2435	1.0290
$\chi^2_{0.2} = 0.064: \hat{\gamma}_{0.2}(\mathbf{h})$	0.0156	0.0660
$Q_{0.2}(\mathbf{h})$ from distbn. of square diff.	0.0600	0.0900
$\chi^2_{0.6} = 0.708: \hat{\gamma}_{0.6}(\mathbf{h})$	0.1725	0.7289
$Q_{0.6}(\mathbf{h})$ from distbn. of square diff.	0.2000	0.2000
$\chi^2_{0.8} = 1.642: \hat{\gamma}_{0.8}(\mathbf{h})$	0.4000	1.6900
$Q_{0.8}(\mathbf{h})$ from distbn. of square diff.	0.4200	0.5800

Note: These are for a lag $|\mathbf{h}| = 1$ in 0° direction.

TABLE A-1   Comparison of Quantile Variogram Values for the Two Models Presented in Fig. A-1

	(3) Gaussian [Fig. A-1(a)]	(4) Non-Gaussian [Fig. A-1(b)]
Semi-variogram value $\hat{\gamma}(\mathbf{h})$	0.369	1.453
$\chi^2_{0.2} = 0.064: \hat{\gamma}_{0.2}(\mathbf{h})$	0.024	0.093
$Q_{0.2}(\mathbf{h})$ from distbn. of square diff.	0.0900	0.110
$\chi^2_{0.6} = 0.708: \hat{\gamma}_{0.6}(\mathbf{h})$	0.261	1.029
$Q_{0.6}(\mathbf{h})$ from distbn. of square diff.	0.270	0.255
$\chi^2_{0.8} = 1.642: \hat{\gamma}_{0.8}(\mathbf{h})$	0.606	2.386
$Q_{0.8}(\mathbf{h})$ from distbn. of square diff.	0.580	1.300

Note: These are for a lag $|\mathbf{h}| = 2$ in 0° direction.

TABLE A-2   Comparison of Quantile Variogram Values for the Two Models Presented in Fig. A-1

	(5) Gaussian [Fig. A-1(a)]	(6) Non-Gaussian [Fig. A-1(b)]
Semi-variogram value $\hat{\gamma}(\mathbf{h})$	0.709	2.523
$\chi^2_{0.2} = 0.064: \hat{\gamma}_{0.2}(\mathbf{h})$	0.046	0.162
$Q_{0.2}(\mathbf{h})$ from distbn. of square diff.	0.100	0.125
$\chi^2_{0.6} = 0.708: \hat{\gamma}_{0.6}(\mathbf{h})$	0.502	1.787
$Q_{0.6}(\mathbf{h})$ from distbn. of square diff.	0.375	0.550
$\chi^2_{0.8} = 1.642: \hat{\gamma}_{0.8}(\mathbf{h})$	1.164	4.143
$Q_{0.8}(\mathbf{h})$ from distbn. of square diff.	0.850	2.600

Note: These are for a lag $|\mathbf{h}| = 5$ in 0° direction.

TABLE A-3   Comparison of Quantile Variogram Values for the Two Models Presented in Fig. A-1

Gaussian nature of the model in Fig. A-1(a). However, the difference between the scaled variogram values and the quantiles of the half-squared increments even for the model in Fig. A-1(a) suggests that the model is starting to stray away from Gaussian assumptions. This could be because of the ergodic fluctuations exhibited by the Gaussian simulation algorithm used to generate the model in Fig. A-1(a).

APPENDIX B

Some Details about the Markov–Bayes Model

Defining the information derived from the secondary source $v(\mathbf{u})$ as:

$$Y(\mathbf{u};z) = \text{Prob}\{Z(\mathbf{u}) \le z \mid v(\mathbf{u})\}$$

The Markov screening assumptions can be written as:

$$P(I(\mathbf{u} + \mathbf{h}) \mid I(\mathbf{u}) = i, Y(\mathbf{u};z) = y) = P(I(\mathbf{u} + \mathbf{h}) \mid I(\mathbf{u}) = i)$$
$$P(Y(\mathbf{u}) \mid I(\mathbf{u}) = i, Y(\mathbf{u} + \mathbf{h};z) = y') = P(Y(\mathbf{u}) \mid I(\mathbf{u}) = i)$$
$$P(Y(\mathbf{u}) \mid I(\mathbf{u}) = i, I(\mathbf{u} + \mathbf{h};z) = j) = P(Y(\mathbf{u}) \mid I(\mathbf{u}) = i)$$

The indicator primary–secondary covariance can be written as:

$$C_{IY}(\mathbf{h}) = E\{I(\mathbf{u}) \cdot Y(\mathbf{u} + \mathbf{h})\} - E\{I(\mathbf{u})\} \cdot E\{Y(\mathbf{u} + \mathbf{h})\} \qquad \text{(B-1)}$$

where the indicator $I(\mathbf{u})$ is defined as:

$$I(\mathbf{u}) = \begin{cases} 1 & \text{if } Z(\mathbf{u}) \le z \\ 0 & \text{otherwise} \end{cases}$$

$$E\{I(\mathbf{u})\} = \text{Prob}\{Z(\mathbf{u}) \le z\} = F_Z(z)$$

The expected value of the soft variable is:

$$E\{Y(\mathbf{u})\} = E\{\text{Prob}(Z(\mathbf{u}) \le z \mid v(\mathbf{u}))\}$$

The expected value of the conditional probability of $Z(\mathbf{u})$ given $v(\mathbf{u})$ is simply the prior probability $\text{Prob}\{Z(\mathbf{u}) \le z\} = F_Z(z)$. The last term in the expression Eq. (B-1) is therefore $F_Z^2(z)$.

We are interested in modeling the noncentered covariance $K_{IY}(\mathbf{h}) = E\{I(\mathbf{u}) \cdot Y(\mathbf{u} + \mathbf{h})\}$, which are finite values only when $I(\mathbf{u}) = 1$. And so we start with:

$$\begin{aligned}
\text{Prob}\{I(\mathbf{u}) = 1 \cdot Y(\mathbf{u} + \mathbf{h}) = y\} &= \text{Prob}\{Y(\mathbf{u} + \mathbf{h}) = y, I(\mathbf{u}) = 1 \mid I(\mathbf{u} + \mathbf{h}) = 1\} \cdot \text{Prob}\{I(\mathbf{u} + \mathbf{h}) = 1\} \\
&+ \text{Prob}\{Y(\mathbf{u} + \mathbf{h}) = y, I(\mathbf{u}) = 1 \mid I(\mathbf{u} + \mathbf{h}) = 0\} \cdot \text{Prob}\{I(\mathbf{u} + \mathbf{h}) = 0\} \\
&= \text{Prob}\{Y(\mathbf{u} + \mathbf{h}) = y, I(\mathbf{u}) = 1 \mid I(\mathbf{u} + \mathbf{h}) = 1\} \cdot F_Z(z) \\
&+ \text{Prob}\{Y(\mathbf{u} + \mathbf{h}) = y, I(\mathbf{u}) = 1 \mid I(\mathbf{u} + \mathbf{h}) = 0\} \cdot (1 - F_Z(z)) \\
&= \text{Prob}\{Y(\mathbf{u} + \mathbf{h}) = y, I(\mathbf{u}) = 1, I(\mathbf{u} + \mathbf{h}) = 1\} \cdot \text{Prob}\{I(\mathbf{u}) = 1 \mid I(\mathbf{u} + \mathbf{h}) = 1\} F_Z(z) \\
&+ \text{Prob}\{Y(\mathbf{u} + \mathbf{h}) = y \mid I(\mathbf{u}) = 1, I(\mathbf{u} + \mathbf{h}) = 0\} \cdot \text{Prob}\{I(\mathbf{u}) = 1 \mid I(\mathbf{u} + \mathbf{h}) = 0\} F_Z(z)
\end{aligned}$$

Employing the Markov screening assumption:

$$P(Y(\mathbf{u}) \mid I(\mathbf{u}) = i, I(\mathbf{u} + \mathbf{h};z) = j) = P(Y(\mathbf{u}) \mid I(\mathbf{u}) = i)$$

$$\begin{aligned}
\text{Prob}\{I(\mathbf{u}) = 1 \cdot Y(\mathbf{u} + \mathbf{h}) = y\} &= \text{Prob}\{Y(\mathbf{u} + \mathbf{h}) = y \mid I(\mathbf{u} + \mathbf{h}) = 1\} \cdot \text{Prob}\{I(\mathbf{u}) = 1 \mid I(\mathbf{u} + \mathbf{h}) = 1\} F_Z(z) \\
&+ \text{Prob}\{Y(\mathbf{u} + \mathbf{h}) = y \mid I(\mathbf{u} + \mathbf{h}) = 0\} \cdot \text{Prob}\{I(\mathbf{u}) = 1 \mid I(\mathbf{u} + \mathbf{h}) = 0\} F_Z(z) \\
&= \text{Prob}\{Y(\mathbf{u} + \mathbf{h}) = y \mid I(\mathbf{u} + \mathbf{h}) = 1\} \cdot \text{Prob}\{I(\mathbf{u}) = 1, I(\mathbf{u} + \mathbf{h}) = 1\} \\
&+ \text{Prob}\{Y(\mathbf{u} + \mathbf{h}) = y \mid I(\mathbf{u} + \mathbf{h}) = 0\} \cdot \text{Prob}\{I(\mathbf{u}) = 1, I(\mathbf{u} + \mathbf{h}) = 0\}
\end{aligned}$$

Let us define the following parameters:

$$m^{(1)} = E\{Y(\mathbf{u}) \mid I(\mathbf{u};z) = 1\}$$
$$m^{(0)} = E\{Y(\mathbf{u}) \mid I(\mathbf{u};z) = 0\}$$

and
$$B(z) = m^{(1)} - m^{(0)}$$

Now we are ready to calculate the expectation:

$$\begin{aligned}
E\{I(\mathbf{u}) \cdot Y(\mathbf{u} + \mathbf{h})\} &= \int_I \int_Y I(\mathbf{u}) \cdot Y(\mathbf{u} + \mathbf{h}) \cdot \{\text{Prob}\{Y(\mathbf{u} + \mathbf{h}) = y \mid I(\mathbf{u} + \mathbf{h}) = 1\} \cdot \text{Prob}\{I(\mathbf{u}) = 1, I(\mathbf{u} + \mathbf{h}) = 1\}\} dY \, dI \\
&+ \int_I \int_Y I(\mathbf{u}) \cdot Y(\mathbf{u} + \mathbf{h}) \cdot \{\text{Prob}\{Y(\mathbf{u} + \mathbf{h}) = y \mid I(\mathbf{u} + \mathbf{h}) = 0\} \cdot \text{Prob}\{I(\mathbf{u}) = 1, I(\mathbf{u} + \mathbf{h}) = 0\}\} dY \, dI \\
&= \int_I 1 \cdot \text{Prob}\{I(\mathbf{u}) = 1, I(\mathbf{u} + \mathbf{h}) = 1\} dI \int_Y Y(\mathbf{u} + \mathbf{h}) \cdot \text{Prob}\{Y(\mathbf{u} + \mathbf{h}) = y \mid I(\mathbf{u} + \mathbf{h}) = 1\} dY \\
&+ \int_I 1 \cdot \text{Prob}\{I(\mathbf{u}) = 1, I(\mathbf{u} + \mathbf{h}) = 0\} dI \int_Y Y(\mathbf{u} + \mathbf{h}) \cdot \text{Prob}\{Y(\mathbf{u} + \mathbf{h}) = y \mid I(\mathbf{u} + \mathbf{h}) = 0\} dY \\
&= m^{(1)} \cdot \int_I 1 \cdot \text{Prob}\{I(\mathbf{u}) = 1, I(\mathbf{u} + \mathbf{h}) = 1\} dI + m^{(0)} \cdot \int_Y 1 \cdot \text{Prob}\{I(\mathbf{u}) = 1, I(\mathbf{u} + \mathbf{h}) = 0\} dI
\end{aligned}$$

Looking at the h-scatterplot (Fig. B-1) between $Z(\mathbf{u})$ and $Z(\mathbf{u} + \mathbf{h})$, it can be seen that:

$$\int_I 1 \cdot \text{Prob}\{I(\mathbf{u}) = 1, \ I(\mathbf{u} + \mathbf{h}) = 0\} dI = F_Z(z) - E\{I(\mathbf{u}) = 1, \ I(\mathbf{u} + \mathbf{h}) = 1\}$$
$$= F_Z(z) - K_I(\mathbf{h})$$

Substituting in the previous equation:

$$\begin{aligned}
E\{I(\mathbf{u}) \cdot Y(\mathbf{u} + \mathbf{h})\} &= m^{(1)} \cdot K_I(\mathbf{h}) + m^{(0)} \cdot (F_Z(z) - K_I(\mathbf{h})) \\
&= K_I(\mathbf{h}) \cdot (m^{(1)} - m^{(0)}) + m^{(0)} \cdot F_Z(z)
\end{aligned}$$

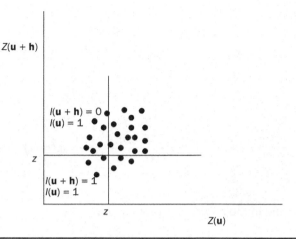

FIGURE B-1 Scatterplot between random variables $Z(\mathbf{u})$ and $Z(\mathbf{u} + \mathbf{h})$ identifying different regions of the joint distribution corresponding to indicator definitions $I(\mathbf{u};z)$ using a particular threshold z.

Now let us look at the definition of the mean $m^{(1)}$:

$$
\begin{aligned}
m^{(1)} = E\{Y(\mathbf{u}) \mid I(\mathbf{u};z) = 1\} &= \int_Y y(\mathbf{u}) \cdot \text{Prob}\{Y(\mathbf{u}) = y \mid I(\mathbf{u};z) = 1\}dy \\
&= \int_Y y(\mathbf{u}) \cdot \frac{\text{Prob}\{I(\mathbf{u};z) = 1 \mid Y(\mathbf{u}) = y\} \cdot f_Y(y)}{F_Z(z)}\, dy \\
&= \frac{I}{F_Z(z)}
\end{aligned}
$$

where $I = \int_Y y(\mathbf{u}) \cdot \text{Prob}\{I(\mathbf{u};z) = 1 \mid Y(\mathbf{u}) = y\} \cdot f_Y(y)dy$

Similarly:

$$
\begin{aligned}
m^{(0)} = E\{Y(\mathbf{u}) \mid I(\mathbf{u};z) = 0\} &= \int_Y y(\mathbf{u}) \cdot \text{Prob}\{Y(\mathbf{u}) = y \mid I(\mathbf{u};z) = 0\}dy \\
&= \int_Y y(\mathbf{u}) \cdot \frac{\text{Prob}\{I(\mathbf{u};z) = 0 \mid Y(\mathbf{u}) = y\} \cdot f_Y(y)}{1 - F_Z(z)}\, dy \\
&= \int_Y y(\mathbf{u}) \cdot \frac{1 - \text{Prob}\{I(\mathbf{u};z) = 1 \mid Y(\mathbf{u}) = y\} \cdot f_Y(y)}{1 - F_Z(z)}\, dy \\
&= \frac{F_Z(z) - I}{1 - F_Z(z)}
\end{aligned}
$$

Therefore:

$$B(z) = m^{(1)} - m^{(0)} = \frac{I}{F_Z(z)} - \frac{F_Z(z) - I}{1 - F_Z(z)} = \frac{I - I \cdot F_Z(z) - F_Z^2(z) + I \cdot F_Z(z)}{F_Z(z) \cdot (1 - F_Z(z))}$$

$$I = B(z) \cdot F_Z(z) \cdot (1 - F_Z(z)) + F_Z^2(z)$$

$$m^{(0)} = \frac{F_Z(z) - B(z) \cdot F_Z(z) \cdot (1 - F_Z(z)) - F_Z^2(z)}{(1 - F_Z(z))} = (F_Z(z) \cdot (1 - B(z)))$$

(B-2)

Substituting in the previous expression for $E\{I(\mathbf{u}) \cdot Y(\mathbf{u} + \mathbf{h})\}$:

$$\begin{aligned}
E\{I(\mathbf{u}) \cdot Y(\mathbf{u} + \mathbf{h})\} &= K_I(\mathbf{h}) \cdot (m^{(1)} - m^{(0)}) + m^{(0)} \cdot F_Z(z) \\
&= K_I(\mathbf{h}) \cdot B(z) + F_Z^2(z) \cdot (1 - B(z)) \\
&= (K_I(\mathbf{h}) - F_Z^2(z)) \cdot B(z) + F_Z^2(z)
\end{aligned}$$

Now we are ready to develop an expression for the cross-covariance $C_{IY}(\mathbf{h})$:

$$\begin{aligned}
C_{IY}(\mathbf{h}) &= E\{I(\mathbf{u}) \cdot Y(\mathbf{u} + \mathbf{h})\} - F_Z^2(z) \\
&= (K_I(\mathbf{h}) - F_Z^2(z)) \cdot B(z) + F_Z^2(z) - F_Z^2(z) \\
&= (K_I(\mathbf{h}) - F_Z^2(z)) \cdot B(z)
\end{aligned}$$

Because: $C_I(\mathbf{h}) = K_I(\mathbf{h}) - F_Z^2(z)$

Consequently: $C_{IY}(\mathbf{h}) = C_I(\mathbf{h}) \cdot B(z)$ (B-3)

B.1   Deriving the Secondary Covariance Model

Within the same Markov–Bayes framework, let us now develop an expression for the secondary covariance $C_Y(\mathbf{h})$. First, let us start with the joint distribution:

$$\begin{aligned}
\text{Prob}\{Y(\mathbf{u}) = y \cdot Y(\mathbf{u} + \mathbf{h}) = y'\} &= \text{Prob}\{Y(\mathbf{u}) = y, Y(\mathbf{u} + \mathbf{h}) = y' \mid I(\mathbf{u}) = 1\} \cdot \text{Prob}\{I(\mathbf{u}) = 1\} \\
&\quad + \text{Prob}\{Y(\mathbf{u}) = y, Y(\mathbf{u} + \mathbf{h}) = y' \mid I(\mathbf{u}) = 0\} \cdot \text{Prob}\{I(\mathbf{u}) = 0\} \\
&= \text{Prob}\{Y(\mathbf{u}) = y \mid I(\mathbf{u}) = 1, Y(\mathbf{u} + \mathbf{h}) = y'\} \cdot \text{Prob}\{Y(\mathbf{u} + \mathbf{h}) = y' \mid I(\mathbf{u}) = 1\} F_Z(z) \\
&\quad + \text{Prob}\{Y(\mathbf{u}) = y \mid I(\mathbf{u}) = 0, Y(\mathbf{u} + \mathbf{h}) = y'\} \cdot \text{Prob}\{Y(\mathbf{u} + \mathbf{h}) = y' \mid I(\mathbf{u}) = 0\} \cdot (1 - F_Z(z)) \\
&= \text{Prob}\{Y(\mathbf{u}) = y \mid I(\mathbf{u}) = 1\} \cdot \text{Prob}\{Y(\mathbf{u} + \mathbf{h}) = y', I(\mathbf{u}) = 1\} \\
&\quad + \text{Prob}\{Y(\mathbf{u}) = y \mid I(\mathbf{u}) = 0\} \cdot \text{Prob}\{Y(\mathbf{u} + \mathbf{h}) = y', I(\mathbf{u}) = 0\} \cdot (1 - F_Z(z))
\end{aligned}$$

Now calculating the joint expectation:

$$\begin{aligned}
E\{Y(\mathbf{u}) = y \cdot Y(\mathbf{u} + \mathbf{h}) = y'\} &= \int_Y \int_{Y'} yy' \cdot \text{Prob}\{Y(\mathbf{u}) = y \cdot Y(\mathbf{u} + \mathbf{h}) = y'\} dy' dy \\
&= \int_y y \cdot \text{Prob}\{Y(\mathbf{u}) = y \mid I(\mathbf{u}) = 1\} dy \cdot \int_y y' \cdot \text{Prob}\{Y(\mathbf{u} + \mathbf{h}) = y', I(\mathbf{u}) = 1\} dy' \\
&\quad + \int_y y \cdot \text{Prob}\{Y(\mathbf{u}) = y \mid I(\mathbf{u}) = 0\} dy \cdot \int_{y'} y' \cdot (\text{Prob}\{Y(\mathbf{u} + \mathbf{h}) = y'\} \\
&\quad - \text{Prob}\{Y(\mathbf{u} + \mathbf{h}) = y', I(\mathbf{u}) = 1\}) dy' \\
&= m^{(1)} K_{IY}(\mathbf{h}) + m^{(0)} \cdot (F_Z(z) - K_{IY}(\mathbf{h})) \\
&= B(z) \cdot (B(z) \cdot C_I(\mathbf{h}) + F_Z^2(z)) + m^{(0)} \cdot F_Z(z)
\end{aligned}$$

Substituting for $m^{(0)}$ [Eq. (B-2)]:

$$\begin{aligned}
E\{Y(\mathbf{u}) = y \cdot Y(\mathbf{u} + \mathbf{h}) = y'\} &= B(z) \cdot (B(z) \cdot C_I(\mathbf{h}) + F_Z^2(z)) \\
&\quad + (F_Z(z) \cdot (1 - B(z)) \cdot F_Z(z) \\
&= B^2(z) \cdot C_I(\mathbf{h}) + F_Z^2(z)
\end{aligned}$$

The required secondary covariance is thus:

$$\begin{aligned}
C_Y(\mathbf{h}) &= E\{Y(\mathbf{u}) = y \cdot Y(\mathbf{u} + \mathbf{h}) = y'\} - E^2\{Y\} \\
&= (B^2(z) \cdot C_I(\mathbf{h}) + F_Z^2(z)) - F_Z^2(z) \qquad \text{(B-4)} \\
&= B^2(z) \cdot C_I(\mathbf{h})
\end{aligned}$$

Index

Note: *Italicized* page numbers refer to figures; **bold** page numbers refer to tables.